ROBUSTNESS DEVELOPMENT AND RELIABILITY GROWTH

ROBUSTNESS DEVELOPMENT AND RELIABILITY GROWTH

VALUE-ADDING STRATEGIES FOR NEW PRODUCTS AND PROCESSES

John P. King
William S. Jewett

PRENTICE
HALL

Upper Saddle River, NJ • Boston • Indianapolis • San Francisco
New York • Toronto • Montreal • London • Munich • Paris • Madrid
Capetown • Sydney • Tokyo • Singapore • Mexico City

Many of the designations used by manufacturers and sellers to distinguish their products are claimed as trademarks. Where those designations appear in this book, and the publisher was aware of a trademark claim, the designations have been printed with initial capital letters or in all capitals.

The authors and publisher have taken care in the preparation of this book, but make no expressed or implied warranty of any kind and assume no responsibility for errors or omissions. No liability is assumed for incidental or consequential damages in connection with or arising out of the use of the information or programs contained herein.

The publisher offers excellent discounts on this book when ordered in quantity for bulk purchases or special sales, which may include electronic versions and/or custom covers and content particular to your business, training goals, marketing focus, and branding interests. For more information, please contact:

U.S. Corporate and Government Sales
(800) 382-3419
corpsales@pearsontechgroup.com/ph

For sales outside the United States, please contact:

International Sales
international@pearson.com

Visit us on the Web: informit.com

Library of Congress Cataloging-in-Publication Data

King, John P., 1944-
    Robustness development and reliability growth: value-adding strategies for new products and processes / John P. King and William. S. Jewett.
       p.   cm.
    Includes bibliographical references and index.
    ISBN-10: 0-13-222551-4 (hardcover : alk. paper)
    ISBN-13: 978-0-13-222551-9
    1. New products—Management.   2. New products—Technological innovations.
    3. Reliability (Engineering)   4. Product management. I. Jewett, William S. II. Title.
    HF5415.153.K535 2010
    658.5'75—dc22
                                                                    2010003456

ISBN-13: 978-0-13-222551-9
ISBN-10:   0-13-222551-4
Text printed in the United States on recycled paper at Edwards Brothers, Ann Arbor, Michigan.
First printing, March 2010

To my wife, Linet, my son and daughter, Jonathan and Jennifer, my grandchildren,
Corey and Natalie, and my mother, Ann, with thanks for all their patience, love, and support.

John P. King

To my wife, Jane, and my daughter and son, Kristin and David. They kept me
encouraged and inspired throughout the project.

William S. Jewett

Contents

Section II   Framework for Reliability Development

Preface

To do anything well, repeatedly over time, requires a disciplined process. This is especially true of product development. People come and go, so your organization needs a process that doesn't depend on having specific people who have internalized your process and know how to practice it well. Your good people and your standardized processes are part of your corporate DNA. Of course your people are the most important ingredient, but it's risky to depend entirely on a veteran staff. Great people following a flawed process won't produce the best results. With a clear call to action, processes can be improved and implemented in ways so that they become a source of competitive advantages.

Ultimately the product development process is about risk management. Failing to meet the expectations of customers for product features, functionality, reliability, usage life, and cost of ownership, for example, can have serious consequences for your business. Products that fall short in the marketplace can threaten both your customers and your enterprise in many ways, such as the following:

- Risks to the health and safety of end users can increase product liability costs.
- Your product development teams may continue to work on urgent problems with current products in production rather than developing your next-generation products.
- Your service organization may have continued demands to correct problems that have escaped the factory because of faulty designs or manufacturing processes.
- Management can be preoccupied with correcting past mistakes and not paying enough attention to future developments.
- In the short run, warranty expenses may increase. This can be the most benign result of quality and reliability problems in products' designs.
- In the long run, customers' dissatisfaction will result in lost revenues and market share.
- If the problems with product quality persist, over time it will become increasingly difficult for your company to attract and retain good people. Work will no longer be an enjoyable or rewarding experience. Eventually people will conclude that making avoidable mistakes over and over is evidence of an organization that does not learn how to improve or to compete to win in the marketplace.

Reliability as Quality over Time

Many books on reliability emphasize the mathematical treatment of the subject. They place considerable focus on topics such as probabilities, life distributions, and the analysis of failure data. These are all important subjects. We too have included them. Statistics provides the mathematical tools and rigor that support the analysis and modeling of system reliability. However, the scope of reliability is broader than statistics. Many reliability problems involve deteriorations in performance over time, not just broken parts. As we will see, robustness is foundational to reliability, and its development should be the early focus of team members responsible for meeting reliability goals.

We have two working definitions of reliability: the traditional definition, focused on the probability of survival, and a more general definition, focused on the stability of performance. Although this second definition is not useful when trying to calculate the percentile for 10% failure, it's a powerful model that is more inclusive and more actionable.

David Garvin's "eight dimensions of quality"[1] include several that are driven by reliability:

1. Performance
2. Features
3. Reliability
4. Conformance
5. Durability
6. Serviceability
7. Aesthetics
8. Perceived quality

When we expand our definition of *reliability* to be "quality over time," reliability has great leverage on customers' satisfaction. Viewed this way, reliability becomes a "quality anchor." If performance drifts off target, customers' perception of quality will follow. When we use the term *reliability* in this book, we are referring to the broader definition of "quality over time."

Developing reliable products requires a good process that develops data that enable the improvement of product performance and its stability under conditions of stress. Many of the stresses affecting performance have a strong dependence on time and how the product is used. Making product performance less vulnerable to these stresses will improve the stability of performance under those conditions over time.

Six Sigma and Reliability

Six Sigma practices have been used for the improvement of business and technical processes for a couple of decades. Their roots go back much further. There are no "magic bullets" in Six Sigma, but the methodology does offer a disciplined process that can be helpful in solving problems. Most technical people accept the scientific method and do not need to be convinced of its value. The Six Sigma process wraps the tools of engineering statistics around the scientific method. We see a problem, form a hypothesis about its cause, and gather data. By analyzing the data we either accept or reject our hypothesis, or conclude that we do not know enough to move forward with an appropriate solution. When efforts are prioritized, the concept of critical-to-quality (CTQ) is central to everything that makes Six Sigma so useful. It offers a nice model for defining what is important to customers and then mapping that information in a hierarchy. The resulting CTQ tree diagrams the linkages among all of the parameters that the system designers have to specify as well as those product outputs that are important to customers. As we describe the process of critical parameter management (CPM), we use the CTQ tree as a way to characterize the flow down of system requirements and the

---

1. David A. Garvin, "Competing on the Eight Dimensions of Quality," *Harvard Business Review* (November– December 1987).

flow up of system performance. We build upon that and other tools as we outline the product development process and best practices for the development of robustness and the growth of reliability.

Typically, Six Sigma practitioners differentiate the tools and practices that are included in DMAIC (Define-Measure-Analyze-Improve-Control) processes from the methods of Design for Six Sigma (DFSS). In the DMAIC process, the focus is on solving existing problems with a product or process. It is a systematic approach to correcting problems. On the other hand, DFSS is a proactive, forward-looking process that is focused on the prevention of problems. Similarly, the applications can be both to new products and to technical and business processes. Because DMAIC and DFSS have different objectives, the tool sets tend to be different. Both processes have best practices that add value to the development of robust and reliable products. In the spirit of Six Sigma, we use what works. Here we speak in terms of the traditional tools wherever possible, trying to avoid the temptation to create new names and acronyms. We decode acronyms when we encounter them. You may already be familiar with many of them. We hope to introduce you to a few new ones.

Robustness Development and Reliability Growth

Our title for this book links two important engineering approaches: the development of robustness in a product's design and the growth of its reliability. Robustness development is focused on preventing problems, while reliability growth is focused on correcting problems. Reliability in the market, or as predicted during development, is a lagging indicator. Robustness metrics in development, however, are leading indicators. The development of robustness in subsystem designs enables the optimizing of robustness at the level of the integrated system. Robustness achieved early in development enables shorter cycle times in the later phases. Fixing problems late in a process may threaten delays in market entry and drive up manufacturing costs as tools are modified and inventories are scrapped. More time and resources spent in early phases[2] has paybacks in the later phases. An early focus on clarifying requirements, selecting superior design concepts, analyzing functional models, and planning designed experiments contributes greatly to "getting it right the first time," which you have heard often. Some problem solving late in the game is inevitable. Robustness development doesn't prevent all problems. We're human, prone to making mistakes. In the context of a project schedule, problems have to be identified and understood quickly and corrected gracefully, with prudent judgments of which problems are more important to correct before market entry and the beginning of production.

Managing the Product Development Process

The product development process can be described simply as a time-phased set of activities and decisions. To ensure a good outcome, certain tasks should precede others. The goal of the process can be simply stated: Get to market as efficiently as possible with a product that offers your customers value that is superior to that of your competition while managing risks to an acceptable level. To the extent that your projects do this better than your competitors', you can be more confident that your products and services will both satisfy and, in some ways, delight your customers. There will be positive benefits to your business. In the long run, superior products delivered on time with acceptable costs will drive superior returns to your business. What are some of the more important elements in your process?

---

2. Sometimes referred to as front-end loading (FEL).

Long-Term: Business and Research Strategies

- Make strategic decisions about your market participation.
- Align your portfolio strategy for new products with your market participation strategy.
- Develop a supporting technology development or acquisition strategy.
- Develop capabilities to deliver superior value to your customers.

These activities are aimed at positioning your company to participate in selected markets and to support this decision with the required development initiatives. Many considerations go into making these decisions. They can include the trends in your business, your core competencies, emerging technological advantages and disadvantages, and the difficulties in repositioning any of your products. Decisions such as these set the foundations under your product development projects.

Short- to Medium-Term: Product Development Process

- Understand the needs of your customers.
- Translate customers' needs into technical requirements.
- Develop better design concepts that satisfy your customers' needs.
- Choose the best concepts from available alternatives.
- Develop the selected concepts into production designs.
- Make the designs robust against sources of performance variation.
- Employ accelerated methods to identify and correct problems to grow reliability.
- Develop robustness in manufacturing, service, and other business processes.
- Prepare for the scale-up of manufacturing, service, and customer support.
- Launch the product into selected markets.
- Provide marketing, sales, and customer support.
- Provide maintenance and repair services.
- Discontinue production at the end of the product's life cycle.

Of course these accomplishments are neither simple nor serial. Development phases overlap. Some activities start as early as possible, while other activities have dependencies on the completion and freezing of earlier work. Still other activities may continue into later phases and impose risks due to their unfinished work. Preparations for manufacturing start long before manufacturing begins. Verification of the product's design is a process that begins while obtaining the validation of requirements from customers and their feedback from their testing of prototypes. Effective organizations seek feedback continually during their progress, from their initial fuzzy ideas through market entry into production. Overlaid on this set of activities are processes to manage the project and to ensure that the business manages its risks.

Key Features of This Book

- Strategies to improve the effectiveness and efficiency of product development, with particular attention to those influencing product reliability
- A comprehensive process for developing new products, with details relevant to reliability development
- Distinctions between the processes for robustness development and reliability growth

- Methods for the development of information critical to the achievement of higher reliability
- Detailed guidance for the planning of experiments and the analysis of data from laboratory tests and field failures
- Methods to optimize performance using both first-principles[3] and empirical system models

In the spirit of *Good to Great*,[4] improvements aggregate over time. Usually we don't make major improvements in anything overnight. Doing things better usually requires working and thinking differently, expecting you to modify your behaviors and eliminate your bad habits. Discipline, persistence, and corporate motivation are essential ingredients to enable positive changes and their implementation. The better process and tools will not help if management lacks the resolve and the desire for better business outcomes. Your "call to action" should be derived from your recent experiences with business results. Ultimately your people, processes, and capabilities make the difference and determine the fortunes of your company.

Topics of Interest versus Functional Responsibility

Of course we believe that our book has value to all who work for the development of new products. We encourage you to read it cover to cover. The book is organized into sections.

The first section, Critical Drivers of Value, provides broad insights about development strategies and tactics. It explores concerns for the consequences of not delivering superior value to your customers.

The second section, Framework for Reliability Development, describes approaches for the development of robustness and the growth of reliability in the context of a product development process with well-functioning governance and decision processes.

The third section, Tools and Methods Supporting Reliability Development, describes selected methodologies with reasonable depth that may enable managers to institutionalize these practices, if they are absent, and to have value-adding conversations with tool specialists. Our intention is that managers set clear expectations for the consistent application of better practices that deliver data with higher integrity to decision processes. For those readers who are not familiar with these tools, we encourage reading further and a pursuit of training to build new capabilities.

In the last section, Integration of Framework and Methods, we discuss the management of critical parameters. An example of a development project ties the strategies, decisions, and methods together in a useful illustration.

With all the demands on your time, we recognize that, depending on your responsibilities, some topics will be of more immediate interest and utility than others. We offer the following suggestions about what to read first.

General Managers: Chapters 1–6, 25

These early chapters should be helpful for understanding the model for product development and how reliability is directly affected by the quality of its implementation. We have given significant

---

3. A first-principles model is one that uses the established laws of science to predict the behavior of a physical system, rather than an empirical model derived from experimental data.

4. Jim Collins, *Good to Great* (New York: HarperCollins, 2001).

emphasis to risk management and decision making in the face of uncertainty. Ultimately, decisions to proceed, as a project moves from phase to phase, are business decisions requiring risks to be identified and evaluated. In the course of running an enterprise, risks are assumed on a regular basis. The likelihood and potential impact of things going wrong in the future have to be understood, and appropriate actions selected and implemented.

Project Leaders: Chapters 1–8, 24, 25

Project leaders are responsible for delivering a quality product on schedule with the right cost structure. There are strong interactions among quality, costs, and schedule. Decisions must be made to select the most appropriate implementation strategies and solution concepts for the product's development. Project management plans must then provide the right resources at the right time to implement the strategies and develop the solutions. Usually projects that are in trouble with their product's performance and reliability are also in trouble with their schedule and resources. The decision that budgets and schedules are all-important should not be made without understanding the associated risks and consequences to product quality. We offer some methods to understand risks.

Product Development Engineers and Scientists: Chapters 2–25

Discovery and understanding are essential to building successful products. As technical people we have a need first to understand how things work and then to make them work better. Usually pride of ownership and an understanding of the impact of shortfalls are drivers in this process. To do this we build both virtual and physical models through thoughtful experiments and first-principles analyses. The models can then be used to improve performance and reduce its variability. In later chapters we offer a collection of tools and practices that, when combined with a well-thought-out process, will help improve your chances of success.

Reliability Engineers: Chapters 2–8, 14–25

Traditional reliability engineering has focused on identifying and correcting failed components. Failed parts that cause the shutdown of a product can be dramatic, and shifts and deteriorations in performance are also a major source of customer dissatisfaction. For this reason we recommend a more holistic approach that broadens the focus to include performance variability and drift. It pulls more involvement by reliability engineers who should be a part of any product development project. Reliability engineers need to be there at the beginning, when specifications are derived from customer requirements, and to be part of the product development team throughout the project.

Quality Professionals: Chapters 2–25

Quality must be designed into a product as well as into its manufacturing and service processes, not achieved by inspection and repair. Quality engineering is a function that can help the organization be proactive in achieving higher levels of quality and reliability. The involvement of quality engineering at one time did not start until a design had been completed. It focused mainly on supply chain activities. This limitation has been recognized by many organizations and corrected. Today, quality engineering has become an important partner in the product development process.

The input of the quality organization is valuable to the planning and execution of reliability design and its verification. It should start early in the process.

Change Begins with You

We intend that our book will provide you with valuable insights. We expect that some will be new and enlightening. It's been an enjoyable project for us to describe the many paradigms and principles that have worked well in our own experience. Time has proven that these strategies, tools, and practices are not just a fad or a theme of the year. They have endured well and work nicely with current initiatives to make product development more lean, flexible, and agile. Certainly their implementation evolves with use. It takes time for new methods to become integrated into a larger set of practices as organizations learn better approaches that can provide major benefits with immediate and lasting value for their companies. It does no good for us to keep them to ourselves. Now it is your job to read, think, adapt, and practice so that you can achieve sustainable competitive advantages. We encourage you and your teams to be students of better processes and methods. Experiment and allow the demonstrated benefits to provide the encouragement that often is needed for institutional changes.

If you are an individual contributor, remember that it is possible to make tangible improvements in the quality of your work without having a formal initiative prescribed by management. If everyone on a team embraced that idea, the organization would reap many benefits in improved product quality and organizational efficiency.

If you are a manager, you can accomplish a lot by setting higher expectations for better information to enable intelligent risk management. By better information we mean developing data that are useful for informed decision making. This means understanding what's important, and developing the relevant performance metrics that must be tracked. Knowing where you are relative to your goals and understanding the problems that must be solved are key to assessing risks and making effective decisions.

The marketplace rewards those who are excellent at execution. Strive to be better than your competitors and serve your customers well. Good luck in your journey!

Acknowledgments

In the early days of our adventure Don Clausing of MIT gave us two fundamental gifts. He gave us a "wake-up call" with the perspective that we were learning well from ourselves but not from others. The processes and practices of our competitors were far ahead of ours, driving us to catch up fast. Second, he introduced us to many thought leaders and practitioners from whom we learned much. That essential learning process continues.

We appreciate the many suggestions that we received from those who reviewed our manuscript. In particular, Dave Auda, Roger Forsgren, and Dave Trimble gave us detailed inputs for all of the chapters. Betty Stephenson, Pat Whitcomb, and Mark Anderson gave us valuable suggestions for the chapters on statistics and design of experiments. They were all very generous with their time and very helpful. Their suggestions have resulted in a better book.

We would like to thank our friend and colleague Skip Creveling, who gave us our first opportunities as consultants.

We owe our publisher, Bernard Goodwin, and editorial assistant, Michelle Housley, of Prentice Hall our thanks for having faith that eventually there would be a book. If at times they had their doubts, they never shared anything with us but encouragement. Thanks to our copyeditor, Barbara Wood, who made many improvements; we learned from her.

John King would like to thank Ron Warner and Dan Rosard of Westinghouse Electric Corporation, who, long ago, had more influence than they ever knew on his development as an engineer.

We have learned from our clients. Our experiences working on their projects have added value to everything we do. Thank you.

We owe our families for their love and encouragement. Without those gifts the project would have taken much longer.

About the Authors

John P. King has made contributions as an engineer, program manager, and laboratory head during a long career in product development and commercialization. His responsibilities have ranged from upstream R&D projects through commercialization and product launch. At Westinghouse, Xerox, Kodak, and Heidelberg Druckmaschinen, John worked as an engineer and manager in developing a wide range of products, including turbo machinery, office products, and medical imaging systems.

As a consultant, mentor, and trainer, John has helped clients worldwide improve both business processes and products. He has worked with project teams for both large and small companies and has written and delivered training programs to clients, followed by project mentoring to ensure the successful completion of their objectives.

John is a member of the American Society of Mechanical Engineers, the American Society for Quality, the American Statistical Association, and the Society of Reliability Engineers. He holds a BME from Rensselaer Polytechnic Institute, an MS in Engineering Mechanics from the University of Pennsylvania, and an MBA from Northeastern University. John can be reached at jpking@jewettking.com.

Bill S. Jewett is a consultant to businesses engaged in the development of new technologies and multidisciplined products. With insights into important paradigms and advancements in practices, he assists improvement teams in upgrading their engineering and management processes, project management strategies, cross-functional teamwork, and governance of their development projects.

For many years Bill worked for Eastman Kodak Company and Heidelberg Druckmaschinen with much of his focus on the development of high-volume electrophotographic copiers and printers. Among his division-level responsibilities were the management of product development projects and of competency centers for mechanical and systems engineering. At the corporate level he was one of the authors of the processes for advanced product planning, technology development, product commercialization, and their related governance. For over a decade he taught the processes and coached teams in their adaptation and implementation. As the process steward, he evolved the process models to incorporate lessons learned from internal practice and external benchmarking.

Bill participates frequently in the conferences of PDMA, Management Roundtable, and INCOSE. He holds a BSME from Swarthmore College and an MBA from the University of Rochester. Bill can be reached at wsjewett@jewettking.com.

The home page for Mr. Jewett and Mr. King is www.jewettking.com.

Critical Drivers of Value

The first four chapters focus on key initiatives, strategies, and plans for improving the robustness and reliability designed into a product.

Chapter 1: Time, Money, and Risks

Products need to be developed without compromising their quality, costs, or market entry schedule. Initiatives to improve these processes depend on a compelling call to action. There are key development principles that should be implemented, with attention to avoid common root causes of poor reliability.

Chapter 2: Reliability, Durability, and Robustness

Reliability represents the probability of a failure, while durability represents the useful life of a product. Robustness is a characteristic of product design that can be developed to improve reliability and durability. All of these measures focus on reducing the deviation of performance responses from their target value. An improved view of performance deviations is in the economic losses that are the consequences to both customers and producers.

Chapter 3: Strategies for Reliability Development

Our book describes strategies for reliability development that you may find to represent improvements to your current approaches. The reliability growth plot shows the problem to be not only one of increasing the reliability in the initial full system prototypes but also growing that reliability at a higher rate leading to market entry. Before development projects can establish project management plans and related budgets, a better strategy must be adopted.

Time, Money, and Risks

The capabilities, performance, and reliability that are achieved in a product prior to mar-ket entry are functions of many activities and decisions during the development process. Physical laws drive product failures. They are also symptoms of implementation problems with plans, resources, methods, and tools. This chapter discusses typical drivers of relia-bility that are inherent in the way product development is practiced. Later chapters address technical concerns.

Inherent in discussions about product development are recommendations for process improvements. Assessment models such ISO and CMMI are effective at driving organiza-tions to define their processes and to follow them well. Benchmarking can help even more by enabling the people in your organization to learn firsthand about better methods that they would not have known from internal experiences. You compete with your processes as well as your products and services.

During our years of working in product development we have experienced many projects that had elements that were executed very well. Excellent design concepts were developed. Manufacturing processes were controlled to be stable and aligned with the specifications of design parameters. Cross-functional teams managed parallel work efficiently across all organiza-tions and partner companies. Customers and suppliers were involved early and throughout the process. Timely decisions enabled the critical path, addressed project risks, and managed cross-project conflicts.

However, few projects were managed to execute all of the necessary elements well at the same time. More often than not, the consequences of good work in one domain were compro-mised by handicaps from other domains. For example, sound designs of robustness experiments were cut short by schedule reductions. The involvement of engineers with customers was can-celed on the assumption that they already knew what customers needed. The transfer of new technology into product development was premature and incomplete, imposing risks on product development. Few projects learned how to repeat the successes of previous projects. As new products approached their market entry, crisis management became the norm. Schedules slipped, costs rose, designs were changed, tools were modified, parts inventories were scrapped, resources were kept on the job. Heroes may have saved the product launch date, but often the product itself limped into the market with less-than-acceptable quality, reliability, and costs. Design changes in production were common. These stories are not good!

We expect that your own experiences confirm this concern about the difficulties of executing product development well and consistently. We've known project teams that proclaimed themselves to be best in class without merit. From the viewpoint of both the business and the technologies, their projects were disasters. Often there was great technical work, but with poor business results. Certainly you have experienced examples when everything important worked well once in a while, but not consistently.

A major lesson is that failures become the call to action for major improvements in the way that product development is practiced. One result is a commitment to a development project that has superior strategies, that stays on its plan, and that adapts easily to inevitable changes. It delivers designs that achieve their requirements and are robust. It delivers the new product system to its market gracefully. The product is viewed to be superior to its competition. What a great experience! The challenge is to repeat those successes in follow-on projects. That's when standardized processes employing better practices and well-managed teamwork need to be institutionalized so that they have lasting value for your corporation.

Quality, Costs, and Schedules

"Bottom-line" business metrics usually relate to accounting parameters such as the costs of poor quality, the costs of development or manufacturing, and the costs of schedule delays. "Top-line" growth metrics usually focus on revenues generated from new products, with insights from customers' satisfaction indices and competitive positions that contribute to pricing. Their forecasts are predictors of revenues due to sales and usage volumes. Broadly, they apply to the value generated by the portfolio of new products evaluated over time.

Higher reliability developed in a product may be perceived as a competitive differentiator. It may also be a significant contributor to reduced service and warranty costs. Higher levels of robustness contribute not only to higher reliability but also to reduced manufacturing costs. To the extent that they are achieved earlier in a development project, they contribute to shorter, more predictable development schedules and reduced development costs.

Many contributors to reliability are within the control of the project teams prior to market entry. During this time, the reliability metrics are forecasts of the level and stability of future performance in the hands of customers. Of course, the ultimate measures of success are derived from the actual performance of the new products in the market, the degree to which higher reliability delivers value to customers, and the new product's contributions to your overall business. For customers of non-repairable products, reliability is perceived as the product's usage life. For repairable systems, it is evaluated by metrics such as failure rate, service frequency, availability when needed, percentage of downtime, or other measures relevant to your customers' business or activities.

With product reliability having the potential for being a significant contributor to business success, your product development teams have to align their technical achievements with those elements of value to be returned to the corporation. To the extent that teams are excellent in the execution of robustness development and reliability growth, they may achieve competitive advantages relevant to their rivals in the market. The more that product development teams work in a productive environment, using efficient methods and benefiting from constructive teamwork, the higher will be their probability of success.

Product Development

Product development is not just a process for engineering teams. "Concurrent engineering" has focused attention on cross-functional activities that develop the product and its manufacturing processes. In addition to engineering leadership teams, effective work groups include representatives of other involved functions to manage the development of their capabilities to launch the product on time, to sell the product, and to provide service and support to their customers. So the management of product development activities must integrate these organizations along with their customers, suppliers, and partner companies.

Product development is a very competitive business. It can be difficult and highly complex. Unfortunately, well-intended processes can be constrained, if not sabotaged, by shortened schedules, late project starts, dysfunctional decisions, inadequate reserves of resources, midcourse changes in portfolio plans, and unresolved conflicts with competing projects. The market rewards good performance but penalizes poor performance, without mercy. If you win, you get to keep your job. If you lose, bad things can happen.

Who differentiates the winners from the losers? Customers do! They vote with their purchases of products and services over time, showing preferences for those companies whose offerings provide superior value in the context of their business, activities, and environments. Customers have options. They choose those products that best meet their needs. Generally, that means those products that provide necessary benefits, in relationship to their costs, better than available alternatives. Everything else follows, such as price, revenues, volumes, competitive position, reputation, and shareholder value.

Companies develop products and services for those business opportunities for which they have competitive advantages. The job of product development is to provide superior solutions to those problems for which customers in target markets are willing to spend their money. Perfection can be elusive, expensive, and time-consuming, but where compromises have to be made, "good enough" must be judged from the viewpoint of customers. Trade-offs represent potential risks for product development, so they need to be made with accurate knowledge of the consequences for customers' value drivers.

The choices that customers have include doing nothing. So there is little forgiveness for not achieving excellence in execution for every functional discipline involved in the project. Customers may be loyal to your company based on a historical relationship. They may cut you some slack, knowing that the products or services are new and understanding that you need to work out some bugs. They may accept your less-than-superior product, knowing that your superior service will compensate, or vice versa. But how long will that last?

The "early bird gets the worm" may be a useful metaphor. Often being "first to market" has its value, since early adopters can set precedents. Products or contracts may commit customers to a long life cycle. Often, however, it's being "right to market" that wins in the end, that is, having the right quality (features, functionality, reliability, usage life) with the right costs at the right time. So if your product is first to market but at a high price or with compromised quality, can your company win in the long run? An old but telling attitude toward product development is "Quality, cost, and delivery schedule—pick two." That implies defeat, a compromise that serves neither your customers nor your company. The business challenge is then to work with a strategy that satisfies all three criteria.

This book does not propose to compromise quality by some clever approach in order to satisfy imposed constraints on development costs or schedule. Customers are not willing to make that trade-off. Likewise, we do not propose that the best product development methods take too long or cost too much for you to adopt them. Companies that are good at product development are also efficient and flexible. They treat product development as an investment in the design of a new or extended business to generate returns that are expected to be superior to alternative investments for the corporation. Otherwise, why would they do it?

There are no magic tricks here. If you lag behind your competitors in a substantial way, we doubt that there are quick fixes that will be adequate. It takes hard work sustained over time with effective implementation of better practices. It takes the right resources at the right time working on technical concepts that are superior to their competition. It takes management working to enable development projects to be successful, having a constancy of purpose over time.

The process does not focus on a specific improvement theme, but rather on the integration of many practices that "winners" do to be successful. Over the past few decades, leading companies have implemented approaches to make their work more efficient, to enable their processes to achieve shorter cycle time, and to develop their technologies to be more robust. Most of these improvements have provided benefits, but none is sufficient alone. They must be integrated into the natural way that people and teams do their work.

The Quality Improvement Movement

Quality training has provided many helpful strategies and methods, from quality circles to an emphasis on analytical problem solving. Many of the approaches use tools that were learned in the past but have been forgotten or were not well applied. Improvement initiatives have encouraged engineers to get closer to their customers and to develop technologies, designs, and processes that are more robust. Project management professionals have pushed sound methods for managing resources and their progress in performing activities and achieving deliverables on predictable schedules. Improved computer-based tools have been implemented to assist analysis, experimentation, and design documentation. Software development has adopted agile methods to embrace late changes that often are imposed by rapidly changing markets. Organizations for software engineering and systems engineering have developed "capability maturity models." Organizational development initiatives have improved the leadership and practices of cross-functional teamwork and of their support from functional management.

Industry studies, conferences, and publications continue to push for improvements. Their agendas have themes such as Design for Six Sigma, lean product development, agile development, risk management, high-performance teamwork, collaborative innovation, and portfolio balance. All of these are important and complementary. The trick is to integrate them in a manner that makes them appropriate for your business model, with implementation that makes them systemic, to become just the way you do business, not a collection of "themes of the month."

It takes "top-down" implementation and "bottom-up" practice. It takes impatience with poor performance and a sense of urgency to improve, with an imperative to achieve the benefits quickly and to remain consistently good over time. It takes creativity to develop new techniques and a flexibility to adapt to changing conditions. It's an ongoing process of improvement.

Companies that "beat" the competition in the market tend to have a knack for integrating resources, processes, and technologies to achieve results that customers judge to be superior. They do it consistently, product after product, to reinforce a reputation in the market for delivering superior value. Often it's that reputation that provides a shortcut to the value judgment, rather than a side-by-side comparison of features or a competitive analysis from an independent assessment. How can the implementation of fundamental improvements be managed?

Management of Process Improvements

The process of change for an organization can be difficult, with nontrivial risks for achieving the intended benefits. Probably you have experienced this yourself. There can be many sources of resistance. The investments that are required may be beyond those that are thought to be affordable or worthwhile. The challenges to learn new methods may be viewed as not contributing to short-term imperatives. People in power may not believe that the problems being experienced are severe enough to justify the changes. Paradigm changes may not be understood easily. We expect that you can add to this identification of barriers easily.

Certain factors have been found to be critical to success in changing the approach to product development. Examples include the following:

- The reasons for the change must be clear and compelling, both for major transformations and for reforms in the context of continuous improvement.
- The adoption of new methods must be developed internally so that ownership is with the people who must practice them.
- The implementation process must be embraced and enabled from the top down. The improvement plans must have objectives that harness the corporate energies.
- The change process must achieve measurable results early to demonstrate the benefits and to build confidence and momentum within the organization.
- The timing of implementation must not be so disruptive that the organization resists it as a "theme of the month." Remember this advice: "The best time to fix the roof is when the sun is shining."

There are several good sources for useful insights and practices for change management. You may find the work of John Kotter[1] and others to be interesting and applicable.

Clear and Compelling Call to Action

With probing, open-ended questions your evaluations of recent development projects can identify those practices that provide competitive advantages and those that consistently put your projects at a disadvantage. Where should you focus? Clearly your business achievements and disappointments will set your priorities. Capability assessment tools can identify the gaps relative to "best in class." Cause-and-effect diagramming can establish probing questions to identify those factors that contribute a larger impact and have a higher probability of occurrence.

For example, suppose your projects miss schedule milestones consistently, with market entries being late. What are root causes that can be addressed? Certainly your projects will have their own specific factors that contribute to these difficulties. If it seems evident that project management practices are deficient, the Project Management Body of Knowledge (PMBOK)[2] can guide probing questions about practices and their integration. If it seems that engineering practices are sound, but cross-functional interactions are too iterative and contentious, the principles of effective product development and of high-performance teamwork[3] can provide probing questions. Clearly there can be complex interactions among the many factors.

---

1. John P. Kotter, *Leading Change* (Boston: Harvard Business School Press, 1996).

2. Project Management Institute, *A Guide to the Project Management Body of Knowledge* (Newtown Square, PA: Project Management Institute, 2000).

What does not work well? What are their causes and effects? Here are some thought provokers that are relevant to our topic of reliability development:

- **Weak design concepts:** Reliability problems caused by immature or inferior design concepts that cannot be controlled well enough can force too many design changes late in the development process. Those that have consequences for tooling and parts inventories, for example, can lead to schedule slips and cost overruns.
- **Delays in market entry:** Major schedule slips increase development costs, threaten revenues, and have serious consequences for those organizations mobilized to launch the product. Long, repeated delays can jeopardize the competitive product positioning and threaten the reputation of the company.
- **Escaping reliability problems:** In addition to dissatisfying customers, design and manufacturing problems that are not prevented prior to production impose higher service costs and the potential need to retrofit design modifications or to recall products.
- **Constrained human resources:** Not having enough of the right skills or experience working on specific problems can cause inferior or delayed solutions. This problem raises the question of whether to hire, to develop internally, or to outsource the missing skills.
- **Constrained prototypes:** Additional prototypes may be necessary to develop robustness and grow reliability, or to obtain customer feedback. That may require additional funding and development time, which management may not tolerate.
- **Overloaded resources:** Centralized functions may be staffed only to maintain a high level of utilization, leaving no reserve capacity for variations in demand. The lack of capacity when needed may be a root cause of conflicts among projects and queues that steal development time. For example, the lack of access to an overloaded testing facility may cause important tests to be skipped in deference to the schedule or delayed to the extent of jeopardizing the product launch date. This problem can be anticipated if there is an integrated approach to project planning when using shared resources rather than a culture of competition, where the project manager having the dominant personality gets satisfied first and everyone else fights for the leftovers. Typically, overloaded resources and top-down pressures lead to multitasking, which actually tends to reduce efficiency and further aggravate the resource problem. For a more detailed discussion, in the context of Critical Chain Project Management, see Leach.[4]
- **Lack of attention to risk management:** Project teams may not be diligent at identifying risks or reducing them. Certain risks may become actual problems that must be resolved, often an expensive process that can delay the product launch. The management of risks takes resources, time, and management attention, all of which may be in short supply.
- **Bureaucratic decision processes:** Management gate reviews and other decision processes may be difficult to schedule or may demand iterations of reviews to answer unexpected questions. Those iterative reviews may actually dilute responsibilities rather than reduce risks. They may tend to add little value for customers and cause significant delays along the project's critical path. They often fail to draw sufficient attention to the rate of reliability growth that lags behind expectations and needs management intervention.

---

3. Jon R. Katzenbach and Douglas K. Smith, *The Wisdom of Teams: Creating the High-Performance Organization* (New York: HarperCollins, 2003).

4. Lawrence P. Leach, *Critical Chain Project Management* (Norwood, MA: Artech House, 2005).

- **Lack of flexibility in a design:** The system architecture of a product may depend upon the robustness of a specific design concept. If that is not achieved early in the project, the inability to change to another, more robust design concept may cause serious consequences for schedules and costs. The capabilities of the entire product and its development schedule can be in jeopardy.
- **Late changes in requirements:** The development of complex, heavy metal products can benefit from requirements being defined and frozen early. However, many products are aimed at markets that can change rapidly, leading to either late-arriving feedback from customers or new expectations set by competitive entries. Design architectures or development processes that cannot accommodate late changes gracefully can be plagued by the consequences to their schedules and costs.
- **Weak investment in early development:** Design concepts that are chosen to be baseline may be found to lack sufficient robustness in the early development phases. This can lead to numerous problems that can force costly corrective actions in the later development phases.

Of course, the list of possible problem-cause scenarios is endless. The challenges for your capability assessment are to identify the business results that are not acceptable and to establish their priorities and their root causes. Process solutions can then be devised, often with assistance from outside specialists who can bring wisdom from their experiences with other projects or companies. The business case for improvement initiatives can then argue for resources, justified by a "call to action" that is derived from the consequences experienced.

Internal Ownership

When new "best practices" and methodologies are introduced, they rarely become "common practices" until lead users inside your projects demonstrate their benefits. It takes learning and experience for people to become believers. With internal experts coaching project work, the consequences of the learning curve for project cycle times can be reduced. Once the benefits are seen to be important to a project's objectives, the new practices can be built into the project's plans. When management understands the intent of the new practices and their technical or business advantages, they can set expectations for their appropriate use. This can be very powerful. Without reinforcement by project leaders and management, new practices with high potential value can be set aside in favor of the more familiar "way we always have done work."

Top-Down Enablement

People in power positions may feel threatened by suggestions that current strategies and capabilities do not work well enough. They may see a recommendation as a reflection on their own personal competence rather than a concern for a corporate process or method. They may not appreciate the long-term value of time off the job for training or of the investment in new software tools. For example, getting customers on your project team may be viewed as too troublesome or a violation of project security. However, pilot projects can show how substantial the benefits can be. The emphasis on the development of superior robustness may take a clear understanding of why a design-build-test-analyze-redesign strategy under nominal conditions is not sufficient. Learning from other companies can help a lot.

It should be in the interest of management to promote changes that will add value. Without their participation, bottom-up initiatives tend to be short-lived, doomed to falling short of their potential impact. It helps greatly to have a high-level champion whose commitment is based on painful experiences. Members of the management team need to be the suppliers of resources, the stakeholders in the expected benefits, and the drivers in the implementation of changes. An effective tactic is for management to teach and coach the new principles and methods. This has a powerful influence on how well new methodologies are understood and integrated. It also ensures that individuals and teams recognize them as being expected in how they do their work.

There is major value in the expectations and rationale being set at the top of the organization. Subsequently, these expectations must cascade throughout the organization and be reflected in individuals' performance. Clear expectations drive behavior changes if they are reinforced in top-down communications and applied in the probing questions and acceptance criteria for project reviews. It is management that enables major changes through strategies, resources, motivation, and a constancy of purpose. It's also management that can sabotage major changes by not making their support deliberate, visible, and consistent.

Bottom-Up Improvement

What if there is no upper-management support for a major overhaul of the product development process? Should you assume that there is no point in trying to improve? Fortunately, even with management reluctance, it is still possible to make improvements. Usually management recognizes when things are not optimal. They would like capabilities to improve, so it is very unlikely that they will prohibit the use of better methods and processes as long as improvements don't jeopardize schedules, budgets, or product quality. If all you can get is indifference, take it. It is better than active interference or prohibition. The problem is that everyone is tired of the "program of the month." A major process overhaul requires the commitment of time and money, making it very difficult to argue that there will be an acceptable payoff. This is especially true if people have been disappointed with past initiatives. Under these conditions, a better way to approach the problem is to scale back from trying to make the "big win" and concentrate on demonstrating success on small subprojects. Shoot for small initiatives that will be "under the radar." It is easier to get buy-in after you have demonstrated improvements. When success happens, publicize it. Before long there will be supporters.

Objectives with Substantial Impacts

What are the objectives of your major improvement initiatives? Examples may include

- Higher levels of customers' satisfaction
- Higher quality in new products and services
- Higher product reliability and longer usage life
- Lower manufacturing and service costs
- Lower development costs
- Shorter, more predictable time to market

These objectives may be obvious. What else applies to your situation? All can have a direct impact on the profitability of the portfolio of new products. They are measurable, but what are their criteria for acceptance? How well are these objectives related to the success of your business? How well can they be reflected in the performance expectations for individuals, teams, and projects? Development time can be measured for a project, but is the real objective a shorter time or the more efficient use of time? Is it increased "leanness" or higher productivity? Possibly a better objective is to start on time.

Is a lower failure rate the objective, or should the emphasis be on reducing those failures with the higher severity levels? Should the project teams care more about constraining development costs to comply with budgets or about getting the right functionality, quality, and reliability to the market on time? Are all development projects comparable, or is each one unique in its challenges? The setting of objectives is not easy, particularly since if it is not done well it can create unintended incentives for the wrong behaviors.

Are these objectives independent? Probably not! Higher product reliability and durability will contribute to higher customer satisfaction. So will features and functionality that are not provided by competitive products. More customer benefits can enable higher prices without jeopardizing the perceived value. Lower manufacturing and service costs can enable lower prices. Lower prices can contribute to higher perceived value and higher manufacturing volumes, which in turn can decrease unit manufacturing costs and increase profit margins. A more efficient product development process should reduce development costs and enable on-time market entry with a more predictable beginning of the revenue stream.

Early Measurable Results

It is through performance expectations, cascaded throughout the organization, that the necessary actions and behaviors can be driven. This is particularly effective if people's paychecks are affected. For example, if engineers are required to shorten product development time, but managers are not expected to start development projects early enough, it will be clear that the engineers are to be punished for a failure in management. There is a powerful message sent when development teams share performance expectations with upper management. Balanced scorecard methods[5] provide a way to make the change process important to everyone's wallet. People then work to the same set of goals. Here are a few suggestions that may be helpful:

- Know what excellence looks like. It will give you benchmarks against which to establish your capability gaps and important projects for corrective actions.
- Evaluate your current experiences with an eye to the business consequences, not just to excellence in the scientific or engineering results.
- Identify lessons learned from past projects and from benchmarking. Incorporate them into the functional and project plans and the standardized business processes.
- Allocate funding, labor, and management bandwidth to the development and implementation of process improvements. Otherwise, not much good will be achieved and the past will also be the future.
- Do not try to fix everything at once. Focus on the "high bang for your buck" opportunities. Prioritize your improvement activities to achieve quick, visible results that build enthusiasm and confidence, and then form the basic foundations for longer-term improvements.
- Keep it simple. That doesn't mean that learning sophisticated methods is to be avoided. However, if a collection of improved engineering strategies and tools is the answer,

---

5. Robert S. Kaplan and David P. Norton, *The Balanced Scorecard: Translating Strategy into Action* (Boston: Harvard Business School Press, 1996).

their direct relevance to "bottom-line" benefits needs to be demonstrated for and understood by those who provide the funding and resources.

Soft Causes of Poor Reliability and Durability

What is reliability? What is durability? Here are useful definitions:

> **Reliability:** *Acceptable quality delivered consistently over time, under specified operating conditions*
> **Durability:** *Acceptable product life, under specified operating conditions*

So we focus on achieving the required quality, as perceived by customers, and on keeping it stable under the operating conditions of the various product applications.

These definitions don't have the rigor associated with being testable and measurable, but they are useful when thinking about reliability and durability at a high level. In Chapter 2 we present a definition for reliability that has statistical rigor.

Fundamentally there are two categories of root causes of reliability problems: "hard" causes and "soft" causes. Hard causes are in the technical elements of the product's design and of its manufacturing and service processes. Soft causes are in the human behaviors of project teams and in the management of the development process.

Many of the tools and processes in this book address the hard causes. Soft causes focus on opportunities for project leaders and management to enable excellence in execution and to avoid being the root causes of failures in project support, cross-functional integration, and teamwork.

Your process capabilities assessment may identify soft causes such as

- Project plans that are not based on sound strategies or cross-functional dependencies
- Production-intent designs specified before robustness and integration have been developed
- Product requirements that are neither complete nor stable (e.g., "feature creep")
- "Seat of the pants" decisions that increase project risks
- Selection of design concepts that lack sufficient capabilities or initial robustness
- Technical deliverables that lack clear expectations at a project milestone
- Development teams that lack process discipline or are complacent about their capabilities
- Project teams that are not able to speak truth to power or ask for help
- Project plans that do not apply lessons learned from past projects
- Development budgets that are reduced by ad hoc decisions external to the project

To the extent that your capability assessment recognizes these or similar problems, effective solutions can be developed. The literature on product development includes references that can provide additional insight into soft causes. For example, Robert Cooper[6] has characterized many things that successful companies do well. An interpretation of these, in the context of your

---

6. Robert G. Cooper, *Winning at New Products: Accelerating the Process from Idea to Launch, 3rd ed.* (Cambridge, MA: Perseus, 2001).

business model, can help you design probing questions for your assessment. Don Clausing[7] describes many fundamental engineering methods and echoes the concern about soft causes in his description of "The 10 Cash Drains." You may recognize many of them as root causes in your own business. Let's look at our list of soft causes in a little detail.

Inadequate Project Management Plans

A development process to achieve higher reliability should follow strategies chosen by the project to have the highest probability of success. It may leverage mature designs. It may be dependent on the development of specific technical concepts achieving prerequisite robustness. It may collaborate with other companies that have necessary and complementary technical capabilities. It may depend on extensive modeling or experimentation in the early development phases. In whatever way the strategies are intended, the project management plans must then reflect their relevant activities, dependencies, timelines, milestones, and resources. The plans have to show *how* the strategies will be implemented. If the resources or funding is not available, the plans may not be achievable and the strategies may not work.

Another concern is that the project management plans must deliver key parameters to the project's business plan. For example, if the plan to achieve the required reliability is not acceptable or achievable, the initiative to achieve the reliability requirement is set up to fail. So if the project plans are not achievable, the business plan also is not achievable and the project should not be chartered.

Inadequate Development of System Integration and Robustness

This is a critical concern for reliability development, being part of the foundation for the familiar guidance to "achieve quality early." To the extent that these criteria are not satisfied, additional risks are incurred and the objectives for quality, reliability, and durability are in jeopardy, as are those for schedules and cost structures.

The evidence of robustness is the specification of critical design parameters that have been demonstrated to control the mean and standard deviation of performance under expected stressful conditions. The specification and configuration control of these set points are the objectives of Critical Parameter Management (CPM). During the product design phase these functional parameters are converted into design parameters suitable for product manufacturing and service. Without them, there is no sound basis for the specifications of a design.

Important actions include these:

- Technical concepts should be developed to be robust for the application prior to becoming baseline for the product development project.
- Subsystems should be developed to be robust within the system architecture prior to being integrated with other subsystems.
- Robustness should be optimized at the system level in order to specify the parameters that must be controlled by the production processes.

7. Don Clausing, *Total Quality Development: A Step-by-Step Guide to World-Class Concurrent Engineering* (New York: ASME Press, 1994).

The strategy for robustness development must incorporate an understanding of the roles and benefits of specific methodologies, many of which are discussed in later chapters. From a project management viewpoint, sufficient time needs to be allocated for the selection of the best design concepts and for the planning of experiments, avoiding a rush to cut metal and to build prototypes prematurely. Too many prototypes, built without the learning from previous generations, contribute to Don Clausing's concern for "hardware swamps." The rush to build-test-fix may satisfy management's desire to see things happen but may be contrary to sound engineering.

Effective product development is analogous to applied systems engineering. It follows a disciplined process of developing linked information and decomposing it to be useful at the various levels in the system hierarchy. Manufacturing and service functions are considered to be part of the system, as are the procedures for users of the product. The development process includes the flexibility to make changes easily within the system and to freeze elements upon which other developments are dependent. Although it is a disciplined process, it should be very adaptable to the technical and business characteristics of a project. The leveraging of existing information and designs is prudent and efficient, while ill-advised shortcuts introduce risks and can sabotage the project.

Changing Product Requirements

Your engineers create solutions to requirements that are understood in the technical language of their design concepts. The origins of these requirements fall into three fundamental categories:

1. The needs and preferences of your customers in selected market segments
2. The standards and mandates imposed internally by your company
3. The government regulations and industry standards for your markets and countries

Ideally these requirements are defined and clarified in the initial development phases and, in the ideal case, approved and frozen. In practice, life is not that easy. Requirements are vulnerable to misinterpretation, internal biases, lack of insight, disagreements, conflicts with each other, risks for achievability, and changing market situations. Feedback from customers may prove the initial requirements to be wrong or ambiguous. Your experience adds to this list. The consequences can range from being trivial to disastrous. If the requirements are vulnerable to changes, engineering is shooting at moving targets with an increased probability of major consequences for their delivered quality, costs, and schedules. If the requirements lack insight into the real intentions of their application, or lack clear differentiation from competitive products, the product and price positioning are in jeopardy. Failure to comply with regulations or industry standards can be a barrier to market entry. There is very little forgiveness.

Good practices are well developed to enable engineering and customers to work together so that designs not only solve the actual problems but also delight customers in ways that establish competitive advantages. Useful tools can enable the thorough and accurate translation of requirements from the language of customers into the technical language of engineering. They can then be decomposed to be applicable at the levels of subsystems, components, and manufacturing processes. Agile strategies for product development enable teams to react to late-arriving changes in requirements and to feedback from customers' tests of prototypes without dramatic consequences for schedules and costs.

It is the job of development teams to design the basic functions of the product, specifying the controllable parameters so that the product transforms the customers' inputs into desirable outputs, in spite of stressful factors that can cause deterioration or failure. Figure 1.1 illustrates

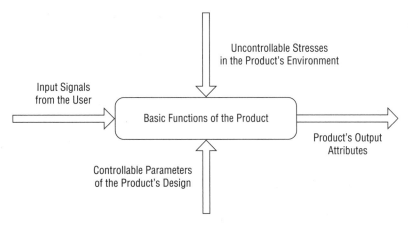

Figure 1.1   The requirements for a product's output attributes are driven by customers (VOC), as well as by internal mandates (VOB, VOT) and regulatory standards (VOM).

this simple viewpoint. It's then essential for development teams to understand the input/output relationships in the applications and the uncontrollable stresses that can affect them.

There are two fundamental concerns that project leaders need to address:

1. There is a tendency for engineers and management to believe that
   a. They already know more than their customers
   b. The process for understanding the voice of the customer (VOC) is too expensive and time-consuming, or it places the confidentiality of the project in jeopardy
2. Requirements can also be driven by "voices" other than those of customers (VOC):
   a. There is the "voice of the business" (VOB), which tends to demand lower development costs, shorter schedules, and lower production costs, but often without comparable concerns for satisfying customers. Lean product development and sound strategic decisions early in the process align with this mandate. Start the project earlier, instead of depending on schedule pressures to compensate for delayed decisions.
   b. There's the "voice of technology" (VOT), which tends to modify requirements to reflect those capabilities that can be delivered within the cost and schedule constraints. Another version of VOT is the "technology push" scenario that assumes, often naively, that customers will benefit from a new technology because it's new.
   c. There's the "voice of marketing" (VOM), which tends to push a wide range of requirements focused on opportunities for current sales and market share. These can be a reaction to the latest market information and to sales challenges with current products.

These additional sources of requirements are not to be ignored. The worst case of technology push assumes that the benefits of the product are needed, that customers will accept and pay for whatever capabilities you deliver, and that there are no competitors. This is often called a "WYSIWYG" product, that is, "What you see is what you get." Of course, in a competitive market that strategy doesn't work.

In many cases there are valid and reasonable strategies to shortcut the VOC process. It may be that a few key people have more insights about future markets than do current customers.

That's a gamble that may pay off. The challenge is to know when a rigorous VOC process must be included in the project plan and how to get more benefits from the investment in its process.

Risk Management

Many decisions are made during a product development project. Good ones clarify the direction of the project, enable the progressive freezing of information, and reduce risks. However, business and technical pressures can promote risk taking, even reward it. Risks are problems that have not yet occurred but might occur in the future. The probability of their occurrence is uncertain, as are their expected consequences for customers or for the company. When problems actually do occur, projects know that they have to be dealt with in a manner appropriate to their consequences. However, potential problems that have not yet occurred tend not to get the same deliberate attention.

Another way to say this is that problem correction is expected and rewarded, whereas problem prevention is a quiet competency that can go unnoticed. Problem correction late in the process tends to cost more than problem prevention. Excellence in achieving schedule milestones and the objectives for reliability and durability, for example, often is rooted in the capabilities of reducing risks, that is, avoiding the problems that consume valuable time and resources when they can be least afforded.

Good practices for managing risks reserve some capacity in resources to identify risks and to implement preventive actions, where justified. Contingency plans are the alternative. These are preplanned actions that will be taken to compensate for the consequences of the problem, when the problem does occur.

Collective wisdom and functional understanding are organized to identify risks and to characterize their expected consequences and probabilities of occurrence. This understanding can be based on experiences with previous projects, on specific technical or business insights, on scenario planning, or on other techniques. Often the foundations for a lower-risk project are put in place with decisions that are made in the early development phases, such as the selections of the development strategies, the system architecture, its design concepts, and their enabling technologies. The concerns increase when significant risks are ignored and risk taking is rewarded. But what are the costs? And who pays for them? For example, decisions based solely on reducing manufacturing costs can result in poor product quality or reliability, higher life-cycle costs, or a loss in market reputation.

Imagine a development project that achieved its requirements for quality, reliability, durability, production costs, and schedule, with no late crises that jeopardized its graceful market entry. Calculate the costs of delayed market entry, and then ask how much you could afford to invest in identifying risks and preventive actions. Remember the wisdom in "Pay me now, or pay me later."

Immature Baseline Technologies

Baseline design concepts are chosen in the early development phases. What criteria are used for their selection? How dependent on them is the product architecture? For new or adapted design concepts, what happens if their capabilities fall short of their expectations for the application?

The concern here is that new technical concepts may be chosen for reasons that are independent of the value they can deliver to customers. It may be that engineers favor a particular design concept for purely technical reasons. Possibly it enables lower costs or an attractive architecture. It may be one for which they have personal preference. It might be the key to major increases in reliability and durability. Fine, but is it ready to be commercialized by a product development project? Does it have functional parameters that control its performance and variability? A disregard for this question tends to place technology development within a scheduled product development project, imposing additional risks. A very difficult situation is created when the product's architecture is entirely dependent on a design concept whose technologies are immature, that is, neither superior to alternatives nor robust for the application.

We remember a disastrous project. The product's architecture was entirely dependent on a design concept that enabled a rather compact system configuration with lower power consumption and reduced acoustical noise. However, the control parameters of the design could not be replicated by available manufacturing processes. In robustness terminology, the process could not be maintained "on target with minimum variability." In addition, the environmental stresses from the product's applications caused shifts in the functional response that customers would not tolerate. Fighting these problems continued well into production. No alternative designs were available, primarily because the system architecture could not be adapted. The root cause of the problem was a decision made very early that "bet the farm" on the hope that clever engineering could compensate for a failure of technology development. It could not. The costs were enormous!

In preparation for its commercialization, a new technical concept must be developed to be superior to available alternatives and to be robust for the range of intended applications. These criteria say a lot. Products often compete with their technologies. What makes them superior to available alternatives? What does it mean to be robust for the application? Good practices include two basic deliverables:

1. A stream of alternative technologies developed in anticipation of the needs of future product development projects
2. An unbiased selection process that places available alternative concepts in competition, with selection criteria that include demonstrated performance and robustness relevant to the intended product applications

Examples of "available alternatives" can include a current design already in production, an analogous mature approach that can be adapted to a new application, a design available from a partner or competitor, or an alternative new technical concept developed in parallel. A selection process that is vulnerable to the loud, bully voice of a concept advocate will lack the objectivity that is necessary.

Lack of Clear Expectations

Management has substantial power to guide a development project and affect its outcome. Certainly they charter projects, review their progress, provide resources, and make decisions. They also have the role of setting clear expectations. These expectations may focus on functional excellence. They may demand rigorous integrity in data that are needed to support decisions. They may express a bias for costs or schedule compliance. They may emphasize concerns for risks, that is, reductions in uncertainty. In the absence of clear expectations, development teams are left to define their own acceptance criteria.

Routine conversations with management, as well as those during project gate reviews, are opportunities for these expectations to be clarified and reinforced. If the discussions are entirely business-oriented, the technical expectations are at risk. If the discussions are entirely technical, the development teams can tend to forget that they are managing an investment that is the generator of a new value stream with expected returns to the business.

The challenges for management are substantial. For the purpose of this discussion, they can include the need to

- Maintain an accessible and interactive relationship with development teams
- Ask probing questions with both technical and business insight
- Recognize good answers and understand their metrics
- Make data-driven decisions so that their customers win
- Set clear expectations for the effectiveness and efficiency of future project work
- Focus on the "critical few" risks that are most important to the business, rather than on the many potentially trivial "issues" that people tend to push on management
- Accept action items to enable the predictable progress by development teams
- Reinforce both flexibility and discipline in standardized development processes

Certainly there are other items for your list, but the important point is that your development teams listen to management. So management needs to be clear about the signals they send.

Lack of a Disciplined Process

Complacency can be a handicap for those companies that have dominant technologies, a high market share, admirable profitability, or a history of successful product launches. With deep pockets, money can be thrown at problems. An unfortunate side effect is that the realities of inefficient processes, growing competitive disadvantages, and customers choosing other companies' value propositions, among others, can remain outside a company's consciousness. Fundamental changes in the marketplace can be overlooked. Your company can become vulnerable to its own successes.

In these situations, the problems that have soft causes become much more fundamental than just product performance or reliability. Your basic business model may no longer be viable. Instead of your capability assessment providing guidance for continuous improvements, it may identify the need for major reengineering projects. Companies that survive over the long run often practice an ongoing reinvention of themselves. Those that do not often find themselves no longer being significant players in the market, or in some cases no longer existing. Although this concern is far beyond the scope of this book, to the extent that it exists it can be a major handicap to the development of reliable new products.

Inability to Speak Truth to Power

In the preceding discussion about the setting of expectations, the advice has been for management to have an "accessible and interactive" relationship with development teams. The absence of this relationship can easily jeopardize the ability of teams to deliver on their value proposition.

Suppose, for example, that management gave clear expectations that they did not want to hear bad news, particularly at gate reviews. This is a "shoot the messenger" scenario. How would development teams behave? Management may intend sternness and ridicule to motivate improved deliverables. On the other hand, such a policy may also prevent bona fide risks from being revealed or significant decisions from being based on the correct data. Project teams may not be willing to acknowledge problems, hoping to correct them before the next gate review. Worse yet, the development teams would be set up to be scapegoats for resulting failures. That situation is neither constructive nor efficient.

The quality leader Dr. W. Edwards Deming[8] was prescient when he urged management to "drive out fear." Rewards should go to those who are open and honest about the status of their project. Management needs to understand the truth early, when there's still time to react without jeopardizing the project's success.

This same intimidation may motivate project leaders to withhold any requests for additional resources or guidance. They may be convinced that it would reflect poorly on their own leadership capabilities. But who pays for this mistake? Imagine how constructive would be the process that brought the right resources to the right problem at the right time, regardless of a naive plan that was established sometime in the past. Independence and self-confidence are healthy attributes, but they can work against you. Certainly it's better to address problems when they are small, when their corrective actions can be managed easily. In his memoir, Harold Geneen,[9] the executive who built ITT into a multinational conglomerate, described his expectations on being open about problems. He felt that hiding problems was a waste of time, time being the one irreplaceable resource that his organization had. He expected his staff, when faced with difficult problems, to be open and to seek help early in the belief that the collective experience, wisdom, and skills of the organization usually trumped those of the individual.

When we work to solve problems, time is our most important asset. Letting too much time pass before getting help on a tough problem makes it more likely that schedules will suffer and that the delivered product performance and reliability will fall short of expectations.

Failure to Implement Lessons Learned

Many organizations engage in the struggle to get a new product to market faster but do not allocate time afterward to learn about what went well and what did not go well. These lessons learned are important contributors to continuous improvement. Without organizational learning, the mistakes of the past tend to be repeated. This can generate inadequate technical and business results across the portfolio. It can also demoralize the workforce who see the same mistakes being made over and over, and who will wonder if product development work will ever become less stressful.

A key element of organizational learning is the clear intention that the evaluation process not be a negative one, that is, one that is perceived to place blame on individuals or teams. To the extent that it is managed to be blameless and clearly intended to build wisdom into follow-on projects, with a top-down commitment to act on the results, it will be a healthy and anticipated process.

Inadequate Development Budgets

If teams are required to shorten development time and also to reduce their resource budgets, they may have objectives that are overly constrained. For example, it is often the case that an overrun in the project's development budget will have only a small impact on a new product's profitability. However, spending more money on resources to commercialize the product earlier, with higher quality at launch, can enable higher revenues and contribute to lower manufacturing and service costs. Unfortunately, budget compliance often gets more attention, probably because it is more easily measured and is a familiar measure of management performance.

This discussion of common soft causes of problems reinforces the notion that product development, although often perceived to be a set of technical activities, is managed in the

---

8. Mary Walton, *The Deming Management Method* (New York: Perigee Books, 1986).

9. Harold Geneen and Alan Moscow, *Managing* (New York: Doubleday, 1984).

context of a wide range of processes and behaviors that have the potential for either positive or negative consequences. That depends on how they are managed. The assessment of your company's capabilities and of the related needs for their improvement must include these soft concerns as well as those more technical hard causes.

Throughout this book we intend to shed light on both categories for improvements, particularly to the extent that they can contribute to the development of improved product reliability.

Key Points

1. The technical development of new products must enable the success of the business.
2. Developing new products is not just a challenge for engineering. Many functions of your enterprise must be involved deliberately in value-adding ways that are appropriate.
3. Ongoing research of markets and technical concepts must identify advantageous business opportunities and evaluate competitive threats.
4. Product development must implement superior design concepts that can be manufactured and serviced economically, with constant attention to risks.
5. Excellence in teamwork has to be achieved across organizations, between teams and management, extending along the entire value chain to suppliers and customers.
6. The processes and practices that are inherent in the way that work is done must deliver effective results in an efficient manner consistently from one project to the next.
7. With the investment in product development being an investment in developing a new or extended business, better methodologies, with excellence in their execution, can deliver substantial competitive advantages in the market and much-improved returns to the company.

Discussion Questions

1. How well do your product development teams understand your customers' value drivers?
2. How well do your new products, after market entry, return value to your company as expected by their business plan?
3. How have risks affected the predictability of your development projects?
4. What types of risks have materialized as problems? What were the consequences?
5. What successes have you had in developing and implementing process improvements?
6. How well have process improvements provided competitive advantages for your projects?
7. Give some examples that show how your development teams practice problem prevention.
8. What are the major difficulties faced by your development projects?
9. How well do your development projects learn from past successes and difficulties?
10. How do you think your product development process compares to those of world-class companies?

Reliability, Durability, and Robustness

Hard failures are easy to detect since things stop working. Soft failures are more the result of performance expectations not being satisfied. Satisfactory performance is often subjective, with variations in judgment among customers. To understand these types of failures better, we introduce the concepts of quality loss and robustness. A concise summary can help readers understand how failures can occur and the tools that are available to prevent hard failures and reduce performance variations.

When purchasing a product, most consumers don't have well-defined or precise expectations for performance and reliability. We can imagine them saying:

"It must not break."

"It has to be there when I need it."

"It should always perform as advertised."

"It can never let me down."

"I expect it to last a long time."

"For that money, it had better work all the time."

As with many customers' expectations, the need for reliability has to be interpreted and translated into specific and measurable attributes that product development teams can address. Designers of product systems need quantifiable requirements.

While customers often refer to parts breaking when discussing product reliability, there are other ways that products can fail. Over time, performance can drift away from its desired operating point, eventually degrading to the extent that customers notice and call for corrective action. An example of this is fuel economy for an automobile. As long as the vehicle is close to the advertised fuel consumption rate, all is well. Drivers may be unhappy with their operating costs, but if the manufacturer of their vehicle has delivered on the promised performance, they really won't complain, at least to the producer. After all, they didn't actually need to buy that fancy "Super Maximus" SUV. However, if the rate of fuel consumption increases substantially from the expected performance level, they'll not be happy and may take their vehicle in for service. Although no parts have failed, they still have the perception of a failure in performance. This is a soft failure, different from the breakage of parts. When developing products, teams must consider both hard and soft failure modes.

Reliability as an Element of Value to Customers

There is little doubt that, for many products, reliability is high on the list of attributes that are important to customers. A well-known consumer guide features reliability and repair data for various automobile models by year. Many of us have consulted this publication when considering a purchase. We avoid the vehicles with black circles and covet the ones having red circles.

If you think in terms of the drivers of value for customers (see Chapter 9 for a discussion of the Kano[1] Model), reliability falls into the category of a "basic need," unless a producer offers a new product with a major leap in reliability or usage life. Often it is taken as a given that reliability will be at least at some acceptable level. The level you need to deliver to avoid dissatisfying customers is set by the marketplace and your competitors. Forty years ago, automobiles needed frequent tune-ups to maintain performance. It was an expectation set by the existing benchmarks. Today it is common for gasoline engines to run well past 100,000 miles without needing anything but fluid changes, filters, and fuel. The bar has been raised, and anything less is unacceptable. This is true regardless of the product, since comparison data are more readily available and well known by consumers. Regardless of the end-user group, the marketplace is interconnected[2] and bad news spreads quickly.[3]

Ask for a definition of *reliability* and you may get a variety of answers. Likely you will expect definitions that sound like those comments listed in the preceding section. The answers all have a common thread. However, in order to apply engineering tools to achieve higher reliability for a new product, you need a definition that will enable you to articulate reliability requirements that are specific and measurable. A more rigorous and useful definition from an engineering viewpoint is the following:

> ***Reliability*** *is the probability that a system or component will perform its intended function for a specified duration under the expected design operating conditions.*

If we parse this definition, we find there are several key elements.

Probability: Probability is the likelihood of an event happening, a number between zero and one. A failure that will never happen within a specified duration has a probability of zero. One that is certain to occur has a probability of one. If you put 100 devices in a test and, after 1000 hours, 90 are still functioning, you would say that the device has a reliability of approximately 90% for 1000 hours. This is the same as saying that the probability of a device surviving until 1000 hours is approximately 90%, also equivalent to the statement that the probability of the device failing before 1000 hours is about 10%.

Assuming that a system is not "dead on arrival" (DOA), the reliability at $t = 0$ is 100%, declining over time, eventually to reach zero. This holds for devices that fail in early life because of manufacturing quality problems, during useful life because of external causes, and at their end of life because of wear-out.

1. Lou Cohen, *Quality Function Deployment: How to Make QFD Work for You* (Reading, MA: Addison Wesley Longman, 1995).

2. Albert-Laszlo Barabasi, *Linked* (New York: Plume, 2003).

3. Emanuel Rosen, *The Anatomy of Buzz* (New York: Currency-Doubleday, 2002).

System or component: In an engineering context, definitions of reliability are usually about the operation of a device or system. We can extend the definition to include the reliability of people or processes: "Jim is not that reliable. His voice mail greeting says that he will return my call before the end of the day, but he hardly ever does." If we wanted to, and had enough time on our hands, we could keep track of how often Jim is true to his word and make an estimate of the likelihood of getting a call by end of day. The probability of getting a call back would be Jim's reliability for this particular task. For a product system with an involved user, what is the probability of the user making a mistake, or not knowing how to operate the product in a particular mode or recover from a specific nuisance condition? The knowledge of the user is a potential contributor to reliability.

Perform intended function: This part of the definition is very broad, including scenarios in which a component or system suffers an abrupt failure, resulting in its no longer being able to function. Another way that a system can fail is for an important output to drift outside its acceptable limits. Either event is a failure, since the system no longer satisfies its requirements. For either, think in terms of the probability that the event will occur within a specified usage duration. If you consider a failure due to drift of an output response, in order to set acceptable limits you have to understand the consequences of the output's variation on customers. In some cases, where the perceived quality of the output has a large amount of subjectivity, you have to perform psychophysical experiments with end users to develop statistics that characterize the range of acceptable performance.

Specified duration: Note that *duration* is used intentionally, rather than a metric such as *time*, since it can have a variety of units. For example, automobile tire life is usually specified in duration of miles. The life of an electromechanical device that is actuated many times, such as a solenoid, would be specified in terms of total actuations. Electronics are examples of devices with failure modes driven by time. So the appropriate units for duration depend on the device and on how the physics of failure determines the failure modes for the device.

Design operating conditions: Having to specify design operating conditions is an acknowledgment that conditions outside the design's control may aggravate failure modes. For some products, such as large production printers, several contributors to performance have expectations that depend on the environmental conditions. For example, paper-handling performance is affected by the bending stiffness of the paper, which in turn is vulnerable to temperature and relative humidity as well as to the way in which the paper was manufactured. So the expectations for the failure rate of paper handling are usually quoted for both nominal and stressful combinations of temperature and humidity and the orientation of the paper.

In fact, there is an entire reliability engineering subspecialty that deals with using increased stresses to force failures to occur sooner. If you have a good predictive model that relates performance to stress level, you can use performance at high stress levels to predict failure rates under normal operating conditions. This is a topic in our later discussion about life testing.

Essential Elements of Reliability

Four essential elements are required for reliable products, each one building upon the other:

1. Designed quality
2. Robustness
3. Durability
4. Manufactured quality

Designed Quality

You might think of designed quality as the degree of compliance with performance requirements under nominal conditions. This can include the basic features of the product system and the functions that they deliver. Generally, product functions have measurable responses that can be compared to requirements and their tolerances. This quality criterion is inherent in the selection of the design concepts with their enabling technologies and the specification of their critical design parameters. Some may call it "entitlement reliability" in that, although it cannot get any better, it can be degraded by the stresses of environments, manufacturing, shipping, rough handling, installation, customer use and abuse, and service maintenance and repair.

An improved concept of quality is the following:

*Customers expect the **quality** of a product to deliver "on-target" performance each time it is used under all intended operating conditions throughout its intended life with no harmful side effects regardless of tolerance limits.*

An obsolete viewpoint is that of quality being represented by the comparison of actual performance variations to their allowable tolerance. That "in-spec" perspective is contrary to the objective of robustness development, which strives to reduce variability.

Robustness

The concept of something being "robust" is familiar. It implies adjectives such as *tough*, *rugged*, *durable*, *long-lived*, and *stable*. Webster's dictionary defines *robustness* as "exhibiting strength or vigorous health." Our engineering definition is similar but more specific and measurable. Controls engineers design robust controls that make system performance less sensitive both to errors in describing the model of the system and to disturbances or perturbations imposed on the system. In the context of product reliability, our definition of *robustness* is to be less vulnerable to the variability in design parameters and to the consequences of stress factors both internal and external to the product. Stress, aka "noise," is a degrading force beyond the control of the design. The end result of robustness development is a system or product that over time in the presence of stresses will, on average, operate closer to its requirements than a system that has not been engineered to be robust. Our operational definition of *robustness* is a relative one:

*A system that is more **robust** is less sensitive to the sources of variability in its performance. Robustness is the ability of a design to deliver its intended function with quality characteristics that are aligned with their requirements ("on target") and that are stable under stressful conditions ("minimum variability").*

Robustness can be measured and normalized. The value proposition advertised by Timex, in the days of analog watches that actually ticked ("Takes a licking and keeps on ticking"), specifically spoke to the robustness that customers expected in their products. Being less vulnerable to stress—that is, being "more robust"—does provide a competitive advantage.

The development of robustness is essential to the objective of problem prevention in product designs and, as such, precedes the problem correction activities. Problem prevention sends its benefits downstream in the development of the new product. Products that are more robust make the transition into manufacturing and service much more easily. The benefits of designs that are proven to be robust for a range of applications must include that knowledge being fed back upstream to the development of follow-on product designs. So first you develop robustness, and then you grow reliability. The same strategy applies to manufacturing processes and, with a little imagination, to business processes.

Durability

The durability of your designs determines how long they will operate before deterioration causes either an abrupt failure or an unacceptable drift in performance. For non-repairable products, durability determines the usage life until the first fatal failure occurs. Generally, increased durability costs money since it requires the use of better components, higher-strength materials, materials that are more resistant to wear, and perhaps materials having special qualities such as corrosion resistance or stability at very high temperatures. In some situations you can achieve increased durability using brute-force approaches where "money is no object." However, for many systems, such as airframes, which are structures requiring a high strength-to-weight ratio, brute force will not work. Nature forces you to select clever design concepts and to develop their robustness for the application. In engineering science you can't ignore nature's laws for even a short time without taking great risks. In cases where nature is not so demanding, you may be able to meet your need for durability without clever design, by throwing money at the problem. Over the long run, however, revenues must exceed costs by some healthy margin. As a result, depending on the product, a business cannot afford the "gold-plated" design for long and must eventually compete with cost-effective solutions.

> ***Durability*** *is the characteristic that enables a product to have an acceptable useful life without significant deterioration.*

This definition is somewhat vague intentionally and requires qualification. Central to the definition is the notion of useful life. How long do customers expect the component, system, or structure to operate without significant deterioration? The answer is different for the Hoover Dam and for the family sedan. Usually, durability is visible to customers. Objects that are highly durable often have the appearance of high quality and substance.

In making design decisions that affect durability, you have to answer questions such as these:

- What are your customers' expectations for the useful life of the system?
- What are the consequences of particular failure modes?
- Is the product or system you are designing meant to be repaired?
- If the answer to the previous question is yes, will the product be serviced at your customers' sites, at a service center, or in your factory? Who should perform the repairs?
- What are the allowable costs for ongoing maintenance and repairs?
- What strategies are your competitors following?

Here are some examples of design strategies for durability:

- Electronic and mechanical components can be derated. If a design requires a capacitor with a 100V rating, you can use one with a higher rating. A bearing can be selected to give a longer life by increasing the load-speed design margin. This buys you an increased cushion between the strength of the device and the applied stresses that would cause its failure. Derating is a common approach that can increase durability, but with a cost.

- You can reduce the sensitivity to degrading stresses, such as by using hardened steel parts to improve their wear resistance.
- Durability and robustness are qualities that are driven by design decisions, by the development of robustness and correction of problems, and by the identification and mitigation of risks during the development process.

Manufactured Quality

When there are defects in a product due to problems with its manufacturing processes, the failures are often seen early in life, at times when the product is unpacked or first used. Poor quality in manufacturing trumps the best and most robust product designs. On the other hand, a design that cannot be replicated with efficient manufacturing processes is not a high-quality design. Manufacturing quality problems must be corrected in order to realize the entitlement of the design for quality, reliability, and durability.

As described for the product's design, the concept of robustness applies also to manufacturing processes. The early Six Sigma movement focused on manufacturing, as did the early work on robustness development. Manufacturing process capability (Cpk) for a design parameter is a measure of how well a process is stable and aligned to the requirement for that parameter. One of the definitions of this metric is

$$Cpk = \frac{Minimum\{(USL - \mu),(\mu - LSL)\}}{3\sigma} \tag{2.1}$$

The design parameter, as reproduced over many products, varies under the stresses that affect the manufacturing process. The distribution of that parameter has a mean of μ and a standard deviation of σ. The product design has latitude for those variations specified by USL and LSL, the upper and lower specification limits. You can see that a higher process capability is gained by increasing the distance of the mean from the closer of the two specification limits and by reducing the variability of the process. Robustness development provides a set of tools and methods to do this. Develop robustness in the process to increase the "capabilities" of the process.

Manufacturing quality problems can be diagnosed with careful failure analysis, as well as by fitting a life distribution such as a Weibull distribution. We explain the basics of the Weibull distribution later in the book. If you see a Weibull shape factor of less than one, it's a sign that you have early-life failures that likely are caused by quality problems in the manufacturing supply chain or the product's assembly process.

The Taxonomy of a Failure

The Bathtub Curve

The "bathtub curve," shown in Figure 2.1, is a useful way to illustrate the types of failures that can affect a component or system throughout its life. There are three types of failures:

1. Failures during the early life of the product
2. Failures during the majority of the product's useful life
3. Failures at the end of the product's life

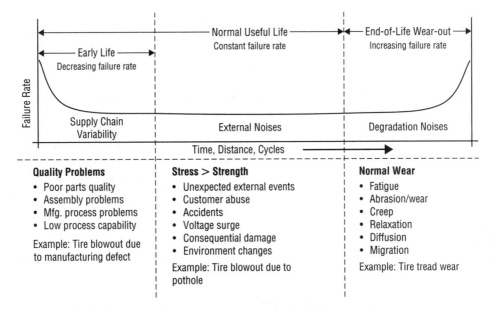

Figure 2.1 The "bathtub curve" shows three phases in the life of a product.

Usually these failures have different root causes. The failure data for most components are mixes of these three types of failure modes. There are two distinct populations: those components or systems having latent quality defects that escape the factory and those that are delivered within specification. Defective components have greatly reduced durability and are likely to fail earlier in life under normal operational stresses. They are also likely to fail from random events where stress exceeds strength, which itself has been greatly reduced by the presence of the latent defects. Most components with latent defects tend to fail early in life under normal operation rather than in high-stress random events.

Early-Life Failures

Hard Failures due to Quality Compromises

Early-life failures are usually associated with quality problems in the supply chain due to flaws introduced during manufacturing, assembly, shipping, handling, or installation. These defects cause the DOA problems. In some cases latent defects escape from the factory and lead to early failures after some limited use. These types of problems can affect both electronic and mechanical subsystems. Examples include cold solder joints or loose connectors causing intermittent faults with electronics assemblies. Early-life failures are distressing because they make your customers very unhappy and usually result in warranty costs. Early-life failures are signs of difficulties that often can be seen in quality audits performed on production samples prior to shipment. Product designs that suffer from these problems are viewed as not being robust against stresses in manufacturing.

A useful tool is the quality-as-received (QAR) audit. In a QAR audit the product is packed and ready to be shipped. Samples are randomly selected to be unpacked and tested. For complex electromechanical systems, historical data show a positive correlation between defects found in the QAR audit and failures observed during in-house product and process verification testing. When defects are observed in QAR audits, you can expect a relatively high failure rate early in life that declines over time as latent defects surface and, for repairable products, are corrected by field repairs.

Unfortunately, customers may come to expect early-life failures. Engineering may treat the opportunity for production modifications as a contingency for low manufacturing process capabilities. It is not unusual to hear it said that "this is a complex system and that's just the way it is." The result is that no one does the hard work required to bring production quality to a level sufficient to drive a major reduction in the early-life failures. The organization lives with the situation, the end result being customers' dissatisfaction combined with high warranty costs and early-life repair costs. The product is vulnerable to competitive products that do not have these problems. Eventually the field service force finds ways to correct the latent defects and the failure rate settles down closer to the "entitlement level," as determined by the durability and robustness inherent in the product's design, and life in production goes on.

Performance Off-Target due to Variations

Another type of failure that can occur early in a product's life is certain performance attributes being off-target. This is a soft failure usually caused by human errors in making adjustments in manufacturing or service, or by manufacturing variations in the fixed parameters of parts, components, or materials. In this case, the product's "out-of-box" performance can be different from its requirements by a noticeable degree. The tools for preventing this type of failure are robustness development methods such as Taguchi's "parameter design" or, if a good empirical model is available, the straightforward application of optimization methods. We cover these methods in later chapters.

Failures during Useful Life

Random Failures

A component that survives its early life without failure can suffer an unexpected failure during routine use. This is a case where the stresses acting on the system momentarily exceed the system's strength. The design is not robust against higher levels of this stress factor. Failures of this type usually occur because of an unusual external perturbation rather than degradation over time.

Examples of random failures include

- Failures of tires due to damage from a road hazard
- Electronic failures from transient surges in line voltage
- Failures due to customers' abuse or to changes in ambient environments
- Failures caused by errors in servicing the equipment
- Damage to structures due to environmental extremes, such as a flood, earthquake, or tornado

Good designs can protect products against some of these failures. To a point, increased durability will reduce the overlap of the stress-strength distributions and reduce the number of failures from transient high-stress conditions. Sometimes the methods of mistake-proofing can reduce the probability of inadvertent damage due to human errors in the use of the product. You can also limit customers' access to components that are vulnerable to damage. Of course, there are limits to the effectiveness of such measures. Regardless of how much durability you design in, it is always possible for stress to exceed strength. It would not be economically feasible to prevent all of the failures of this type. In the case of customers' access, it is difficult to prevent people who are determined from gaining access to components they should not touch.

Performance Drifts due to Degradations

Soft failures take time to develop. Driven by the lack of robustness that would have reduced the rate of degradation in a critical-to-quality parameter, performance can decrease until customers notice it. How quickly this failure moves from its latent state to being noticeable depends on the robustness of the system against the variability in its critical parameters. These failures are similar to wear-out failures, except that they occur before the expected end of life and may be correctable by preventive maintenance or repair.

Wear-out Failures at End of Life

Many components degrade over time. Eventually they wear out. Abrasion, wear, fatigue, creep, corrosion, and inelastic strain are examples of mechanisms at the root cause. The device or component can survive its early life and be very useful until this deterioration causes its eventual failure. This is a lack of sufficient robustness against long-term causes of degradation in the components or materials.

You have a couple of options to extend life. You can increase the design's strength to have more safety margin against failure. This may require some design trade-offs. For example, in the design of tires there is a trade-off between stopping capability and wear rate. Good stopping favors softer materials, but they tend not to have the wear resistance of the harder, longer-wearing materials. You can also consider increasing the design margin by reducing the stresses or loads where possible. Another option is to shield the system against the external stresses contributing to degradation. For example, rolling element bearings operating in an environment that is contaminated with dust can be sealed. Generally, extending useful life costs money. You can implement preventive maintenance to correct a deteriorating condition before it jeopardizes the life of the product system. What constitutes useful life for a product is a business decision that must consider customers' expectations, the possible opportunity to introduce a superior product concept, and, of course, where the competitive market is heading.

Reliability Metrics

To analyze contributors to reliability, two important considerations must be clarified:

- What are you going to call a "failure"?
- What measures are you to use to characterize reliability?

Is a failure anything that causes the product system to stop functioning, even if correctable by customers? Will you count only those stoppages that require a service call because your customer cannot correct the situation? Or are you going to count only problems that create a safety concern? It depends on your intent for the analysis.

If your product is not repairable, a failure that your customers cannot tolerate is an end-of-life scenario. If you are worried about warranty costs, you may choose to count just those failures that are related to the terms of the warranty.

If the product system is repairable, you may choose to count just those failures that require repair or parts replacement. That would exclude stoppages that customers can correct themselves without significant downtime or annoyance. If you are worried about customers' perceptions of reliability, you may choose to count any noticeable deterioration that causes customers to stop using the product and take some action. So your teams will have to define clearly what they mean by a "failure."

"Reliability requirements" can have several different meaningful dimensions:

- **Mean time between failures (MTBF):** For repairable systems, MTBF is a useful metric. Valuable insights can be gained from the distribution of failures versus time to understand the costs of failure to your customers' business. The criteria should reflect how customers are most affected.
- **Service frequency:** The inverse of MTBF is the number of failures over a given time and, potentially more important, the number of service calls during that time. The number of "unscheduled service calls per month," for example, may be a more significant metric than MTBF. A moderate increase in MTBF may be measurable by engineers but not noticeable to customers, while seeing the service person less frequently may be very noticeable. Also, as duty cycle increases, customers might expect that the frequency of service calls should not increase. That would be a critical requirement to understand.
- **Mean time to failure (MTTF):** For non-repairable systems this is a design life requirement. As is the case for repairable systems, you benefit more from knowing the shape of the life distribution.
- **Warranty costs as a percentage of sales:** Know the shape of the life distribution, since MTBF alone does not provide sufficient information.
- **Mean time to repair (MTTR):** For repairable systems, customers place high value on the system being available for use when they need it. The time to repair includes both response time and service labor time.
- **Availability:** If you know MTBF and MTTR, you also know the availability:

$$Availability = \frac{MTBF}{MTBF + MTTR} \tag{2.2}$$

With both reliability and repair time having distributions, availability will also be described by a distribution. You may be interested in the distribution of "availability" across your population of customers to identify customers who may be getting service that is unacceptable. You have an opportunity to probe this question when gathering VOC data. Graceful failure modes that allow your customers to continue to use their system while waiting for service can increase the availability.

Reliability Is a Probability

One measure of reliability is the probability of a period of operation being free of failure. That probability depends on an assumed duration, such as operational time, cycles of operational usage, or possibly a warranty period. For example, the reliability for a car may be stated relative to 100,000 miles of driving. The probability depends on how the product is used, the stressful nature of the usage environment, and the specific expectations for that product. So the reliability of a passenger car will be much lower when it is used in a competitive or off-road environment. Clear assumptions must be made about how it is driven and how often preventive maintenance is performed.

This raises another concern. Do you want to count scheduled outages for preventive maintenance or only those that are unscheduled? There is no universal answer. You have to choose your metrics and do the bookkeeping with a strategy that makes sense for your business and customers. Certainly it is important to make the assumptions clear and visible, and the measurements consistent.

In this case, "reliability" is a probability, a number between 0 and 1. Assuming that the product either works or does not work, the probability of success is 1 minus the probability of the failure $P(f)$.

$$R = 1 - P(f) \qquad (2.3)$$

If the product is not repairable, the reliability becomes the probability of survival (zero failures) over a period of time under specific conditions, given that the end-of-life failure mode is clearly defined.

A second characterization for repairable systems is the number of failures over a specified period of time or usage duration. Usually this is determined easily by counting failures, yielding a metric such as

- Number of failures per 100,000 actuations
- Number of service calls per month of operation, or per 100,000 miles of driving

The "failure rate" is often given the symbol lambda (λ).

A third characterization is the time or usage duration between failures. This metric, MTBF, is the inverse of the number of failures over a usage period.

$$MTBF = 1/\lambda \qquad (2.4)$$

You'll then have a metric such as

- Mean images between stoppages, as an example for a printer
- Number of miles driven between warranty service calls

A fourth reliability metric, MTTF, represents the average usage life for a population of non-repairable products. MTTF and MTBF are often used interchangeably in reliability literature. However, a more precise definition of MTBF is

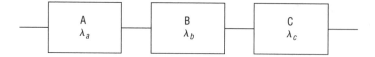

Figure 2.2   Series system

$$MTBF = MTTF + MTTR \tag{2.5}$$

where *MTTR* is the "mean time to repair." If the product cannot be repaired, *MTBF* equals *MTTF.*

The MTTF then is the "mean life" of the product, the average time to the first fatal failure of a component in the system. If the product is covered by a warranty, a metric that is more important to your business than MTTF can be the percentage of the population that may fail before the end of the warranty period. When products under warranty fail, they have to be repaired or replaced and may be subject to warranty claims. So a plot of the population's life distribution indicates, as a function of time, the number of failed products generating warranty costs.[4]

A population of complex repairable products will have a distribution of failures, as a function of time and for a given application, that covers the range of early-life failures, usage failures caused by random events, and wear-out failures. The overall population can be approximated by an exponential distribution with a constant failure rate. The system-level reliability (probability of being operational) will be the product of the reliabilities for the contributing elements. Suppose you have a series system composed of three major elements, A, B, and C, each with its independent failure rate $(\lambda_a, \lambda_b, \lambda_c)$, as shown in Figure 2.2.

The probability of a system failure is the product of the probabilities of the potentially contributing failures:

$$R_{sys} = R_a \times R_b \times R_c \tag{2.6}$$

The system reliability, as a function of time, is given by the exponential function

$$R(t) = exp\big[-(\lambda_a + \lambda_b + \lambda_c)t\big] \tag{2.7}$$

You can see that the exponential contains the sum of the failure rates of the major contributors. So with a series system, the failure rates are additive. This is equivalent to adding the reciprocals of the MTBF of each element to calculate the reciprocal of the system MTBF, as shown in equations (2.8) and (2.9):

$$\lambda_{sys} = \lambda_a + \lambda_b + \lambda_c \tag{2.8}$$

$$\frac{1}{MTBF_{sys}} = \frac{1}{MTBF_a} + \frac{1}{MTBF_b} + \frac{1}{MTBF_c} \tag{2.9}$$

---

4. Paul Kales, *Reliability for Technology, Engineering, and Management* (Upper Saddle River, NJ: Prentice Hall, 1998).

The Anatomy of Quality Loss

In this section we explore three concepts with major benefits to product or process development:

1. **Performance variations:** During development it's more important to focus on variations in product performance than on failures.
2. **Quality loss:** Performance variations are losses in quality that impose costs on customers as well as on the producer.
3. **Quality loss function:** The economic consequences of performance variations are important criteria for decisions about the funding of product improvements, either through additional development or by the tightening of tolerances in manufacturing and service.

Consider a technical "function" to be a process that transforms an input signal into an output response for which there is a requirement and a tolerance for variations in the response. The amount of variation in the output is a measure of the quality of the function.

Consider a nonengineering analogy that may help you to understand: the kicking of field goals in American football, illustrated by Figure 2.3. Getting the ball through the uprights is a measure of quality for the kicking function. Traditionally, its requirement is viewed as a "go/no go" criterion. If you meet the requirement, you get 3 points; if not, you get no points and give the ball to the other team—a cost for being out of spec. This example is useful for discussions in this and later sections.

The concept of a "product" can have a wide range of applications. For example, something as static as a parking lot is a product. Its design has a set of requirements derived from the needs of its customers. It incorporates specific architectural elements and technologies for road building, coating, drainage, and marking. It's produced by a manufacturing process that itself was designed to replicate the product with a high degree of predictability. The performance of a parking lot is subjected to a wide range of stresses, from the many elements of weather to the abrasive effects of traffic and snowplowing, for example. Although customers may have various degrees of tolerance for deterioration, they can be sensitive also to small variations in quality. For example, faded markings may be annoying, but worn-off markings are not acceptable. A cracked

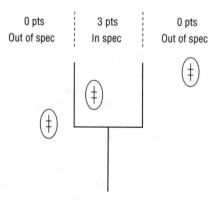

Figure 2.3   In football, field goal kicking can be a traditional go/no-go view of tolerances.

surface is annoying, but potholes need to be repaired. Rainwater puddles are annoying if small, although they may not be acceptable if deep and along walkways. The costs associated with these situations of poor quality are reasonably easy to imagine.

This example illustrates that deterioration in performance is often noticeable to customers prior to an actual failure event. It may be a cause of dissatisfaction without creating the need for repair. For example, a cracked road surface is a predictor of a future pothole. So variations, not just failures, are important to both customers and producers.

Reliability as a Subjective Concept

How do people perceive product reliability or durability? Engineers and managers may view "reliability" simply as the usage interval between those failures that require repair. They may think of "durability" as the usage life between replacements or scrap. For how many seasons does the parking lot remain useful before potholes have to be repaired or the parking lot has to be reconstructed? But wait, are all parking lots the same? To what stresses are the parking lots subjected? Are all customers' vehicles the same? Are the tolerances for deterioration common among customers? Customers don't make it easy on engineers and managers.

A performance failure is whatever customers perceive it to be, and the consequences of a failure can vary quite a bit among customers. A failure can be considered to be an event that causes the product to no longer operate as intended, or at least to no longer provide acceptable performance. Certainly there can be a range of interpretations of what constitutes a failure. Is it a loss in appearance that bothers them? Is it a condition that imposes an unacceptable cost on the customer? You'd expect that at least it's the consequences of the failure that impact them. A loss of appearance for a VIP parking lot may harm the corporate image. A drainage puddle may damage a visitor's shoes. A pothole may damage a car's steering mechanism. Certain customers may have more stressful applications or operating conditions, such as, in the parking lot example, heavier vehicles or more severe weather. Cracks and potholes that may not be acceptable for a VIP parking lot may not be noticed on an access road to a warehouse. Are customers more sensitive to the number of these situations, to their frequency of occurrence, or to the loss of use during the repairs? Is it the hassle that the situation imposes on them or the absence of a work-around? Is it the cumulative effect on their cost of ownership? There are no easy answers, since all of them and more are correct. So how can we add some value to the engineering process?

Performance Variations

Variation in a performance characteristic can be measured and thought of as a leading indicator of a failure. For example, in an office printer the stoppage of paper requiring clearance by the operator would be an observable failure. Just counting the number of paper stoppages over time would tell you something about the customers' view of the printer's quality, although nothing about its predictability. However, the variation in the timing of the paper's arrival at a paper path sensor not only would provide information about the variability of the paper's movement, but would also provide a prediction of the paper's timing exceeding a control limit and causing the stoppage. This variation would not be observable to customers but would be important for engineering. For other factors, such as the quality of the printed image, the variation itself may be observable to customers and be a major element of their dissatisfaction. The absence of performance variation, particularly under stressful usage conditions, would then be a competitive advantage!

The Concept of Quality Loss

The concept of quality loss was introduced in the 1980s with the methods of Dr. Genichi Taguchi. It emphasizes that the variations in performance characteristics are more important to measure than are product failures. Variation is a measure of poor quality, that is, the loss of quality. However, not all variations are the same, and their unit of measure depends on what is being measured, so how can you compare them and use this information? The premise is that all quality losses should be viewed as unnecessary costs. What are these costs?

The consequences of a failure may be the costs to replace parts, make adjustments, replace consumable materials, or clean the system. In some cases the failure may end the useful life of the product, incurring the costs of replacing the product and scrapping the failed one. However, that's a little too easy and misses an important concern.

If you are a customer, you value a product, in part, by the comparison of its benefits to its costs of ownership. You suffer economic loss when those benefits are not available or the costs are too high.

Here are some examples of losses in value to customers:

- Expectations that the product design cannot provide
- A vulnerability to stresses in the routine operation of the product, either all the time or intermittently over extended periods of use
- Observable differences from product to product due to flaws in the manufacturing processes
- Lost productivity, and even profits, due to the loss of use of the product
- Time lost in working around minor failure modes
- The need to provide a backup capability to protect against downtime
- The need to train key operators to perform routine replacements and maintenance as a way to reduce the cost of waiting for service
- Excessive replacement of consumable materials and their costs
- Costs of maintenance and repair, if borne by the customers

Quality loss cuts both ways. In addition to the losses sustained by consumers, producers are also affected. Examples of losses in value suffered by producers include

- Lost sales due to delayed market entry caused by late design changes
- Late market readiness of consumable materials, replacement parts, or software
- Features, functionality, performance, or reliability being less than those of market leaders
- Inspection, scrap, rework, and verification in manufacturing
- Excessive costs of repair and maintenance, if borne by the producer
- Logistics costs in moving the product to a service center for repairs
- Product recalls or replacements, or field modifications of products
- Distraction of resources from developing the next generation of products
- Warranty costs or, in extreme cases, legal costs
- Loss of customer acceptance and competitive advantages, with the resulting losses of revenues and market share and potential downgrading of customer expectations for your products
- Scrapping of parts and tools due to late design changes
- Pollution of the environment from unintended or uncontrolled product wastes
- Lost corporate respect in the marketplace

Is there a cost when the loss of quality is seen as a variation in an important quality characteristic, but not yet a failure? Remember our parking lot example. Stability in performance may be critical to your customers' applications. In certain cases, any customer-observable variation may not be acceptable or may be viewed as a leading indicator of a failure. Customers may notice that variations in one product are greater than those in a second example of the same product. It is not the case that the relationship between performance variation and costs is zero until there is a broken part. And the relationship is not at all linear.

These consequences all have the waste of money as a common measure. If you are the producer of the product, quality losses will affect you both in the factory and in the field. Quality losses are economic losses both to the customers of the product and to the producers of the product, so the total quality loss is viewed as a "loss to society."

The Quality Loss Function

In the Taguchi methodology, the relationship between performance variation and costs is simplified to be a quadratic function. Reality checks confirm that this is a reasonable assumption. It is characterized by small losses when deviations are close to the target requirement and losses that increase at an increasing rate as deviations from the target increase. In fact, it's often called the "quadratic quality loss function."[5] The mathematics based on this assumption can then be useful for engineering and management decisions, since they convert a range of variations in performance into a universal economic metric, such as money lost versus deviation from quality target.

Actually there are three basic forms of the quality loss function, illustrated in Figure 2.4. If variation from the target value of a performance response can be in either direction, the loss function is centered on the target value. The scenario is that of "Nominal is best." It can also apply to the dynamic situation in which the "target" is the slope of the relationship between the output response and the input signal. Consideration must also be given to variations that can be in only one direction. If the performance response can vary only greater than the target value, it's "Smaller is better," whereas the opposite situation would be "Larger is better." In our example of a parking lot, the variations we discussed were one-directional, depending on the functional response chosen. There is a fourth situation, an "asymmetric loss function," for which one direction of variation has a higher quality loss than the other.

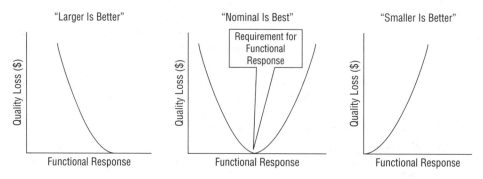

Figure 2.4   The quality loss is a quadratic function of the deviations from requirements.

5. Madhav S. Phadke, *Quality Engineering Using Robust Design* (Englewood Cliffs, NJ: Prentice Hall, 1989).

The application of the quality loss function is for a population of products and customers. It represents the average of the consequences to customers in a market segment due to deviations in a particular functional response of the product. For any specific customer the relationship might be considerably different. The approach was developed to contribute to the many economic decisions that must be made during the development of new products and of their manufacturing and service processes.

The mathematics of the quality loss function represents those situations easily, since the function is simply that of a parabola, as shown in Figure 2.5 for the nominal-is-best scenario:

$$Quality\ Loss = k(Y - T)^2 \tag{2.10}$$

where

Y = measured response

T = target value

k = economic proportionality constant

Consider a contrived example. Suppose, when characteristic performance response (Y) deviates from its required value (T) by the average amount of $\Delta = 10$ millimeters, a service call is generated at an average cost of $500 for each repair.

Then, using equation (2.10), we can solve for the economic proportionality constant k:

$$k = \frac{Quality\ Loss}{(Y - T)^2} = \frac{\$500}{(10)^2} = \$5/mm^2 \tag{2.11}$$

When the response deviates by a larger amount, say, 15 millimeters, a 50% increase, the quality loss is

$$Quality\ Loss = k(Y - T)^2 = 5(15)^2 = \$1125 \tag{2.12}$$

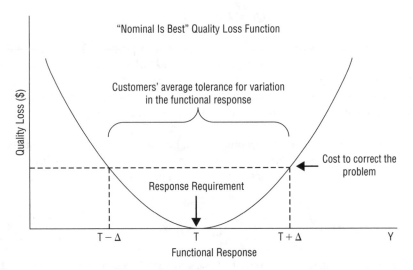

Figure 2.5 When the tolerance for a deviation is exceeded, a cost is suffered.

This is an increase of 125%!

Suppose this same problem causes the repair five times per year, for the average deviation of 10 millimeters. That problem would cost $2500 per product per year. How much of an increase in manufacturing cost would you accept to save $2500 per unit per year?

How else can the quality loss function be used? A typical example would be to answer the question "How much money would you be willing to invest to change the product design as a way to eliminate the problem?"

Suppose that 5000 products are in the market, all with the same problem occurring on average five times per year. The total quality loss due to the problem in question would then be 5000 × $2500 = $12,500,000 per year. So this is a very big problem! Depending on the complexity of your design, the consequences for its manufacturing, the particular implementation plan for the design change, and possibly the implications for marketing and corporate communications, your business case could justify an improvement project up to that $12.5 million with a one-year break-even time. It would be beneficial for you to invest in this product improvement as long as the investment is less than the total savings in quality loss.

When considering the broader implications for the level of investment required to correct a design problem or to improve a manufacturing process, the total quality loss across the population of customers and products must be considered. Across those populations, the total quality loss would be the sum of the individual quality losses over time. This is one way to determine the constant k. If the total cost to repair a specific problem, for the population, is known from your service database, that service cost can be assumed to be the quality loss for the mean square of the deviations that triggered the service call.

For a population of products, the mean square deviation (MSD) from target can be shown[6] to be

$$MSD = \sigma^2 + (\overline{Y} - T)^2 \tag{2.13}$$

The quality loss for the population is

$$QL = k[\sigma^2 + (\overline{Y} - T)^2] \tag{2.14}$$

Being able to use this expression to calculate quality loss presumes knowledge of the population mean and standard deviation. Solving this equation for k will give you the shape of the quadratic loss function for that particular problem. You can then use the function for subsequent decisions.

The quality loss function is analogous to the concept of the cost of poor quality (aka "cost of quality"), but it has several advantages for engineering and management:

- It focuses the discussion on economic terms, instead of on the frequency of occurrence, on the consequences of deterioration, or on the population statistical parameters.
- It provides a mathematical relationship that can be useful in decisions by engineering management.

---

6. Naresh K. Sharma, Elizabeth A. Cudney, Kenneth M. Ragsdell, and Kloumars Paryani, "Quality Loss Function—A Common Methodology for Three Cases," *Journal of Industrial and Systems Engineering* 1, no. 3 (Fall 2007).

- It broadens the scope of costs that are included, beyond just those costs of repair and maintenance.
- It focuses attention on variations in performance that are closer to the target value of the functional response.
- It changes the objective of product design and that of manufacturing from an "in-spec" model to one that controls performance to be "on target, with minimum variability," particularly under stressful conditions.

The Reduction of Performance Variations

There is a valuable strategy to reduce variations in functional performance in order to improve reliability and durability. Parameter design is the foundation of Taguchi's techniques for Robust Design.[7] We devote Chapter 7 to the topic of robustness development and also address it in Chapters 18 and 19. For now we'd like you to understand some of its key principles and appreciate the valuable role that the methodology can play in problem prevention and correction.

We have pointed out that we are concerned not only about variations beyond "specification limits" but also about small variations from requirements targets, those that are leading indicators of larger variations or of failure events. Even "within-spec" variations in performance may be observable to customers and cause some consequences, such as costs. Variations in the internal functions of components and materials can lead to permanent deterioration or breakage. Regardless of how the concepts of reliability or failure are described, at their root are the variations in the characteristic responses of functions that are specific to the performance of interest.

So variation is the enemy of reliability and durability. However, it is also the friend of engineering. Consider the implications of these key points:

- To the extent that you can measure variations in experiments, you can anticipate failures during product development.
- To the extent that you can use experimental stresses to force variations to occur deliberately under laboratory conditions, you can find more problems faster and prove their solutions to be better.
- To the extent that you can identify the controllable design parameters that reduce the vulnerability to stresses, you can devise solutions that are more effective at increasing reliability and durability.
- To the extent that controllable design parameters are successful without the tightening of their manufacturing tolerances or the increase in their material costs, increased manufacturing costs are avoided.

Remember our example of field goal kicking in football? Is it really true that this is just a concern about specification limits? Do we care about variation in the ball's flight? If you are the kicker and worried only about scoring, you might just think with an "in-spec" model. However, if you are the coach worried about how predictable the kicking game is, particularly under poor

---

7. Phadke, *Quality Engineering Using Robust Design*; Clausing, *Total Quality Development*; William Y. Fowlkes and Clyde M. Creveling, *Engineering Methods for Robust Product Design Using Taguchi Methods in Technology and Product Development* (Upper Saddle River, NJ: Prentice Hall, 1995).

weather or field conditions, the variability of the kicking function will be your concern. How can the kicking function be improved? The rate of success at scoring does not provide information that is useful for improving the kicking game. When a kicker places the ball just inside the goalpost on one side, then misses wildly to the other side on the next attempt, this wide variation in performance means a reduced predictability for future kicks.

Can you see the analogy with engineering? The development of a product or process is focused on identifying and specifying those parameters that control the performance of the function in spite of the degrading factors that cannot be controlled.

Process of Parameter Design

How can you improve the reliability of field goal kicking? Parameter design is a two-step optimization process, as illustrated in Figure 2.6:

1. Reduce the variability in the distribution of performance.
2. Shift the mean of the distribution to the required target for that performance.

A product system performs basic functions that provide value to customers. This system-level function, illustrated by Figure 2.7, can be decomposed into a hierarchy of contributing sub-functions, down to the level of the basic science and engineering of the technical concept.

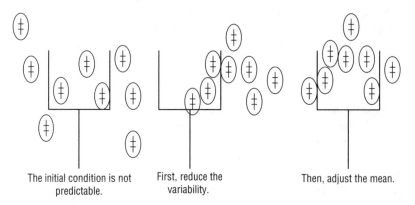

Figure 2.6   How do you make the function of field goal kicking more robust?

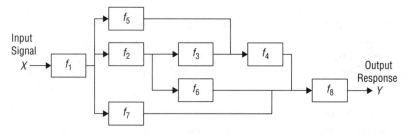

Figure 2.7   System-level functions can be decomposed into lower-level functions for subsystems, assemblies, and components.

Basic Function

Figure 2.8 shows a functional model known as a "parameter diagram" (P-diagram)[8] that is very useful for many of the discussions in this book. Think of it as the core element of a functional flow diagram, but with its forces described in a way that helps you deal with the issue of reliability improvement. As a functional parameter diagram, somewhat akin to a free body diagram in engineering mechanics, it can be applied to the design of either a product or a process.

Mathematically, this graphic illustrates that the output (Y) of a basic function (f) is derived from parameters (X) that control the behavior of that function, such as input signals, functional design parameters, and stress factors. Simply, $Y = f(X)$.

The basic function is to be understood at its fundamental level in the physics (or other science) of the technical concept that is implemented by the design. Taguchi Methods consider the role of the basic function to be the transformation of energy from one state to another. This is an engineering description, decomposed from that of customers. For example, customers may define the function of a copier as reproducing the image of their original. Engineers must consider the technical functions of the system design to have many sub-functions, such as the transfer of a physical image into discrete light pixels, the movement of toner particles onto a photoconductor, or the bonding of toner particles to paper. It's at the level of these sub-functions that development and design are performed.

The function has an input signal. If that signal is movable during operation, you have a dynamic situation. The turning of a steering wheel in a car is a simple example of a dynamic input signal that is transformed by the steering system into the radius of turning for the car. If the input does not change, the case is static, as in our parking lot example.

The function has a response that is a direct measure of the function's technical output, a "critical functional response" (CFR). That response has a requirement—a target value—derived from the high-level system requirements, which themselves were derived from the needs and preferences of the product's end users. In the terms of Taguchi Methods, these responses are often called "quality characteristics."

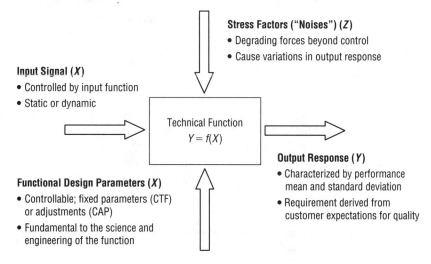

Figure 2.8   A parameter diagram defines the major elements of a design function.

---

8. Phadke, *Quality Engineering Using Robust Design.*

Now, what controls the ability of the function to deliver the response that achieves its requirement? There are two groups of forces:

1. Parameters that can be controlled by the design
2. Factors that either cannot be controlled, or that you choose not to control

Design parameters that are effective at controlling the variability of a basic function, or of a manufacturing process, are called the "critical functional parameters" (CFPs). There are two basic types of CFPs:

1. Certain of these parameters may be fixed in the design, as is a dimension or shape that is achieved by tooling or by the manufactured properties of a material. Traditionally these parameters have been called "critical-to-function" parameters (CTFs). In the simplest of cases, the CTFs play a strong role in reducing the variability of the function.
2. Other CFPs may be adjustments, designed to be modified by the product's user, by manufacturing, or by service personnel. This, of course, depends greatly on the strategy incorporated into the design's architecture. These are often termed "critical adjustment parameters" (CAPs) or "tuning factors." In the simplest of cases, the CAPs adjust the mean of performance to the required target value.

The design concept that is selected and developed has in its architecture these controllable parameters that are found to be effective at controlling the functional output response. If those controllable parameters are not effective, particularly under stressful conditions, the architecture of the design concept is flawed and must be improved or replaced. Remember that the best selection of set points for design parameters cannot transform a bad concept into one with superior performance and robustness.

The second group of parameters that influence a function consists of stressful factors that are unavoidable, that is, beyond the control of the design. Alternatively, you may choose not to control them because of considerations for their economics or practicality. Jargon in Taguchi Methods calls these factors "noises," a terminology derived from communications technologies for which undesirable variations in audio output are clearly noises. For some readers it may be more understandable to think of these factors as stresses. They are deteriorating forces beyond control.

In our football example, wind, rain, and mud are stress factors beyond the control of the team, unless they choose to invest in a roof over the stadium. In many cases, that would be judged to be offensive to the purity of the game or too expensive. However, there are controllable parameters of the kicking function that reduce the vulnerability of the process. These may be in the equipment (shoe cleats) or in the human movements that are practiced. Look for them the next time you watch a game. Certainly there are other stresses in the tactics of the opposing team and the tension of the game that affect the ability of the players to focus on their tasks.

Robustness Development

In the world of robustness development, there's a motto: "To improve reliability, don't measure reliability," measure the robustness of the design. What does that mean?

Don't measure the rate of failures as a method to understand the misbehavior of engineered systems or to determine how to improve them. Measure the variability in their characteristic functional responses and compare it to the function's requirements.

The premise here is just that the counting of failures and their frequency of occurrence are not very useful either to engineering or to management. Failures are events. They can be counted when they occur. However, when they happen, they tell you almost nothing about their root causes or what to do to correct them. If you just count failures, you are not measuring parameters that contribute to or anticipate the failures. Certainly failure rate is important to customers and service, but it is a lagging indicator for engineering.

Building upon the application of sound science and engineering, robustness is achieved initially without depending on tightened tolerances, higher-grade materials, or other tactics that drive up the costs of manufacturing or service. Under stressful laboratory conditions, problems can be found faster and better solutions can be developed. If done in the earlier development phases, more costly changes in designs and tooling can be avoided. This is very important to achieving shorter cycle time, a tactic of "lean" product development. When robustness is focused on variability under deliberately stressful conditions, it is a leading indicator of deterioration under naturally occurring stress conditions. Consequently, it is a more sensitive leading indicator to foster design improvements.

So focusing on variation is more important for reliability development than is a focus on failure events and their rate of occurrence. Later we describe the measurement of reductions in variability in the context of measuring the improvements in robustness under stressful laboratory conditions. Customers may perceive it as designing to achieve a more reliable or more durable product. It's the same thing. Making the design of a product more robust is the way to achieve a product that is more reliable and more durable. The same strategy applies to processes. Figure 2.9 is an overview of how components, parts, and materials fail. CPM (critical parameter management) is covered in detail in Chapter 23.

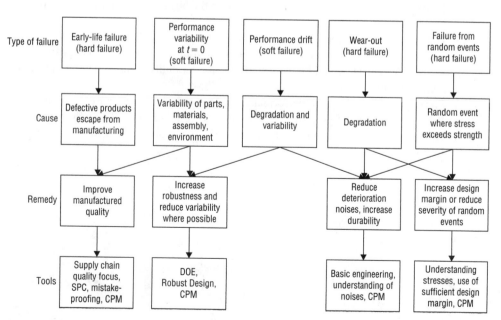

Figure 2.9 Summary of how things fail.

Key Points

1. Losses in quality are unnecessary costs, that is, wastes.
2. Quality losses are caused by deviations from requirements, not just nonconformance to specification limits.
3. High quality can be viewed as freedom from the costs of poor quality.
4. Making decisions based solely on reducing manufacturing costs can result in lower quality and reliability, higher life-cycle costs, and losses in market reputation.
5. For the development of improvements in product designs, a focus on reducing variations is more useful to engineering than is a focus on reducing failure rate.
6. The investment in improvements for products or manufacturing processes can be justified as long as the investment is less than the savings in quality loss.
7. The fundamental contributors to the loss of product reliability, either perceived or measured, are variations in a design's functional responses.
8. Stresses are the enemy of reliability. They cause variations in performance that are the contributors to unacceptable reliability and the resulting quality loss.
9. Stresses are the friend of engineering. Stressful test conditions enable development teams to find problems faster and to prove that their design improvements are superior to alternatives.
10. Systems can fail in a number of ways. There are specific tools and methods to prevent the failures.

Discussion Questions

1. What characterizes failures in your products?
2. Pick one of your products that is important to the business. What are the top three problems with it?
3. How do your customers perceive reliability and durability?
4. How do reliability and durability contribute to your competitive market position?
5. For your products, would higher reliability and durability command higher prices or earn larger market share?
6. For your products, how do performance variations and failures cause economic loss?
7. What are the costs of performance variations to your customers? To your business?
8. How well do your development teams use effective tools to evaluate the causes of failures?

Strategies for Reliability Development

A common experience is that the reliability development process is an afterthought, done in an ad hoc manner as a flurry of problem solving near the end of the project. Achieving a better outcome to the process requires the recognition that reliability development is a subproject within the context of the larger product development project. Developing reliable products requires the selection of effective strategies and the development of improved capabilities. Without this investment, the results may be highly variable. In this chapter we consider strategies that can add value as part of your process. A capability assessment tool is illustrated that is easily adapted to the nature of your business.

Are Better Strategies Important?

"Fix as many problems as you can before time runs out!" That's an approach to reliability development that may characterize your experiences. The engineering misery in that statement reflects an absence of understanding what it takes to achieve higher levels of reliability. A more effective strategy is necessary to achieve levels of reliability that not only are satisfactory to your customers but also are superior to levels delivered by your competitors. Through management decisions and project tactics, strategies are transformed into plans with resources. They are in turn implemented as tasks with dependencies, enabled by effective tools and efficient methods.

What would you expect to see in results from an internal assessment of your capabilities for reliability development? Are you already "best in class"? How do you know? Do you have notable examples of excellence? Or does feedback provide clear indications that serious attention to improvement is needed?

What comments might you expect?

"Reliability metrics are included in our managers' balanced scorecards."

"Management will intervene with necessary resources if progress is too slow."

"Development budgets are subordinate to product quality and project schedules."

"At a gate review, a product in development will be kept in its current phase if its reliability growth is not acceptable."

Or would you hear comments like these?

"There are more rewards for fixing problems than for preventing them."

"Our gate reviews do not focus on reliability."

"In spite of our requests, we don't have enough resources, prototypes, or time."

"Reliability plans are viewed as threats to schedules and budgets."

"Field reliability data are of no value to product design."

At the end of this chapter we offer an example of a tool that can be adapted to enable your internal assessment. The ratings themselves can be helpful if there is good evidence of systemic behavior. However, recognize that you should expect a range of ratings when assessing a large number of development projects and business units. There can be significant differences between intentions and achievements. Specific case studies, root cause analysis, and anecdotal evidence can also provide valuable insights. As mentioned in Chapter 1, a serious "call to action" must be compelling to provide the energy behind your assessments and their follow-on action plans. Are the gaps that you choose to close important to your project's business case?

Reliability Growth Plot

One way to visualize reliability development is provided by the reliability growth plot shown in Figure 3.1. Also known as the Duane plot, it is a useful way to evaluate progress in developing the reliability of a repairable system. Technically, it's a log-log plot of cumulative reliability as a function of cumulative problem solving. Reliability growth is an iterative process of testing, identifying problems, analyzing their causes, designing and verifying solutions, and implementing them in the prototypes being tested. Based on the initial reliability and on the slope of the curve, the growth of reliability over time can be predicted, with its consequences for management decisions and for the project's schedules and resources.

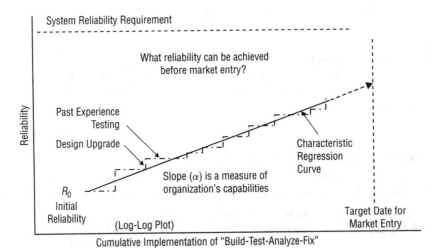

Figure 3.1   The reliability growth plot maps the improvement of system reliability as a function of the cumulative process of "build-test-analyze-fix."

Another reliability growth model, the Crow-AMSAA model,[1] builds on and extends the Duane model. It makes it possible to do statistical estimates of model parameters using maximum likelihood estimators (MLEs). Duane's approach, on the other hand, uses a least-squares fit of the growth model[2] to the failure data. In this book we focus on Duane's model because it is simple, does not require any advanced statistical methods, and has proven to be useful. We return to the Duane model in Chapter 8.

Okay so far, but a few strategic questions are raised. Look at Figure 3.2.

- How can the initial reliability (R_0) be increased?
- How can the slope (α) of the curve be increased?
- Is reliability expected to grow at a constant rate?
- How should the plot reflect the acceptance criteria at various project milestones?
- If the growth of reliability is too slow or underfunded, what are the consequences for the market entry date?
- What are the costs of delayed market entry?
- What would be the economic benefits of an earlier market entry date?
- What are the consequences of the product entering the market with its reliability being less than required?

An objective of our book is to provide guidance to effective strategies, plans, methods, and tools in answer to questions like these. By thinking of the problem first as one of an integration of

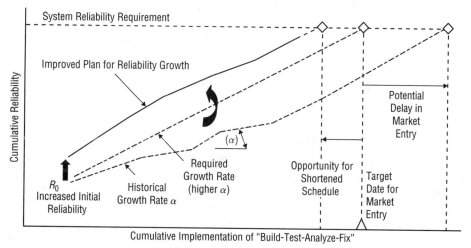

Figure 3.2 How can you improve the shape of the reliability growth plot to achieve higher reliability with less development time?

1. Larry H. Crow, *Reliability Analysis for Complex Repairable Systems* (Aberdeen, MD: U.S. Army Materiel Systems Analysis Activity, 1975).

2. William J. Broemm, Paul M. Ellner, and W. John Woodworth, *AMSAA Reliability Growth Handbook* (Aberdeen, MD: U.S. Army Materiel Systems Analysis Activity, 1999).

development strategies, you have an important initial point. The foundation for higher reliability in a new product system is built upon

- Advanced development of technical concepts prior to the beginning of a product development project
- The improvement of your organization's capabilities for problem prevention and problem solving

With a solid foundation in place, it is then the job of your product development teams to make decisions and to invest resources in efficient ways to develop the product and its manufacturing, service, and customer support processes. Development has to be managed to achieve growth in reliability at a rate that not only is sufficient to achieve its requirements prior to market entry, but also establishes sustainable competitive advantages. That sounds easy to say, but how can it be done with limited time, labor, skills, and funding?

A review of several strategies should provide useful approaches for both ongoing processes and for product development projects. There are no magic bullets here, just sound processes and expectations for excellence in execution. Certainly, to the extent that the achievement of superior results can be expensive and time-consuming, the ability to do it well and efficiently provides competitive advantages. These mandates may also be barriers to market entry.

Strategies for Ongoing Processes

An important foundation for reliability development is the ongoing work that prepares market information and develops new or improved technical concepts in anticipation of the needs of product development. Without this foundation, that work becomes part of the product development project or is neglected, adding time, costs, and risks to the project. What is this up-front and ongoing work? Let's look at four elements: customers, competitive markets, technologies, and information.

Voice of the Customer (VOC)

VOC has been popularized as a way to establish requirements that reflect the real needs of customers. The principles have much depth in the involvement of customers with engineering. The intent is that customers and users of the product be integrated into every phase of the development process, including its advanced planning. In the context of strategies to achieve higher reliability, customers have a lot to tell development teams. There are two fundamental approaches to that conversation:

1. Customers can be directly involved with development activities. Their concerns for reliability can thereby be articulated more accurately in the value proposition, in its translation into specific requirements, and in the selection of product architectures and design concepts to deliver that intended value.
2. Engineers can be involved in their customers' activities and business. This way they can understand well the perceptions of reliability problems and their context, both as spoken by customers and as observed in their actions and frustrations.

In the absence of this direct involvement, reliability requirements and design concepts can be vulnerable to the biases of people acting as surrogates for customers, or acting with an internally driven agenda. Probably the customer will not win.

In a competitive market, the job of product development is to provide solutions (products and services) that satisfy unfulfilled needs better than competitive alternatives, and to delight customers in ways that they could not have anticipated. The resulting value proposition will then have much more impact than can be related just in words.

Prior to a specific product development project, the VOC process should be focused on the needs of customers who are grouped into logical segments with similar value drivers. There are many ways to segment markets, so your company needs to articulate its participation strategy and then decide how market segments are defined and which segments it chooses to serve. The VOC processes can then identify the more important value drivers and translate them into requirements that serve many customers. Customers have many problems, some of which they choose to spend money to solve. For which ones do you choose to develop solutions? How do they map into families of future products? This has two significant benefits:

1. It enables the mapping of a sequence of value propositions and related requirements to be introduced over a series of evolving products.
2. It enables the mapping of new or improved technical capabilities to be developed or acquired in anticipation of the needs of the development projects for those new products.

Competitive Market Situation

It does no good to begin a development project when the competitive context of the chosen market is not understood.

- What features and requirements will add differentiating value to your new products and services?
- For a selected market, what does "higher reliability" mean in practice?
- What functionality, failure rate, usage life, and so on are expected in new products from your competitors? When?
- How will the obsolescence of technologies or product platforms, or the evolution of applications for existing technologies, influence your selection of design concepts?
- How will the economics of the marketplace influence the product's price and thereby the required manufacturing costs?

Product development cannot wait for lengthy studies of the market. So these studies have to be conducted prior to the beginning of the development project. Likewise, product development cannot forget to keep its eyes open for changes in the market situation, so market studies must be ongoing.

There is an important point for the information that is gathered to represent the competitive market situation. It must be collected, analyzed, and reported with insights that are most useful to the consumers of that information. Certainly they are not just one group, such as the business unit managers. What good is that? This information needs to be one of the basic sources for

- Product family plans and product line plans
- Technology development road maps

- Strategies for competitive product and price positioning
- The portfolio of investments in product development and technology development
- Value propositions and the derived requirements for new products and services
- The selection of the best design concepts

Consequently, the scope, depth, and integrity of that information face high expectations. They must then be matched by comparable investments in the efforts to acquire and disseminate the information to people who need it, when it's needed. A major source of risks can be heard in declarations such as "We already know it all," "We know better than the customers what they need," and similar statements that rationalize the deliberate exclusion of ongoing market and customer studies from the project's plans and decisions.

Technology Development

The role of technology development is to prepare a consistent stream of new or improved technical capabilities in anticipation of the needs of future development projects. What new capabilities are needed? When? Technology road maps establish the direction and the timing of the needs for robust new capabilities that are required by the product line plans. Technology development plans establish the required timelines, resources, and investments. If your company develops its own new technologies, these plans are yours to fund. If you expect to acquire new capabilities from your suppliers or partner companies, these plans contribute to the partner selection process and to the detailed expectations built into the relationship. What are the sources of these needs? One answer is the two activities previously described, namely, the study of the needs of target customers and the analysis of the competitive market situation.

The primary objective of technology development is to develop those new or improved functional capabilities to be robust for a range of intended applications. These "new or improved capabilities" will become part of the product, its manufacturing processes, its service processes, or some other element of the value delivery system. The new capabilities may be acquired from another company, either purchased or via a collaborative development arrangement. Most important are the activities that identify their needs, with a high level of integrity, early enough that the capabilities can be developed or acquired without being on the critical path of the product development project.

A second objective is to identify experimental stresses ("noises" in the language of Taguchi Methods) that are effective at forcing variations in a functional response. They enable the identification of controllable design parameters that reduce the vulnerability to those stresses. These experimental stresses can be surrogates for stresses in real applications of the product. Temperature variation is an example of an experimental stress since it can cause material distortions and other deteriorations that in real use actually have other root causes. The experimental stresses that are effective at forcing variations in a technical function should be understood to enable the building of development fixtures and the planning of experiments. Stressful test conditions enable problems to be identified earlier and their solutions proven to have superior robustness.

Information Management

Data about existing designs are critically important to the development of new products. They may be collected from, for example, in-house testing, feedback from service or customer support, sales wins and losses, competitive assessments, and consumer reports. For reliability development there are many questions to ask of these data, such as these:

- How are deteriorations (soft failures) apparent to customers?
- What hard failures occur and under what conditions?
- What problems are the most costly to detect and repair?
- How long do consumable materials last?
- What parts have to be replaced and how often?
- What are the opportunities to reduce the consequences of deterioration in a product?
- What are the difficulties in diagnosing problems in the field?

Often this type of information is neither collected nor retained. For example, existing databases may be designed to support the service organizations but not to provide inputs to product development. For repairable products, they may collect information such as the inventories of maintenance and repair parts or materials, repair and maintenance labor hours, service call and repeat call frequency, and service center workloads. They tend not to be motivated to collect information that would be useful to the next product's development. That's unfortunate, since the population of products with customers represents a large test with a much wider range of stressful conditions than can ever be simulated in development experiments.

This emphasis on the ongoing work prior to product development has significant potential benefits. You should expect higher performance, reliability, and durability to be achieved with a shorter time to market than would be expected if the results of this work were not in place. You might also expect lower risks in unanticipated problems or in the integration of new design concepts. Lower risks and shorter development times both contribute to lower development costs.

Probably you have heard the wisdom:

"Reliability is a characteristic inherent in the product design."
"Good design cannot compensate for a flawed technical concept."
"Good manufacturing cannot compensate for a poor design."
"Good service cannot compensate for poor manufacturing."

And so on. The point is that the foundations are set very early. Their absence places product development in jeopardy. In your company, when quality, costs, and schedules are in conflict, what wins?

Strategies Specific to Product Development

Once a product development project is chartered and staffed, the job of the development team is to develop the product and/or services so that they not only perform as required, over time under stress, but can be replicated efficiently over and over with the same acceptance criteria. This is not the time for lengthy studies, and certainly not for inventions. It's a process of timely decisions, putting stakes in the ground (aka progressive freezing), efficient management of risks and corrective actions, and moving deliberately toward market entry, with increasing predictability.

System Architecture and Design Concepts

From a reliability viewpoint, the architecture of the product has several requirements to satisfy. Among these are the following two:

1. Subsystems should be defined so that the significant risks are within subsystems, rather than between them. This means that the interfaces are robust. Assuming that different teams are assigned to each subsystem, the risks then are within the work of teams rather than between teams.
2. Systems architectures should enable easy changes without highly interactive consequences. This means that the full system configuration is not dependent on a specific design concept for each subsystem. It can tolerate an initial design concept being replaced by another one later if the first one is proven to be inferior. This enables alternative design concepts to be developed in parallel, with the winner being chosen later in the process. Carrying forward alternative design concepts is a way to reduce the risk of being dependent on a difficult-to-develop technology, a principle of "set-based design" described in Chapter 4.

Reliability Development Strategies

So what is your strategy to achieve reliability that is superior to your expected competition prior to market entry? To help remove the glaze from your eyes, let's think of the implications of two ways to describe reliability.

Reliability is derived from the robustness inherent in the product's concept and specified designs.

Reliability is derived from developed robustness against the stresses and human mistakes that can reduce the inherent reliability, such as are found in its

- Intended use and abuse by customers, over time
- Manufacturing, packaging, storage, shipping, and rough handling
- Installation, maintenance, diagnostics, and repairs

These viewpoints should help you toward strategic decisions about how to

1. Select the best system architecture, design concepts, and enabling technical concepts for the intended applications.
2. Analyze the functional designs to identify opportunities for improvement and to predict field reliability.
3. Design experiments under deliberately stressful conditions in order to identify problems earlier and to develop better solutions.
4. Assure that product designs and the processes for manufacturing, installation, service, and use are less vulnerable to degrading factors beyond their control.

Not only does this viewpoint force discussions to be broad in scope and cross-functional, but it has a strong influence over the development plans that require decisions, guidance, resources, funding, time, and management attention.

So what are useful strategies for the development of reliability? You have many choices to integrate into your engineering and organizational development plans. What approaches will be used to achieve the required reliability? Will they be efficient in using resources and time?

Examples of candidate strategies are described in Tables 3.1 and 3.2, applying to the engineering of the product as well as to the organization approach adopted by the project.

Certainly a mix of these approaches will be more effective than any single approach. Given your viewpoint of the reliability equations, any or all of these are strategies that need to be

Table 3.1 Examples of reliability development strategies for design engineering

Example Development Strategies	Rationale
Reuse mature, robust components and design modules; avoid "state-of-the-art" technologies.	Having fewer new design concepts and technologies avoids the introduction of unnecessary risks.
Apply new design/manufacturing technologies only where they can provide the most value.	Efforts to reduce risks will be more related to customers' perception of value.
Choose simpler design concepts with fewer adjustments.	Less complexity and fewer adjustments reduce the opportunities for failures.
Partition the system architecture so that risks are within subsystems rather than at their interfaces.	Risks will tend to be within development teams rather than between them.
Emphasize stressful test conditions.	Finds problems faster; verifies solutions.
Develop subsystem interfaces to be robust.	The absence of degrading stresses or variability in signals crossing interfaces enables system integration and the changing of modules.
Develop subsystems to be robust within the system architecture early.	Subsystems that are robust reduce the risks in system integration.
Optimize robustness at the full system level.	The achievement of on-target, stable product performance may require selected compromises at the level of subsystems.
Decompose and deploy requirements for reliability throughout the system architecture.	Thorough requirements contribute to criteria for the selection of the best design concepts.
Make selected adjustments available to customers.	Customers can correct problems themselves.
Bias set points away from expected deterioration.	Increases tolerance for deterioration.
Design for graceful failure modes.	Customers can use the product while awaiting service.
Emphasize "design for service"; design diagnostic capabilities into the product.	Enables diagnostics and repairs to take less time and be more accurate.
Apply the quality loss function to select tolerances to tighten or to component/material quality to increase.	Increases in manufacturing costs will be aligned to expected savings in quality loss in the market.
Employ scheduled preventive maintenance.	Customers will be able to anticipate downtime.
Extend usage intervals for preventive maintenance.	Over time customers will experience less downtime.

(*continues*)

Table 3.1 Examples of reliability development strategies for design engineering (*continued*)

Example Development Strategies	Rationale
Enable customer-performed maintenance and repair.	Customers can control their downtime.
Design the system architecture to be accessible with independent modules.	Flexibility in the architecture enables late design changes to one module without affecting others.
Develop alternative design concepts and parameter set points in parallel (set-based design).	Available design options enable changes with less impact on project schedules and costs.
Develop design concepts to have "additivity" in their functional parameters (see Chapter 7).	Design parameters that act either independently or constructively are much easier to implement.

Table 3.2 Examples of reliability development strategies for your organization

Example Development Strategies	Rationale
Develop core competencies in robustness development methods and tools.	Robustness development acts to prevent problems and thereby reduce risks. Benefits include reduced cycle times and costs.
Benchmark and reverse-engineer the designs in competitive products.	Design teams need to define superior performance. Select the best design approaches. Beat the competition's design or use it.
Develop design concepts and technologies to be robust for a range of applications prior to being transferred to product development projects.	Technology development reduces risks that can jeopardize product performance or development schedules and costs.
Conduct research on experimental stresses.	Stressful conditions known to be effective at forcing variations accelerate the finding of problems.
Involve customers, suppliers, service, manufacturing, and others early in the concept selection process.	"Street smarts" are critical inputs to the clarification of requirements, the selection of design concepts, and the planning of stressful experiments.
Develop alternative design concepts in parallel.	The best design concepts must be selected from alternatives available within the constraints of the development project.
Develop capabilities for rapid prototyping.	Excellence in the rate of reliability growth depends on fast build-test-fix iterations.
Locate engineering teams close to test labs.	Increases the capabilities for fast corrective actions.
Lead teams with a steady focus on problem solving.	Instills a bias toward fast response to problems.
Apply in-process quality control and/or 100% manufacturing burn-in early in production.	Correcting more problems in the factory reduces the probability of manufacturing defects escaping to the market. However, you have to "design in" quality, not "inspect in" quality.
Develop user training and certification.	Failures due to operator error and unintended abuse can be reduced.

decided upon early in the development project. Certain ones may represent opportunities for improvement. Of course you can always take a brute-force, expensive approach by designing sophisticated control systems, by using redundant designs or high safety margins, or by isolating your product from stressful conditions. Their implementation may have many cross-functional consequences and dependencies and certainly may have major impacts on the project management plans and costs.

Design engineering works within the larger framework of the company's capabilities and capacities. In our experience, improvements in an organization's operations can enable design projects to be much more successful. Table 3.2 gives some suggestions.

Reliability Modeling

Analytical approaches to reliability improvement can benefit greatly from functional models of the interactions among subsystems and modules. With the models reflecting the system architecture and its functional dependencies,

- The development challenges for tough reliability requirements can be understood.
- Analyses can explore more thoroughly the probability that the drivers of failure events and their interactions will cause degradations or failures.
- The system-level consequences of alternative subsystem design concepts and their vulnerabilities to failures can be evaluated.
- Feedback from previous products can focus attention on opportunities for improvement.
- Predictions of system reliability can be based on improvements for subsystems and modules.

Your reliability development has to include the resources to develop models that are representative of the system architecture and to exercise them early enough that their guidance can focus the development activities. We discuss empirical modeling in Chapters 16 and 17.

Project Management

Time, money, and skills are the food of development teams. If they are not available when needed, the achievements of the projects are in jeopardy. A reliability development plan that is not backed by an enabling project management plan is no plan at all. For example:

- The project management plan, integrated across all of the involved organizational and partner functions, must provide the funding and the right resources at the right time.
- Those resources must be organized to make effective impacts on the product's design.
- Sufficient numbers of prototypes of representative configurations must be planned and funded.
- The schedule and resource allocations must allow the necessary time to build and apply analytical models, to plan thoughtful experiments, and to conduct sufficient iterations of corrective action.
- Certainly, iterations of problem solving need to be prioritized by the impact of those problems that are found and their probability of occurrence in customers' applications.

Our earlier discussion illustrated that the application of resources to drive up the slope of the reliability growth plot should also serve to shorten the development schedule. Similar thinking places the progress of reliability growth on the agenda for routine project integration meetings so that slow progress can be identified early and corrected by intervention.

There are substantial interactions of project management with portfolio management, and with the management of centralized resources that support development projects. You want your people working on the right projects (those with higher expected value) at the right time, without being so overloaded that none of the projects are enabled to be successful. Similarly, you don't want your centralized resources to be without reserve capacity. The absence of capacity when needed can lead to process queues and the resulting schedule delays. Worse, they may force decisions to compromise the design in order to meet the product launch date.

Risk Management in Product Development

A technical risk is a potential quality or reliability problem with the product or its development, one that has not yet occurred. So it has an uncertainty derived from the probability that its root causes will occur and from the probability that their effects will result in a local deterioration or failure. If the problem does occur, it imposes consequences, viewed as economic losses. So the risk is the "expected value" of a negative event. This definition of *risk* is more actionable than vague terms like *issue* or *concern*. Preston Smith's work on risk management[3] adds to your assessment the uncertainty about how the local failure will affect the system-level performance. That depends on how the problem fits within the product's architecture or its related service or customer support. This viewpoint is very useful both for understanding the nature of risks and for developing preventive actions to reduce the probability of occurrence and/or the impacts.

Risk management in product development must be an ongoing, visible, and funded process. This can be a challenge because it's fundamentally problem prevention rather than problem solving. Corrective actions driven by demonstrated problems tend to receive a higher level of attention since the problems and their consequences have actually been experienced. It is more difficult to gain support for preventive actions against potential problems that have not yet occurred. There are opportunities here.

From the viewpoint of the customers of a product, reliability assessment is very much a risk assessment. If problems were known to occur at regular intervals under specific conditions, they would become objects of preventive maintenance. The problems would be controllable by deliberate actions that prevent their impacts from affecting the product's use. But for uncertain problems, ones for which the stressful conditions and the vulnerability of the design are less predictable, customers can ask:

- What is the probability of something going wrong?
- If it did, what would be the consequences for the product's use?

---

3. Preston G. Smith and Guy M. Merritt, *Proactive Risk Management: Controlling Uncertainty in Product Development* (New York: Productivity Press, 2002).

- Are there some actions that can be taken to reduce the probability of occurrence?
- If the problem occurs, what can be done to correct it?

So reliability growth is not just about solving the problems experienced in development, but also about preventing problems from occurring in the first place. The strategies that we discuss here are very much about preventing problems. By selecting superior design concepts and their enabling technologies, and by developing them to be inherently less vulnerable to stresses, you reduce the probability that a risk event will become an actual problem, and you reduce the consequences of that problem to the end user of the product.

By limiting the number of new, risky design concepts to just those that add essential value to customers, you keep the probability of system failure aligned to your capabilities to deliver value to your customers. Smith and Reinertsen[4] point out that if your new system has a number of independent subsystems $(a, b, c, . . .)$ each with a probability of success of less than 100%, the probability of success of the full system is the product of those probabilities:

$$P_{system} = P_a \times P_b \times P_c \times P_d \times \ .....\ \tag{3.1}$$

The more new concepts you add, the lower is the probability of succeeding. This advice is analogous to that of leveraging mature and robust design concepts as much as possible.

Capabilities of the Organization

We describe here two methods to recognize the capabilities of your organization for reliability development: the demonstrated growth rate for reliability and a capability assessment using a tool specific to reliability development.

Reliability Growth Rate

Earlier in this chapter we discussed the reliability growth plot. The challenges included how to start the curve at a higher point (R_0) and how to increase the slope (α). The ability to increase these metrics is essential to the management of reliability development. One perspective interprets the slope of the reliability growth plot as a measure of your organization's capabilities. Table 3.3 provides examples of these definitions. You may choose to adapt them to the character of your business. This reliability growth model is discussed in detail in Chapter 8 along with its implications for project management.

Assessment of Capabilities for Reliability Development

Table 3.4 is an example of a capability assessment tool specific to reliability development. You can adapt it easily to apply to your business. We used categories relevant to the development of reliability and criteria relevant to improved results. An interesting way to map the results is with a radar chart, shown in Figure 3.3. We've illustrated a difference of opinion about the ratings, which might be expected if people feel the assessment to be a report card affecting their pay.

---

4. Preston G. Smith and Donald G. Reinertsen, *Developing Products in Half the Time* (New York: Van Nostrand Reinhold, 1991).

Table 3.3   How can you increase your organization's capabilities to grow reliability?

Rating	Demonstrated Characteristics
$\alpha = 0.4$–0.6 *Excellent*	• The organization has a high level of dedication to preventing failure modes. • Development teams are responsive and efficient at analyzing and correcting failures. • Accelerated test methods are being used.
$\alpha = 0.3$–0.4 *Good*	• Reliability improvement initiatives have reasonable priority. • Analyses and corrective actions are prioritized and managed well. • Laboratory tests employ effective environmental stress levels.
$\alpha = 0.2$–0.3 *Fair*	• Reliability improvements receive only routine attention. • Corrective actions are developed for the more important problems. • Environmental stresses are lacking from test plans.
$\alpha = 0.1$–0.2 *Poor*	• Reliability improvement receives little attention. • Failure analysis and correction receive little deliberate attention.

Table 3.4   Assessments of your capabilities for reliability development should be adapted to your business model.

Characteristic	Beginning	Better	Excellent
		Design Execution	
Concept selection	• System architecting and integration are formal processes. • Concept selection balances risks of new concepts and technologies against maturity of leveraged approaches. • Decisions implement a consistent strategy for reliability development.	• System architecture keeps risks within subsystems, not at interfaces. • System architecture enables design changes. • Reliability data are valued in the selections of concepts and components.	• Concept selection criteria include the full scope of "RAM" (Reliability, Availability, Maintainability) design guidelines. • Design concepts are selected to be superior to available alternatives. • Design concepts and technologies are robust for the application when transferred.

Characteristic	Beginning	Better	Excellent
Reliability development	• Peer reviews value collective wisdom • Analytical reliability methods and tools support product development • Development capabilities enable reliability requirements to be achievable.	• DFSS and DMAIC methods and tools are used. • Design guidelines are employed consistently. • Bias for action is high for failure reporting, analysis, correction, implementation. • Subsystems are robust prior to system integration.	• Robustness is optimized at the system level. • Critical parameter management is formal. • "FRACAS" is effective. • Reliability growth methods are applied to software development.
Reliability testing	• Plans have cycles of build-test-analyze-fix. • Prototype configurations are current. • Test planning has high technical integrity. • Objectives for tests differentiate "learning" from "proving."	• Designs are stressed beyond expected usage and environmental conditions. • Test designs and surrogate stresses are demonstrated to be efficient and effective. • Data acquisition and analysis systems enable high integrity in test results.	• Test planning, execution, and analysis are efficient and enable timely decisions. • Full range of stressful tests accelerate problem identification and correction. • Decisions derived from tests have proper statistical conclusions.

Functional Involvement

Characteristic	Beginning	Better	Excellent
Customer involvement	• VOC studies are the basis for reliability requirements for products and services.	• Advanced customers provide routine advice on reliability value drivers. • Customers validate reliability requirements.	• Engineers are involved with customers' value drivers. • Customers evaluate design concepts. • Customers test production prototypes.
Manufacturing involvement	• Manufacturing engineers provide DFMA guidance to early product development. • Development prototypes provide feedback from manufacturing methods and personnel. • Suppliers reduce infant mortality.	• Suppliers provide advice to early design decisions and component selection. • Demonstrated manufacturing process capabilities are included in program decisions.	• Supplier selections include demonstrated manufacturing process capabilities. • Suppliers work as development partners. • Demonstrated manufacturing quality is a pre-production decision criterion.

(continues)

Table 3.4   Assessments of your capabilities for reliability development should be adapted to your business model. (*continued*)

Characteristic	Beginning	Better	Excellent
Service involvement	• Service engineers provide "design for service" guidance to early development. • Strategies define roles for customer-performed maintenance.	• Serviceability tests improve designs. • Demonstrated service capabilities are included in project decisions. • Reliability demonstration tests are used to estimate parts and materials usage, maintenance schedules.	• Customer acceptance tests include field service feedback to improve designs.

Management Support

Characteristic	Beginning	Better	Excellent
Expectations	• Importance of reliability to customers is visible and consistent, with targets. • Need for improved development capabilities is communicated.	• Funded development plans and metrics include focus on reliability achievement. • Expectations and funding are consistent for reliability resources and skill building.	• Reliability requirements are not compromised by schedules and funding budgets. • Gate review agendas focus on reliability, risks, and business impacts. • Management teaches reliability courses.
Resources	• Some engineers are trained in methods and tools of reliability development and problem solving, e.g., DMAIC Six Sigma.	• Development engineers are trained in reliability methods and tools. • A few reliability specialists are available for highest-priority problems.	• Reliability specialists are available early. • Responsibilities and advice are respected. • Organizational capabilities are improved continuously.
Accountability	• Development teams are rewarded for fast problem correction. • Teams must use lessons learned from previous and competitive products.	• Development teams have incentives to avoid compromising reliability to meet schedules. • Teams are rewarded for problem prevention.	• Balanced scorecards include reliability. • Development teams remain responsible for production and field failure correction.

Characteristic	Beginning	Better	Excellent
Research	• Market research has funding that is sufficient and consistent. • The drivers of markets' value for reliability are current.	• Product plans drive technology development. • Market situation analyses are current. • Competitive product benchmarking is current.	• Technologies are developed to be robust. • Experimental stresses are studied. • Customers' value drivers are current. • Databases of field failure analyses are useful and accessible to development teams.

Project Management

Project management	• Project plans include significant time for reliability growth testing. • Sufficient prototypes with correct configuration are available. • Reliability growth is tracked routinely.	• Funded plans incorporate specific strategies for reliability development. • Responsibilities for specific reliability requirements are clear and reinforced. • Facilities and equipment for stressful testing and data analysis are accessible.	• Management intervenes when the progress of reliability growth is not acceptable. • Reserve capacities for analysis, testing, and problem solving are available.
Reliability requirements	• Engineering and marketing define value drivers for market segments. • Stressful conditions are defined for product operation, storage, handling, etc.	• Reliability requirements are defined in detail, superior to previous products. • Reliability requirements are allocated to subsystems with balanced achievability. • Requirements also cover service and support.	• Reliability requirements are based on value to customers and competitive product benchmarking. • Reliability requirements reflect the intended product and price positioning.
Decision making	• Failures and risks are prioritized by the severity of their impacts on customers. • Management acts early to enable reliability growth.	• Risk management and reliability growth have time on the agenda at routine project reviews. • Reliability specialists have a voice in program decisions.	• Decision criteria include reliability status and program risks. • Interpretations of reliability growth data are proper and understood.

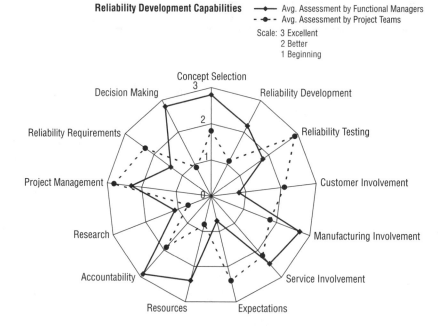

Figure 3.3   The assessment of capabilities can lead to prioritized improvement initiatives.

Facts and root cause analyses from past projects can resolve the differences. In this example, the term *reliability* is meant to be a broad representation of metrics such as failure rate, usage life, availability, maintainability, safety, and other terms related to the ability to prevent problems from affecting customers' use of your product. Use metrics that are meaningful to your organization and actionable.

What levels of capabilities are needed by your organization or by a specific development project?

- Studies of past projects may identify certain characteristics that, if improved greatly, could have major benefits to the probability of success for development projects.
- Possibly the business case for a development project may require dramatic improvements in particular capabilities.
- By conducting benchmarking studies of successful companies or organizations, your teams can identify opportunities for significant gains in competitive advantage.

In these assessments, the capabilities gaps that are identified are more important than the raw scores that are assigned. The potential efforts for and consequences of closing particular gaps can be evaluated to determine the justifications for specific improvement initiatives.

Key Points

1. The achievement of higher levels of reliability is not just an engineering problem. It depends upon substantial and consistent support from management, sufficient resources with the necessary capabilities, ongoing processes for understanding markets and developing new capabilities, and many other contributing factors.
2. First understand where your projects need to improve.
3. An internal assessment of your capabilities for reliability development can identify gaps between your intentions and your practice.
4. External benchmarking can identify additional opportunities derived from understanding what other companies do well.
5. Continuous improvement can ensure that the competitive advantages that you do achieve are sustained from product to product.

Discussion Questions

1. How effective are your strategies for increasing product robustness?
2. What new strategies have you learned from this chapter?
3. What are your projects' current capabilities (α) for growing reliability?
4. How important is the slope (α) of the reliability growth curve considered to be in judging your projects' progress and your expectations for reliability at market entry?
5. How well can your projects control their reliability growth rate?
6. What barriers to reliability growth do your projects face?
7. How well do you rate your projects for reliability development capabilities?
8. Do you believe that reliability is a much lower priority than meeting schedule in your organization? What evidence is there to support your belief?

SECTION II

Framework for Reliability Development

Chapter 4: A Process Model for Product Development

The work to develop reliability is within the overarching context of the product development process. We describe such a process that has strong engineering and business characteristics.

Chapter 5: Road Map for Reliability Development

How can you recognize critical elements of your project management plan that are essential to reliability development? It doesn't just happen. You have to plan specific activities with the right resources at the right time. Relevant chapters in our book are identified for their contributions to these plans.

Chapter 6: Effective Decisions

Product development is, in part, a process of making and implementing many decisions. We describe several types of decisions, in addition to gate reviews.

Chapter 7: Robustness Development for Product Designs

The strategy promoted in our writing is first to develop robustness to reduce the probability of a problem occurring. Then, after those best efforts at problem prevention, grow reliability with fast problem solving. Robustness development reduces the vulnerability to stresses that cannot be controlled. Applied both to product designs and to manufacturing processes, it is the heart of the strategy to increase manufacturing process capabilities.

Chapter 8: Reliability Growth and Testing

The slope of the reliability growth plot is a measure of an organization's capabilities for fast problem solving. The familiar DMAIC Six Sigma methods provide clear advantages. The growth rate for reliability benefits from accelerated testing and effective corrective action for problems whose consequences have a higher level of severity for customers.

A Process Model for Product Development

Most companies that enjoy consistent and repeatable outcomes in product development follow a standardized process that is adapted to the characteristics of their business model. Whether or not their process is a good one may be an open question. Likewise, whether or not a company actually achieves competitive advantages in following its process is also a question for an assessment. In this chapter we outline a descriptive model for developing products with enough detail to illustrate elements that are important to the development of product reliability. The process is scalable and flexible, intended to be practiced with the principles of lean and flexible product development. It can be adapted to highly leveraged designs as well as to complex, clean-sheet product concepts.

What Is a Product Development Process?

Product development is a flow of activities across organization functions, partners, and suppliers to create solutions to the needs and desired outcomes of target customers. The process follows fundamental steps that are reasonably universal:

1. Establish requirements that represent a differentiated, superior value proposition.
2. Select the best architectures and design concepts from available alternatives.
3. Develop, design, and specify solutions for
   a. The design concepts integrated into a product system
   b. The production processes to replicate, deliver, and maintain the solutions
4. Validate the value proposition, requirements, and designs with feedback from customers.
5. Verify that the designs of the product and its production processes satisfy their requirements.
6. Develop the value delivery system.
7. Manage smooth transitions
   a. Into product development from advanced planning, portfolio management, and technology development
   b. Out of product development into production, service, sales, and customer support

As an investment in the creation of a new or extended business, a development project applies resources, knowledge, capabilities, capacities, and decisions effectively over time to meet its objectives in a predictable manner. A value-adding development process incorporates the best wisdom and practices of the company. As a source of competitive advantages, it embraces core principles and paradigms that enable it to drive higher quality, lower costs, and shorter, more predictable cycle times.

The goal is to achieve product designs and production processes that are better than those against which they will compete in the market. In parallel, the model also develops the capabilities and capacities to sell, deliver, install, service, and support the new product in order to deliver superior value to customers while achieving acceptable returns to the company.

When designed to be suitable to your internal culture and technical challenges, your development process should establish consistent, more predictable methods of accomplishing work. That enables people on projects to know what is expected of them, when it is needed, and what the acceptance criteria are at particular milestones. It frees development teams to be more creative and efficient in how they do their work. When work is done in collaboration with partner companies, it may be either the common process with which everyone works, or the basis for other companies to see how their deliverables integrate with your work.

For the development of robustness and the growth of reliability, the product development process provides a clear framework. With it your project teams can establish project management plans for those activities, with resources and timelines necessary to develop sufficient robustness in the design concepts and to achieve the necessary growth in reliability prior to market entry. Those plans implement the strategies chosen for the development of the particular product.

Phases and Gates

Phases are logical groupings of cross-functional activities that develop information to answer key business and technical questions. The plans for these activities with their dependencies enable progress and risks to be monitored, enabled, and controlled intelligently.

Gates are decision processes (see Chapter 6) supported by project-level reviews. They are positioned at natural points of convergence of critical, integrated information and the need for project-level decisions. The concept of a gate implies a respected barrier to be opened rather than a "speed bump" to be disregarded. Gate reviews separate the development phases, although typically the beginning of the work for the next phase does not wait for the gate. The decision to be made is whether or not the project is prepared to do the work of the next phase with acceptable risks and more predictable progress. Through this process, management leads product development, implements the strategies in the product development portfolio, and empowers the project teams. It may also be necessary for management to remove barriers, supplement resources, or resolve conflicts to enable more predictable progress.

What's the Problem?

We expect that your project teams follow a development process, either written or at least understood. To determine whether or not your process reflects how work actually gets done and is a source of competitive advantage, ask questions of your organization such as these:

- Do development teams know what to do, when, why, and how well? Are people who really understand the process no longer with your company? How well does management understand your process?

- How repeatable, scalable, and adaptable are your processes? How lean and flexible are projects' applications of your process?
- Do all development projects follow the same basic process?
- How responsive are projects to changing conditions? How often is your process improved with lessons learned internally or by benchmarking?
- Are the criteria for management decisions known before the information is developed?
- How well are functional organizations, partners, and suppliers integrated? How well is product development integrated with business development?
- How vulnerable are development projects to degrading factors they cannot control?

If you do not like the answers, it may be a good time to upgrade the guidelines your development teams follow. Your improvement objectives may be technical, such as higher levels of customers' satisfaction, increased product reliability and usage life, or lower manufacturing costs. They may be more business-oriented, such as shorter, more predictable time to market, increased product differentiation, or reduced technical and business risks. An assessment of your past projects can identify the opportunities.

Why Should Your Organization Have a Standardized Development Process?

We expect that no two development projects are the same. There are ranges in technical complexities, cross-functional integrations, business challenges, and project risks, regardless of whether projects involve major platform changes or highly leveraged designs. However, we do not advocate the use of a variety of processes. An example promoted by some would be a fully detailed process for a major platform change and a simpler or reduced process for an easily leveraged product. It makes more sense to us to have a single standardized process written so that all organizations can understand the relevance to their particular functional challenges and contribute effectively to the technical and business coordination. Then the project leadership team can decide which expectations of the process are either not relevant to their project or are already satisfied, as is often the case with a leveraged design. They also can decide where processes can be accelerated or started earlier and how phases can be overlapped. Likewise, the need for more rigorous discipline can be identified, which may place certain reviews and decisions on the project's critical path. Although a design may be highly leveraged, a single new technical concept may determine that extended development in the early phases is necessary.

Since a development project is an investment in a future business, the stakeholders responsible for those business returns should approve of those modifications to the standard process that are proposed for a specific project. It then becomes the basis for the project management plans.

What benefits can be expected from having a single standardized process that is adapted to the character of each project? Here are some of them:

- Development teams can devote their time to working on solutions rather than determining what work to do.
- People new to the company will have the same process to learn, regardless of their assignment in the organization. People transferred between units internally will not have to learn a different process.
- Adaptations of the standard process to a range of projects can have a common basic model. The process knowledge can be transferred from project to project, across business

units and collaboration partners, with a common understanding of the expectations. Lessons learned from a variety of projects can contribute to continuous improvements that can help all projects.

• Standardized documentation can set expectations for the contents and facilitate the interpretation of those details. Project stakeholders can provide governance to a variety of projects across business units with a common understanding of the process, its principles, decision criteria, paradigms, and key questions.

Published Development Processes

Your development process may be derived from one of several popular processes that have been published. All have their merits, and there's no claim here that one is better than the others.

• Stage-Gate, developed by Robert Cooper,[1] and PACE, from PRTM,[2] employ three phases or stages to move from an initial screening and project approval to product launch. Don Clausing[3] also uses three phases to describe the development of new products, although the scope of his first two differs from PACE and Stage-Gate. In his NexGen Stage-Gate Process,[4] Bob Cooper uses an additional phase up front to describe the early work of getting a project off to a good start, improving its alignment with portfolio plans and ongoing market research.

• CDOV[5] describes the process in four phases, with subphases to differentiate activities for the product design from those for their manufacturing processes.

Their value is in how they are adapted to guide the work of your product development teams. Your process then will have details that relate to steps that your teams find to be necessary and that fit with your development strategies. Pressures on development timelines, costs, and out-of-box quality can vary greatly among companies. Some businesses are highly seasonal, with little forgiveness for delays in market entry, and others demand more of competitive performance and costs. Your process should guide the integration of insights into those elements that promote, for example, superior achievement, reduced risks, efficient project management, and the effective investment of resources. There are no magic remedies here. The processes all expect rigorous involvement with customers and suppliers and expect functional excellence to achieve improved business performance.

1. Cooper, *Winning at New Products.*

2. Michael E. McGrath, Michael T. Anthony, and Amram R. Shapiro, *Product Development: Success through Product and Cycle-Time Excellence* (Boston: Butterworth-Heinemann, 1992).

3. Clausing, *Total Quality Development.*

4. Robert G. Cooper and Scott J. Edgett, *Lean, Rapid, and Profitable New Product Development* (Canada: Product Development Institute, 2005).

5. C. M. Creveling, J. L. Slutsky, and D. Antis, Jr., *Design for Six Sigma in Technology and Product Development* (Upper Saddle River, NJ: Prentice Hall, 2003).

Process Documentation

We do not propose that your development process be described with large amounts of documentation. A combination of concise descriptions, checklists, flowcharts, value stream maps, dependency maps, skeletal project plans with milestones and decisions, and other devices may be found to be useful. Documentation could be in paper form or on your company's intranet, reflecting how your people prefer to learn and receive guidance. Descriptions of activities and deliverables should not be so prescriptive that they inhibit flexible practice and creative approaches. The objectives of phases and the criteria for decisions must be clear, but without bureaucratic demands for compliance with every detail. Above all, the documentation must be understood by a wide range of people and kept evergreen as experience and benchmarking identify improvements. It is a foundation for your organizational learning.

The process descriptions are a high-level framework with expectations that can be understood by all disciplines, organizations, partners, and management. They are not thick volumes that are burdensome to read and that collect dust. They may include subprocesses for specific disciplines that need guidance with more relevant references, practices, and jargon. This has been our experience for the development of software and consumable materials.

Integration of Key Development Principles

How can competitive advantages be achieved in ways that consistently deliver higher value to your customers and return higher value to your company?

DMAIC Six Sigma

The classical tools of DMAIC[6] Six Sigma have been used for years to identify and solve existing problems with processes and product systems. The approach applies disciplined methods of building prototypes, testing them to identify problems, analyzing the causes of the problems, and correcting them with solutions that are verified prior to implementation. It's a "design-build-test-analyze-fix-verify" process derived from Deming's "Plan-Do-Check-Act" strategy.

Six Sigma methods have saved a great deal of money for organizations by reducing the variability in manufacturing processes that causes deviations in product performance. For product development the focus has been on problem correction, particularly for product problems found later in development. The methods have also found valuable applications to nonengineering business processes. Walter Shewhart,[7] an early quality pioneer at Western Electric, provided sound guidance to identify root causes and to decide whether or not corrective actions were necessary. Statistical process control was then focused to correct those problems that have assignable causes and to increase the ability to tolerate the consequences of random events. In Chapter 8 we discuss DMAIC methods in the context of efforts to grow reliability.

---

6. *DMAIC* stands for "Define, Measure, Analyze, Improve, and Control." Variations on this acronym include an extra I for "Innovate."

7. Walter A. Shewhart, *Economic Control of Quality of Manufactured Product/50th Anniversary Commemorative Issue* (Milwaukee, WI: ASQ/Quality Press, 1980).

Design for Six Sigma (DFSS)

Design for Six Sigma (DFSS)[8] extends the concepts of Six Sigma as applied to product development. While DMAIC strategies are focused on problem solving, DFSS strategies are focused on problem prevention. Certainly the approaches overlap. DFSS is an umbrella methodology that addresses the broad scope of activities in product development. This includes the ongoing work of market research, benchmarking, customer studies, advanced product planning, and technology development. During product development, the practices of DFSS bring improved rigor to the development of robustness in products and processes, the management of requirements and critical design parameters, the mitigation of risks, and other valuable practices through product launch into production.

Management should be involved early to provide insight, guidance, and focus. Their governance processes can enable efficient progress by clarifying the value to be developed for customers, by helping to resolve significant risks in the project, and by taking steps to improve the predictability of the project's future achievements. It's the early decisions that have the most opportunities to set the path for successful development of better solutions to a product's requirements. Their foundations are derived from

- The ongoing studies of markets, customers, and competitors
- The development of new and improved technical capabilities in anticipation of future projects
- The continuous improvement of the organization's resources, knowledge, and processes

With DFSS, the interactions among management and project teams are more deliberate and structured with deliverables, acceptance criteria, reviews, and decisions. Decisions are based on data, verified to have high integrity, and backed by methods and thought processes that support sound engineering and science. With DFSS-enabled processes, management implements the portfolio for product development.

Lean Product Development

Competitive and financial pressures emphasize the need for shorter development times and reduced development costs. For many types of products, such as those for rapidly changing markets, development speed is vital to success. The best defense against changing markets is to get to market fast.[9] Strategies for lean product development enhance the DFSS approach by reducing the waste of time and resources during the process and by increasing the focus on engineering productivity. Lean principles[10] ask us to question the relative value of each activity and the efficiency with which it is accomplished. The objectives are to reduce cycle time and costs without compromising quality or failing to mitigate risks. Resources that may become available can then be applied to higher-value activities.

---

8. Creveling et al., *Design for Six Sigma in Technology and Product Development*.

9. Clausing, *Total Quality Development*.

10. Allen C. Ward, *Lean Product and Process Development* (Cambridge, MA: Lean Enterprise Institute, 2007); Ronald Mascitelli, *The Lean Product Development Guidebook: Everything Your Design Team Needs to Improve Efficiency and Slash Time-to-Market* (Northridge, CA: Technology Perspectives, 2007); Donald G. Reinertsen, *Managing the Design Factory: A Product Developer's Toolkit* (New York: Free Press, 1997).

Although lean methods were developed in the context of manufacturing processes, their generalities are very relevant to product development. For example, lean principles minimize batch size. That guides strategies to evaluate prototypes with shorter test cycle times, providing faster feedback. Problem prevention reduces the costly rework of designs and tooling during preparations for production. Lean organizations strive to minimize queues that often arise when plans for resource allocation do not address the potential variability in project schedules or when centralized functions have insufficient reserve capacity to handle surges in demand. A key strategy is to maintain planned resource utilization to be less than 100%. The avoidance of resource overload is a sound principle, reflected in development portfolios managed to avoid too many simultaneous projects.

Large benefits should be expected from delivering the right quality at the right price in products provided to the market at the right time. Lean principles emphasize that a demand for compliance with naive funding budgets has lower value when compared with the benefits of avoiding the costs of delayed market entry. Wasted time means the waste of resources that could have been used to develop functionality further, to improve reliability, to prevent problems that customers care about, or to fix them. How can you afford not to do this? You should expect that your competitors are doing it.

Set-Based Design[11]

The strategy of set-based design (aka "Set-Based Concurrent Engineering"), exemplified in Toyota's product development process, is to develop thorough knowledge of the behaviors across the design space and its alternative configurations. This is a series of maps of the relationships among design parameters and the output response of the technical function. Sets of acceptable parameter set points are defined. Adjustments to controllable parameters can then be made later in the process, during production preparation, with full knowledge of their impacts on performance and its stability.

Set-based design fosters the parallel development of not only alternative set points for parameters, but also alternative design concepts. It starts with technology development and extends well into product development. A change in the design from one subsystem concept to a better one is an example of the strategy that can be applied during system integration. The key is to have alternatives ready to be used and design space understood well enough to enable the changes. The methods that we describe contribute to that knowledge. A flexible product architecture that is receptive to these changes enables this strategy.

Agile Development

Traditional development processes have tended to depend upon the freezing of requirements in the early development phases and on the completeness of information in deliverables. However, for products aimed at rapidly evolving markets, this may be neither wise nor practical. In fact, there can be noteworthy benefits in delaying decisions until dependencies require them. These milestones can be determined by reverse-engineering the project schedule, from market entry back toward the project beginning. Premature decisions can commit the project to a path too early, blocking the ability to respond to late-arriving information that may be critical to customers.

---

11. Durward K. Sobek, Allen C. Ward, and Jeffrey K. Liker, "Toyota's Principles of Set-Based Concurrent Engineering," *Sloan Management Review* (Winter 1999).

Demands for shortened cycle times often force development teams to begin activities before their prerequisites have been satisfied. Teams have to learn how to work with partial information. For example, your customers may not know what they really need until they have opportunities to test your prototypes. They may change their minds when one of your competitors launches a new product with unexpected capabilities. Rather than resisting design changes, as might be prudent when substantial tooling is involved, those projects working toward dynamic markets will have advantages if they have systems and processes that can embrace changes gracefully. This can have implications for the

- Development of architectures for the new products in order to enable changes more easily
- Development of the knowledge to adapt the designs (set-based design)
- Funding of testing and feedback activities to gather inputs from customers
- Processes for change management being selective and efficient in the later phases

The principles of agile development[12] were established for software design, embracing the principles of Extreme Programming. For software, late-arriving feedback is a routine part of the process, and fast design changes are expected. Software capabilities enable this strategy since they can be changed more easily than can hardware or materials. The challenge is to adapt these principles to the design of multidisciplinary products so that late changes can be less disruptive.

The strategy implies deliberate plans to engage customers to obtain their feedback about both product requirements and design concepts. Within the process described here for product development, there are validation steps throughout most of the phases. They depend on an architecture for the product's system and an attitude for the project team that both enable and embrace late design changes. The benefit is an improved ability to be responsive in delivering competitively superior products to your customers.

So DFSS emphasizes sound methods and tools with a notable emphasis on problem prevention. Lean emphasizes efficiency, and agile places the focus on flexibility. Set-based design encourages alternative solutions being developed in parallel, with the best one chosen later in the process. DMAIC Six Sigma enables your development teams to correct problems more quickly with better solutions. Combined, these principles give your projects a set of powerful strategies and tools that can prevent a wide range of problems and achieve much-improved results. Better yet, they use resources wisely and are adaptable to a wide range of project characteristics. It's a crime to waste resources and time when they can be applied to those tasks that are more important to your customers and to your company. This is far better than depending on heroic and costly efforts to fix problems in a crisis mode late in your project.

Figure 4.1 illustrates that strategies aimed at learning how to control and optimize performance and reliability apply in the early development phases, while later phases are focused on fast problem solving.

---

12. Preston G. Smith, *Flexible Product Development: Building Agility for Changing Markets* (San Francisco: Jossey-Bass, 2007).

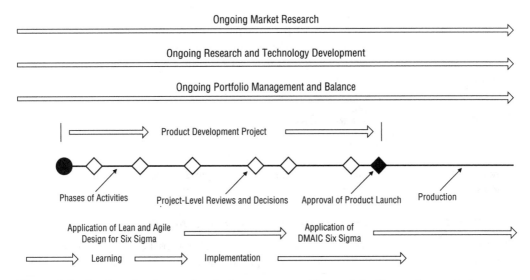

Figure 4.1 Product development projects are managed in parallel with ongoing market research, technology development, and portfolio management.

Portfolio Management

Although it is not a topic for this book, these processes fall under the umbrella of portfolio management. Portfolio plans serve to balance the loading of resources across projects and align the selection of projects with your business objectives and organizational synergies. They define the long-term investment plans that enable a range of projects with various timelines. Lean practices make them more predictable. Large and uncertain variations in resource loading and project timelines can jeopardize not only a new product's market entry but also the other projects in the portfolio that compete for resources and management attention.

Each product development project is chosen for its potential value to customers and for its expected returns to the business. The basic foundations for these forecasts are in the portfolio plans and in their competitive response to the needs of chosen markets. The ability to achieve these products is in the functional capabilities of the organization, its partners, and suppliers, and in the technical concepts that can be developed, or acquired, and commercialized. The fundamental decision about what to include in the portfolio is a stochastic optimization problem, where development costs and financial returns have uncertainties. Projects compete for limited resources. The challenge is to select the portfolio that maximizes the risk-adjusted expected return.

These ongoing activities can have dramatic benefits for product development projects. They provide the foundation information to start projects on time with strategies, knowledge, and concepts that reduce their development cycle times. They improve the chances of development projects being successful. As market or technology factors change or as other projects compete for resources and priorities, the project's direction might have to be changed.

The importance of these processes to management cannot be overstated. They are not just processes for the engineers. The value streams that are developed are the basis for future profits. So management should own your processes. In their interactions with development teams, managers have the opportunity to set clear expectations for excellence in execution and an appropriate level of discipline in following the process. By asking better questions and recognizing good answers, they can foster respect for the strategies behind the process and the criteria supporting decisions. By accepting action items to enable projects, such as by allocating resources and removing barriers to progress, they act in partnership with project teams rather than being bureaucratic barriers themselves.

A Process Model for Product Development

In this chapter we describe a somewhat more detailed development process to enable the elements of robustness and reliability development to be illustrated. Figure 4.2 shows that its framework extends to include advanced product planning and technology development, which provide solid foundations under development projects.

What Principles Affect the Layout of the Phased Activities?

The architecture for the product development process has characteristics that may be evident if you look closely at the details. The guiding principles include

- Technology development before product development
- Product requirements before concept selection
- Subsystem development before system integration
- Development before design, for both the product and its production processes
- Product development in parallel with process development
- Problem prevention first, problem correction later
- Design completion before design verification

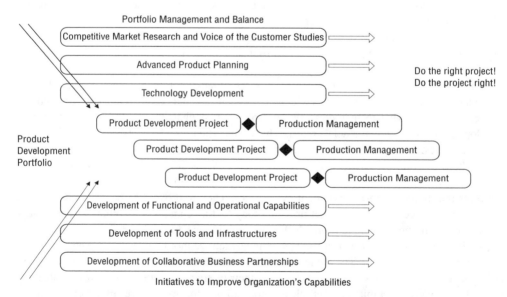

Figure 4.2   A development project benefits from ongoing studies of markets, advanced planning, technology development, and continuous development of organizational capabilities.

- Business plan approval prior to commitment to product design and production preparations
- Certain long-lead expenditures potentially approved prior to a completed business plan
- Phased beginning to production preparations, as design information is stabilized
- Process activities, expectations, and milestones derived from the market entry date and acceptance criteria

In practice, competitive advantages may be found in how the phases and gates are practiced. It is important, then, for guidance to convey the intent of the phases so that project teams can strive to comply with their objectives rather than to satisfy checklists. The major initiatives in product development expect that. We encourage your teams to be students of strategies such as

- Project management
- Risk management
- Concurrent engineering
- Decision making
- Knowledge management
- Design for Six Sigma
- Lean product development
- Flexible product development
- Agile development
- Collaborative innovation
- High-performance teamwork
- Collaborative product development

Advanced Product-Planning Process

What has been thought of as the "fuzzy front end" does not need to be fuzzy at all. It encompasses specific activities to be staffed and managed, plans to be developed, and decisions to be made. The deliverables include long-range product family plans (portfolios) that give birth to

- Product development projects started at the right time
- Technology development (or acquisition) projects in anticipation of the needs of those product development projects
- Long-range, integrated investment plans that enable the development or acquisition of technologies, products, and manufacturing processes at the right time

It starts with basic decisions about the participation strategy for the business and the markets within which the company chooses to participate. It incorporates conclusions from evaluations of the

- Value drivers for selected market segments
- Competitive situations in those markets
- Trends in technologies and their obsolescence
- Unfulfilled needs and preferences of target customers
- Strategies for product and price positioning

- Business opportunities for which you can have sustainable competitive advantages
- Creative ideas for new and differentiated products and services to offer to those markets

Your business should have a market participation strategy with which the plans and core competencies need to be aligned. A series of differentiating value propositions can then characterize the tactics the company intends to use to serve those customers in ways that will be perceived as being better than available alternatives. This takes a lot of insight, based on knowledge rather than guesswork. The winners include those who see the market first.[13]

Figure 4.3 illustrates a linkage of activities and plans that are the responsibilities of those people assigned to advanced product planning and to technology development. Product family plans are maps of conceptual solutions to high-level requirements that are derived from value propositions. The capabilities that the products are challenged to commercialize are the sources of the need for new product or manufacturing technologies that must be developed or acquired. They are also the source of plans to leverage existing designs and platforms for applications to new market needs. The mapping of these plans serves as the basis and timing for the technology development (or acquisition) projects. That determines when an identified new or improved technical capability needs to be ready for transfer into a product development project. The passage of Gate 0 provides the project with a charter and the necessary allocation of resources. This must take place early enough that the "window of opportunity" for market entry can be achieved with the value proposition implemented by an acceptable development effort.

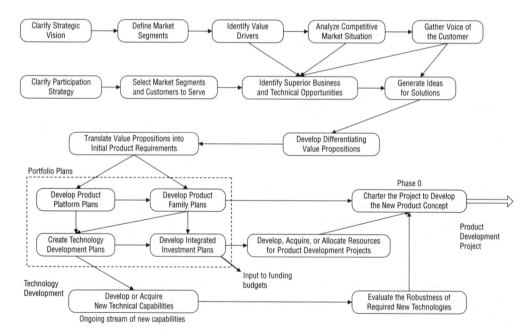

Figure 4.3   The front-end foundations of the process do not need to be "fuzzy," but they do need to be funded and managed.

---

13. Gary Hamel and C. K. Prahalad, "Seeing the Future First," *Fortune*, September 5, 1994.

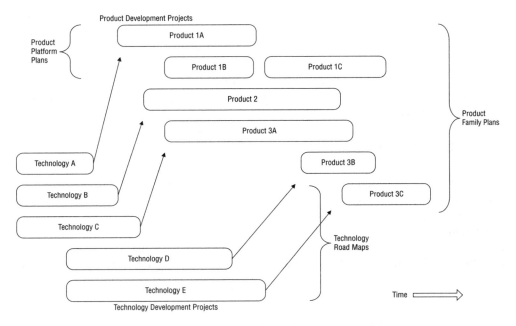

Figure 4.4 Technology development plans are derived from the needs of product family plans and product platform plans.

You can see that these are highly dependent activities. Given the risks that they embody, alternative approaches and contingency plans will be valuable, such as the funding of more than a single new concept to be developed in parallel to answer a particular new requirement. As illustrated in Figure 4.4, projects should start at times that depend on their development cycle time, working backward from their target date for market entry. They depend on resources being available and on the required new technical concepts being ready for transfer. The road maps for technologies are intended to plan the migration of new capabilities into products and the availability of alternative solutions from which the best ones are chosen for the products' applications. Each new product represents a differentiated and superior value proposition.

Technology Development Process

The technology development process applies to what you might call "advanced research and development." It is not intended for basic research or invention activities. Those tend to be more open-ended processes without timelines that can be managed or predicted. The objective of technology development is to prepare a stream of new or improved technical capabilities to be robust for a range of planned applications to new products or manufacturing processes, in anticipation of the needs of future development projects. Those applications are identified by the product family plans and their related technology road maps. You could even consider the process to apply to the capabilities for product service or customer support. For the development of robustness and reliability into product designs, let's stay focused on product technologies and the ability to replicate them.

The process can have four phases, shown in Figure 4.5. As with product development, "gate" decisions judge whether or not phase-specific acceptance criteria have been satisfied to enable the technical concept to move into its next phase. Ideally, the results of technology development will be sufficient to enable the product-specific development to focus on tuning the adjustment parameters to be appropriate for the subsystem design and later to optimize system integration. In reality, further development of the concepts in the context of the product-specific configuration and stresses may be needed.

Technology Transfer

The best way to transfer the knowledge of a new technical concept is to transfer the key people and their development equipment into the product development project. That can have important benefits for the careers of technology development people, making them much more aware of the needs for quality and content in their deliverables. People who make this transition and then return upstream often acknowledge it to be one of the better experiences of their careers. However, this is not a reason to avoid providing thorough documentation. Here's a brief checklist of deliverables that can be helpful to ensure a complete and sufficient transfer of a new technology.

Technology design specifications:

- Decomposition of subsystem requirements at the level of the technical function
- Controllable functional parameters, both fixed and adjustable, with recommended set points
- Sensitivities of the controllable parameters to the range of product applications
- Stresses ("noises") found to be effective at forcing variations in functional responses
- Maps of the design space, including how the response of a function varies with changes in its controllable parameters
- Risk assessments and recommended preventive actions
- Recommendations to enable product-specific system integration

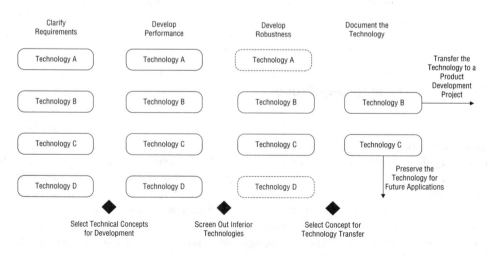

Figure 4.5   A phase/gate construct can be applied to the technology development process.

Technology development reports:

- Rationale for the selection of the technical concept for development
- The range of applications and stress levels that have been tested, with conclusions
- Assessments of strategic fit with product family plans and technology road maps
- Conclusions from the competitive benchmarking of technologies
- Conclusions from key analyses and trade-off studies, such as failure modes, cause-and-effects, design limits, performance cliffs, costs, risk assessments
- Reports from tests, the application of stressful conditions, exercises of analytical models

Resources for technology transfer:

- Personnel for transfer to product development, at least for the initial phases
- Robustness development fixtures and other development breadboards
- Instrumentation, with data collection and analysis equipment
- Analytical and empirical models

There is an important distinction between technology development and product development. It is not the job of product development to understand the detailed science of why a technical function is vulnerable to particular stresses or how certain controllable parameters actually affect that function. That is the job of technology development when the time pressures are not so critical and the context is that of a range of applications. During product development the focus is on making the specific design work with repeatable performance, quickly and at low cost. Schedule pressures tend not to allow the time for the development of basic understanding. If controllable parameters are reproducible and can be shown to have predictable benefits to the output response, they are good enough.

Another distinction is in the mind-set of management and work groups. In technology development the focus is on forcing failures to occur in order to learn how to control the performance of a function. In product development the focus is on developing value in the product and keeping it stable so that it will be successful in the market. The nature of the questions, analyses, and experiments are different. It's best for their organizations and management to be different. When technology development is mixed with the work of product development, product development can be jeopardized.

Product Development Process

We expect that the process we describe here is not a sufficient statement for your product development process. Our objective is to describe enough detail that you can understand the framework and see how the plans for reliability development map to it. To be useful in practice, it must be adapted to the character of your business, with details and flexibility added to be relevant to the challenges in your functional and project work. For example, a development project that is software-only will have a production preparation process that is quite different from an electromechanical/materials system design that might have substantial tooling or facility modifications. Although the language tends to be more familiar to mechanical and electrical engineers, it should be easily interpreted for the development and integration of software, materials, and the other technical elements of a multidisciplinary product system. The process must be understandable to management, not viewed just as a tool for engineering.

It may be that the real challenges in a product development project are for new manufacturing processes or facilities. Consequently, the cross-functional guidance must be clear and viewed to be just as important as that for the product design. For that matter, the challenges may be for new marketing or service capabilities. Many subordinate value streams are developed and integrated to create the overarching value delivery system. Any way of looking at it, the process has to be inherently cross-functional in its objectives, activities, dependencies, deliverables, acceptance criteria, and decisions. That creates high expectations for the structure of the development teams and for the methods of their leadership.

For now let's look at the structure of the process. A quick study of Figure 4.6 shows that

- Phases are expected to overlap, with activities starting when they are enabled by information and resources, rather than when a formal gate review is passed.
- Not all gate reviews need to be of equal importance.
- Not all gate reviews need to be on the project's critical path.

Flexible Phases

The intended flexibility contradicts the criticisms that phases are strictly sequential and inflexible, that the work of one phase cannot begin until there's approval of the work of the previous phase. There may be value in combining phases or in managing less important gate reviews by informal means. You may choose to skip certain gate reviews while reinforcing others.

Two examples from our experience are the reviews prior to the "demonstration" Phases 4 and 6. Initially we found those gates to be critical to ensure that configurations were complete and frozen but later made them informal as the organization's behaviors improved. The gate prior to the verification of the product designs was found to be necessary to ensure that the prototypes submitted for verification testing were complete representations of the production-intent design. A misbehavior was the implementation of late-arriving design changes during the verification tests, tending to invalidate the previous test results. The gate at the end of Phase 5 served to judge whether or not the preparations for production were complete and satisfactory. This was a "manufacturing readiness"

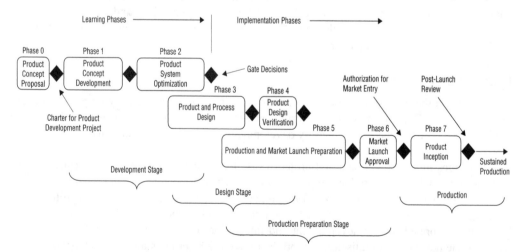

Figure 4.6   Once chartered, product development projects follow a phase/gate process through market entry into production.

review. As the project teams learned to incorporate these decisions into the agendas of their routine meetings, that management review became no longer necessary. The product samples submitted for the final tests that supported the market entry decision were selected from the production line that was preparing products to be shipped to revenue-bearing customers. All design improvements that were required prior to market entry were verified to be installed.

The benefits of overlapping, compressing, and even eliminating phases are in the reduction of cycle times and development costs. The difficulties are in the potential for additional risks. There's no universal right answer. It must be the judgment of the project leadership and their governance committees that agree upon those adaptations that provide an advantage and are acceptable for a specific development project. It's best not to promote these shortcuts casually, since the risks can be very significant. Clear acceptance criteria can be established to ensure that the objectives of your process steps are satisfied and that mistakes of previous projects are not repeated.

For leveraged product designs, as illustrated in Figure 4.7, certain phases may be shortened, not because they have to be, but because much of the work is already completed. However, the development of a single new design concept may determine the critical path.

For an extremely leveraged project there may be no need for the development of design concepts or of new manufacturing processes. That may allow phases to be combined and gate reviews to be eliminated. A design enhancement or extension may be a useful example of that situation. A phase flow such as the one illustrated in Figure 4.8 may be appropriate.

For some business models, the work of the next phase is not permitted to begin until the previous phase has been completed. In this case, the gate reviews will be on the critical path. This is not efficient but may be required by contracts and their funding model. The example in Figure 4.9 can illustrate a policy that purchase orders for tooling or production parts cannot be placed until the design has been verified.

Within each phase there are iterations of negotiations between requirements and capabilities, plans and resources, experiments/analyses and design changes. Typical process descriptions look to be linear. They are not. Envision that within each phase are spirals of "design-build-test-analyze-redesign-verify" scenarios that characterize development work. There are iterations of movement from the system level to subsystems, components, and manufacturing/service processes and back as

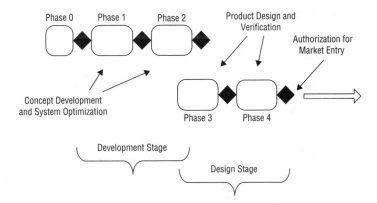

Figure 4.7 Highly leveraged products can have shorter phases because much of their necessary information has already been developed.

characterized by the "V" model (Figure 4.10) familiar to systems engineers. Project management plans may include interim milestones and frequent reviews when phases are long, not just the phase-ending gate review.

Each phase has a unique objective, from which acceptance criteria are derived. The activities within the phase have dependencies based on the value of the information developed as prerequisites. The information is an input to start the activity. It has to be planned and managed, often in the form of a map of information flow. An activity has the task to transform input information into output information that is needed by downstream activities. This "pull" of information is the basis for the process defining what information is needed when, by whom, to do what. It is a much more efficient viewpoint than a "push" model, which can develop excessive amounts of information just in case it is needed. Often this is in response to concerns about unanticipated questions from management at gate reviews. It is risk-averse, but not lean.

So the flow of information tends to be highly linked. This is a source of many dependencies among project documents. In Figure 4.11 you see negotiations among the information and decisions. This is expected as the project teams strive to develop a product that will provide differentiated

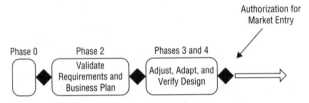

Figure 4.8   Projects for design enhancements can combine phases and eliminate selected gate reviews.

Figure 4.9   For certain business models that are controlled by contracts, rigid processes for gate reviews may be needed, and overlapping phases may not have acceptable risks.

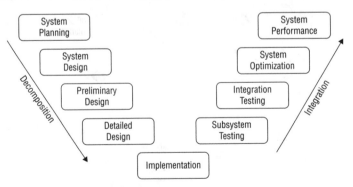

Figure 4.10   The "V" model guides the decomposition from the system level to components, and then the integration back to the system level.

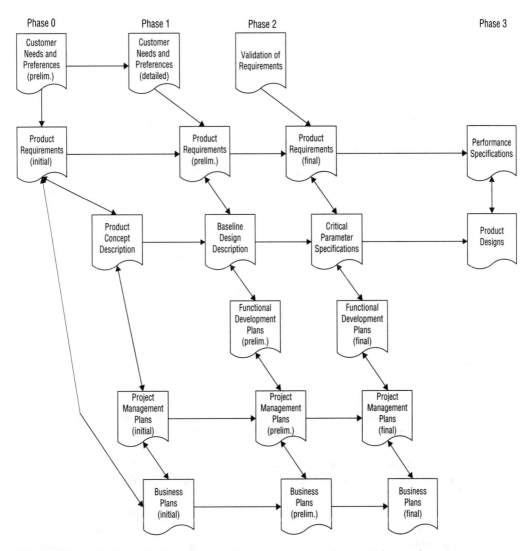

Figure 4.11   The linked information in critical documents is integrated over phases.

value both to customers and to the company, value that can be accomplished by achievable plans. This information matures over the early phases from "initial" to "preliminary" to "final." It illustrates the point that the early development work has to be accomplished with partial but stable information. For example:

- The "preliminary" requirements are those necessary to enable the selection of the base-line architecture and design concepts.
- The "preliminary" business plan needs inputs from
  - The "preliminary" project management plan, which describes necessary resources and timelines

- The "preliminary" functional plans to replicate the design concepts, to deliver services at a cost, or to achieve the expected revenues
- The baseline design descriptions, with their estimated abilities to achieve those preliminary requirements that drive prices and the estimates of revenues and costs

The "final" level of maturity reflects completeness, built upon the earlier levels. The contents and integrity have to be sufficient to obtain approvals, based on negotiated commitments to the expectations of the project.

What Questions Are Answered by Major Deliverables?

Each project document has a purpose and expectations for its contents. There is no expectation that project documents be thick bureaucratic documents written "for the record." One page may be sufficient in some cases, as is the construct of the "A3 charts"[14] familiar to lean product development. The format can include spreadsheets, drawings, flowcharts, scenario descriptions, and other tactics to convey information, clarity, and insight. Most important, they have to be useful and be used.

The authorship of many of the documents is with cross-functional teams, such as for the product requirements. Some may be combined, such as functional implementation plans and project management plans, and others may be kept separate to serve the needs of a functional organization. Here are a few examples:

- **Requirements:** What specific technical requirements, with target values, must be satisfied in order to deliver the selected value proposition and to satisfy the relevant constraints of the project?
- **Baseline design description:** What are the selected architectures, design concepts, and enabling technologies for the full system product, its modules, accessories, subsystems, and components?
- **Critical parameter specifications:** What controllable design parameters, with set points and tolerances, must be replicated to enable the full system product to satisfy its requirements?
- **Marketing plan:** What are the strategies, tactics, resources, and plans to prepare for market entry and to achieve the stream of revenues in the project's business plan?
- **Project management plan:** How will the project apply available resources, partnerships, and capabilities over time to achieve its requirements for quality, costs, cycle time, and capacity, while reducing risks to an acceptable level? What are the critical milestones and deliverables?
- **Business plan:** As an executive summary, how will superior value be delivered to target customers, and how will superior value be returned to the business with predictable processes and acceptable timelines?

---

14. James M. Morgan and Jeffrey K. Liker, *The Toyota Product Development System: Integrating People, Process, and Technology* (New York: Productivity Press, 2006); John Shook, *Managing to Learn: Using the A3 Management Process to Solve Problems, Gain Agreement, Mentor, and Lead* (Cambridge, MA: Lean Enterprise Institute, 2008).

Key Principles for Robustness Development and Reliability Growth

The principles of robustness development and reliability growth should be reflected in your product development process. We have identified some of the more important principles below. In our process model, the development of robustness is the focus of technology development and of product development Phases 1 and 2. It also contributes to the correction of problems found later. Reliability growth is the focus in Phases 3 to 5, when prototypes reflect production-intent designs that implement the critical parameters specified earlier by robustness development. In Chapters 7 and 8 we elaborate on these principles. It would be good if you saw them in action within your product development process.

1. The goal of product development is to create specifications for design parameters that enable the functional responses of the product to
   a. Satisfy its requirements that are customer-driven, externally driven, internally driven
   b. Achieve "on-target" performance, not just "in-spec"
   c. Minimize variations in performance due to stressful conditions beyond control
   d. Enable efficient manufacturing, testing, service, and distribution
2. Reliability must be designed into the product early in development. Responsive corrective actions resolve problems found later. Dependence on corrective actions in manufacturing and service is neither effective nor efficient.
3. Robustness development methods are both analytical and experimental. They
   a. Build upon basic scientific and engineering methods
   b. Use stresses to find problems faster and to prove that solutions are better
4. Superior robustness is a major criterion for the
   a. Selection of new design concepts and their enabling technologies
   b. Completion of development phases
   c. Proof of competitive superiority
5. If achieved early, superior robustness is a major advantage for
   a. Improved quality, reliability, and durability
   b. Reduced risks for system integration
   c. Reduced manufacturing and service costs
   d. Shorter, more predictable cycle times, with related reductions in development costs

Product Development Plan

Product development plans embody the design strategies chosen for the project. They are the basis for the commitments to achieve the requirements for the product. The work to manage the technical development and integration of design concepts and to translate those results into production-intent designs that may need substantial coordination among in-house design groups, advanced manufacturing and service support, partner design organizations or companies, and suppliers of outsourced components. These activities with their dependencies, resources, prerequisite information, deliverables, and timelines are core elements of the project management plans. The plans to determine how the required reliability will be developed are subsets of the product development plan.

The product development plans

- Describe how the selected strategies, such as those described in Chapter 3, will be translated into tasks, dependencies, and decisions
- Define the specific resources that will be needed and when they will be needed
- Define how the achievement of performance and reliability will be developed, integrated, measured, monitored, and managed
- Describe how design documentation, key project deliverables, product configurations, and the like will be approved, released, controlled, changed, and archived
- Define the design deliverables, specific to the particular product system, with any adaptations from the corporation's standard process

To the extent that collaboration among various organizations and companies is part of the plan, the methods for aligning that distributed work and integrating the deliverables must be carefully articulated. A lack of diligence in managing collaborative work can place the full system at risk. We've experienced several cases where deliverables from a partner company were consistently late and fell far short of our expectations.

The integration elements of the plan describe the methods, sequences, decisions, and schedules by which the partitioned elements of the design will be united, debugged, optimized, and verified. The identification, analysis, and correction of system failures have to be managed without confusion about the process or conflicts in individual responsibilities. Risk assessments, preventive action management, and contingency planning must be reflected in the planning for resources and timelines.

Resources, acting as a coordinated system of capabilities, enable the project when they are needed. Examples of resources to consider in your plan, with their timing, include the following:

- **Human:** labor (full-time and part-time), skills, subject matter experts, analysts, coaches, organizations, teamwork, incentives, management support
- **Centralized functions:** testing services, prototype fabrication, benchmarking capabilities, information technology, patent attorneys, facilities
- **External capabilities:** testing labs, development partners, critical suppliers, appraisers of regulatory compliance
- **Funding:** product development labor, materials, equipment, nonengineering labor (e.g., marketing), nonrecurring engineering costs to partners or suppliers, administration costs, overhead rates
- **Capital investments:** tooling, development instrumentation, test equipment, development labs, environmental chambers, EMI/EMC test labs, shipping/rough handling test labs, facility construction or modifications
- **Major prototypes and their allocation:** robustness development fixtures, architectural concept mock-ups, engineering models, early production models
- **Time:** specific timelines and milestones for major tasks and deliverables

The development plans must also address key steps to make these resources available and ready when needed. This can be a source of significant risks when, for example,

- Dedicated project or shared resources are overloaded or under conflicting mandates.
- Centralized resources are overloaded or under competition from other projects.

- Acquisition timelines are outside the control of the project's leadership team.
- Training plans have significant learning curves.
- Timelines for equipment acquisitions to enable new people to work are not aligned with the need for that work.
- Decision responsibilities are not clear, particularly among partnering organizations or between the project leadership team and upper management.

Certainly wisdom from your teams' experiences can identify other sources of risks.

The timelines and milestones of the development teams are integrated with other functions to create a schedule that is consistent with manufacturing lead times. It is "lean" to develop these timelines by working backward from the target date for product launch. Their consequences are thereby included in the project management plan and business plan. For a project with a mandatory product launch date, the consequences for achieving the gates and other project-level milestones will be much more critical, placing additional emphasis on risk reduction as a routine element of project management.

Reliability Development Plan

What would you expect to see in a product development plan specific to reliability development? You could argue that everything is relevant to reliability development. What strategies will be employed? How will they benefit the growth of reliability?

In Chapter 3 we describe several strategies that are relevant. Certainly there may be others that are appropriate for the nature of your products, technologies, processes, organizations, and markets. Please think about how these strategies can have strong influences over the details in your project management and business plans. Remember that a product development plan, without being adequately reflected in resources, funding, time allocations, and so forth, is no plan at all. It cannot serve its purpose as being the basis for committing to the achievement of requirements.

The reliability development plan describes how the robustness will be developed and the reliability grown to achieve the graphical plan depicted by the reliability growth plot, illustrated in Figure 4.12. The justifications for resources can benefit from the avoidance of costs in delayed market entry, from the reduced risks of customers' dissatisfaction, and even from the benefits of a shortened development schedule.

The Model for a Product Development Process

Tables 4.1 through 4.5 describe the phases of our process model for product development. Each phase focuses on creating value. It starts with Phase 0, which determines whether or not the project should be chartered, with resources allocated to it. Gate 6 is the approval to release the product to revenue-bearing customers. Gate 7 is a formal post-launch review to evaluate how well the project achieved its business and technical objectives and to summarize feedback that can be useful to future development projects. The gate reviews should be designed to focus not so much on the completeness of deliverables but on the value developed for customers and for the company, and on the readiness to accomplish future work with less risk. They are not just quality control checkpoints for the project.

Phase 1 deserves particular attention. There can be much uncertainty about a project when it is begun. Phase 1 plays a critical role in developing new information that helps to design the business plan and to determine whether or not the decision to charter the project

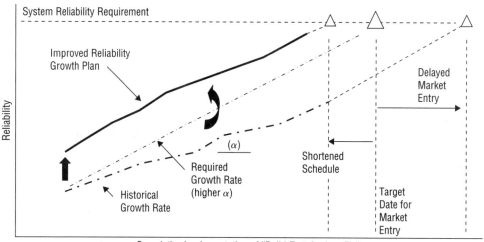

System Reliability Requirement

Reliability

Improved Reliability
Growth Plan

Delayed
Market
Entry

(α)

Shortened
Schedule

Required
Growth Rate
(higher α)

Historical
Growth Rate

Target
Date for
Market
Entry

Cumulative Implementation of "Build-Test-Analyze-Fix"

Figure 4.12   How can you achieve the required reliability within the constraints of an accelerated project schedule?

was wise. Can the core requirements of the project be supported by available design concepts and controllable technologies? Can the technical concepts be developed into reproducible designs by achievable implementation plans with available resources? What are the strategies for manufacturing, or for repair and maintenance? What manufacturing and service costs can be expected? Can the desired value proposition be brought to market when desired by the new product portfolio? If not, the project should be redirected, postponed, or canceled. The earlier this decision can be made, the sooner resources can be reallocated to a new plan. It is a major source of waste for teams to fight to keep their project alive when it is neither achievable nor of sufficient value.

Note that there is not yet enough information in Phase 1 for a commitment to a specific market entry date. However, there is a "window of opportunity" for market entry that determines the approximate time available to the project and guides the plans to be achievable within that time frame. Each phase develops information that enables the planning of the next phase to be more predictable. Phases beyond that will have rough plans, whose details are improved when their time is closer and the prerequisite information is clarified and more complete.

During the initial production phase, responsibility for the product is often shared between product development and production management as remaining problems are solved. After sufficient experience with the product in production and customers' use, a post-launch review is quite valuable. Phase 7 leads to that review, after which the product is completely in the hands of product and production management.

Table 4.1 Phase 0

Objectives: Decide whether or not to charter the proposed development project and to allocate necessary resources.

Activities/Deliverables	Deliverables for Reliability Development
• Clarify the business opportunity. • Clarify the differentiating value proposition. • Describe the vision for the product. • Propose design concepts, enabling technologies, and functional capabilities. • Evaluate the strategic fit of the product. • Develop achievable project management plan-swith an estimate of the time to market entry. • Recommend adaptations to the standard product development process. • Develop an initial business plan, with a project-level risk assessment.	• Value proposition addressing reliability intent • Initial customer-driven requirements for reliability and durability • Initial requirements for failure rate, usage life, availability, downtime, cost of ownership, etc. • Availability of technologies and design concepts to achieve the reliability requirements • Consequences of the challenge for reliability growth to the project management plan • Consequences of the reliability requirements for the business case

Acceptance criteria:

The decision to charter a development project should be based on analyses and forecasts that show that

- The new products and services can be expected to deliver value to target customers better than alternatives expected to be available to them after your product's market entry.
- The project is expected to return value to the business better than the next best alternative investment.
- The project is prepared with knowledge, technical concepts, resources, funding, for example, to achieve its objectives within an acceptable timeline.

Remarks:

- The selection of a new product concept for development is an implementation of the portfolio plans for new products. If it doesn't fit, it should not be chartered. Ad hoc development projects that are established "under the table" can be a waste of resources.
- Information that is prerequisite to Phase 0 is derived from the work of advanced planning and technology development. If that is not available, the Phase 0 team has to do it, with higher risk consequences for cycle time.
- A "go" decision at Gate 0 triggers an investment in Phase 1 to develop and evaluate the product to the point that the engineering, marketing, and business parameters can be understood better. It is not a commitment to the entire process leading to market entry. That commitment is at Gate 2.
- It is better to start a project on time than to focus on reducing development cycle time as a way to compensate for a delayed project start decision.
- In the early phases there can be much uncertainty. Activities need to be managed to work with partial but stable information and to adapt to changes or refinements in that information as the project matures.

Table 4.2   Development Stage (Phases 1 and 2)

Objectives:
- Complete the development of subsystem-level robustness.
- Complete system-level integration and robustness, with specifications of critical functional parameters.
- Develop an achievable business plan derived from the product requirements, the selected design concepts, the approaches for manufacturing and service, and the project management plan.

Activities/Deliverables	Deliverables for Reliability Development
<ul><li>Validate the "voice of the customer."</li><li>Translate the needs and preferences of customers into requirements that are achievable within the resources and constraints of the project.</li><li>Decompose system-level requirements to the level of subsystems and components.</li><li>Select the system architectures, subsystem design concepts, and enabling technologies.</li><li>Complete the development of new or improved technical concepts to be robust for the new product's applications.</li><li>Validate system architectures and design concepts with feedback from customers.</li><li>Optimize full system performance and robustness, with the specification of critical functional parameters.</li><li>Complete the project management plan committing to a market entry date.</li><li>Complete and commit to the business plan, with updated financial evaluations and risk management plans.</li></ul>	<ul><li>Customers' perceptions of reliability and durability</li><li>Requirements for reliability and durability allocated to subsystems and components</li><li>Detailed stresses from customers' environments and product applications</li><li>Conclusions from analytical and/or empirical models of system reliability</li><li>Conclusions from reliability risk analyses</li><li>Conclusions from initial life tests</li><li>Consequences of the challenge to reliability growth for the project management and business plans</li><li>Specifications for critical functional design parameters</li><li>Reliability growth plan with consequences integrated into the project management and business plans</li><li>Initial predictions of system reliability, of service intervals for replacement parts and consumable materials, and usage lives</li></ul>

Acceptance criteria:
- Critical requirements are complete, approved, frozen.
- System-level performance and robustness are optimized.
- Project management plans are approved and supported through product launch and production start-up.
- Product launch dates for various markets are acceptable and committed to.
- Project-level risks and their management plans are judged to be acceptable and supported.
- Project business plan is complete, approved, and committed to.

Remarks:
- It is natural that there are conflicts between "market pull" and "technology push." These need to be resolved early and represented in the product's requirements.
- Prototypes of the system configuration need to be flexible to enable rapid iterations of testing, learning, and corrective action.
- The thorough development of critical functional parameters enables the follow-on product designs and production preparations to require less cycle time and to involve lower risks.

- There are substantial dependencies upon the effectiveness of these controllable parameters, particularly for product quality and reliability, for the quality and costs of manufacturing and service, and for the remaining development cycle time and costs.
- The project management plan now should have sufficient information about the path forward and its predictability for the project leadership team to commit to the market entry date.
- The business plan has integrated all of the consequences and forecasts of the information developed through the early phases.
- While it is logical that the approval of the business plan triggers the release of funding for production preparations, such as tooling and parts procurement, certain expenditures with long lead times may need to be approved earlier. An example might be the funding for a new production building or a major piece of production equipment.
- At Gate 2 the project team is expected to commit to deliver the expectations of the business plan, e.g., achieve the product requirements and deliver the product to the market entry process on time.

Table 4.3   Design Stage (Phases 3 and 4)

Objectives:
- Complete and release the production designs of the product system.
- Validate that the product's features and functionality are acceptable to customers.
- Verify that the product designs satisfy their requirements.
- Develop the manufacturing and service processes to replicate and maintain the designs.

Activities/Deliverables	Deliverables for Reliability Development
• Develop product designs with feedback from internal reviews and from trade-offs to achieve required manufacturing and service capabilities.	• Reliability analyses to improve designs
	• Empirical results to optimize performance and robustness
• Test and refine design prototypes to comply with product requirements and to be robust, with feedback from customers' tests.	• Feedback from customer acceptance tests
	• Tracking of failure reports and corrective actions, prioritized by problem severities
• Manage prioritized corrective actions.	• Tracking and analysis of reliability growth
• Complete the specification and release of the product designs with cost-effective manufacturing tolerances.	• Specifications of the fixed and adjustable critical design parameters with their manufacturing tolerances
• Develop the required production and supply chain capabilities and processes.	• Demonstration of product reliability and the usage lives for consumable materials
• Begin the procurement of production parts, components, and tooling.	• Determination of whether or not the reliability growth rate is acceptable, with related plans
• Develop capabilities and processes for product service, customer support, and marketing.	• Forecasts of reliability, durability, usage lives, etc. after market entry
• Develop plans to support the market launch.	• Consequences of the expected reliability and usage lives on customers' applications, product and price positioning, launch and sales plans, and on the business plan
• Test the full system product prototypes to demonstrate compliance with their requirements.	

(continues)

Table 4.3   Design Stage (Phases 3 and 4) (*continued*)

- Identify product design improvements that are required prior to market launch.
- Ensure that the expectations for the verified product are aligned with the business plan.

Acceptance criteria:

- Product designs are complete and released, with validation by customers.
- Manufacturing suppliers, processes, and materials have been selected.
- Verification tests have demonstrated that the released product designs
  - Are acceptable to customers
  - Satisfy product requirements over the range of expected applications and operating conditions
  - Have a reliability growth rate that is acceptable
  - Satisfy requirements for manufacturing and service quality, costs, and cycle time
  - Are superior to their expected competition

Remarks:

- The work of product design transforms the controllable functional parameters into critical design parameters that can be reproduced with efficient and stable manufacturing and service processes.
- Customer acceptance tests expose lead customers to prototypes of the production-intent designs while there's time to react to their feedback.
- An important purpose of verification testing is to identify those design problems that must be corrected before market entry. Whether or not they jeopardize the project schedule depends on how quickly resources can solve the problems and on the consequences of the implementation plans.

Table 4.4   Production Preparation Stage (Phases 5 and 6)

Objectives:
- Prepare for market launch and for sustained production and product support.
- Authorize the delivery of products to revenue-bearing customers.

Activities/Deliverables	Deliverables for Reliability Development
• Develop and verify the production and supply chain capabilities to be stable, of acceptable quality, with required capacities.	• Implementation of design corrections to enable increased reliability and durability
• Manage reliability growth and corrective actions.	• Demonstration that manufacturing processes deliver acceptable "quality as received"
• Develop and verify corrective design changes that are required prior to market entry.	• Demonstration that service and customer support capabilities accurately maintain product performance and reliability
• Manage the regulatory and statutory approval processes.	• Acceptable inventories of repair parts, maintenance supplies, consumable materials, tools, media, etc. where needed
• Develop product service and customer support with feedback from early customer trials.	• Service and support personnel trained and organized where needed
• Develop readiness for marketing, sales, and distribution.	

Activities/Deliverables	Deliverables for Reliability Development
• Complete preparations for the product's launch into selected markets.	• Plans for field feedback about customers' satisfaction, replacement usage intervals and frequency for parts and consumable materials, service labor and parts costs, customer support and training
• Demonstrate the product's performance, reliability, consumable usage life, etc. to provide updated expectations to customers and to service/support organizations.	
• Confirm that company requirements are satisfied for target markets.	• Demonstration of the product's reliability, the usage rate of consumable materials, repair parts usage, etc.
• Confirm that the product and its supporting processes are prepared to achieve the financial and other expectations of the business plan.	• Forecasts of customers' satisfaction with perceived reliability, replacement lives, costs of ownership, etc.
• Obtain executive authorization to release the product to revenue-bearing customers.	

Acceptance criteria:

- Corrective actions required prior to market launch have been completed, verified, and implemented.
- Products built with production tools and methods satisfy their requirements.
- Factory and supply chain operations are stable, satisfy launch criteria for quality, costs, and cycle time, and have capacities to meet the demands of production scale-up.
- Support organizations are prepared to provide service and customer support in the initial markets.
- All regulatory and statutory requirements have been satisfied.
- All company requirements have been satisfied.

Remarks:

- Many activities in preparation for product launch are planned to start with sufficient lead times prior to market entry; they "pull" the development of prerequisite information.
- Customer acceptance tests in this phase expose lead customers to product service and customer support processes when there's still time to react to their feedback.
- Regulatory approval processes are based on released designs.
- Testing in Phase 6 has the objective to provide the best information available for product service and customer support, and to clarify the expectations of marketing, sales, and customers.
- Approval for market entry is a commitment by the company to deliver products to revenue-bearing customers in volumes and quality expected to sustain the introduction, and to support the product with its resources, warranties, and reputation. It may be appropriate for approvals to be obtained from company officers.

Table 4.5   Phase 7

Objective: Manage the product transition into stable, routine production and product support.

Activities/Deliverables	Deliverables for Reliability Development
• Manage the scale-up of production and supply chain operations. • Develop and verify improvements to quality, costs, and cycle times for production and the supply chain processes. • Manage improvements to product service and customer support. • Analyze feedback from product service and customer support to guide improvements to the current product and to requirements for future products. • Evaluate how well the product in production has satisfied the expectations in its business plan. • Identify lessons learned that can benefit future product and development projects.	• Analyses of data for product failures, usage lives, stressful conditions, usage of parts and consumable materials • Feedback on the effectiveness of preventive maintenance and the efficiency of problem diagnosis and repair • Conclusions from QAR audits

Acceptance criteria:

• The production processes are mature and stable, meeting their requirements for quality, costs, and cycle time.

• Feedback from the market and customers no longer demands improvements to the current product.

Remarks:

During the early months of production, sales, service, and customer support, there is much work to resolve start-up problems and to scale up the required capabilities and capacities. Feedback during this phase is important not only to trigger corrective actions but also to determine how well the enterprise is meeting the expectations of the business plan. Recommendations can benefit follow-on product development projects. Given these objectives, post-launch reviews should be held not only after processes have been stabilized, but at interim milestones to ensure that feedback is received when it's needed and can be acted upon.

For electromechanical/materials systems, design changes after launch may involve the modification of products already in customers' hands. That can be extraordinarily expensive and embarrassing. It should be viewed as a failure of the product development process, which should be effective at preventing the escape of significant problems to the marketplace.

For software-only products, design changes after launch may be more easily implemented. In fact, your customers may expect that process to be routine and advantageous. The challenge is to do it efficiently without jeopardizing your customers' continued use of the product.

Lessons-Learned Reviews

The building of capabilities to grow reliability can benefit greatly from the many lessons available to be learned from the experiences of previous projects. The key is to identify those recommendations in ways that make them useful to future projects. We suggest that your teams conduct lessons-learned reviews at the end of their development project. It's noted in the table for Phase 7.

It might also be appropriate at the end of significant phases of work. These are after-the-fact evaluations of the process for achieving requirements or implementing plans.

- What went well? What actions enabled those successes?
- What did not go well? What were the root causes of those difficulties?

In both cases the objective is to identify specific recommendations that can enable future teams to repeat successful approaches and to avoid difficulties. Priorities can be assigned. You can also include suggestions for the best development phase for implementation. We've found that facilitation by a process specialist can be useful to maintain the team's focus on opportunities for improvement rather than on the placing of blame. This step is essential to organizational learning, particularly if it identifies improvements to the standardized processes. Table 4.6 is an example of a useful template.

Table 4.6 This template for a lessons-learned review should be edited to provide transferable recommendations.

What Did Go Well?	Weight	Key Enabling Factors	Recommendations	Phase for Earliest Action

What Did Not Go Well?	Weight	Process Root Causes	Recommendations	Phase for Earliest Action

Older versus Better Methods

You may ask what could be wrong with your current processes. One way to judge that is to look at the principles and strategies that are incorporated to see how well they serve your objectives. Here is a quick view of some ways better approaches can differ from older, less effective ones.

How can customer satisfaction be improved?

Older Methods	Better Methods
Reactive feedback from customer surveys and complaints	Understand value drivers for market segments (see Chapter 9). Identify a superior value proposition. See the future first.
Voice of management, sales, or service	Involve customers early, often, and proactively throughout the process. Spend a day in the life of your customers. Validate requirements with customer feedback.
Engineering specifications focus on what you can deliver	Identify requirements that can establish competitive differentiation.

How can product quality be improved?

Older Methods	Better Methods
Develop performance under nominal operating conditions.	Develop robustness in performance under stressful operating conditions (see Chapters 7, 18, 19).
Quality is manufacturing's conformance to design specifications.	Specify controllable design parameters with tolerances that enable high manufacturing process capabilities.
Tighten tolerances to improve quality.	Apply tolerance design selectively after parameter design (see Chapters 7 and 20).
Quality is subordinate to costs and schedules.	Develop quality and costs in parallel with achievable schedules and efficient methods, e.g., Robust Design
"Seat of the pants" engineering	Expect excellence in functional execution. Develop alternative solutions in parallel (set-based design). Select the best architectures and design concepts from available alternatives (see Chapters 11 and 12).

How can product reliability and usage life be improved?

Older Methods	Better Methods
Failure rate is a design metric.	Use signal-to-noise ratio (S/N) as a design metric.
	Use stresses known to be effective at forcing deteriorations and failures.
Depend on reliability tests under nominal usage conditions.	Develop robustness to be better than that of competitive designs under the same stressful conditions.
Optimize subsystem designs.	Optimize specifications of design parameters at the system level.
Incorporate many new technical concepts.	Leverage mature, robust designs.
	Limit the risks inherent in new, complex design concepts to those that can truly add value for customers.

How can the costs of manufacturing and service be reduced?

Older Methods	Better Methods
Focus on reducing manufacturing and service defects.	Develop design latitude for manufacturing and service variability.
	Develop robustness in manufacturing and service processes to reduce their variability under stressful conditions.
Develop manufacturing and service processes and tools after the product is designed.	Design processes and tools in parallel with the product design.
	Incorporate manufacturability and serviceability in the design.
	Involve suppliers, manufacturing, and service experts early in cross-functional development teams.
Minimize manufacturing costs.	Minimize life-cycle costs.

How can product development costs be reduced?

Older Methods	Better Methods
Development costs are vulnerable to rework, staff overloading, late starts, non-value-adding tasks, project risks, changes in requirements, etc.	Develop higher quality and robustness earlier.
	Leverage proven, superior design capabilities.
	Build fast feedback loops into the design process.
	Develop a flexible design that can accommodate reasonable changes later in the process without severe consequences.

How can time to market be reduced and made more predictable?

Older Methods	Better Methods
Heroic crisis management to compensate for late starts, late changes, and unanticipated problems	Start projects on time with the right resources. Manage project risks. Incorporate the principles of "flexible" product development.
Developing products the way we always have done it	Incorporate the principles of "lean" product development.
High-risk activities as a way of life in product development	Separate technology development from product development.
Technologies developed within product development projects	Impose rigor to the decisions and process of technology transfer.

Key Points

1. A structured process should be commonly understood by all development teams and by management. Its expectations integrate technical development with business development to facilitate the management of cross-functional activities and dependencies.
2. It serves as an important framework for understanding the elements of reliability development, emphasizing that reliability development is not managed in isolation from other initiatives.
3. A well-designed process is an important basis for organizational learning and for the achievement of sustainable competitive advantages.
4. One of the roles of leadership is to intervene when, for example, good processes are not being followed, when the rate of reliability growth is not adequate, or when risk reduction plans are not being implemented. Without consistent and proactive attention, reliability growth will become subordinate to project costs, schedules, and other imperatives.

Discussion Questions

1. How well do your projects implement lessons learned from past projects?
2. What elements of your development process have been learned from benchmark companies?
3. How well do the principles and practices expected by your process provide competitive advantages?
4. How clear are the objectives for each phase of your development work?
5. How well do deliverables enable decisions at project milestones?
6. How well do deliverables, such as information and prototypes, enable future work?

7. By what approval process do your projects adapt your standard development process to their specific needs?

8. How well does your development process reflect how work actually gets done?

9. How well do project plans reflect the work to achieve higher reliability and durability?

10. How predictable are the achievements of requirements for quality, costs, and schedules?

11. How well do new products in production satisfy the expectations of their business plan?

Road Map for Reliability Development

The design for reliability starts early in product development and continues into production, service, and customer support. The details of your project initiative depend on the nature of the product and its technical challenges. Certainly, designs must be based on customers' requirements as well as on the needs of other stakeholders such as manufacturing, service, and your supply chain. Competitive markets have strong voices also. In this chapter we discuss what must be accomplished to get the reliability results required for your business and your customers. Robustness development and reliability growth activities are mapped to the phases of the product's development process.

Build a Compelling Case for Your Reliability Project

When formulating plans for an aggressive reliability initiative, it is important to focus on the expected business and financial benefits. For example, what are the potential impacts of failing to meet reliability goals on time, such as the cost of delayed market entry or the loss of customer acceptance for your new product? Often reliability development plans are challenged as being expensive and time-consuming. If you have not made a compelling case for the project, you share the responsibility for a project plan that has a lower probability of success. At the heart of the concern may be a lack of understanding of the benefits of higher reliability for your particular product as well as a lack of appreciation for the serious engineering efforts necessary to achieve improvements that are significant enough to provide a competitive advantage. The minimum scenario is that you design, build, test, and fix the product and ship it when you run out of time. We expect that your competitors have a more productive approach.

So a responsibility of the project team is to outline clearly the required project management plan, with the associated risks and benefits. Teams responsible for achieving a much improved reliability growth curve, described in Chapter 3, have to identify those improvements that will be needed. Then plans to accomplish them can address the root causes of known failure modes and prevent other problems that can be anticipated or discovered.

With a clear and urgent call to action you expect a constructive response from management. How well do they understand the value of the project? It's natural for stakeholders to argue for reductions in resources, prototypes, funding, and timelines. Not all products need higher reliability or a longer usage life. In many cases there are diminishing returns. So the starting point is to develop a justification that is backed by high-integrity data. Throughout this book we suggest strategies that can accelerate the process. What capabilities do your projects actually have to develop in order to increase product robustness and to achieve a higher reliability growth rate? How capable are your manufacturing processes? Do these capabilities need to be improved? Is the path to higher reliability to be achieved in the next product or over a sequence of new products? Can the timing of market entry be delayed, or does the rhythm of product launches dominate the plans?

Linking Activities to the Product Development Phases

All of the activities described are managed in the context of the product development process, illustrated in Figure 5.1. Remember that activities and phases can overlap, depending on their preconditions. You may need to start certain activities earlier. Life testing, for example, is time-consuming, so it has to be started as early as production-intent designs are available. The news may be good: You will have higher integrity in data to justify necessary improvements to the product design before market entry. If the product is dependent on new manufacturing processes, you may need to develop the robustness of those manufacturing processes in parallel with the work on product robustness.

Key Initiatives of a Reliability Development Project

Several initiatives are important to reliability development. Each should be reflected in your project management plans in achievable ways.

- Defining realistic requirements for the improvement of reliability and durability
- Selecting and developing a superior product concept
- Identifying potential failure modes and their levels of severity
- Developing analytical and empirical models to evaluate designs
- Developing robustness to prevent problems in the product designs
- Growing reliability and durability with efficient problem solving
- Balancing performance and costs through the optimization of tolerances
- Improving manufacturing process capabilities

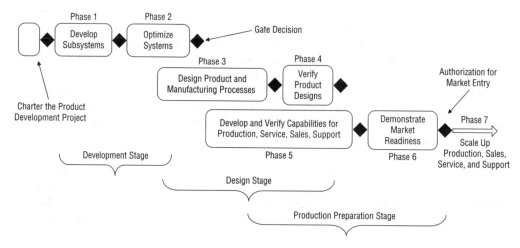

Figure 5.1   The phases of product development.

Defining Realistic Requirements for the Improvement of Reliability and Durability

First your leadership team has to understand how much improvement is needed. What value does your market place on increased reliability? There are numerous ways that you can segment markets for products. These categories represent a framework for assessing their value drivers related to reliability and durability.

Single-Use, "One-Shot" Products

Certain products have to work only once. Examples cover a broad range, including medical products such as blood collection devices and disposable syringes, rocket boosters, military ordnance, and consumer products such as single-use cameras. Some of these products are not meant to be repaired. Others are designed to be refurbished if storage causes deterioration. For example, the U.S. stockpile of nuclear warheads is routinely evaluated to detect deterioration and to plan potential replacement. A cruise missile is another example. It's a complex system that could be tested periodically to evaluate the functionality of certain subsystems and to implement service to restore its expected performance. Low-cost consumable medical products, such as blood collection devices, are usually subject to accelerated life tests, where functionality may be determined after storage at an elevated temperature for an extended period of time.

Clearly, the enemies of single-use products are time and the environment. Shelf life, defined by the time required for component parts or materials to deteriorate to an unacceptable level, is a measure of durability. In addition, you have to ensure that the population of devices provides consistent performance "out of the box." This requires robustness against the stresses both in manufacturing and in their storage, shipping, and handling. The cost of single-use devices covers many orders of magnitude, ranging from fractions of a dollar to fractions of a billion. The dilemma here is that the more costly the system, the more important is its durability but the fewer system tests you can afford. Here the reliance is on subsystem tests and on excellent system integration.

Repairable Products with a Long Expected Life

Low-cost products: Some lower-cost products have high complexity. They include devices such as MP3 players, cell phones, window air conditioners, small kitchen appliances, desktop printers, laptop and desktop computers, simple instruments, and many others. These products have to perform consistently over an extended duration of time or actuations. Some of them are serviceable and, depending on their remaining value, can be either repaired or replaced. Some are not inexpensive, but their value may decline quickly because of technical obsolescence. Customers of these products value higher reliability and durability.

Moderate- to higher-cost systems: This also is a broad category that includes both consumer and industrial products. Automotive products, high-end copiers and printers, complex instrumentation such as medical instruments for blood analysis or genetic analysis, centralized HVAC equipment, refrigerators, clothes washers and dryers, and dishwashers are just a few members of this class. A common denominator is a fairly high level of complexity coupled with the need for service. Generally, service is driven by the troublesome electromechanical and fluid-handling components. Often the process physics of the device is vulnerable to environmental stresses and to the deterioration of material properties. Clearly, these devices can benefit from higher levels of robustness.

High-cost capital assets: Systems in this category are very expensive and high levels of durability are assumed. Examples of this class include aircraft, power generation equipment, large machine tools, locomotives, offset printing presses, chemical process equipment, custom machines for automated high-volume or heavy manufacturing, and many others. For these complex systems preventive maintenance and effective problem diagnostics are expected.

Relevant chapter:

Chapter 9, Understanding and Managing Customer-Driven Requirements

Development phases for focus: Advanced product planning and development Phases 0 and 1

Tools: Customer interviewing, KJ analysis, Quality Function Deployment (Chapters 9 and 10)

Areas for focus:

- Voice of your customers (VOC): It all starts with understanding those elements that drive value for your customers and how your customers make trade-offs. What are the consequences of product failures to your customers' business or activities?
- Voice of your business: Your projects are chartered to deliver value to your customers at an acceptable cost to your company. Those costs reflect your organization's capabilities for developing higher reliability as well for maintaining it during production and customer support.
- Voice of your technologies: Different technical concepts can offer different advantages and costs. Project teams have to understand their capabilities, risks, and trade-offs at a basic level in order to make the best choices of design concepts.
- Voice of the market: What's going on in the marketplace that you choose to serve? How do competitive alternatives affect the expectations of your customers? Ignore this and your projects risk developing me-too products having a much lower probability of success.

Key results:

- Critical product functions that support customer requirements: When developing an understanding of reliability requirements, capture all of the important customer needs. What types of failures cause the more severe consequences for your customers?
- Historical data for previous products: These data can be an excellent source of realistic requirements for reliability and maintainability. How have previous products fared in service? These data will be invaluable in developing your reliability growth strategy, serving to highlight shortfalls that need to be corrected in new products. Keep in mind the view of some Japanese manufacturers: "A defect is a treasure."
- Benchmark data for competitive products: When gathering VOC data, there are valuable opportunities to understand how competitive products currently satisfy customers. How do customers perceive the reliability or durability of the other products that they use? The planning matrix in the House of Quality (Chapter 10) is a great tool for documenting this type of information.
- Requirements for reliability metrics, as examples:
  - Mean time between failures (MTBF)
  - Customers' perspective for service frequency (e.g., service calls per month)
  - Mean time to failure (MTTF) for non-repairable components and products

- Warranty costs
- Mean time to repair (MTTR)

$$Availability = \frac{MTBF}{MTBF + MTTR}$$

- Total cost of ownership (TCO): TCO includes first-time costs, ongoing costs of operation and maintenance, and the costs of disposal or decommission. This is an important metric for your more sophisticated customers, particularly for repairable systems expected to have long lives. While not a reliability requirement per se, it is strongly affected by system reliability and service costs.
- Average repair costs: These can be a concern for both your business and your customers, depending on how your service policy is structured. The frequency of service contributes to this value driver. Do your customers pay for maintenance and repairs? If your customers pay a usage fee, are the costs of service built into that pricing? Do your customers purchase service contracts?
- Service strategy: Who performs the repairs and/or maintenance? Do your customers do this or are highly trained mechanics required? Is it done in a service center or on your customers' premises? How is your strategy reflected in your product designs?

Selecting and Developing a Superior Product Concept

A decision critical to the reliability of a new product is the choice of the product concept. Broadly, the concept includes the system architecture and the design concepts for the subsystems, with their enabling technologies. In some cases these decisions may have strong implications for the manufacturing and service processes, with their technologies.

Often done without sufficient thought, the selection of the baseline product concept casts a long shadow. Inherent in how well the technical functions can be controlled is the upside potential for performance and robustness. Choose wisely and the development process will be easier and have a better outcome. Make poor choices and your projects may struggle constantly to overcome the flaws inherent in the weak concept. The old saying "You can't make a pig fly" is highly relevant to the selection of the baseline design concepts. If a design concept does not have parameters that are practical and effective in controlling its performance and robustness, it should be redesigned or replaced. Likewise, if its control parameters cannot be replicated by efficient manufacturing and service processes, it is a flawed concept. In a rush to design, the pressure to show tangible progress in the form of drawings and prototypes often results in teams taking shortcuts, such as not evaluating alternative architectures objectively or depending on an attractive technical concept that has not yet been developed to be robust.

Ironically, this activity has a large payback in benefits for the amount of time spent. The activity is not intensive in its use of resources, except that it requires some time, information, and patience up front. However, the investment here can generate dividends later in fewer struggles to integrate the product system, to satisfy its requirements, and to avoid problems.

Relevant chapters:

Chapter 11, Concept Generation and Selection
Chapter 12, Axiomatic Design
Chapter 13, Tools to Reduce Risks

Development phases for focus: Phases 0 and 1

Tools: Functional modeling, axiomatic design, concept generation and selection

Areas for focus:

- Identifying system and subsystem functions: The requirements of the system and sub-systems flow from the outside in. The features and functions of the product support those requirements, with the higher-level functions being decomposed to the functions of the subordinate subsystems, components, materials, and software. A function is what the system or subsystem does, involving the transformation of energy, material, or information. An excellent tool for visualizing the flow of requirements and supporting functions is the functional model.
- Defining product architecture to minimize functional interdependencies: Axiomatic design (see Chapter 12) is a tool that can be very helpful for identifying and reducing dependencies. When you have coupled designs, it can be difficult to optimize system performance. Coupling makes the effects of the critical design parameters non-orthogonal, where a change to one design parameter can affect more than one system function.
- Using the design matrix (DM) to select better concepts: Depending on the complexities of the product architecture or on the sequence of project activities, the development of a system can become excessively iterative. Often assumptions have to be made about one part of the system to enable the design of another part of the system. The DM can help teams choose better concepts that require fewer design iterations, enabling the development process to be more productive (see Chapter 12).
- Choosing a superior design concept from available alternatives: Your project plans should contain deliberate activities to compare alternatives objectively. Flexible product architectures enable concept changes more easily, reducing the consequences of premature decisions that lack sufficient information.

Key results:

- Several available strong concepts competing to become baseline: Your teams need more than one option and fewer than too many. A balance can be struck between enough alternatives to provide a rich source of ideas and so many that teams bog down in an excess of detail. The objective is to choose the best available design concepts for the product's intended applications.
- The selection of the best concept for the application: Stuart Pugh's method[1] for concept selection is an iterative process of generating new ideas or concepts, filtering out weak ones, creating hybrid concepts that combine the best attributes of others, and converging on the one that is best suited to satisfy its requirements.

---

1. Stuart Pugh, *Total Design: Integrated Methods for Successful Product Engineering* (Reading, MA: Addison-Wesley, 1991).

Identifying Potential Failure Modes and Their Levels of Severity

In what ways can your product fail? What is the likelihood of each failure mode? What are the consequences of each failure to your customers? These are key questions to be answered when making an assessment of the risks posed by potential reliability being less than acceptable. Of course, depending on how the system fails, there can be a broad range of consequences for your customers, such as wasted time, lost production, annoyance, inconvenience, additional costs, property damage, or personal injury.

Chapter 8 provides criteria that you may find useful for judging the levels of severity for problems. Their financial impact can range from very little, through lost revenues and warranty costs, to large judgments with significant financial penalties. Large costs may invalidate the project's business plan or, in the extreme, jeopardize the future of the company. Publicity associated with product failures can be very damaging to a company's reputation. Do you recall the rollover problem with unstable SUVs, triggered by tire failures? Many viewed the tire manufacturer as liable. A systems approach might have avoided the problem and its consequences. If the development project had made the vehicle design robust against tire blowouts and also eliminated the manufacturing defect for the tires, that problem could have been avoided. Less dramatic problems can also damage your business. When the results of product comparisons are published, there are winners and losers. A poor rating of reliability, relative to that of your competitors, may have serious consequences for sales and your corporate image.

Relevant chapters:

Chapter 13, Tools to Reduce Risks
Chapter 17, Building Empirical Models Using DOE
Chapter 18, Developing System Robustness Using DOE
Chapter 23, Critical Parameter Management

Development phases for focus: Starts early and continues throughout the project

Tools: Cause-and-effect diagrams, failure modes and effects criticality analysis (FMECA), fault tree analysis (FTA), analytical and empirical modeling, assessments of manufacturing process capabilities and design capabilities

Areas for focus:

- Identifying failure modes for the product system: At the early stage of concept development, teams depend heavily on engineering and scientific insight. They may not yet have either first-principles or empirical models. However, they may have past experience with similar products or technologies.
- Assessing the causes of failure modes, the severity of effects, and their probabilities: All this is part of the application of FMECA. The insights derived from the analysis help to guide your improvement activities.
- Assessing potential problems due to poor design capabilities: Failures due to performance drifting outside acceptable limits can be caused by a design that is too vulnerable to stressful conditions or to variations in its design parameters. So the process of risk assessment continues, using logical tools such as cause-and-effect diagrams, FMECA, and FTA with efficient corrective actions.

- Assessing potential problems due to poor manufacturing process capabilities: The best designs cannot have consistent performance over a population of products unless the manufacturing processes can replicate their critical parameters with sufficient accuracy and precision. The process capability for manufacturing a design parameter is a combination of the design's latitude for variations in the parameter and the ability of the process to set the parameter on target and to minimize its variability. Figure 7.3 in the later chapter illustrates this.

Key results:

- A proactive corrective action program to prevent failures from recurring: FMECA has high value when applied to prevent problems. Once you have prototypes, the cost of solving problems rises quickly. In production, the cost is orders of magnitude higher still. The most difficult scenario is that of correcting problems in the field after products have been shipped. It is so much more cost-effective to prevent problems before they happen in the development process. You should plan for liberal investments in problem prevention in the early phases. The costs can be relatively small compared to potentially large costs later.
- An understanding of the weaknesses in design and manufacturing capabilities: With this understanding leadership teams can make business decisions based on likely yields in manufacturing, warranty expenses, and service costs.

Developing Analytical and Empirical Models to Evaluate Designs

When projects are fortunate enough to have a system that can be modeled analytically, development teams can get early insights into system or subsystem performance and their design space before creating physical prototypes. Often, however, systems are too complex for analytical models to yield meaningful results. Notwithstanding the "fun factor" of building analytical models, your development teams have to identify the value of the model. Is the physical system amenable to building a realistic and useful analytical model? Will it help to improve the understanding of its behaviors? "First-principles" models apply the fundamental laws and equations of science and engineering. There are many systems for which they are extremely useful, saving time and resources.

The subject of modeling brings to mind the tongue-in-cheek comments of a former boss who was a very proficient analyst and an expert modeler: "If you can possibly avoid it, never run an experiment to test your analytical model. Chances are that your experimental results won't agree with your model's results. If you're forced to run an experiment, never repeat it, because the chances are that the second experiment will deliver a different result from the first. Then you'll really have some explaining to do!" Of course he was joking, but the concern is whether or not an analytical or empirical model is appropriate. It is really about whether the time spent building the model can yield valuable insights. In cases where the validity of the model has been demonstrated and results are easily obtained, "just work the model."

Once you have physical prototypes, empirical models can be derived from experimental results. That may be the best strategy, since the validity of the model can be determined easily.

Relevant chapters:

> Chapter 16, Building Models Using ANOVA and Regression
> Chapter 17, Building Empirical Models Using DOE
> Chapter 23, Critical Parameter Management

Development phases for focus: Phases 1 and 2

Tools: First-principles analysis, designed experiments (DOE), cause-and-effect diagrams, functional modeling, analysis of variance (ANOVA), critical parameter management

Area for focus:

- Relating product functions and critical design parameters: Establish linkage among variations in functional performance at the system level with changes to critical functional parameters at the levels of subsystems, components, and manufacturing processes.

Key results:

- Analytical or empirical $y = f(x)$ models relating functional parameters and product performance: An improved understanding of the design space enables parameters to be specified to set functional responses on target and to reduce their vulnerability to stressful conditions.
- Understanding the relative importance of design parameters: Which parameters are most effective at adjusting the mean of the distribution of performance? Which parameters reduce the vulnerability to stresses and increase the latitude for variability in manufacturing or service?
- Initial assessment of design capability: This is the system performance before its robustness has been improved. It is the baseline for developing the robustness of the product designs and, later, for growing the reliability.
- Critical parameter management plan: A plan for documenting and managing design parameters has the objectives of managing design changes, maintaining configuration control, and enabling real-time access to accurate specifications. The system that implements that plan applies throughout the development phases into production, including the supply chain.

Developing Robustness to Prevent Problems in the Product Designs

The performance of a product with increased robustness is less sensitive to sources of variation. Its design builds upon excellence in applied science and engineering with additional methods of analysis, experimentation, and optimization. Important questions that development teams need to answer include the following:

- What characteristics of system performance are most important to your customers?
- How do you measure system performance and what are its metrics?
- Which critical functional parameters have the most influence over the mean and variability of system performance?

- What are sources of variation affecting those design parameters?
- Are there any customer usage factors that can degrade system performance?
- How can environmental factors, such as temperature, humidity, and vibration, degrade performance?

To answer questions like these, development teams have to understand their design space, that is, how variations in the controllable parameters cause variations in system performance. This understanding comes from developing a model that describes these functional relationships. The model can be derived either from first-principles analysis or empirically from experimentation with physical prototypes. Typically, robustness development depends heavily on building empirical $y = f(x)$ models.

Relevant chapters:

Chapter 18, Developing System Robustness Using DOE
Chapter 19, Robustness Optimization Using Taguchi's Methods

Development phases for focus: Phases 1 through 3, building upon technology development

Tools: Taguchi's methods of robust design using P-diagrams, noise diagrams, orthogonal array experiments, response surface methods, classical and stochastic optimization methods

Areas for focus:

- Understanding the consequences of performance variability to your customers: When considering the consistency of performance, a logical question is the relative importance of on-target performance versus its variability. It's possible to have average performance on target over time but still have customers dissatisfied with the lack of stability in performance. Product users may notice variability before they are aware of a shift in average performance. This is especially true when performance is somewhat subjective, in the "eye of the beholder." As we discussed in Chapter 2, the quality loss depends both on average performance and on the variability of performance around its mean.
- Specifying the critical design parameters that affect performance and its variability: This is the beginning of the critical parameter management process. It continues throughout product development into production.
- Identifying experimental stresses that affect system performance: Empirical stresses can force performance variations and failures in the laboratory to occur deliberately and sooner. Problems can thereby be found faster and designs proven to be better than earlier designs.

Key results:

- Specified set points for critical design parameters: In the most general case, design parameters will affect both nominal performance and performance variability. Your challenge is to find those critical design parameters that will mostly adjust nominal performance and those that will mostly reduce performance variability, particularly as aggravated by stressful conditions.

- Reassessment of design capability: The purpose of making the system robust is to ensure on-target performance over time as environments change, components degrade, and manufacturing processes vary, all conspiring to push performance off target. An assessment of the design capability after robustness development can determine how much improvement has been achieved.

Growing Reliability and Durability with Efficient Problem Solving

For reliability, what are the expectations for repair frequency and costs? For durability, what is the target design life of the product? Some products can serve indefinitely as long as repair parts are available and the costs are justified. Examples include automobiles in low-stress climates, printing presses, machine tools, and many others. The expectations for repair frequency, costs, and design life are best answered by considering the type of product and by evaluating its competitive market. A number of questions need to be answered:

- Will technical obsolescence signal the end of life long before the product wears out? A good example of this is the cell phone, often on the way to technical obsolescence at the time of purchase. Since most people replace their cell phone at the end of its contract, a target for a longer design life would add costs yet return little value to most customers.
- Is it an expensive product that customers will expect to be repairable?
- What are the competitive benchmarks in the market?
- What are the practical limitations to making all components last for the life of the product? Are there inherent technical limitations or is it more a cost constraint?
- What is the cost of service for the product? Components that are very expensive or time-consuming to repair may be required to last the life of the product.

Relevant chapters:

Chapter 3, Strategies for Reliability Development
Chapter 8, Reliability Growth and Testing
Chapter 9, Understanding and Managing Customer-Driven Requirements
Chapter 13, Tools to Reduce Risks
Chapter 23, Critical Parameter Management

Development phases for focus: Phases 0 through 5

Tools: Quality Function Deployment (QFD), FMECA, FTA, DMAIC problem solving, durability analysis, life testing, accelerated stress testing

Areas for focus:

- Understanding stressful loads imposed during product life from customers' use or abuse and from the product's storage, shipping, and handling
- Understanding the physics of failure for system components
- Understanding the "rare event" that can damage the product
- Fast problem solving and implementation

Key results:

- Requirements for design life for all system components
- Design concepts chosen to be inherently long-lived
- Conclusions from failure analysis and corrective actions for every failure that occurs during product development
- Results of life tests to improve or confirm durability
- Results of accelerated experiments, such as highly accelerated life testing (HALT), described in Chapter 8
- Decisions that enable prioritized corrective actions

Balancing Performance and Costs through the Optimization of Tolerances

Tolerance design follows the development of robustness. The tightening of a tolerance drives up its manufacturing costs. Think of tolerance design as a way to balance increased manufacturing and service costs with the benefits of increased stability of replicated parameters. It is not a tool to make your design work. In general, tighter tolerances should be avoided unless the increase in manufacturing costs is less than the expected decrease in quality loss. If you are in a situation where you need to tighten tolerances to get the design to achieve acceptable performance, particularly under nominal conditions, you have a flawed design concept. Is it ever justified? It may be if that's the only way to get the product to market on time. However, it will be an improvement opportunity to change the design so that the tolerances can be relaxed. If your competitors solve that problem before you do, they will have cost, robustness, and performance advantages.

Relevant chapter:

Chapter 20, Tolerance Optimization

Key phases for focus: Phases 2 and 3

Tools: Taguchi's methods of Robust Design using P-diagrams, noise diagrams, orthogonal array experiments, response surface methods, classical and stochastic optimization methods, methods of empirical tolerancing using designed experiments, quality loss function

Areas for focus:

- Translating performance variations into costs
- Using the quality loss function to balance performance and costs with an optimal allocation of tolerances for critical design parameters

Key results:

- A set of design tolerances that can minimize the cost of poor quality
- An understanding of the costs associated with quality loss
- An evaluation of the impacts of tighter tolerances on manufacturing and service costs

Improving Manufacturing Process Capabilities

Will the manufacturing process produce a stream of products with little variability in quality "out of the box"? It's clear that low manufacturing quality can degrade the replication of a design's inherent reliability and durability. If the product is not manufactured properly, it might be DOA, or perhaps the outputs of the system can be out of spec or close to being off target at the beginning of the usage life.

At times the design organization can find itself overly constrained. The product concept may depend on a new manufacturing technology that is not yet robust and that requires tighter tolerances for dimensions or material properties. In this case, as the design goes into production, supply chain partners may have a difficult time maintaining the tolerances without costly efforts. One possible path forward is to accept higher manufacturing costs and lower yields. Is this really a manufacturing problem? Good manufacturing cannot compensate for a poor design. This is an excellent reason to insist on the participation of manufacturing engineering and suppliers in the cross-functional teamwork throughout the product development process, particularly in the early phases. Often problems of inherently weak manufacturability can be headed off by the selection of better design concepts.

A good way to determine if a tolerance is too tight is to understand the Cpk for the given dimension or material property. A low capability is evidence of lower yields, more waste, and a higher probability of escaping defects. Poor control of manufacturing processes may contribute to degraded product reliability. Factors contributing to manufactured product quality include

- Critical process parameters that control the mean and variability of the replicated design parameter under stressful conditions in manufacturing and its supply chain
- Production strategies and systems that minimize human errors and potential damage to the product during manufacturing

What are some of the manufacturing actions that can favorably affect these factors? Here are a few that can have great impact:

1. Standard operating procedures (SOPs): SOPs provide guidance for a disciplined and repetitive process. More than most other approaches, SOPs are essential to making acceptable products consistently. Without SOPs you can have chaos. If production uses SOPs, even if they are not optimal, at least there is a basis of repeatable processes and starting points for making improvements.

2. Robust manufacturing processes: Just as product designs need robustness, so do manufacturing processes. A good indicator of process robustness is its long-term Cpk, where the stresses affecting manufacturing have had time to demonstrate their effects on the process mean and its variability. One way to evaluate process robustness is to introduce stresses deliberately, using a designed experiment, and to measure the changes in the variability of reproduced design parameters.

3. A bias toward lean manufacturing: Why is "lean" important? Lean processes have
   a. Fewer process steps and movements of materials
   b. Reduced work in process (WIP) and smaller, less costly inventories
   c. Less waste of time and resources
   d. Shorter, more predictable cycle times

   Certainly, the opposite of lean sounds neither reasonable nor desirable. It's clear that "not lean" presents more opportunities for error and a greater chance for accumulated process variability. There is an increased potential for damage because you have more products and components to store, handle, and move more often from point to point.

The objectives of lean manufacturing are often viewed as the reductions in cycle time and costs. However, these steps are not to jeopardize quality or risks. The concerns for the consequences of failures, and their probabilities, apply to manufacturing process steps as they do to the product system. It may be that additional investments in manufacturing costs can increase the probability of consistent manufacturing quality and cycle times.

4. Mistake-proofing of the manufacturing and assembly processes: Manufactured quality depends on being able to build the product correctly. The design of the production processes is as important as the design of the product itself. It's a strong motivation to involve suppliers, manufacturing, and assembly personnel in building engineering prototypes.

Relevant chapter:

Chapter 23, Critical Parameter Management

Key phases for focus: Phases 3 through 6

Tools: Lean manufacturing, excellent SOPs, process capability studies, mistake-proofing, robust design applied to manufacturing processes

Areas for focus:

- Development of robustness for critical manufacturing processes
- Review of manufacturing SOPs for the completeness and accuracy of in-line quality controls
- Elimination of waste and superfluous process steps using lean principles
- Mistake-proofing[2] of the manufacturing process

Key results:

- Accurate specifications for critical process parameters for manufacturing
- Updated SOPs that guide manufacturing processes
- Reduced costs and cycle times from lean improvements
- Fewer human errors requiring additional costs for corrective actions

The Role of Testing

The old quality adage "You can't inspect quality in" can be extended to include testing. Testing has always been part of the "build-test-fix" triad. In the absence of concept selection and robustness development, this trial-and-error approach can be expensive, inefficient, and ineffective.

The importance of problem discovery should not be underestimated. Developing reliable products requires a balance of analysis, experimentation, and improvement. Analytical models enable you to explore the design space with behaviors that you already understand. Experiments enable teams to explore design space of which they have little knowledge. To obtain more information in less time,

---

2. C. Martin Hinckley, *Make No Mistake! An Outcome-Based Approach to Mistake-Proofing* (Portland, OR: Productivity Press, 2001).

accelerated test designs can be efficient strategies. Even with a high-quality design process, many problems may not surface until subsystems and systems have been subjected to rigorous testing. Among the starting questions in a test design are these:

- Is the purpose of the test to improve or to demonstrate?
- Are the operating conditions to be nominal or extreme?
- Are all expected scenarios to be tested or just a fraction of them?
- Is the configuration to be fixed or updated periodically with improvements?

Stress Tests

Stress tests are conducted under conditions that do not represent normal use. They are deliberately stressful to force inherent problems to occur sooner and to prove that solutions are effective. The stresses used can be expected external factors, or their surrogates, set to extreme levels that may occur infrequently. Examples of stress tests, described in Chapter 8, include

- DOE-based tolerance stress tests
- Highly accelerated stress testing (HAST)
- Highly accelerated stress screen (HASS)
- Manufacturing burn-in

Reliability Growth Tests

In project management plans, reliability growth tests contribute to the improvement of product designs, while the reliability demonstration tests contribute to verifying the completed designs. For both tests, the operating conditions are selected from those that are expected in customers' applications. However, in growth tests the product's configuration is upgraded with block changes as corrective actions respond to problems that are found.

Although they are based on build-test-fix iterations, growth tests should be planned after robustness development has given the product design a higher initial reliability (R_0). With some test "mileage" under their belts, teams can use the data to forecast how long they need to continue the build-test-fix iterations in order to achieve the reliability requirement. Reliability growth testing is one activity where there is clear payback for additional investments. More resources and responsive corrective actions can increase the slope (α) of the growth curve, while more time enables more failure modes to be corrected along that slope. Development teams have fitted growth models, such as Duane and Crow-AMSAA, to data successfully and used them to evaluate the rate of progress. If the forecast of the reliability expected at market entry is too low, they have justifications to increase the intensity of the reliability improvement.

Life Tests

Life testing is an essential tool for increasing a product's durability. As designs are improved to endure a long useful life, the longer it takes to get enough failure data to make a reasonable statistical inference. For this reason, accelerated testing has become a staple of design for durability. Two examples are

- Accelerated life testing (ALT), described in Chapter 22
- Highly accelerated life testing (HALT), described in Chapter 8

Reliability Demonstration Tests

A reliability demonstration test uses expected operating conditions and applications, but with the production-intent designs complete and frozen. If the system is repairable, planned preventive maintenance and repairs should be performed by trained personnel with formal procedures and production tools. When the product design is complete, a demonstration test can verify the achievement of the requirements for reliability, which might be a contractual mandate for certain projects. In addition, the demonstration test can identify the rates of use for consumable materials, the incident rates for repairs, and their consumption of labor and spare parts. This information can enable service organizations to plan parts inventories and personnel staffing more accurately. It may also be useful to marketing and sales as they clarify expectations for customers. The data can contribute to judging whether or not the product is prepared to achieve the expectations of its business plan.

Reliability Development Process Map

Figure 5.2 illustrates a high-level process map that suggests a logical order and relationship of activities to develop reliable products.

Of course the process is flexible and appears to be straightforward. In reality it may not be so easy. There are interactions, overlaps, and iterations that defy being illustrated. Certainly

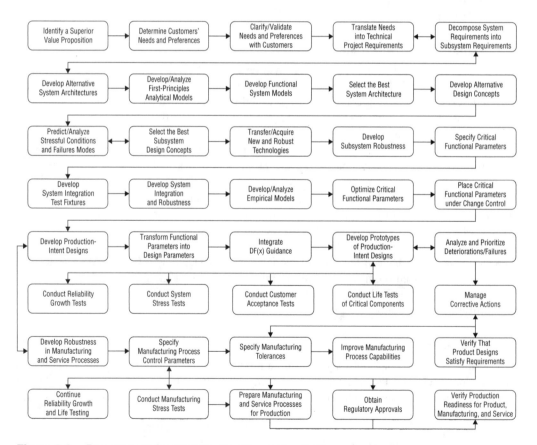

Figure 5.2   Process map for reliability development activities.

the process must be adapted to the particular challenges of the project. An early task for your project team is to convert the strategies that they select into plans with decisions, dependencies, resources, and timelines.

Key Points

1. Reliability development is an integral part of the development of product designs and of manufacturing and service processes. It involves robustness development followed by reliability growth.
2. Critical design parameters must be replicated by efficient manufacturing processes and maintained by effective repair and maintenance processes.
3. It all begins with the ability of technology development to provide a stream of alternative design concepts that have been developed to be robust for future applications prior to the beginning of product development. If your company does not manage the development of new technologies, your technology acquisition process should have demonstrated robustness as a selection criterion.
4. Robustness is focused on preventing problems by specifying critical parameters that reduce the vulnerability of performance to stressful factors that cannot be controlled.
5. Reliability growth is focused on finding problems and correcting them at a high rate.
6. Durability is developed by extending the life of components until the first failure that can end the product's usefulness.
7. The ability of your development processes to establish strong competitive advantages in your products is a measure of your organization's capabilities.

Discussion Questions

1. What capabilities have your projects demonstrated in developing products with higher reliability? How can these capabilities be improved?
2. What benchmarks are there against which your project teams can judge their competitive position and identify opportunities for significant process or product improvement?
3. How well do your teams understand the effects of stressful conditions on product performance?
4. How can effective stresses be incorporated into your experiments?
5. What trade-offs do your teams have to make among reliability, development costs, and schedules?
6. How much value do your customers place on product reliability and/or durability?
7. How can failure modes be designed to be more graceful, keeping your product useful while awaiting repair?
8. What competitive pressures in reliability do your products face in their markets?
9. What barriers do your project teams face in planning for predictable reliability development?
10. How well does your product development process incorporate the basic elements of robustness development and reliability growth?
11. What opportunities to reduce manufacturing costs are available by reducing design tolerances or by specifying less costly components?

Effective Decisions

Efficient product development depends upon many decisions being made by the right people with the right information at the right time. Each decision has to be supported by the analysis of data, the evaluation of alternative proposals, and the comparison of expected outcomes, with criteria that are relevant to the intended consequences of the decision. Project-level decisions are scheduled in the project management plan when they are on the critical path. They enable the future work to be more focused with clear guidance and lower risks, with follow-up implementation of the decision being managed as a systematic agenda for project management.

Decision makers in product development face many challenges. People involved may have different interpretations of data, recommendations, and decision criteria. The benefits, costs, and risks associated with options may not be well understood. No one may feel accountable for the consequences of the decision. Decisions made too early in the process may prematurely commit the project to design concepts or implementation strategies with constrained consequences. So decision processes are challenged not only to choose the best available option based the best information, but to do so at the right time. However, development teams will not know whether or not their decision was truly the best until it is implemented and the actual consequences are seen in practice.[1]

Product development is not just a process of turning engineers loose on challenging problems. It is an investment in the creation of a new or extended business, a value stream to be delivered by the new or improved products and services. So it has the expectations of populations of potential customers who look for products and services to be better than alternatives available to them. Similarly, a development project has investors—managers of corporate funding who are responsible for portfolios of investments in development projects and who are challenged to achieve superior returns from those investments. It has the minds of people who have to integrate technical and business capabilities into a well-functioning value delivery system, available to the market at the right time.

---

1. Dr. David G. Ullman, *Making Robust Decisions: Decision Management for Technical, Business, and Service Teams* (Victoria, BC, Canada: Trafford Publishing, 2006).

Many choices have to be made. They include whether or not to approve the charter for the development project in the first place or, later, to cancel it in favor of a better investment. Within project management plans are many decisions that need to be preceded by the activities that prepare information for them. For example:

- Target markets and customers have to be identified and characterized. Their needs and preferences have to be reflected in those value propositions selected to focus on the problems for which customers have money to spend for solutions. Development teams have to determine whether or not certain requirements are achievable with the capabilities available and within the timelines and constraints of the project.
- Design concepts, production processes, infrastructures, manufacturing facilities, and marketing channels, for example, have to be selected based on comparisons of alternatives. Their capabilities need to be specified and verified to represent the best efforts and wisdom of the collective enterprise that is the corporation and its partners, suppliers, and channels.
- If the team agenda is focused on reliability growth, the strategies, methods, and timelines are selected for that initiative. Whether or not the value proposition represents a challenge for reliability is an early decision reflected in the project requirements. The specific requirements are the results of negotiations between the demands of the market and the capabilities of the organization. The particular architectures, design concepts, and technologies are chosen as the best available solutions to those requirements. The accelerated test and corrective action strategies are selected to achieve required results within the allocated time and resources.

The quality and timeliness of decisions and the effectiveness of their implementation have dramatic consequences for a project. Decisions made at the right time can enable teams to have a higher probability of success. However, a company's inability to make decisions effectively or when needed can be a critical weak link. Decisions that are flawed because data have low integrity, or that are not supported by resources, can be very costly. Decisions that lack functional support or follow-up implementation are no decisions at all.

Given all of this, what should be expected of decision processes within your development projects? Certainly, the judgments of many wise people have to be integrated. People experienced with past projects need to contribute the lessons they learned to current projects. The solution-deep development activities need to benefit from business-wide guidance from those not so immersed in the technical details.

Qualities of a Good Decision-Making Process

When you look at *how* the more effective decision processes ought to work, you can expect a few important features:

1. Comparisons of alternative approaches and expected outcomes, with decision criteria that are unambiguous and appropriate for the point in time of the project
2. Core principles, representing the values and strategies of the company
3. Evaluations of work in progress to identify the risks to future activities, not just to appreciate the achievements of individuals and groups

4. Routine cross-functional teamwork that enables many decisions based on the consequences to the various business units and partners, the open-minded evaluation of trade-offs among approaches, and the planning of implementation activities
5. Decision milestones within development phases to enable progress to be expedited by the progressive freezing of information that is reasonably stable while leaving open those decisions that can be delayed until supporting information is stable and has sufficient integrity
6. Project-level gate reviews at critical milestones to make major business and technical decisions that enable the forward progress of the project

Product development projects integrate the wisdom of people who have decision-making responsibilities with those doing the project work. Without these elements, decisions probably will fail to meet their objectives and their implementation can fail to achieve the benefits expected. Risks may increase, customers may reject the solutions, and the investment of resources may be squandered. Those are avoidable wastes that should not be tolerated.

Core Principles Supporting Decisions

Certainly alternative proposals for decisions need to be compared with facts, supplemented by the judgment of people experienced in the development of similar products. Are decision processes that objective in your business? In many cases, core principles can provide tie-breaking criteria. It's easy for teams to be biased by the short-term pressures of resources, budgets, and schedules. Company politics and strong voices may tend to distort the logic.

It can be helpful for those organizations responsible for product development to articulate core principles that they accept to guide their activities and decisions. You would expect that they would reflect company values. What else? That's an excellent thought activity for organizations, since the principles would represent the fundamental relationships among the company, its suppliers, and customers. You need to develop your own principles. Here are some suggestions to stimulate your thinking:

- Make decisions so that your customers win.
- Don't pass the buck. Avoid "pass-through" reviews that distribute accountability or seek ratifications.
- Coordinate partners and suppliers to act as a single enterprise delivering to customers.
- Develop solutions that are expected to deliver more value to customers than alternatives available to them after market entry.
- Return value to the corporation better than the next-best alternative investment.
- Recognize empowerment as a "three-legged stool": responsibility, accountability, and authority.
- Make decisions at the right level in your organization, where the knowledge and empowerment coincide. Develop the necessary information to have high integrity to enable decisions when needed.
- Practice an inverted organization structure: management working to enable the cross-functional work groups. Management accepts action items to implement decisions, just as do project teams.
- Avoid short-term reverence for budget compliance, which can jeopardize the opportunity to deliver the right product at the right cost to the market at the right time.

Each of these statements represents an agenda. So it may not be reasonable for all of them to apply to your organization, unless they do. What other principles are relevant to your business model and its drivers of success?

What Is Value?

The concept of *value* can be a thought provoker. Customers tend to view value as benefits compared to their notion of costs, which might mean purchase price, the broader sense of cost of ownership, or a combination of objective and subjective criteria. Corporations may view value in financial terms, such as a stream of revenues minus a stream of costs over time. They may see value in other ways, such as a product being a test of a market to provide feedback to a follow-on product. They may see a product as a device to establish or maintain corporate presence in a market. Certainly there are other views. They should be articulated clearly in those project objectives agreed to in the product development charter.

Who Makes Which Decisions?

Decisions are made at different levels in the organization. Many decisions are the responsibility of the development work groups, cross-functional integration teams, and the project's core leadership team. It's best that decisions be made by those who have access to the breadth of information necessary and who are doing the work. That empowers work teams to affect the progress and outcomes of their project. In certain cases, decisions can be made by customers, such as the validation of requirements or the judgment of the acceptability of solution concepts. That level of involvement depends on the nature of the business.

Certain decisions are the responsibility of the corporate or business unit's managers. For example, those might apply to strategic objectives specific to products, portfolio balance, and funding levels. Other decisions are the responsibility of the project's governance committee, such as the project-level gate decisions. There are two principles that are helpful in determining decision responsibilities:

- In accordance with the principle that decisions should be delegated as much as possible, they should be made at the lowest level in the organization where both the knowledge and the responsibility for implementation coincide.
- In accordance with the principle of empowerment, decisions should be made by those who have the responsibility, authority, and accountability for their consequences.

A "RACI" diagram can clarify responsibilities. Table 6.1 shows how it can work. The acronym *RACI* refers to a mapping among relevant managers and teams, for each type of decision, to indicate who is

- **Responsible** (R) for making a decision
- **Accountable** (A) for the consequences of its implementation
- To be **Consulted** (C) to provide critical inputs to the decision
- To be **Informed** (I) of the results of the decision

These agreements need to be worked out early, with a clear understanding of the implications for the organization. Phase 0, when the charter for the project is agreed upon, is a good time

Table 6.1 A "RACI" diagram clarifies roles to be played for a range of development decisions

Organizational Role	Decisions about Product Design	Project Gate Review Decisions	Performance Appraisals of Core Team Members
Business unit manager	C	R	C
Technical or business functional managers	I	C	R
Project leaders/core team leader	A	A	R
Core team members	I	A	I
Development team leaders	R	A	C
Technical, business, and process specialists	C	I	C

for this. If a manager does not have the resources or other capabilities to act on the decision and be responsible for the outcome, then he/she should not have the accountability for the decisions.

Since there is a range of decisions to be made, the roles that various organizations and individuals play will vary. It is most important that decision responsibilities be very clear for a project, particularly when the organizational implications are complex, such as, for example, with collaborating business units, with external development partners or supplier companies, or with sales and distribution channels. We've seen confusion and conflicts in these responsibilities cause major delays and poor implementation of decisions, and even their reversal.

A "RAPID" analysis, illustrated in Table 6.2, is another tool to clarify the key decision-making roles. Try both methods to determine which works better for your business model. It can be dysfunctional and disruptive to have a power struggle within the project organization or, even worse, organizational indifference.

The table presents an example for RAPID that differentiates among those who have responsibilities to

- **Recommend** the best decision proposal
- **Agree** to the consequences of the decision's implementation
- **Perform** the work of implementing the decision
- Provide technical or business information as **inputs** to the decision
- **Decide** on the best option for the decision

Decision responsibilities may differ among organizations, since the depth of knowledge in specific details and the accountability for business and technical results can vary quite a bit.

Rules of Engagement

There are many ways for decision processes to become dysfunctional. One of them is rooted in how people act in decision meetings. How often have you witnessed behaviors that are

Table 6.2 RAPID analysis clarifies who decides, who has input, and who gets it done

Recommend (R)	Agree (A)	Perform (P)	Input (I)	Decide (D)
Project Leadership Team Members	**Functional Managers**	**Project Core Team Leader**	**Technical and Business Specialists**	**Project Governance Body**
Coordinate the work during the development phases Develop valid data relevant to the objectives of each phase Derive conclusions from analyses and experimentation Make sound recommendations	Ensure that functional plans are achievable with excellence in execution Participate in gate reviews with knowledge and perspective about recommendations Can commit resources or intervene to stop a decision	Coordinates the interactions with upper management Explains recommendations and rationale to management Is responsible for implementing the decision and making it stick Coordinates plans and decisions by the project team	Consulted by the project core team to ensure data integrity and the soundness of recommendations Consulted for advice to support the decision, with no obligation to act on the advice	Facilitates a high-quality decision process Asks probing, ended questions Recognizes good answers Sets a climate for open discussion and evaluation of data Achieves a decision with consequences that are acceptable

contrary to an open and honest assessment or to a supportive management process? Certainly time pressures place stress on all of us. That may be reflected in key people not showing up or being distracted by side conversations, cell phone calls, or emails. They may not be prepared with current knowledge of the project's major concerns. They may choose to "play gotcha" and withhold critical information from the project team, or to distract the convergence to a decision with low-probability "what-if" questions.

An effective tactic to combat these risks is for decision bodies and project leaders to establish rules of engagement for themselves. The implication is that participants agree to improve their behavior since they themselves define the rules, agree to their rationale, and want the mutually desirable benefits. Here are some examples that are broadly applicable:

- Hold the review meeting when it was planned and on managers' calendars, even if the work has not yet been completed or the recommendations not yet prepared.
  - The urgency of a review may force late work to be accomplished more quickly.
  - Virtual meetings may be sufficient to resolve remaining concerns.
- Participate with attentiveness, ownership, and commitment.
  - Be on time to establish the required quorum.
  - Permit substitute attendees to carry decision responsibilities.
  - Be prepared with perspectives about the risks and relevant questions.
  - Turn off laptops, cell phones, PDAs, and other electronic distractions.
  - Resist disruptive side conversations.

- Plan the agenda to be concise, cross-functional, and value-adding.
  - Be receptive to bad news or unexpected results. Don't kill the messenger.
  - Invite "truth to be spoken to power."
  - Reward open, honest, and responsible discussions.
  - Improve the meeting design with feedback from previous reviews.
- Set expectations at the beginning of the phase.
  - Avoid new objectives or acceptance criteria at the last minute. The project teams will not have time to develop solutions within the planned timeline.
  - Resist disruptive or "gotcha" questions that are out of scope or beyond expectations. They tend to add more risks than value. Hidden agendas, unspoken criteria, or surprise attacks are not fair to the project teams and tend to be self-serving.
  - Notify the project team of major concerns or potential showstoppers in advance.
- Maintain a business perspective.
  - Avoid submitting a request to management unless it is appropriate for them to act upon it.
  - Be willing to compromise at the functional level to optimize at the business level.
  - Focus on value development rather than on process compliance.
  - Support conclusions with facts rather than opinions. Ask probing questions to verify the integrity of data and assumptions.
  - Base decisions on available information, perspectives, and acceptance criteria.
- Ensure the rigorous implementation of decisions.
  - Communicate decisions with their reasoning to the project team face-to-face.
  - Document decisions with their qualifiers, action items, and implementation plans. Manage the implementation of the decisions.
  - Sign a decision letter to demonstrate buy-in to a common understanding of the decisions and their implementations.

These are just examples. We expect that you have experienced the benefits of many of them. The key is for participants to make the rules for themselves, to agree to them overtly, and to evaluate routinely whether or not they are following their own rules. This can be very helpful for fostering improvements.

Decision Processes

Generally, decisions that are prudent and timely are the result of a structured process. A standardized process should be repeatable from project to project, improving with use and feedback. The benefits are not only in better decisions being made at the right time, but also in people knowing how decisions are made and how to contribute to them. Belief in a decision and its process enables effective implementation to be achieved as intended.

Here are some characteristics of an effective decision process that you may find useful:

- The decision schedule should be critical to the advancement of work and the management of risks, although it is not necessarily on the project's critical path.
- Acceptance criteria must be relevant to the objectives of the phase of development work and to the follow-on activities.
- Large decisions should be decomposed into smaller, incremental decisions.

- Selected noncritical decisions should be delayed until the need for their benefits is on the project's critical path and their prerequisite information is mature and stable.
- Background information, data, analyses, and perspectives relevant to the acceptance criteria must be available and summarized for evaluation.
- Standardized contents and formats improve the ability of stakeholders to interpret information that they evaluate from a range of development projects.
- The setting and atmosphere should be conducive to unbiased and honest evaluations. For example, the layout of the room and seating should foster the interactions among the presenters and deciders.
- There must be open discussions among knowledgeable and responsible people, with ample opportunity for responses to probing questions. Ask for a response with rationale from each decision maker.
- A decision must have alternative outcomes, from which the best one is selected.
- There should be a managed process for the implementation of decisions, with responsibilities, resources, verification, and follow-up.
- There must be agreement among those managers responsible for implementing the decision to honor the rationale and live with the implications. This does not necessarily require unanimous agreement, which would give too much power to a single person.

Decision processes are thoughtful, cross-functional processes, involving empowered representatives of the business. They are not competitions among powerful voices. The consequences of flawed or late decisions can be severe for a development project, particularly those decisions made by management.

Standardized Data

Many companies have found that by standardizing the content and format of information, such as project-level metrics and the results of analyses and tests, their managers have an easier time comparing data, recognizing the trends, and understanding the underlying implications. The illustration in Figure 6.1 is sometimes called a "dashboard" or a "4-up chart," an example of a way to display data that are monitored routinely. The data are selected to be most important to understanding the current state of progress. Showing the relevant acceptance criteria or target values enables judgments about the expectations for future progress or the need for intervention to be made more easily.

Alternatives for Decisions

A decision is a selection of the best approach among available alternatives. Often it is the case that the alternatives require different levels of investment, different timelines, or different allocations of resources. The team may have different degrees of knowledge about the alternatives and their risks, how to implement them, or even their requirements. Their expected outcomes may be different, as might be the ability to implement the decision. One option may be to cancel the project and redirect its resources to another project with higher potential value. Each decision may have uncertain outcomes that are outside the control of the team. For example, the probability of customer acceptance may vary between two alternative design approaches.

Uncertain outcomes with their probabilities of occurrence can be mapped by a decision tree, such as the one illustrated in Figure 6.2. Analyses can then identify the approach with the higher

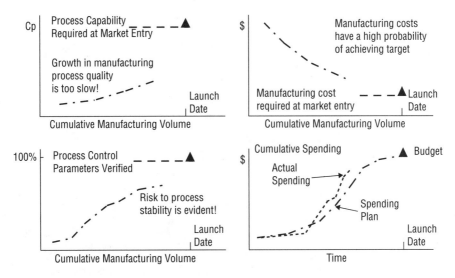

Figure 6.1   What data should be displayed? Project data are easier to interpret if they are standardized in content and format, with relevant acceptance criteria.

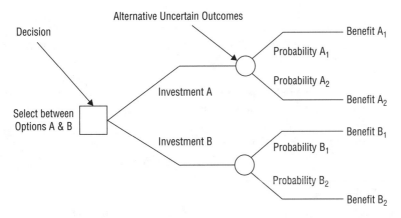

Expected Value of Investment A = (Benefit A$_1$ × Probability A$_1$) + (Benefit A$_2$ × Probability A$_2$) − Investment A

Figure 6.2   The decision tree maps the alternative approaches (A and B) with analyses of the potential benefits and their probabilities of occurrence.

expected value. That's one way to do it. Although the values for outcomes and their probabilities will be just approximations, the process by which the team thinks through the alternatives, maps them, and estimates their business values can be quite enlightening. Teams will understand the elements of the decision and more easily agree to the best choice.

Figure 6.3 shows another way to use a decision tree. Suppose that the financial forecast for a project is judged to be extremely vulnerable to a number of factors beyond its control. For example, there may be little tolerance for competitive price actions, uncertain customers' acceptance,

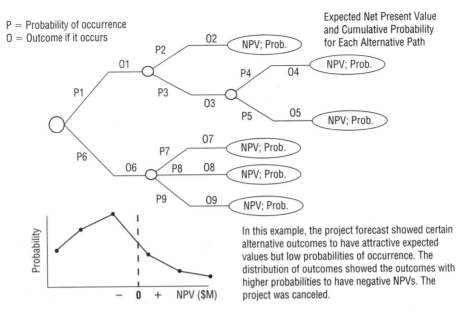

Figure 6.3   A decision tree evaluates a range of alternative outcomes with their various probabilities of occurrence.

or the introduction of competitors' products. When the possible outcomes are mapped, a clearer understanding of the situation can be gained. The net present value (NPV) is the stream of future revenues and costs over time, converted to the value of money in the current year, using an acceptable discount rate.

The project scenario shown in Figure 6.3 is one for which there were a number of potential outcomes that had negative expected returns and higher probabilities of occurrence than those with positive expected returns. Once the leadership team understood that there was little within their control that could correct these handicaps, they recommended that the project be stopped and the resources redeployed to a higher-valued product concept.

Risk Assessments

Assessments of risks, their drivers, and mitigation plans serve important roles in the routine coordination of projects and in their reviews. With selected intervention by management, actions to reduce risks can be enabled further by higher-level decisions, reserves of resources, and other approaches outside the control of the project core team.

A useful tool to evaluate risks is an adaptation of failure mode effects and criticality analysis (FMECA), described later in Chapter 13. Table 6.3 shows a template for the application of FMECA to risk management.

Remember that a risk is a problem or failure that has not yet occurred. It could happen in the future, but there are uncertainties about when or how often a problem may occur, about the behavior of its drivers, and about the severity of its consequences. A problem can have several root causes, each with its probability of occurrence. So problem prevention is a process of reducing the probability of occurrence and/or mitigating the consequences if the event does occur. On the other hand, contingency plans are steps to compensate for the consequences of the event

Table 6.3 The FMECA model is useful for risk assessment and management.

Risk Situation	Probable Root Causes	Probability Occurrence	Consequences to the Project or Customer	Preventive or Contingency Plans

after the problem has materialized. If the problem has already occurred and is severe enough to justify correction, you have a corrective action process. All three of these approaches (assessment, prevention, and correction) require decisions to allocate resources, time, and leadership attention.

Expectations for the Product Development Core Leadership Team

The development project is coordinated by a core team of representatives of the various organization functions involved in the project. Some companies may use other terminology, such as project leadership team (PLT). We have characterized the actions of this leadership team as being inherently cross-functional and acting within their empowerment. The process places important expectations on its members. Here are useful suggestions that can be adapted to your culture:

- **Own** the development project:
  - Negotiate achievable requirements.
  - Select the best strategies and solution concepts.
  - Integrate development activities into plans and capabilities across functions.
  - Manage the actions to reduce risks.
  - Close the gaps between "actual" and "plan."
  - Commit to the project's success; don't pass the buck.
- **Deliver** superior value to customers:
  - Make decisions so your customers win.
  - Practice functional excellence.
  - Embrace the behaviors of high-performance, empowered teams[2] across your enterprise.
- **Return** superior value to the corporation:
  - Be efficient, lean, and agile in all practices.
  - Leverage proven capabilities.
  - Implement new capabilities only where their benefits add high-priority value.
  - Manage deliberate actions to reduce risks.
  - Respect the uncertainty due to factors beyond your control.

---

2. Katzenbach and Smith, *The Wisdom of Teams*.

Types of Project Reviews

Our perspective is that the purpose of a review is to make a decision. Otherwise a review is vulnerable to being a non-value-adding activity demanded to update those who have not been paying attention. Because decisions are made at various organizational levels and at various times along the process, it makes sense that there should be appropriate levels and timings for reviews. So what reviews should be reflected in project plans? What are the intended consequences of each review? We suggest that a few basic types of reviews be reflected in your development strategy:

- Cross-functional progress reviews
- Out-of-bounds review
- Peer reviews
- Design reviews
- Functional reviews
- Portfolio reviews
- Gate reviews

Each serves a different purpose, so their participation, agenda, format, and formality should be tailored to their objectives. With the exception of the "out-of-bounds" review, their timing should be planned to coincide with the convergence of the prerequisite information and the need for the direction that results from the decision. Their value to your business should increase with practice, coaching, and feedback.

Cross-functional Progress Reviews

These are the routine meetings that involve representatives from the various organizations contributing to the development project. At the level of the project's leadership, these gatherings of functional representatives are important methods to coordinate and integrate activities across the technical and business disciplines of the company and its partner companies. In some situations customers or suppliers may participate when the agenda is appropriate.

The project's core team is chartered by its governance body to manage the investment in product development. Their work is in the context of the portfolio of product development projects, technology development initiatives, and strategic business alliances, as examples. The project's core team and its leader are empowered to get the work done. It is then their responsibility to make the best use of the allocated resources (time, funding, labor, partners, knowledge, functional capabilities) to achieve the project objectives.

> *Resources can be thought of more constructively as investments to enable progress rather than budgets to constrain it. So use them wisely.*

By describing these meetings as progress reviews, we do not intend them just to focus on sharing status. To add value they need to enable progress. The many elements of progress can be measured and compared to acceptance criteria. Lagging progress may just need some improved tactics or motivation. Possibly some additional resources or extra efforts are needed for a short period. The "investment" budget should not rule.

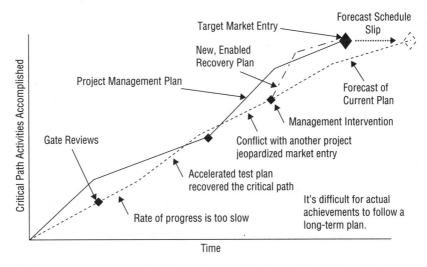

Figure 6.4   Progress that lags behind the "critical path" may need intervention by management with additional resources or a new plan.

Out-of-Bounds Review

It may be that the current plan is not working well enough and a new plan is needed, with immediate implementation. Progress has drifted beyond the acceptable boundaries, as defined by the project's charter. If intervention is necessary, actions can be taken within the scope of empowerment provided to the core leadership team. If the correction of progress is beyond the control of the core team, the situation must be escalated to a higher level of empowerment as soon as possible, outside the scheduled gate review process. This unscheduled intervention is often called an "out-of-bounds" review. For example, if the achievement of deliverables from tasks is far behind schedule, deliberate new steps must be taken to recover, such as an increased focus on higher-priority problems or the allocation of specialized expertise. Figure 6.4 illustrates a situation where management intervention invests in more accelerated testing with additional resources allocated to avoid the slip in the market entry date.

Another example, shown in Figure 6.5, is the case for which the rate of growth in reliability is too low. Decisions may be necessary to reconsider the design concepts or technologies being implemented, to accelerate the test methods being applied, or to stimulate the responsiveness to problem reports. If corrective actions are beyond the control of the core team, the situation may need a mid-phase allocation of additional funding or specialized outside resources. Some of the strategies from Chapter 3 may need specific direction from upper management.

Project leadership teams may have to deal with major threats to the project's existence. For example, there may be situations when the project is redirected or canceled, perhaps because of changes in the portfolio balance. The core team will then have to develop new plans to react to negotiated adjustments to its commitments or to redeploy its resources.

Peer Reviews

A peer review is an informal gathering of knowledgeable people offering wisdom based on their past experiences, specialized knowledge, or unique insights. By "peers" we mean people doing

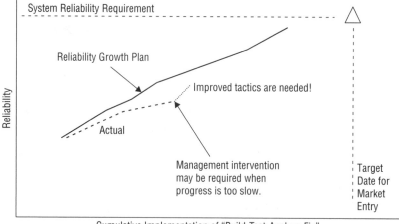

Figure 6.5   How can you accelerate the rate of growth in reliability?

similar work, but in other work groups or on other projects. If you involve people in higher-level positions, you may invite the risk of their being judgmental or unintentionally sabotaging the desired openness. You can see peer reviews all the time at athletic events in team huddles, individual coaching, and advice from veterans to rookies. In engineering you can see it in corridors, around CAD terminals or test fixtures, near the coffeepot, and in impromptu office gatherings. When suppliers or subject matter experts need to be included, they may have to be invited. More often a peer review can be spontaneous, when the particular problem and relevant wisdom collide. You could argue that knowledge derived from a lessons-learned database fits the model, but the face-to-face discussions, clarification, interactions, and brainstorming provide valuable benefits. Knowledge management plays an important role in making peer reviews more valuable; Toyota's standardized documentation (A3 charts)[3] is an example.

Feedback is requested when there's time to react to it. The objective is to solicit recommendations that can strengthen solutions while the work is in progress. If it's too late—for example, after a design is released—the feedback may be either just a report card or the beginning of a potential crisis. This process is important to the principle that customers pay for solutions that represent the best that can be produced by the enterprise, not just by an individual.

A key ingredient is the "pull" of input by people responsible for the design. They may focus on creative options to improve reliability or on clever stresses to impose in a designed experiment. Advanced manufacturing or service engineers may provide detailed guidance by reviewing design concepts in depth to enable more capable processes. Reviews of software code fit the model, with the objective of finding defects for correction while the development continues. These are not inspections or performance appraisals merely to develop a score. They are specific design improvement steps to integrate the best advice from others.

Effective peer reviews can be characterized as

- Being early, frequent, short, and informal
- Having little preparation, just evidence from work in process

---

3. Morgan and Liker, *The Toyota Product Development System;* Shook, *Managing to Learn.*

- Demonstrating "sleeves rolled up" involvement in details
- Bringing together the collective wisdom of peers, experts, suppliers, customers
- Generating recommendations that are nonjudgmental and nonbinding
- Being facilitated by a focused agenda, with emphasis on effective follow-up
- Being documented sufficiently to ensure responsible action and to avoid losing the suggestions or their context

Design Reviews

Certain development processes may mandate a sequence of structured reviews of the product design, its manufacturing processes, and the achievements of other functional elements of the project. They may be driven by management, by customers, or by their surrogates. If necessary, the objectives of these reviews ought to be very clear for the business, particularly as they iterate from phase to phase within the process. For example, how do the acceptance criteria for a design review in an early phase differ from those later in the process? Design reviews can be given names, such as "Preliminary Design Review" or "Final Design Review," that do not necessarily convey their reason for existence. There may also be internal reviews prior to external reviews.

Unfortunately, it has been our experience that often those who participate may do so because it's required but may not understand the purpose of the review, its expected content or criteria, or the consequences of the decisions that are expected. It makes sense to us that the acceptance criteria for a design review should reflect the objectives of the development phase within which it is scheduled. If it is a decision process, rather than just an information exchange with some advice, that should be clear to the project team, along with the intended consequences of that decision. So care should be taken to clarify the decisions to be made and to ensure that there's time to implement the actions that are decided upon.

Some development processes may need to manage technical or business risks by imposing project milestones as "inspection points," requiring criteria to be satisfied and approvals to be obtained. Contrary to the description of peer reviews, this process has an "after-the-fact" character which implies that actions derived from the review would be either going back to rework something already developed or adding work to future activities to reduce risks in designs. Care is needed to ensure that the process does not become dysfunctional. We've observed iterations of additional work to satisfy unexpected criteria or changing requirements, or to accommodate weak achievements in other parts of the system.

Contractual agreements may mandate post-work approvals. For example, in situations where the approval of funding for the next phase is based on deliverables satisfying particular requirements of the current phase, the design review could become very formal with much due diligence. Participation would have to include those with specific decision rights.

Effective design reviews can be characterized as

- Formal, with feedback at significant milestones
- Assessments of progress, forecasts, or methods
- Comparisons of plans, concepts, and developed results to requirements
- Focused on identifying
  - Corrective actions to meet specific acceptance criteria
  - Preventive actions to reduce risks

Functional Reviews

Often companies have strong functional organizations supporting development projects. While functional excellence is achieved every day on the job by the people in the work groups, there may be tangible value in selected functional managers periodically reviewing and guiding the work specific to their function. The managers of those organizations may even share responsibilities for the success of the projects and participate in the governance processes.

Functional reviews held prior to gate reviews can enable the gate reviews to benefit from summaries of technical information and thereby concentrate more clearly on business decisions. A functional manager may see these reviews as valuable opportunities to contribute personally to technical advancement or risk reduction. Although they may be an extra burden on the project core team, functional reviews can be efficient ways for functional managers to determine how to support the core team's recommendations and how to ensure that there are no major surprises in the presence of his/her peers. Routine involvement with projects is far better, but functional reviews may be necessary and even valued in some situations.

For example, functional reviews can add value by ensuring that

- Functional implementation plans are achievable.
- The best functional tools and thinking are being applied.
- Subject matter experts are available when needed.
- Centralized capabilities are not overloaded, so that demands with unpredictable frequencies and durations can be handled without generating excessive queues.
- The right functional resources are applied at the right time.

Effective functional reviews can be characterized as

- Decision processes themselves, not "pass-through" reviews merely for advanced awareness of concerns
- Resulting in recommendations based on evaluations of achievements, risks to future activities, and implications of higher-level decisions
- In-depth evaluations of technical concerns specific to the function
- Being focused on the achievement of functional and technical excellence
- Enabling resolution of cross-functional or cross-project conflicts
- Identifying business problems or conflicts that need to be escalated to gate reviews
- Being managed with objectives that are clear and do not contradict the responsibilities of the project's leadership team

Portfolio Reviews

Although beyond the scope of this book, portfolio reviews are mentioned here because they have interactions with gate reviews. External to a specific development project, the portfolios of projects are developed, balanced, and optimized for the business. Long-range investment plans are determined and contribute to the funding and timing for each new project.

However, these plans are not stagnant. Competitive markets change. New ideas for products are recommended. Feedback from customers can indicate the need for different value propositions. New technical developments "push" concepts for products or features that have not yet been considered.

So management needs to review the portfolio periodically to ensure that it has the best balance for the business. A standardized schedule of quarterly portfolio reviews is a familiar practice. Similarly, the development projects need to be monitored to ensure that they remain aligned with the portfolio. It may be that changes to the portfolio can have significant consequences for ongoing new product projects. Some may need to be redefined, others may need to be accelerated, and certain others may need to be canceled or postponed.

Management is responsible for the implementation of the portfolio through the development projects. That has deliberate consequences for the development objectives and for the allocation of funding and other resources. So the agenda for the project gate reviews should ask how well the project is aligned with the portfolio. Similarly, the portfolio reviews should ask how well the collection of projects is enabled to achieve the objectives of the portfolio.

Gate Reviews

A gate review is a decision process. It oversees, if not enables, the transition from one development phase to the next, although it is not necessarily on a project's critical path. The development phases overlap. A gate review is a process by which the company makes incremental investment decisions for projects developing new products with improving clarity as information matures.

At each gate review, a governance committee has the option to modify the direction of the project or its allocation of funding and resources. For example, a product proposed by advanced planning may look good in Phase 0. At Gate 0 a development project can be chartered to determine if the requirements are achievable within an expected timeline or with an acceptable financial return. However, development work in Phase 1 may determine that this is not the case because the enabling technical concepts are not yet controllable. The project can be canceled or postponed. The investment decision at Gate 0 was to "buy a look," with the option to back away. As markets change and a project looks to be more valuable, resources can be added if an accelerated strategy is required. Similarly, a project with high potential value but unacceptable risks may require increased funding to reduce those risks. When management is routinely involved with teams, expectations can be clarified and an appropriate sense of urgency can be reinforced. This decision process establishes an agreement between the recommendations of the project's leadership team and the support from the governance body.

The concept of "options" being available is important to the financial analysis in the business plan. Estimates of expected value for a project should recognize that the investment committed at Gate 0 can be modified later in the project as new information is developed. That new information should also reduce the uncertainties associated with the streams of future revenues and costs.[4]

Gate reviews are an important way for management to share responsibilities for the success of the project. The process balances project-specific concerns with those of the overarching business. As we mentioned at the beginning of this chapter, product development is driven by decision making that determines what products to develop, how resources are to be assigned, how development is to be conducted, and many other important concerns. Through this process senior management leads product development, implements the specific strategy derived from the new product portfolio, and empowers the project leadership team to manage the investment.

---

4. Robert G. Cooper, Scott J. Edgett, and Elko J. Kleinschmidt, *Portfolio Management for New Products* (Reading, MA: Addison-Wesley, 1998).

Gate reviews are positioned in the development process at natural points of convergence of business and technical information and the need for project-level decisions. They are major integration events, a valued part of work processes. For example, the milestone to approve a business plan depends on valid requirements, acceptable solutions, and achievable implementation plans having similar levels of maturity and integrity. The schedule is not necessarily based on the passage of time. An exception would be when a phase is very long, for which interim project-level reviews might be warranted. However, unless there's a major decision to be made, the objective of a mid-phase review would relate more to risk management.

The concept of a "gate" implies that there's a respected barrier to be opened, rather than a "speed bump" to be negotiated. However, this does not mean that gate reviews cannot be combined or treated as being off the critical path. In the latter case, individual activities of the next phase begin when the prerequisite information or resources are available, and the consequences of this early start do not jeopardize lagging activities. In fact, principles of lean product development encourage that. It becomes a matter for the objectives of each gate review and how the consequences of the decisions are reflected in the work of the teams. The agendas have higher value when they are focused more on improving expected value and reducing risks than on the inspection of deliverables or the compliance with work standards.

Not all gates need to have the same weight. For example, Gate 3 may be an important way to ensure that engineering models reflect the production-intent design and that late design changes do not compromise the verification tests. With improved discipline within the project team and its communications with management, this gate review may be achieved by a review within the project leadership team rather than by a formal meeting with the governance committee. However, it may have to be more formal if critical modules of the system to be verified are being delivered from partner development companies.

At a development gate, management has options. As for any good decision, there are alternative outcomes, such as these:

- Continue the development project with the work of the next phase.
- Go forward, but with specific actions required and follow-up checkpoints.
- Remain in the current phase to resolve specific risks and revisit the gate at a planned later date.
- Redirect the project: accelerate, retarget, reconfigure, or place it on hold.
- Stop the project and redeploy the resources to a higher-valued project; document the results for the benefit of future projects.

Often projects are faced with a common dilemma, as shown in Figure 6.6. Suppose you had to decide whether to enter the market with the design difficulties identified by your failure-reporting process, or to delay market entry until those problems were corrected. There are costs for the delay in market entry, such as the extended development costs and the delay or loss of revenues. Your reputation in the market may either suffer because of the delay or benefit because of the improved product. What is the right answer? It depends on how the team evaluates the effects of the delay and its consequences and what it would take to avoid those consequences. How severe are the failures in question? Is being "first to market" the key to success, or is being "right to market" the better strategy? Can problems be corrected easily after the product is launched, or would expensive recalls or service-installed modifications be required? Can the project efforts be bolstered to correct its

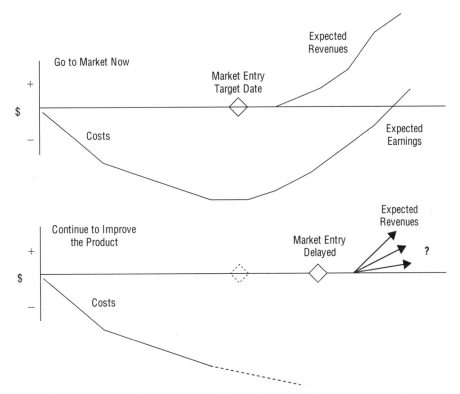

Figure 6.6   Which choice is better: Go to market now, or continue to improve the product?

shortcomings and save the market entry date? Certainly there is no universal answer. It depends on the particular circumstances both in the marketplace and within the development project.

A selection matrix is a useful tool for decisions. It is described more fully in Chapter 11. The process is to compare decision options against a set of criteria for the decision. The objective is to select the option that is better than the available alternatives at satisfying all of the criteria. You do not care how much better it is. The method compares each option to a referenced option (datum), counting only whether or not each option is better than, the same as, or worse than the reference. Promising options can be improved by integrating good features from other options to create hybrid new ones for the selection process. Table 6.4 is derived from Pugh's "concept generation and selection process."[5] If no decision option is the clear winner, as illustrated in this example, the review process needs to modify the decision concepts, creating a hybrid that is the best for all the criteria.

As mentioned earlier, decisions are agreements, with acceptable rationales, so that all involved organizations can live with their consequences and support their implementation. The process does not require everyone to favor the decision based on his/her own functional interests. Some organizations may have to compromise their own position for the benefit of the whole business and its customers. By being involved in the project issues and collaborating in the decision

5. Pugh, *Total Design.*

Table 6.4 A decision is a selection of the best available alternative. Can a better option be defined that will be superior to the reference (datum) for all of the criteria?

Decision Criteria	Option 1	Option 2	Option 3	Option 4
The customers "win"; the product will be superior to the competition's	D	+	S	–
Enabled by available resources	A	–	+	+
Acceptable to functional organizations	T	–	+	S
Aligned with technology portfolio	U	S	+	+
Acceptable impact on market entry timing	M	–	–	S
Summary	Referenced option	+ = 1 – = 3 S = 1	+ = 3 – = 1 S = 1	+ = 2 – = 1 S = 2

process, functional managers will more likely suggest better approaches, accept the compromises, buy into the implications, and contribute to the implementation.

Some projects may not be designed well enough to deliver sufficient value. The gate review is a forced opportunity for the project to be refocused. Possibly the product's value proposition needs to be more competitive or the scope of the project constrained. It could be that the achievable cost structures would not provide sufficient profit margin. The market may be changing so quickly that the product would be obsolete by the time it is launched. Particularly if submitted early, a recommendation to stop an ill-fated project should be rewarded. People at all levels should be motivated to avoid wasting resources and time on a project that has a low probability of delivering acceptable technical or business value.

Although not its main purpose, a gate review serves as a quality control tactic for a stage of development. However, the conclusions may be "fuzzy" since it's rare that all work is complete or that all acceptance criteria are satisfied. So people involved in reviews need to develop capabilities to make important and timely decisions when faced with partial information and uncertainties about the future.

Agenda for a Gate Review

With a focus on business concerns, the agenda of gate reviews should be concise, for example, with durations of one to two hours, with attention aimed at improving the quality of future work rather than just evaluating work completed. We like agendas that devote about 20% of the time to review and 80% to planning, with plenty of time for questions and discussions. The consequences should enable future work to be more predictable. The topics should enable the decision to be the right one, within that business day. For example, agenda topics can

- Clarify the objectives for the current phase and its acceptance criteria.
- Validate the key project assumptions.
- Evaluate the recommendations of the project leadership team.

- Compare the significant achievements and forecasts against expectations for the phase, ensuring that they are supported by clearly interpreted data.
- Evaluate the business consequences of technical data, functional capabilities, key assumptions, and competitive market forces.
- Evaluate how well the project is expected to deliver value to customers, as well as to return value to the corporation.
- Understand the integrity of the data contributing to the decision, such as the methods used for data gathering and analysis (see Chapter 15).
- Evaluate business and technical risks, such as the consequences of unfinished work or of criteria not satisfied, or risks in work of the next phase started early.
- Demonstrate how the project is prepared, with functional resources and partners committed and aligned, to achieve the work of the next phase with fast, deliberate implementation and with a high degree of predictability.
- Verify alignment with the development portfolio and its improvement initiatives.
- Set clear expectations for future work.
  - Clarify the objectives and deliverables for the next phase.
  - Identify specific actions and milestones necessary to make future work more predictable and of higher value both to customers and to the company. Assign actions to gatekeepers, if appropriate.
  - Approve high-level action plans for the next phase, including short-term actions to reduce current risks or to resolve conditions imposed by the gate decision.

Some companies focus the review agenda just on questions and answers derived from the prerequisite reading of advance documentation. In this case there are no presentations. Often this is not practical, since data gathering and analysis can be ongoing up to meeting time. Other companies depend entirely on reacting to evidence presented during the review. There's no best way to do it. Local management style probably rules, as long as the end results are value-adding, decisive, and efficient.

Traffic-Light Forms

One way to focus the agenda on those topics that represent risks is to use a checklist of deliverables with assessments against their acceptance criteria. The "traffic-light" form, illustrated partially in Table 6.5, is developed by the project leadership team to identify weaknesses that deserve attention. The question asked is not whether or not a deliverable has been completed. The acceptance criteria are derived from the objectives of the deliverable. Remember that, depending on its consequences, a single shortfall can jeopardize the project.

The criteria are reasonably straightforward and familiar:

Green = The acceptance criteria are satisfied and the work of the next phase can proceed without risk.

Yellow = The acceptance criteria are not completely satisfied. The work of the next phase is handicapped, but the risks are acceptable. Corrective actions should be completed with near-term target dates and a high level of predictability.

Table 6.5   Traffic-light forms help to focus discussions on the most critical risks. Suppose this was the situation at Gate 2. Would you approve this recommendation?

Deliverables	Owner	Achievement versus Criteria			Project Recommendation			Remarks about Risks
		G	Y	R	G	Y	R	
Complete, approve, and freeze project requirements					○	●	○	Risk will be reduced by customer acceptance testing in next phase.
Validation of superiority of value proposition		●	○	○				Value proposition has been validated by focus group.
Project requirements document		○	●	○				Requirements have not yet been accepted by marketing.
Validation of project requirements		○	○	●				Funds are not budgeted for customers to validate requirements.
System requirements document		●	○	○				Requirements deploy the assumed value proposition.
Subsystem requirements documents		●	○	○				Requirements deploy the assumed value proposition.

Red = The acceptance criteria are not satisfied. The work of the next phase is handicapped to the point that the risks of proceeding are not acceptable. Additional work with uncertain timelines is required to reduce the risks to an acceptable level.

Action Items for Gatekeepers

A gate review is also an important opportunity for management to enable the project leadership team further. To improve the probability of success of the project, it may be very appropriate for management to take action items, such as

- Provide specific resources or funding to recover the critical path or to reduce project risks.
- Maintain sufficient capacity and clear priorities at sources of queues.
- Resolve conflicts over resource allocations or project priorities.

- Approve major purchase orders when needed by the critical path, such as for tooling, facility upgrades, or long-lead production parts.
- Participate in the redesign of the project, if necessary.
- Foster the reengineering of processes to increase efficiencies or to reduce wastes.
- Provide guidance derived from lessons learned by previous projects.
- Delegate decisions to the appropriate level of knowledge and responsibility.
- Make expectations specific and clear prior to the work.
- Recognize systemic problems with the process and correct them quickly.

Resources and funding are usually high-priority topics. Are they viewed as constraints or, more important, as investments? Contrary to typical management incentives, larger financial gains can be achieved by getting products to market on time with higher quality than by complying with budgets for development funding. Unfortunately, individual performance expectations for managers foster the opposite behavior. A manager's reserve of discretionary funding is one way to overcome this conflict. If the product isn't right, customer acceptance, revenues, competitive advantages, market reputation, and so forth can suffer. If the product is late to market, the costs of delay can include those just mentioned, as well as extra development costs, lost window of opportunity, and many others specific to the project. These costs tend to far outweigh the additional development costs to get the right product to market at the right time.

Gatekeepers

Development projects should have a governance body composed of specific managers committed to the life of the project. These leaders serve in this role in a routine and consistent way. Who are these "gatekeepers"? Usually they are senior managers who act as a small decision-making body for all gate reviews, for all development projects in their business unit. Because they are responsible for the portfolio of development projects, you can think of them as investment bankers. Although some may be functional managers, collectively they are worried more about how well the business elements of development are progressing than the specific technical nuances. There may be a challenge for development engineers to provide presentations in the language of management, free of unfamiliar acronyms or jargon.

Gatekeepers can be

- Stakeholders, accountable for the success of the portfolio of development projects
- Linked to higher levels of governance bodies within the company
- Knowledgeable of the business, its markets, and its technologies
- External to the project, but internal to the business
- Interested in the project and in guiding it to be successful
- Able to understand the metrics, ask better questions, and recognize good answers
- Familiar with best practices, as well as their implications
- Owners of resources that can enable a project's progress to be more predictable
- Possessors of position authority to remove organizational or bureaucratic barriers and to resolve cross-project conflicts

Expectations of Gatekeepers

To emphasize their role, some companies call these managers "key enablers." Similar to the commitment of a project leadership team, members of this governance committee are stable for the project and participate with commitment to the business. The project's success is part of their balanced scorecard. Some companies also include expertise external to the business unit to act as a corporate conscience or "devil's advocate." Since a gate review without the right body of gatekeepers would be flawed, we suggest that you establish a defined minimum set of gatekeepers as a quorum required for the review and its decision process.

As with the core team, there are expectations that the process places on these managers:

- Ensure business success.
  - Promote understanding of the business of the company and that of its customers.
  - Ensure alignment with business and technical strategies.
  - Set high, yet achievable, expectations for excellence in execution.
- Enable the progress of all selected projects.
  - Provide the skills and funding required for the project's success, when required.
  - Remove excessive constraints and unnecessary bureaucratic barriers from the project's path.
  - Maintain focus and a constancy of purpose.
- Be decision makers.
  - Exercise independence in the judgment of recommendations.
  - Ensure the integrity of the data that influence decisions.
  - Work for the success of the whole business.
  - Manage the follow-up implementation of decisions.
- Own the product development process.
  - Reinforce the discipline and expectations of the process.
  - Encourage flexible process adaptations to improve effectiveness and efficiency.
  - Institutionalize process improvements to sustain its competitive advantages.

Key Points

1. Decision making is a structured process that is critical to the technical and business success of new product development.
2. Projects are driven by many decisions made at the right organization levels throughout the development phases. Their acceptance criteria and intended consequences must be relevant to the objectives of the phase of development.
3. Empowerment for a decision requires clear and relevant responsibility, accountability, and authority.
4. Decisions are enabled by tasks that employ the best tools and processes to deliver data with high integrity at the right time.
5. Decisions require options among which the best path forward is chosen. However, they are ineffective without commitment to their follow-up implementation.
6. These processes set high expectations for the leadership team and work groups within the development project, as well as for the management team acting as a governance body for the project.

Discussion Questions

1. What difficulties do you have getting a quorum of decision makers to gate reviews?
2. Are your projects' gate reviews on the critical path? What would happen if they were not?
3. What tactics have you found to be effective in getting managers to understand the objectives of reviews and to add appropriate value?
4. How well do your managers accept action items and follow up on them to enable projects to have a higher probability of success?
5. How clear are the decision responsibilities for your projects?
6. How well empowered are your project teams? What does that mean in their daily work?
7. What behaviors create difficulties for your projects' reviews being value-adding?
8. How clear are the objectives for key decisions in your projects?
9. How well can your projects speak "truth to power"?
10. How well do your peer reviews bring collective wisdom to your project when needed?
11. What are the objectives of your design reviews?
12. Can you see how your weaker projects could have been canceled earlier?
13. How well do your decisions consider available alternatives objectively?
14. What decisions in past projects could have enabled reliability to be increased at a faster rate?
15. How standardized are your decision processes?

Robustness Development
for Product Designs

Methods for the development of product robustness are essential steps to increase the quality and reliability in new designs with reduced risks for their integration, manufacturing, and service. To the extent that engineering teams apply the best practices of Robust Design[1] in the early development phases, they prevent failures in later phases. This problem prevention strategy contributes to reduced cycle times and related project costs, making them more predictable. Robustness development applied to manufacturing can reduce the variability of those processes with valuable benefits to manufacturing yields, cycle times, and costs. Since the methods optimize designs rather than requiring tighter tolerances or higher cost components, increases in manufacturing costs can be avoided. Robust Design methods applied to the development of new technologies for products or for manufacturing processes prepare those capabilities to be more easily integrated and commercialized for a wider range of applications.

This chapter is intended for those of you who manage or oversee development projects, helping you to have value-adding conversations with practitioners of robustness development. We cover the fundamental principles and practices of robustness development, building upon the descriptions of performance variation and quality loss in Chapter 2. In Chapters 17, 18, and 19 we describe the experiment methodologies for those of you interested in learning those details.

Technology and Product Development

Product development is a process of creating new or improved solutions for customers. Customers judge the results. In addition to the features, functionality, and nominal performance, customers evaluate how well the capabilities remain stable over the range of applications and environments, regardless of how harsh those situations are. In the language of robustness

---

1. The best practices of Robust Design were made practical by the methods of quality engineering developed by Dr. Genichi Taguchi. The references in this chapter describe these methods in detail.

development, that can be stated as how well the new product's performance remains "on target, with minimum variability" under stressful conditions.

Among the deliverables of product development, from the viewpoint of your business, are the specifications of product designs that are expected to

- Provide functionality that delivers superior value to customers and that implements the selected value proposition
- Achieve performance that will satisfy its various requirements (customer-driven, externally driven, and internally driven)
- Be reasonably insensitive to stressful factors that cannot be controlled in its operation, or that you choose not to control
- Be replicated by economic and efficient processes for manufacturing, testing, distribution, installation, maintenance, and repair and can be tolerant of variations in those processes
- Be superior to their competition in the market, as judged by feedback from customers and competitive benchmarking

Robustness development improves many of those attributes. The benefits are greater when robustness development is applied during the phases of technology development and in the early phases of product development. The earlier that quality and reliability can be achieved, the less vulnerable is the project to the higher costs and delays caused by late design changes.

Technology Development[2]

Technology development has two objectives relevant to product development. One is to develop the robustness of a technical concept for a range of future market applications. Those applications are derived from long-range plans in the portfolio of product development projects.

An example is the fusing of toner to paper for an electrophotographic printer. The increasing of system speed represents the range of applications in future products. Customers' requirements for the bonding of toner to paper are independent of speed. However, without robustness the functional performance of the fusing function deteriorates, as illustrated in Figure 7.1, with reduced mean and increased variability in the distribution of fusing quality.

This is a two-step optimization process across the range of applications. For each application, first specify controllable parameters within the design concept that can reduce the variability. Then specify parameters that adjust the mean back to the customers' requirement. If those design parameters cannot be found for future applications, or are no longer effective enough, the design concept is not appropriate for those applications and should be replaced with a more controllable concept.

A second objective of technology development is to study stresses (aka noises). Analyses and experiments identify stressful conditions that are effective in engineering experiments. Their purpose is to excite failure modes deliberately by forcing extreme variations in the functions being developed. These stressful conditions would be either representative of those expected in the market or surrogates for expected conditions. They may work independently or have

---

2. Don Clausing and Victor Fey, *Effective Innovation: The Development of Winning Technologies* (New York: Professional Engineering Publishing, 2004).

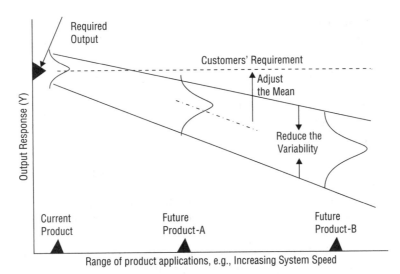

Figure 7.1 During technology development, robustness is developed for a range of applications, as represented by system speed.

cumulative effects as "compound stresses," such as high temperature and humidity representing high moisture levels in the air.

Product Development Phases 1–2

During the early phases of product development the objective is to optimize the functional parameters that control the performance, robustness, and integration of the design concepts. These "critical functional parameters" are documented and under change control. In Phase 1 the focus is on subsystem robustness, and in Phase 2 development teams integrate the subsystems and refine the parameters to optimize full system performance and robustness.

Product Development Phases 3–5

During product design, in Phase 3, the engineering activities convert the functional parameters into design parameters with set points and tolerances. Here the focus is on refining configurations and their specifications, not only to satisfy the product requirements but also to enable efficient manufacturing and service. The CAD files and other forms of design documentation provide these parameters to manufacturing, with production-oriented change control and product data management (PDM).

Here's an example of the difference between a functional parameter and a design parameter. In early development experiments, a functional parameter might be a force applied within a subsystem. The objective would be to specify the force, without necessarily defining how that force would be implemented in the production-intent design. During the product design phase, the mechanism to implement the force would be designed in the form of springs, levers, pneumatics, or other suitable devices.

During the preparations for production, in parallel with the refinement of the product design, engineering teams develop the manufacturing and service capabilities to reproduce and maintain the design parameters. As for the product, they specify controllable process parameters that enable those processes to be robust.

Functional Diagram

In general, a product performs many functions. Based on the product architecture, the system-level functions can be decomposed into many sub-functions, illustrated previously in Figure 2.6. Each basic function can then be described by a unique parameter diagram,[3] illustrated by Figure 7.2. The design concepts that are selected have the job of accomplishing the basic functions that are then integrated to achieve the system-level functions.

The parameter diagram illustrates several important elements that are fundamental to our discussion of robustness development:

- The basic function is part of a system of functions that comprise the product.
- It has an input signal that may be static or dynamic and may have variability ("noise").
- The basic function transforms an input signal into an output response.
- The output response is directly related to the physics of the basic function.
- The output response has a requirement, derived from the full system product requirements.
- There are stressful factors that are not controlled by the design but that cause undesirable variations in the functional response of the design.
- There are critical parameters within the design concept that are found to be effective in controlling the output response of the function. Their purpose is to
  - Reduce the sensitivity of the function to the stresses imposed on it

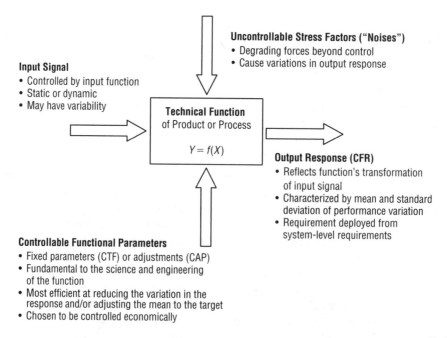

Figure 7.2   The parameter diagram illustrates those parameters that affect the mean and the variations in the response of a function.

---

3. Phadke, *Quality Engineering Using Robust Design.*

- Adjust the mean of the output response to its required target
- Or both

In the early development phases, the critical functional parameters should be adjustable so that their set points can be changed easily and their design space explored. In the production design, decisions must be made to either leave a parameter as an adjustment—a "critical adjustment parameter" (CAP)—or have it be fixed as a "critical-to-function" (CTF) parameter created by tooling or some other element of the manufacturing process.

The parameter diagram is an important model to keep in mind. It helps you to visualize performance situations and to focus on the basic functions of the solution to be developed. A design concept may have several functions to perform, each one described by a unique parameter diagram. Failure modes can then be analyzed in terms of cause-and-effect relationships among controllable parameters and stress factors. Asking again the definition of the basic function of a device can clarify the important attributes of its design concept and the requirement for the function. It simplifies your discussions and sets the framework for thoughtful experiments.

Stressful Conditions

The language of Taguchi Methods uses the term *noises* instead of *stresses* for the degrading factors. Taguchi Methods also use the term *noise* to refer to the variations in both the input signal and the output response of a function, in the sense that an output of one function is an input to another function. This may be a source of some confusion, but it is familiar in the context of electrical or audio noises being unwanted variations in the output of speakers and communications products, or electromagnetic interference that deteriorates performance. For the purposes of this chapter, we're using the more familiar term *stress* to refer to a degrading force that acts on a function and causes variations in its output response. We hope that will provide some clarity in our explanations.

The performance of a product is subjected to stresses from three fundamental sources:

1. Among products in production: Degrading factors in manufacturing, assembly, distribution, installation, and service contribute to differences among units of the same product. They also reduce process yields and out-of-box quality. Examples of these stresses can include
   - Controllable parameters reproduced off target for adjustments, part dimensions, material properties, and component characteristics due to degrading conditions within the production operations or those of suppliers
   - Mechanical wear in tooling, increasing the variability in part dimensions and shifting the mean off target
   - Environmental factors in production, such as contamination or varying temperature or humidity
   - Infant mortality problems with purchased components
   - Errors in human procedures, such as those used in assembly, handling, and service
   - Inadequate process controls in manufacturing
   - Inadequate attention to preventive maintenance of the product

2. Internal to each product: Factors internal to the product degrade performance during its use. In the context of the system architecture, teams can map damaging interactions among subsystems (a "noise map"[4]) to determine how to reduce their effects. Examples can include
   - Unintended variations in the input signal from an upstream function, transmitted to the variations of the output response for the function of interest
   - Interactions of subsystem failures or the cumulative effects of variations across subsystem interfaces
   - Material wear, distortions, corrosion, and fatigue
   - Distributed contamination, such as dust, oils, acids, heat, moisture, electromagnetic interference, vibrations, shocks, stray light
3. External in the use of the product: Factors external to the product can degrade its performance because of shipping, handling, storage, installation, use, and service. Examples can include
   - Operator misuse and abuse; inadequate training
   - Extreme usage conditions or applications, such as from loads, job streams, consumable materials, duty cycles, operation time
   - Extremes or variations in ambient environments, such as temperature, humidity, atmospheric pressure, water depth, vibration, power sources
   - Contamination from external dirt, liquids, particles, vapors, stray light, electromagnetic interference
   - Damaging environments in shipping, handling, storage and display, such as temperature, humidity, atmospheric pressure, vibration, shock

These examples of stressful factors present three basic challenges to development projects:

1. The identification of those stresses that have the highest expected consequences for the performance of the new product
2. The selection of the most efficient strategy for dealing with those stresses
3. The design of methods to introduce the most significant stresses into engineering development, either analytically or empirically

Expected Consequences of Stresses

Certainly not all stresses are important, and certainly not all products have the same vulnerability. As an example, gasoline-powered vehicles may be vulnerable to the variations in fuel quality found across global markets, if that is their application. Off-road vehicles encounter vibrations, shocks, dust, temperatures, and other conditions not expected for passenger cars. The tractors for 18-wheelers experience loads that impose much more severe stresses for acceleration and braking than do smaller, local delivery trucks. It is best to study the stresses from these applications prior to the beginning of your development project to avoid delays by last-minute efforts to identify these factors.

---

4. Creveling et al., *Design for Six Sigma in Technology and Product Development.*

Effective team methods, such as the cause-and-effect diagram, FTA, and FMECA, provide excellent ways for people to pool their wisdom and experiences about potential root causes of problems. Experiments can then prioritize the causes, thereby making the development process more efficient.

Strategies for Handling Stress

In the context of your product systems, your development teams may have several ways to handle the potential consequences of stressful conditions. Here are some common ones. You may think of other examples for your types of products.

1. Eliminate or control the source of the stress.

   This can be very expensive and can limit the application of the product. An example is in the processing of photographic film or paper. Light sources over most of the spectrum are not tolerated by the photosensitive emulsions, so the film and paper are processed in the dark or under safe light conditions. Darkrooms can be expensive and limiting, but processing mini-labs enabled this work to be distributed to pharmacies and discount stores. However, the strategy for the mini-lab was not to eliminate the sources of the light, but to create a design architecture that is not vulnerable to it. A more extreme example is the domed stadium that some teams use for American football. That strategy eliminates the rain, snow, and wind, but at what a cost! Not only are the structures expensive, but natural grass is eliminated from the game conditions, the players have to worry about new types of injuries, and the spectators may swelter in the heat if the stadium is not air-conditioned. Whether or not the benefits are worth the costs is a trade-off for management.

2. Inspect the product prior to shipment.

   The inspection of manufactured products can contribute to preventing quality defects from escaping the factory, although it is not entirely foolproof. What happens to a unit that does not pass the inspection? Is it scrapped, reworked, or sent to a more tolerant market? Inspection is expensive and its necessity acknowledges that the processes do not create sufficient or consistent quality. Quality cannot be inspected into a product; it has to be designed in.

3. Depend on additional service, preventive maintenance, or customer support to compensate for manufacturing variability.

   This strategy can be very expensive and can be perceived as a desperate tactic to get a flawed product into the market prematurely. A decision to continually repair escaping problems that are inherent in the design can drive up service costs and contaminate customers' perception of the product's quality. A decision to depend on additional training of customers may modify customers' behaviors and procedures to compensate for a flawed design, but it may place your product at a disadvantage in side-by-side comparisons with competitive products.

4. Depend on feedback from customers' use to define the "real" problems in need of correction.

   This strategy can address problems demonstrated to have higher frequencies of occurrence and more serious consequences for customers. However, it delays the time for proper solutions to reach customers, leaving them suffering the quality loss while waiting. With hardware solutions, the modification of products in production and those already placed with customers can be very expensive. This may not be so much the case with software products, which can be upgraded more easily. In some cases, customers may be offended not only by having to wait for improved performance for which they have already paid, but also by being treated as a testing function after market entry. In other cases, such as software products, customers may expect upgrades after purchase.

5. Incorporate feedback or feed-forward control systems to compensate for the consequences that stresses have on the functional performance.

   The instrumentation and control systems add costs and complexity to the design, as well as additional sources of failure. A cynical view of a product that is advertised as employing feedback control is that the product is not inherently stable and requires those compensating systems to maintain its performance. However, if robustness development has been exhausted, the additional design content may be justified as a method to extend usage life further. You'd want to be certain that the added manufacturing cost will provide noticeably higher reliability and large reductions in service costs. Our discussion of the quality loss function in Chapter 2 provided the criterion that the increase in manufacturing costs should be less than the savings in quality loss.

6. Incorporate redundant features or increased safety margins.

   This is a brute-force approach that may be acceptable in many circumstances. However, in a design that has challenges to be efficient in its use of resources and to be capable in its achievement of quality, it's a waste of costs. Additional features may also be additional sources of failure.

7. Optimize the design by developing it to be robust.

   This strategy is the application of good engineering practices, including Robust Design. Developing and integrating the designs so that the full system performance is on target with less variability under stressful laboratory conditions means the product system will be less vulnerable to whatever stressful conditions are encountered in actual use. This is economically prudent because money is not spent on the elimination of the sources of the stresses or on additional features to compensate for their consequences. In its competition with other products, a design that is more robust than its competitors in the same stressful factory tests is expected to be more robust in the market under customers' use conditions.

Develop Methods to Introduce Experimental Stresses

We mentioned that one of the activities of technology development is the study of stresses. For laboratory experiments, not all stresses need to be introduced. Stresses having the largest effect on product performance can be identified by screening experiments. This will reduce the complexity and costs of the subsequent robustness experiments. When it is known which stresses, or combination of stresses, are most effective at forcing performance variations, test plans can introduce the critical few. When it is known how easily applied stresses can be surrogates for others that may be more difficult to apply, test designs can be more efficient. The consequences of those factors may represent their direct effects on a material, such as its electrical properties. For example, temperature and humidity may cause changes in material dimensions, just as do variations in manufacturing processes.

During development, extreme variations in experimental temperatures, humidity, and atmospheric pressure can be achieved with environmental test chambers. A paper-handling device may be vulnerable to static when the moisture levels are low, or to the loss of paper stiffness when the moisture levels are high. Devices that depend on the properties of air can be vulnerable to altitude changes. A range of consumable materials can provide artificial stresses. Certain types of fuel may be more stressful for an engine. For a printer, if papers with extreme properties can be handled without jamming, more normal papers should be handled reliably. A particular job stream or application might be designed to exercise a system more severely than would be expected under normal operation.

It is not possible to duplicate accurately the frequency, intensity, or combinations of stresses that a product will experience in the market. For the development of a product, this does not need to be a concern. The objective is to improve the designs and to make them better than their competitive designs. To that end, the stressful test conditions need to be extreme enough to excite each failure mode that is inherent in the design and to force performance variations in each critical function. The tests need to be measurable and repeatable to enable the comparisons of alternative designs. However, they should not be so extreme that they force new failure modes that are not representative of the intended applications. The objective is not to demonstrate current reliability or to predict the reliability after market entry. Those are the objectives of reliability demonstration tests, with the concern remaining about the relevance and accuracy of stresses imposed.

Parallel Development of Products and Processes

The basic concept of concurrent engineering is the parallel development of product designs and their manufacturing processes. The teamwork and guidelines that do this efficiently have been promoted for decades. They focus on best practices such as the collaboration of manufacturing engineers with design teams, the early involvement by production suppliers, design guidelines for manufacturing and assembly (DFMA), and design prototypes being built by people and methods that are transferable to those to be used in production. The principles apply to your suppliers of components and materials as well as to your internal manufacturing and assembly operations.

Suppliers' Involvement with Product Development Teams

When you invite key suppliers to be involved in early development discussions, what do you talk about? Certainly the topics of manufacturability and costs are important, as are lead times, production capacities, the availability of raw materials, and other mutual concerns. More important, particularly in the early phases, suppliers may have very good ideas about design concepts. They may be making similar components for your competitors or for your own existing products. That enables them to provide valuable suggestions toward identifying concepts and technologies that are both superior and achievable within their process capabilities.

Product Development Teams' Involvement with Suppliers

In the early development phases, manufacturing costs are probably not as important a topic as are the capabilities to reproduce a design concept. If a concept has design parameters that are expected to be difficult to manufacture with existing processes, process development may be necessary. The development of improved manufacturing capabilities would be the responsibility of your suppliers, but they may need help from your own experts in robustness development. They help you and you help them, to your mutual benefit. An improved process capability (Cpk) is aimed at both higher quality and lower costs in manufacturing.

This mutual involvement can be particularly beneficial for suppliers of manufactured materials. Those materials perform functions within your product system, but how well are their design parameters known or specified? It may be that the material's behavior depends on particular equipment with which it is made, or on a raw material whose own control parameters are not well defined or controlled. We expect that you may have experience with suppliers who have changed an ingredient to save costs or to reestablish a dependable supply chain and inadvertently changed the behavior of the material in your product. That's very difficult because the critical parameter that was actually changed may never have been identified. You might say that it was a "lurking parameter" that affected performance but whose role was never understood.

An example that we've heard about is in the production of a particular single-malt Scotch. It had a unique flavor that gave it a competitive advantage. A dispute among the owners led to the start-up of a competing distillery in a different location. However, that new operation was never able to replicate that differentiating flavor. Why? The recipe was essentially the same. It turned out that critical parameters for the process were in its building and equipment being located next to the sea and thereby being saturated with moisture from the salt water. The new building and equipment, located inland, did not have those parameters. They were doomed by the absence of the "lurking parameters."

You may experience similar problems when scaling up a manufacturing process. Smaller equipment or slower processes may have contributing parameters that are not the same in larger or faster equipment. You will then be faced with the need to optimize the process again without jeopardizing production schedules. Good luck!

Manufacturing Process Capability

How do you measure the capability of a manufacturing process? It's all about the ability to replicate a required parameter consistently. Figure 7.3 illustrates the strategy. Process development, to align the mean and reduce the variability of manufacturing for a critical design parameter, parallels the work to increase the latitude in the design for those variations. When placed under control in production, the resulting alignment and variability, compared to the design latitude, is the metric described as the "manufacturing process capability" (Cpk) in Six Sigma practices. It's defined as

$$Cpk = \frac{Minimum\left\{(USL - \mu),(\mu - LSL)\right\}}{3\sigma} \tag{7.1}$$

USL and LSL are the upper and lower specification limits for the design parameter, its latitude for manufacturing variation. The distribution of the parameter's manufacturing variations has a mean (μ) and a standard deviation (σ). The smaller of the two values in the equation is

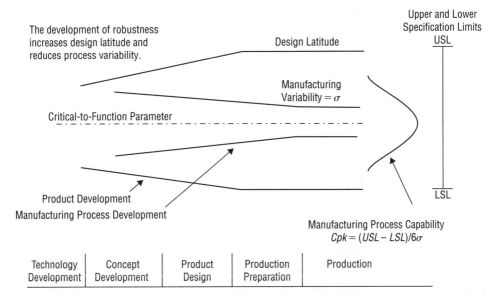

Figure 7.3   The improvement of manufacturing capabilities involves both product development and process development.

the process capability index. The index accounts for both the variability in the process and the position of its mean relative to the tolerance limits.

In Figure 7.3 you can see a simpler form of the capability index:

$$Cpk = \frac{(USL - LSL)}{6\sigma} \tag{7.2}$$

This expression assumes that the mean of the distribution is aligned with the target, which might be the case initially. Over time, manufacturing stresses can cause the mean to drift and the variability to increase, a situation reflected by a decreasing Cpk. For example, the internal dimensions of a mold for a plastic part may increase because of mechanical wear. In addition, there can be dimensional variations due to changes in material properties, process temperatures, pressures, and other root causes.

Figure 7.3 illustrates that the demonstrated or forecasted Cpk can have an acceptance criterion that increases during development to the start of production. When quality problems and the need for process corrections are reduced, an increased robustness in manufacturing contributes to higher out-of-box quality for the product, reduced unit-to-unit variations, reduced need for process control and problem solving, higher manufacturing yields, shorter manufacturing cycle times, and lower manufacturing costs. Those benefits sound pretty good! Higher production quality does not have to require higher manufacturing costs.

The resulting distribution of a part dimension would then be described by its standard deviation (σ) and mean (μ), which may drift off target, as illustrated in Figure 7.4. The work to develop the robustness of the manufacturing process would identify those parameters to change

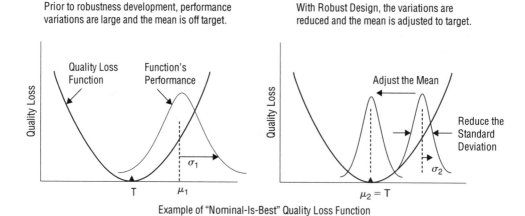

Example of "Nominal-Is-Best" Quality Loss Function

Figure 7.4   The improvement of robustness reduces the quality loss. In manufacturing it increases the process capability index.

in order to reduce the variations and return the mean to its target requirement, the two-step optimization mentioned earlier.

Robustness Development

Another goal of robustness development is the improvement of the reliability of the product when used by customers. That increases the usage interval before repairs are needed by making the design less vulnerable to the other two categories of stresses:

1. Stresses from degradations within the product's configuration during its use
2. Stresses imposed upon the product from external sources

Figure 7.5 illustrates the advantage to cycle time via the reliability growth curve described in Chapter 3. Robustness achieved during technology development and the early phases of product development provides a higher initial reliability. Robustness development methods during product development contribute to faster problem solving and thereby a higher rate of growth for reliability, with the potential of enabling an earlier market entry.

During product development, the reliability of a design may be less than adequate in the early project phases. It's an attribute that is improved through engineering processes. The same can be said for manufacturing process capabilities. Time and resources usually are in short supply, so the rate of progress in development needs to be high to meet their requirements prior to market entry. Chapter 3 offers many suggestions for increasing the growth rate.

Traditional Product Development

At a high level of description, product development is a logical sequence of steps, although they are usually iterative and overlapping:

1. Define and clarify product requirements.
2. Select the best available product architecture.
3. Decompose the functions of the full system into layers and networks of sub-functions.

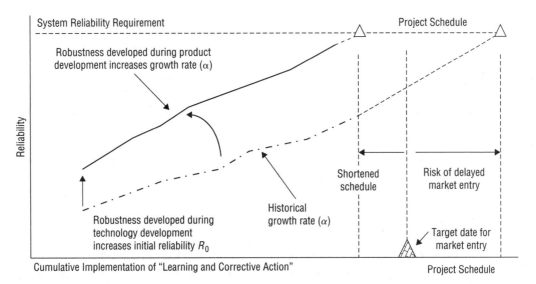

Figure 7.5 The methods of robustness development are effective at increasing both the initial reliability and the rate of growth for reliability, contributing to a shorter project cycle time.

4. Decompose the system requirements to the level of the sub-functions.
5. Select the best design concepts and enabling technologies to achieve those functions.
6. Develop analytical, empirical, and physical models of those functions.
7. Apply scientific knowledge and engineering best practices to develop the subsystems.
8. Identify and specify critical design parameters to control those functions.
9. Integrate subsystems into the full system product.
10. Optimize the design parameters to achieve the required full system performance.
11. Test the full system product to identify overlooked problems.
12. Manage corrective actions through iterative problem solving and implementation.
13. Verify that the product designs satisfy their requirements.

Scientific and engineering methods are familiar practices that emphasize the development and evaluations of analytical models, CAD models, virtual and physical prototypes, and other ways to represent the design concepts. Many good practices have been integrated into the engineering tool kit, such as those derived from systems engineering, quality engineering, reliability engineering, statistical analysis, and other sources of design guidance. Those practices are very valuable and should be used to their best advantage prior to incorporating the methods of Robust Design. Traditional approaches work well for exploring known design space. Robust Design adds value in exploring the unknown design space.

Robustness Development

The methods of Robust Design provide a simple but powerful set of tools that build upon the traditional development methods. The principles have many nuances to appreciate:

- **Fundamental function:** The parameter diagram, illustrated in Figure 7.2, places emphasis on defining the basic functions within a design concept, not just its parts and assemblies. It requires definitions of the basic function, its input signals and output responses, and those controllable parameters and noncontrollable factors that contribute most strongly to the behavior of the function. The requirements for the output response are decomposed from the system-level requirements through a process such as QFD, described later in Chapter 10.

- **Superior solution:** Robustness development emphasizes the achievement of a better solution. This does not mean that it is essential to have a deep understanding of the scientific relationships among root causes and functional misbehaviors. Technology development has the charter and time to do that. However, time and cost pressures on product development force an emphasis on preventing or fixing problems fast in ways that are repeatable and reproducible. It's not essential to understand deeply why a solution works. The object is to improve the design and move on to the next problem.

- **Excite failure modes deliberately:** Robust Design methods deliberately introduce stressful conditions to excite failure modes that are inherent in the design. The measurable variations in a function's output response enable problems to be found sooner and solutions proven to be better than previous designs.

- **Stressful conditions:** Robustness development is not concerned about the test conditions being an accurate reflection of degrading forces expected in customers' applications. The role of the stresses is to excite failure modes deliberately. When the stresses are the same within an experiment, the alternatives that are tested can be compared. As long as the stresses are extreme enough to force variations, solutions developed to be less vulnerable to those extreme stresses should be less vulnerable to whatever stresses are experienced in customers' applications.

- **Explore unknown relationships:** Analytical modeling can contribute well to design optimization as long as the model is an accurate representation of the function and of the ways in which the controllable parameters and noncontrollable factors act upon it. However, in many cases these relationships are not known very well. The methods of Robust Design are well suited to explore the unknown behaviors and develop empirical models, guided by good engineering judgment about potential cause-and-effect relationships.

- **Use of designed experiments:**
  Orthogonal arrays: Taguchi's methods are based on using orthogonal arrays as the experiment design of choice. Orthogonal arrays are Resolution III experiments in which main effects are aliased with two-way interactions.

  Response surface designs: Response surface designs are available as central composite, Box-Behnken, and optimal designs. They can be used to develop empirical models having unaliased two-way interactions.

  Orthogonal arrays and response surface designs are discussed in more detail in Chapters 17, 18, and 19.

- **Better concepts:** The strategy of robustness development is focused on identifying and specifying those controllable parameters that are most effective at controlling the mean and variability of the output response. If they cannot be found, the design concept is flawed and will never be optimized. Determining that very early in the development process enables a better concept to replace the flawed one.

- **Technology development:** If a better concept and its enabling technologies can be developed to be robust during technology development, prior to the beginning of a product development project, the cycle time and risks for product development can be reduced.

- **Metrics for minimizing variability:** Signal-to-noise ratio (S/N) is normally used when applying Taguchi's methods. The S/N has several forms, depending on whether you want the response to have a target value or to be maximized or minimized. Chapter 19 discusses the S/N in more detail.

  Variance is generally a response that is co-optimized with the mean after you have the transfer function from a response surface experiment. Usually the objective is to put the mean response on target while minimizing the variance. We show how to co-optimize the mean and variance in Chapter 18. Another optimization strategy is to maximize Cpk, which will minimize defects per unit (DPU).

- **Robust subsystem interfaces:** A key attribute of a product's architecture is the robustness of the interfaces among its subsystems. The output responses of one subsystem are inputs to the next subsystem in the functional flow. If those responses have little variability and if there are no extraneous stresses being "exported" downstream, no significant stresses cross the interface. The interface is robust, with some important implications. To the extent that there are no stressful interactions, inadequate subsystems can be replaced by superior ones without jeopardizing the system integration. This enables the benefit of set-based design, described in Chapter 5. Likewise, subsystem replacements in production or service can be made with fewer complications. Subsystems with robust interfaces require less development effort for system integration, since there are no degrading interactions among the subsystems that have to be corrected in later phases. To the extent that the responsibilities of a work group are aligned to the definitions of subsystems, the risks will be within the work of the team rather than between teams.

- **Additivity:** A design concept whose critical functional parameters either are independent or interact constructively is much easier to implement in production. The effects of the parameters add to each other, rather than counteract each other's effects on the response. In an empirical model there would be no complicated cross terms. Smaller experiments can be designed since interactions can be neglected. This property, called "additivity," is a characteristic sought in a design concept. It allows one parameter to be changed without requiring a companion change to another. This also enables the benefits of set-based design. If additivity is seriously lacking, the possible causes are control factor and/or response factor choices. Worst of all, the architecture of the concept may be flawed and in need of redesign.

- **Predictive empirical model:** If interactions in a system are small compared to the main effects, the analysis of the data from an orthogonal array experiment provides an efficient model for Robust Design. The model enables the development team to select the best set points for those parameters it chooses to control. Subsequent confirmation tests, using the optimal set points, should demonstrate the performance and S/N that are predicted by the model. If the prediction and results of the confirmation experiment differ significantly, the process may have strong interactions, and additional development, or another experiment design, is needed. To be clear, the objective of this model is not to predict field reliability. Its construct is focused on selecting the best set points for parameters and on predicting the resulting performance and variability under the test conditions. Systems having significant interactions that cannot be reduced or eliminated are handled better by an experiment design from the response surface family, combined with classical or stochastic optimization.

- **Parametric relationships:** The tactics to reduce the sensitivity to stress depend on relationships that are found to exist among the input signal, the controllable parameters, the uncontrollable stressful factors, and the output response. In particular, the strategy to reduce sensitivity depends on nonlinearity between certain controllable parameters and the output response, as illustrated in Scenario 1 in the next section. Choosing set points for the controllable parameters where the functional response is less sensitive to the sources of variation reduces the variability in the output response. Parameters that have linear relationships with the response are effective at adjusting the mean.

Chapter 18 covers the use of traditional DOE tools such as response surface methods to improve robustness. In Chapter 19 we cover Taguchi's methods and how they fit with the goal of improving system or product robustness.

The following scenarios of robustness development may clarify these key points.

Examples of Robustness Development

Scenario 1

Suppose that an environmental stress factor acts on design parameter A, which in turn has a nonlinear effect on the output response (Y). Figure 7.6 illustrates this scenario, showing the following:

- Control parameter A is set where the output response (Y) is less sensitive.

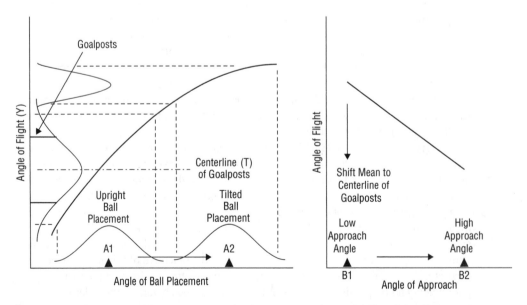

Figure 7.6   Scenario 1: Tilting the placed football reduces the variability of the flight angle.

- There's no attempt to reduce the effect of the stress on parameter A.
- Adjustment parameter B is set to shift the response back to its target (T).

In the case of a stress affecting a manufacturing process, the nonlinearity of the function would enable the tolerance on parameter A to be relaxed to reduce manufacturing costs and still have reduced variability in the output response.

Consider our friend the field goal kicker, discussed in Chapter 2. The kicking process includes the ball holder, who receives the ball from the center and places it to the best advantage of the kicker. For soccer-style kickers, that placement is with the ball laces facing away from the kicker's foot and the ball tilted slightly off vertical. The parameter in question is the angle of tilt that works best for the particular kicker. Once the holder finds the right angle to minimize the variability in the ball's flight, the kicker can adjust his approach to the ball to shift the distribution back to be aligned with the centerline of the goalposts. The critical enabler is the nonlinear relationship between the angle of the ball, when positioned by the holder, and the variability of the angle of flight.

Scenario 2

Suppose that external stresses act on the output response (Y) in an unknown way.

Experiments may identify two control parameters that interact in their effect on the output response. Figure 7.7 may help you understand this story.

- One control parameter (B) is found to have variability due to the influence of the stress.
- The second control parameter (A) is not vulnerable to the stress but is found to have an interaction with parameter B, reducing the variation caused in the output response (Y).
- Parameter A is specified with a set point that reduces the vulnerability to those variations in parameter B that are caused by the external stresses.

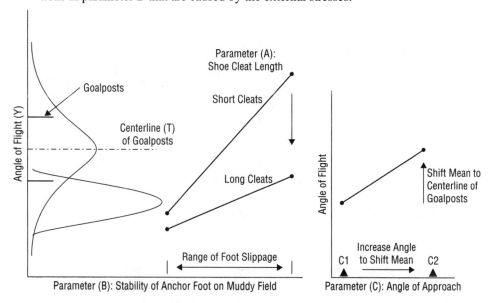

Figure 7.7 Scenario 2: Mud cleats reduce the variability of field goal kicks.

- The influence of the stress on parameter B is not reduced.
- Parameter C is set to shift the response back to its target (T).

In our field-goal-kicking example, consider the increased variability when the natural grass playing field is slippery and muddy. The anchor foot that is planted just prior to the kick needs to be firmly positioned, so a slippery footing will increase the variability of the ball's flight. By changing the cleats of his shoes to longer ones, the kicker can increase the stability of that foot and reduce the variability in the kick under slippery conditions. It may shift the mean, which he can correct by adjusting his angle of approach.

Scenario 3

External stress acts directly on the output response (Y).

- Control parameter A is found to desensitize the output response (Y) from the effects of the stress.
- Adjustment parameter B is set to shift the response back to its target (T).

In Figure 7.8, Scenario 3 illustrates another problem for field goal kicking: the wetness of the ball due to rain. The problem is more for the holder, who has to handle the ball, grab it again when it slips in his hands, and still get it placed in the right position before the kicker's foot hits the ball. The wetter the ball, the greater the variability of its flight. This scenario depicts the case

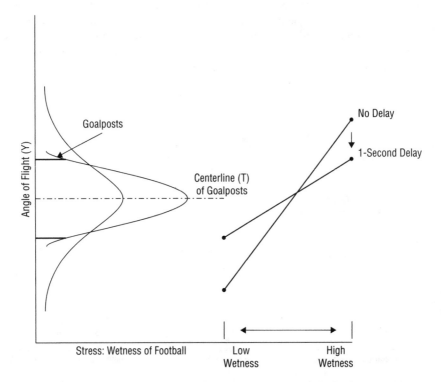

Figure 7.8   Scenario 3: A slight time delay allows time for the holder to place the ball.

for which a slight delay in the start of the kicker's move forward would give the holder a little more time to handle the ball and correct its position. There is a limit to this design window, since the defensive team is charging in to block the kick.

Robust Design

Enhancing the familiar steps of product development, the methods of Robust Design guide product development teams to perform two specific steps—parameter design and tolerance design—instead of an ad hoc design-build-test-analyze-fix process. The following are overview descriptions. More detailed explanations can be found in the works of Clausing,[5] Phadke,[6] Fowlkes,[7] Creveling,[8] Launsby,[9] and others.

Parameter Design

Parameter design can be performed with analytical or empirical models and with physical prototypes. The models have to be reasonable representations of the functional relationships, and the physical prototypes must be constructed to embody those relationships. With parameter design, development teams select and specify the critical functional parameters. In production, these parameters may be designed as fixed parameters, such as dimensions replicated by tooling or a material property set by a chemical formula. They may also be specified as adjustments available to assembly, service, and/or users. The manufacturing tolerances for these parameters are relaxed to levels easily achieved by efficient manufacturing processes. The steps of parameter design are as follows:

1. Clarify the objective of the basic function for its role in the architecture of the product.
2. Define the output response to be most relevant to the objective of the basic function. Preferably this is a continuous variable.
3. Clarify the requirement and best metric for the output response, as decomposed from the full system requirements.
4. Identify candidate functional parameters that have a high potential for influencing the response.
5. Define alternative values for each functional parameter to represent the range of the design space.
6. Select experimental stresses to force variations in the output response or to excite failure modes.
7. Instrument the test fixture to measure the output response, input signal, controllable design parameters, external stresses deliberately imposed, and internal stresses occurring naturally.

5. Clausing, *Total Quality Development.*

6. Phadke, *Quality Engineering Using Robust Design.*

7. Fowlkes, and Creveling, *Engineering Methods for Robust Product Design, Using Taguchi Methods in Technology and Product Development.*

8. Creveling et al., *Design for Six Sigma in Technology and Product Development.*

9. Stephen R. Schmidt and Robert G. Launsby, *Understanding Industrial Designed Experiments* (Longmont, CO: CQG Ltd Printing, 1989).

8. Select an experiment design that will be appropriate.
9. Follow the test procedure dictated by the chosen experiment design.
10. Analyze the effects of the parameter values on the output response.
11. Identify constraints to the design space of controllable parameters and the output response.
12. Identify which parameters are most effective at reducing the variability in the output response.
13. Identify which parameters are most effective at adjusting the mean value of the output response.
14. Evaluate the interactions among controllable parameters.
15. Evaluate the additivity that is evident in the behavior of the parameters.
16. Develop a predictive model for the functional behavior.
17. Specify the best set points for the critical functional parameters.
18. Run an experiment to confirm that the specified parameters produce the predicted performance and cause a significant increase in the S/N, or a reduction in the variance of the response if using a response surface design.
19. If the interactions or additivity is not acceptable, reconfigure the design concept. Just because strong control factor interactions can be modeled effectively and included in an experiment design does not mean they are desirable. If they can be eliminated, that is good.
20. If the predicted performance is not confirmed or the increase in S/N is not sufficient, expand the design space of controllable parameters and their potential set points.

Tolerance Design[10]

The processes of tolerance design determine for which parameters to impose tighter constraints on the upper and lower specification limits for manufacturing. Otherwise the general tolerances remain at economically efficient levels.

The setting of tolerances always involves trade-offs. A tighter tolerance might force a more expensive manufacturing process to be used or secondary operations to be included. Higher-cost materials or purchased components might be required. The quality yield of the manufacturing process might be reduced, with potential consequences for production capacities and delivery schedules. If design teams specify a tighter tolerance for a parameter, you should expect its manufacturing cost to increase.

Robustness experiments can identify those parameters with major influences over the output response. In tolerance design, experiments with selected parameters varied by small amounts can determine the effects of tightened tolerances. Where a small reduction in the variation of a parameter causes a significant reduction in the variation of the response, the increase in the S/N will show the benefit. The reduced deviation will enable the reduction in quality loss to be calculated. The acceptance criterion for the proposed constraint on manufacturing is that the resulting increase in manufacturing cost must be less than the expected decrease in quality loss.

Increasing the robustness of a manufacturing process will increase its process capability index (Cpk) and reduce its associated quality loss. The Cpk leads to a forecast of the frequency of out-of-spec parts. The quality loss places a monetary value on that benefit and helps to justify the

10. C. M. Creveling, *Tolerance Design: A Handbook for Developing Optimal Specifications* (Reading, MA: Addison-Wesley, 1997).

investment. As a caution, depending on the particular data, it's possible to have an increased Cpk but also an increased quality loss. That could be the situation of a distribution with a small standard deviation but a mean that's off target, close to the specification limit. The quality loss would then provide more important information to drive corrective actions.

We cover tolerance design in detail in Chapter 20.

Robustness Development Fixtures

Usually, physical prototypes are needed to explore the unknown behaviors of a function. To be useful for experimental development, physical models should have characteristics that enable testing, analysis, and configuration changes.

There are three fundamental types of test beds for product development:

1. Breadboards at the level of subsystems or assemblies
2. System-level development fixtures with various elements of subsystem integration
3. Full system design prototypes

At the subsystem or assembly level, breadboards are designed to test alternative design concepts. These can be employed during the phases of technology development as well as in the early phases of product development. The objective of this work is to identify and specify controllable parameters that are most influential over the subsystem performance. The experimental evidence will determine whether or not the chosen design concept will be robust enough against the expected stresses when it's integrated into the larger product system, and whether or not it is better than available alternative design concepts. These are among the acceptance criteria for the end of Phase 1.

A system-level test fixture can have a variety of forms. It can be an integration of the several subsystem breadboards that comprise the system, or it can be more product-like for those subsystems not subject to development. An example of the latter configuration is an adaptation of a current full system product attached to a subsystem under development. The modified product can be an inexpensive way to represent elements of the future system's design, particularly in ways that impose valid inputs and stresses on the subsystem being developed. You may have heard of these adapted products being called "test mules," since they support the subsystem of interest. They are particularly applicable in Phases 1 and 2. The objective for their use is the identification and specification of critical functional parameters that both integrate the subsystems well and optimize the system-level performance and robustness.

Full system engineering prototypes can also be useful for robustness development. However, the more they represent production designs, the less flexibility they have for making changes. Also, production designs may have reduced access for the detailed instrumentation required to evaluate designs. An engineering model, constructed from production-intent designs, may be useful for demonstrations of robustness. However, the ability to make changes may be limited to its designed-in adjustments. A full system prototype designed specifically for robustness development will be more applicable to Phase 2 of product development. In Phase 3 a production-intent prototype will be more focused on translating the functional parameters into configurations and specifications for manufacturing and service. It will also be suitable for reliability growth testing.

To be most useful for robustness development, a prototype or test fixture should be designed specifically to enable

- Design concepts to be changed easily
- Key factors to be measured and adjusted easily:
  - Input signals to the functions under development
  - Controllable functional parameters
- External stresses to be imposed deliberately in controllable, measurable ways
- Internal stresses from subsystem interactions to be measured
- Output responses of the function and their variations to be measured

These features are not necessarily intended to be replicated in production designs. As we mentioned earlier, certain critical functional parameters may be fixed by tooling in production designs but configured as adjustments in development fixtures. The architecture for the development prototype may enable one subsystem concept to be easily replaced by an alternative concept, while in a production design the mounting would comply more with guidelines for durability under the shock and vibration of handling or for the ease of assembly and service.

Key Points

1. Variability is the enemy of product performance and reliability. The methods of robustness development can improve the specification of those design parameters that control performance to be "on target, with minimum variability" under stressful conditions.
2. Stress is the ally of engineering. Deliberately introducing stressful conditions to analyses and experiments enables problems to be found faster and solutions proven to be better than earlier designs.
3. The design optimization process has an efficient strategy:
   a. Select the best available design concepts and technologies for the application.
   b. Develop the concepts with the current knowledge of the functional behaviors and parametric relationships.
   c. Explore unknown relationships using specialized statistically designed experiments employing stressful conditions.
   d. Identify and specify the critical functional parameters that minimize the variability caused by stresses.
   e. Tighten tolerances only where they add value, such as by reducing the quality loss more than they increase manufacturing costs.
4. Robustness achieved early in a development process is a major advantage for
   a. Increased customer satisfaction
   b. Improved product reliability and usage life
   c. Reduced risks for system integration of the product design
   d. Increased quality of the product's manufacturing and service
   e. Reduced manufacturing and service costs
   f. Increased range of operating and environmental conditions that can be tolerated, contributing to lower operating costs
   g. More predictable and reduced cycle times, with related reductions in development costs

5. Product designs that are more robust than their competitors under the same stressful laboratory conditions can be expected to have higher reliability in the market under whatever conditions are in their operating environment.

Discussion Questions

1. How well do your engineers understand how stressful conditions can improve their ability to develop robust products?
2. How well do your experimental prototypes lend themselves to easy changes in parameters and design modules?
3. How well do your development teams use effective data collection and analysis equipment to monitor functional parameters as well as input signals and output responses?
4. How much training will be necessary for your development teams to become competent with Robust Design methods?
5. How much in manufacturing costs could be saved if many parameters were specified with less-constraining tolerances?
6. How receptive is your organization to nontraditional statistical methods of designing and analyzing experiments?
7. How receptive are your managers, during development, to asking for assessments of relative improvements rather than for predictions of performance in the field?
8. How much training will be necessary for management to understand Robust Design well enough to
   a. Enable thoughtful test planning?
   b. Provide useful test and stress equipment?
   c. Ask better questions of development teams?

Reliability Growth and Testing

The development of robustness in design concepts is a major step toward achieving higher reliability in a new product. However, time can be a scarce resource. Certain failure modes may not be identified, and critical design parameters may not be optimized or implemented properly. Prioritization in development can result in the probability that some problems may not be prevented. Many problems in products under development may not be identified until prototypes of production-intent designs are subjected to rigorous testing.

During the design and production preparation phases, when the time remaining to product launch is short, it is necessary to find problems fast and correct them. The challenge is to be efficient, so management has to set clear expectations that the path to satisfying the requirements for market entry is predictable and that investments in problem correction are prioritized to get the most benefits.

The aggressive development of reliability is focused on both problem prevention and problem solving. Robustness development, the topic for Chapter 7, helps to prevent problems. Reliability growth, the topic for this chapter, depends on fixing problems quickly and thoroughly. These concepts are reflected in the title of our book. Consequently, the strategy for development should include planned activities focused on finding problems quickly so that their priorities can be judged and their correction managed.

The objective of fast and early problem identification presents strategic questions such as these:

- What test strategies should be implemented?
- How much time should be allocated for testing in each development phase?
- How can tests incorporate stresses that are effective at forcing problems to occur?
- How should test results be interpreted if the test conditions are artificially stressful?
- What will be the configurations of those prototypes constructed for testing?
- For specific tests, is the configuration fixed or subject to design upgrades?
- When will prototypes of the production-intent design be available?

Your product development process should provide those details. More important, the project management plans have to reflect the necessary resources and timelines with commitments to achieve them. Specific to our focus in this chapter, the growth of reliability to achieve its requirements prior to market entry depends upon these processes being planned, implemented, and managed.

Many critical problems must be corrected quickly and their solutions verified to be effective. For these efforts to be predictable and of sufficient capabilities, important elements of a development system should be in place and functioning well. What should such a system include? We discuss several elements of a productive system for reliability growth:

- An appreciation for the roles of analysis and testing
- A model for evaluating the growth in reliability
- Organizational capabilities that drive the reliability growth rate
- Strategies for accelerated testing to find problems faster
- An understanding of the relationship between stress and strength
- A systematic process for problem solving
- A centralized failure-reporting and corrective action system
- Criteria for differentiating problems based on their consequences for customers
- A method of reporting with metrics that are easily interpreted and linked to decisions
- A variety of tactics for testing to be incorporated into project management plans

Product Testing

Testing to identify new problems is an important element of any product development process. In spite of the best efforts to develop robust designs, as long as humans are designing and building the product, errors will creep into the process. Even if robots are building the product, humans are giving the orders in the form of software instructions. So there is almost 100% certainty that mistakes are inevitable. There is an entire body of knowledge about mistake-proofing[1] focused on manufacturing. Of course, some errors in design do escape to manufacturing.

Some organizations have suffered a backlash against testing. Management, tired of the build-test-fix methods, may push hard for a more analytical approach. However, the choices are not limited to either analysis or build-test-fix. The real question is whether or not to make the product more robust and durable before build-test-fix begins. While more analysis can provide value, usually it will not replace testing completely. Structured tests such as designed experiments, covered later in the book, and the accelerated testing and reliability growth testing we discuss in this chapter continue to add great value even as design teams get much better at modeling their way to success.

Another reason for testing is to develop statistics about the expected product reliability and component life. Component life is a random variable that is described by a statistical distribution. Developing enough data to describe these distributions requires testing.

Field Data

Another source of information is the reliability data from customers' use of the product and the resulting maintenance and repairs. The conditions that products experience during customers' use are uncontrolled, causing more variability than a laboratory life test. Once a product is in the hands of customers, the possibilities for increased stresses are greater, translating into an increased number of root causes for failures and an increased variability in product life. Noisy

---

1. Hinckley, *Make No Mistake!*

data are also caused by the variability in maintenance and repairs and in their documentation. It may be expected that the incentives for gathering that information are more aligned with the needs of the service organization than with those of product development.

You might agree that real life is messy and sometimes surprising. In spite of the "untidiness" of field data, they can never be replaced by laboratory testing. Good field data must come from a process that has high integrity so that design teams can depend on the data to guide product improvements. Inconsistency in the data collection or reporting process adds errors in the difference between what is reported and the true state of problems and their context.

Very often, unexpected failure modes will surface in customers' use. One reason is that important stresses were neither anticipated in the development phases nor included in any life or stress testing. These stresses could be noises emanating from other subsystems or stresses created by unforeseen applications or environments. A way to prevent these surprises is to perform in-house system testing that duplicates as much as possible how the product will be used. Of course, those tests depend on development teams understanding how customers will use the product. The occurrence of unexpected failures in customers' hands emphasizes the value of design teams being students of the stresses that can be applied to products during customers' use.

Some of the testing can apply a broad range of stresses, forcing failures to occur deliberately. Other strategies aim at demonstrating reliability growth over time. Reliability growth testing requires a well-functioning corrective action process where project teams are committed to identifying and resolving problems quickly.

Getting the Most Out of Testing Programs

Testing produces data that must be acted upon. What is the most effective way to organize and interpret these data? What causes the observed performance problems? Why did the designs fail? These questions have to be answered to make progress.

Any testing program should include the following elements:

1. **Test design and procedures:** A test must follow a written plan with specific objectives.
2. **Failure reporting and tracking:** There has to be a disciplined process for capturing each problem and tracking progress toward resolving it. Typically this requires a database that can be used to generate different views of the data as well as to deliver reports showing the current status, the projected dates for solutions being implemented, and estimates of the expected effectiveness of solutions.
3. **Problem solving:** Once a problem is identified, engineering teams determine what corrective actions to take. This is critical to any reliability growth program.
4. **Data analysis:** Testing generates data that must be analyzed. Two important data, time to failure and cause of failure, are required in order to fit a distribution to the life data, discussed in some detail in Chapters 21 and 22.
5. **Problem resolution and closure:** As problems are solved, they need to be crossed off the list. There are two good reasons for this. There is a morale boost from the visible accomplishment of resolving problems. Second, resources are scarce and should be spent only on solutions that clearly add value.

Reliability growth testing is a key part of any efforts to improve the reliability of a new product.

Reliability Growth Testing

Reliability growth testing has characteristics of the build-test-fix approach that we have recommended against. However, there is a major difference: Reliability growth testing should not be performed on a system that includes subsystems whose robustness and durability have not been developed.

Once subsystems are available, there tends to be pressure to integrate them quickly and begin system-level testing. This can lead to inefficient "build-test-fix" if the system-level testing includes underdeveloped subsystems. Multiple subsystem problems can make it very difficult to understand the real system integration problems. The process can be slow, complex, and painful. Reliability growth testing works best on systems whose subsystems have achieved some minimal level of robustness, with control parameters specified, although subject to refinement at the system level. It is important for management to give the development teams enough time to debug and upgrade subsystems before starting full system integration and testing.

Duane Growth Model

How can you evaluate progress due to reliability growth activities? The Duane Growth Model,[2] developed for that purpose, was leveraged from work done in the 1930s on the learning curve model for manufacturing costs when, among other companies, Boeing was interested in the relationship between manufacturing costs and cumulative production volume. The learning curve concept is based on the assumption that there is a power-law relationship between cumulative production volume and unit cost. Intuitively this makes sense. We know that the more you do something, the better you get at doing it, but the more difficult it is to make additional improvements. The law of diminishing returns applies. The Duane Growth Model, built on that idea, applies to a wide variety of complex systems, evaluates the rate of growth in reliability, and forecasts future progress. The reliability growth law can be viewed either as the decline in the failure rate (λ) or the growth in MTBF.

Cumulative Failure Rate (λ)

Duane's paper uses the declining failure rate, where the cumulative failure rate at any time is

$$\lambda_C = \frac{n}{T} \tag{8.1}$$

The model expresses cumulative failure rate as a power-law model, shown in equation (8.2):

$$\lambda_C = kT^{-\alpha} \tag{8.2}$$

where

T = cumulative test time
n = total failures

---

2. J. T. Duane, "Learning Curve Approach to Reliability Monitoring," *IEEE Transactions on Aerospace* 2, no. 2 (April 1964).

λ_C = cumulative failure rate = $\dfrac{n}{T}$

k = constant

α = exponent determined by slope

It is useful to take the log of both sides of equation (8.2), producing

$$\log \lambda_C = -\alpha \log T + \log k \tag{8.3}$$

This equation has the form $y = ax + b$, where

$a = -\alpha$

$b = \log k$

$x = \log T$

There will be variability in the cumulative failure rate. As you gather more data points, you can use least-squares regression and solve for the unknown model coefficients α and $\log k$. Initially, with little test data, there will be much uncertainty in the model, so the calculation should be updated as test data accumulate.

Using equation (8.2) you can express the negative growth of the failure rate another way:

$$\lambda_C = \lambda_0 \left(\frac{T}{T_0} \right)^{-\alpha} \tag{8.4}$$

where

λ_0 = cumulative failure rate at the beginning of the test

T_0 = start time of the test

By running for some time before starting the test, you can establish T_0 and λ_0.

Cumulative MTBF

An alternative to working with the failure rate is to use the cumulative MTBF, which is

$$MTBF_C = \frac{\text{total test time}}{\text{total failures}} \tag{8.5}$$

The Duane Growth Model, in terms of the growth of cumulative MTBF, is

$$MTBF_C = (1/k)T^\alpha \tag{8.6}$$

Equation (8.6) can be derived directly from equation (8.2). Again, it is useful to take the log of both sides, giving

$$\log MTBF_C = \alpha \log T + \log(1/k) \tag{8.7}$$

If you want to work with MTBF instead of failure rate, you can fit a least-squares regression model to the data and solve for the model coefficients. Again, as data accumulate, the regression model should be updated.

Another form of the growth model that comes from solving equation (8.6) is

$$MTBF_C = MTBF_0 \left(\frac{T}{T_0} \right)^{\alpha} \qquad (8.8)$$

where

$MTBF_0$ = mean time between failures at the start of the reliability growth test

T_0 = test time that was used to establish $MTBF_0$

Using equation (8.8), you can calculate the total test time required to meet the reliability requirement:

$$T = T_0 \left(\frac{MTBF_C}{MTBF_0} \right)^{\frac{1}{\alpha}} \qquad (8.9)$$

Establishing MTBF at the Start of Reliability Growth Testing

The Duane Growth Model requires an estimate of MTBF at the start of the test. There are three ways to do this:

1. Use historical data as a starting estimate, with appropriate modifications to reflect design improvements in the new product.
2. If there is little historical basis for an initial system MTBF, an estimate must be made analytically. Tools such as cause-and-effect diagrams, FMECA, and FTA (explained in Chapter 13) can be helpful for identifying failure modes and their causes. In-house component life tests and supplier data can be helpful for estimating the lives of components in the context of the intended application, usage, and environments. Usually there are historical data at the component level, unless the component is new and untried. That can be a risky situation.
3. Estimate starting MTBF by running a relatively short reliability demonstration test. When failures occur in this test, failed components are replaced, but design improvements are not incorporated.

Estimating MTBF Using Results from Reliability Demonstration Tests

Three factors are required to calculate the confidence interval for MTBF: the total test time, the number of observed failures, and the acceptable α error. Total test time and number of failures are outcomes of the test. Acceptable α error is a business decision that depends on the amount of uncertainty (risk) that is acceptable. The confidence interval for the system MTBF can be estimated[3] using the chi-square distribution. It's given by

---

3. William Q. Meeker and Luis A. Escobar, *Statistical Methods for Reliability Data* (New York: John Wiley, 1998).

$$\left[\frac{2T}{\chi^2_{(1-\alpha/2,v)}} \leq MTBF \leq \frac{2T}{\chi^2_{(\alpha/2,v)}}\right] \tag{8.10}$$

where

T = total test time (machine hours, not clock hours)

χ^2 = chi-square statistic

α = acceptable level for a Type I error when evaluating confidence limits (not to be confused with the slope of the reliability growth plot)

v = number of degrees of freedom

For a failure-truncated test, $v = 2n$, where n = number of failures

For a time-truncated test, use $v = 2n + 2$

Equation (8.10) and its parameters require some understanding of statistical concepts, which we cover later in Chapters 14 and 15.

Instantaneous Failure Rate and MTBF

Because the cumulative MTBF is calculated using all the failures and the total time, it underestimates the current system MTBF. Some of the problems causing failures that are included in the calculation will have been resolved. A better metric to represent the effect of problem solving is the instantaneous MTBF. It represents the MTBF with some or all of the corrections implemented for those problems that have been identified. Starting with the cumulative MTBF, you can derive the instantaneous MTBF.

The cumulative $MTBF_C$ is the total time divided by the total failures, given by

$$MTBF_C = MTBF_0 = \left(\frac{T}{T_0}\right)^\alpha = \frac{T}{n} \tag{8.11}$$

where

n = total number of failures identified over time T

The instantaneous failure rate is

$$\lambda_i = \frac{dn}{dT} \tag{8.12}$$

If you solve equation (8.11) for n and substitute it into equation (8.12), you get the following for the instantaneous failure rate. It plots as a line parallel to and above the plot for the cumulative failure rate, as shown in Figure 8.1.

$$\lambda_i = (1 - \alpha)\lambda_C \tag{8.13}$$

Once you have the instantaneous failure rate, the instantaneous MTBF is

$$MTBF_i = \frac{MTBF_C}{1 - \alpha} \tag{8.14}$$

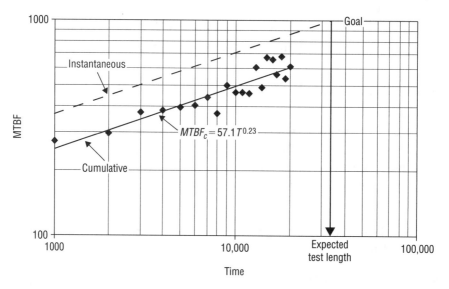

Figure 8.1   Log-log plot of MTBF versus time.

As shown in Figure 8.1, the instantaneous MTBF is useful for estimating test time required to reach a given reliability goal.

The slope of the reliability growth curve is directly related to the rate at which problems can be identified and solved. This depends on the size and membership of the development teams and the responsiveness and quality of their corrective actions. Experience with complex systems has shown that a value of α between 0.3 and 0.4 is indicative of an effective problem-solving organization, as explained in Table 3.1, although not an excellent one.

Ideally, reliability growth tests have durations long enough to give failures a chance to occur. In the next section we discuss what to do when long tests are not practical.

Dormant Failures

Test Length versus Component MTBF

When running reliability growth tests, consider the effects of potential failures that have not yet occurred and may not have time to occur given the planned test length. These failures are called "dormant" or "latent" failures.

Failures may be dormant for a couple of reasons. If test length is significantly shorter than the expected MTTF for a component, it's likely that the component will not fail during the test. Test conditions such as severe environmental stresses can be a factor also. If the failure of a particular component is driven by conditions that are not duplicated in the reliability growth test, failure during testing is less likely. However, given enough test time and the right test conditions, these failures can occur. So if the product is repairable and the MTTF for the dormant failures is expected to be within the useful life of the product, those long-life components having dormancy should be considered when setting targets for reliability and deciding on the length of the test.

Setting Reliability Growth Test Targets When There Are Dormant Problems

As we've mentioned, the best estimates of system failure rate and MTBF are the instantaneous failure rate and MTBF. This is the failure rate and MTBF you should observe if, at some point, you terminate the reliability growth test and run a reliability demonstration test where failures are repaired but no further design improvements are made. System failure rate and MTBF can be calculated using equations (8.13) and (8.14). If there are dormant problems that have MTTFs considerably greater than the system MTBF, and if you don't plan to run the test long enough to allow the long-life components to fail, you have to calculate an adjusted failure rate or MTBF that includes the effect of the dormant failure components. Consider an example that asks for an estimate of the required test time using the Duane Growth Model.

Example: Setting Test Goals and Estimating Test Time

Suppose you are developing a product that has an MTBF goal of 1000 hours. In addition to those components that have the most influence on system MTBF, there are three long-life, non-repairable components in the system. Based on supplier data and component life testing, the long-life components have MTTFs of 6500, 7500, and 10,000 hours. They can be replaced upon failure and are expected to fail before the end of the expected useful life of the product.

After 500 hours of initial testing to establish a baseline MTBF, there were two failures, so the expected MTBF at the start of the test is 250 hours. Based on historical data for reliability growth testing of similar products, it is reasonable to expect a slope of $\alpha = 0.40$ for the plot of log (cumulative MTBF) versus log (cumulative test time). Using the Duane Growth Model, what is the best estimate for required test time, and what test goal should be used for system MTBF?

You can calculate the system goal as follows and solve for $MTBF_{si}$:

$$\frac{1}{1000} = \frac{1}{MTBF_{si}} + \frac{1}{6500} + \frac{1}{7500} + \frac{1}{10,000} \tag{8.15}$$

where $MTBF_{si}$ is the instantaneous MTBF for the system, but excluding the long-life components.

$$MTBF_{si} = 1632 \text{ hours} \tag{8.16}$$

Using equation (8.14), you can solve for required $MTBF_C$, which includes all elements except the long-life components.

$$MTBF_C = (1 - \alpha)MTBF_{si} = (1 - 0.4)\,1632 = 979 \text{ hours} \tag{8.17}$$

Use equation (8.9) to calculate total test time.

$$T = 500 \left[\frac{979}{250} \right]^{\frac{1}{0.40}} = 15{,}173 \text{ hours} \tag{8.18}$$

Various combinations of clock time and number of test machines can be used to achieve this test duration. For example, five systems, each tested for 3035 hours, is one combination that can work. Even though there is reciprocity between clock time and the number of systems in the test, you have to consider the issue of problem dormancy when planning the test. You should have a balance between efficiency and effectiveness. It does no good to put many machines in a test for a short time, observe very few failures, and "declare victory." There has to be enough time to allow failures of those elements with shorter lives, at the same time without testing for longer than necessary. This requires you to consider dormancy and plan the test duration accordingly.

Suppose that the expected slope of the growth plot is $\alpha = 0.30$ instead of 0.40. What effect does that have on the required system-hours of test time?

$$T = 500 \left[\frac{1142}{250} \right]^{\frac{1}{0.30}} = 79{,}078 \text{ hours} \qquad (8.19)$$

It increased by over five times! There are two contributing factors. A reduced slope means longer time, as does a higher required cumulative MTBF (1142 hours versus 979 hours) to yield the same instantaneous MTBF. So test time is very sensitive to the rate of reliability growth that you can achieve. This means that the intensity of the efforts to find and solve problems is a critical factor in determining test productivity and total required test time.

Reliability Growth Test Considerations

Clock Time and the Number of Systems in a Test

The downside of longer clock times and fewer systems in the test is that the test program takes longer. The benefit is that there will be fewer dormant problems. There may be some surprises with longer test times, where unexpected failure modes are identified. This also is a benefit since it reduces the number of surprises when the product is in the hands of customers.

Shorter test time with more systems in test can be advisable as long as the test program is effective in helping the team improve the product. However, shorter tests that don't identify failure modes are not useful.

Adequate Resources Available for Solving Problems

Resource allocations can contribute significantly to determining the effectiveness of a reliability growth test. Several types of resources are required. The most labor-intensive activities are the analysis of problems, the identification of root causes, and the development of effective corrective actions. Solutions to problems should be bench-tested and validated off-line by subsystem design teams before being installed in the systems undergoing reliability growth testing.

Shared Resources

Reliability growth testing can require resources from multiple disciplines. For example, with complex electromechanical products, establishing root causes of failures can involve electrical, mechanical, and software engineers as well as manufacturing process experts. Solving problems that occur in products using consumable materials can require material scientists and engineers in both development and production. When planning a reliability growth test, it is essential to anticipate those resource needs.

Drivers of Reliability Growth Rate

What are organizational attributes that can drive the rate of reliability growth? Once you have increased the initial reliability using problem prevention methods, how do you increase the rate of growth of reliability to be fast enough to achieve the requirement prior to product launch or, better yet, to shorten the development cycle time? Without a deliberate effort, you cannot expect reliability to be developed fast enough to satisfy the project requirements by the time of market entry. Within a product development project are many potential difficulties that can handicap the reliability growth rate and present the project leadership team with dilemmas, such as to trade off requirements in favor of schedules or cost budgets. Your customers will then suffer the consequences. Figure 8.2 shows the situation. It's a simplified version of those figures we used in Chapter 3 to discuss strategies. What drives the slope of the curve? In Chapter 3 we describe the slope to be a characteristic measure of the capabilities of the organization. The implication is that the drivers are not just engineering methods.

What would you do if, at a gate review, you observed that the rate of growth in reliability is not high enough to achieve the requirement in time for product launch? Management intervention would be expected, but often it is expensive with results that fall short. It would be better to improve the basic capabilities for rapid problem solving so that they provide benefits throughout product development. Table 8.1 illustrates some ideas that you may find to be applicable to your business. What strategies can you identify to improve the capabilities of your projects?

Typically, reliability growth tests are performed under conditions of expected customer usage and normal stress levels. The objective is to analyze and correct problems with conditions duplicating, as much as possible, those that the product should experience in the market. In the following section we describe accelerated testing, a strategy to employ stressful conditions to force problems to occur deliberately so that they can be corrected in less time.

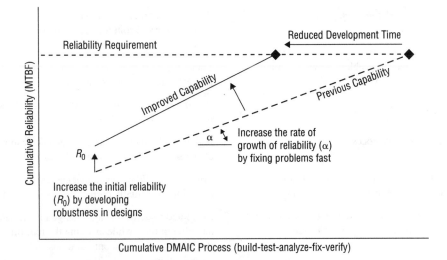

Figure 8.2 The growth of reliability is a sustained effort to correct many problems quickly with solutions that are verified to be better.

Table 8.1   Systems and behaviors that drive reliability growth

Drivers of Higher Reliability Growth Rate	Rationale
Responsive behaviors toward fast problem solving	A bias for action fosters creative ways for teams to use time and resources wisely while preparing the product for use by customers.
Excellence in cross-functional teamwork	Project teams have the responsibility to optimize at the system level with the best capabilities of the organization.
Colocation of cross-functional teams with their facilities for development and testing	Engineers who work near the testing tend to respond to problems faster and achieve higher accuracy of problem analysis.
A standardized system for managing corrective actions	Problem analysis and correction need strong project management with clear responsibilities and visible tracking.
Management expectations for an efficient, consistent process for corrective action	A well-understood, predictable process helps progress to be enabled, tracked, and managed.
Documentation system for failure reporting and corrective actions	The solutions to problems must be understood accurately by production and service operations and for the development of future products.
An effective risk management process	The investment in problem solving should be based on the severity of the consequences of the problems to customers.
Time, labor, and capacity allocated to sufficient iterations of test planning, execution, failure analysis, and corrective actions	Reliability at market entry is the result of efficient problem identification and solving with much iteration.
Stressful and accelerated test plans	Accelerated processes help problems to be found faster and solutions to be judged superior.
Workforce trained in problem-solving methods and tools, such as DMAIC Six Sigma	Standardized approaches provide a consistent process across development teams.
Rapid prototyping	Rapid prototyping enables faster build-test-fix iterations.
Agile development process (see Chapter 5)	Agile development capabilities enable design changes in later development phases.
Independence in responsibilities for testing and verification of analyses and solutions	Priorities should be set so that customers win and the solutions actually correct the real problems.
System for PDM integrated with the production enterprise resource management system	Integrated systems manage change implementation and effectivities, while ensuring the real-time visibility of accurate configurations and specifications.

Accelerated Tests

Stress testing is an important part of any product development project. Many books[4] have been dedicated to the subject, so we offer a high-level summary to point you in the right direction. There are two classes of stress tests: those that are designed to develop predictive models and those that are designed to find problems faster.

Accelerated Life Testing (ALT)

The main benefit of ALT is that it can be used to develop a model for predicting product life. Typically the model is developed using data taken under conditions of higher-than-normal stress, causing failures to occur in a shorter time. The model can then be used to predict life under conditions of normal stress. The process requires less test time to generate failures, resulting in increased test productivity. We cover ALT tests in more detail in Chapter 22, after some topics in statistics and probability distributions.

Accelerated Stress Testing (AST)

AST is a family of tests employing stressful conditions to find problems faster. There are several different tactics for this test design, each with advantages and disadvantages. An improved design should demonstrate degradation rates that are less than those of its predecessor or of competing designs under those same conditions. That sets an expectation for longer usage life under actual conditions. An excellent reference that covers the broad subject of AST is Chan and Englert.[5]

Highly Accelerated Life Testing (HALT)

HALT is performed to uncover weaknesses in the system rather than to develop predictive models. A HALT test identifies opportunities to make the product more durable.

There are a number of ways to apply stresses in a HALT test. The best way is dependent on what you are trying to learn and on what field stresses you are trying to mimic. Since you are searching for the threshold of failure, you gradually increase the stress until a failure occurs. As shown in Figure 8.3, the stresses can be applied in several ways, from cyclic to constant. When a failure occurs, design teams analyze the failure and decide whether or not to correct the problem.

Stresses in a HALT test are more severe than those expected during customers' use. Consequently, the recommendation to invest in design improvements to increase product durability can be met with resistance. Opponents may argue that a design improvement represents a waste of time and money since the failures are precipitated by stresses that are much higher than normal. While it may be tempting to terminate the test and declare victory, experience has shown that many of the failures at elevated stresses can be caused by the same failure modes that cause field failures under nominal conditions. Why is this? We discuss why later in the chapter, in the section about the stress-strength distribution.

---

4. Gregg K. Hobbs, *Accelerated Reliability Engineering* (Chichester, UK: John Wiley, 2000); Alex Porter, *Accelerated Testing and Validation* (Oxford, UK: Newnes-Elsevier, 2004); Patrick D. T. O'Connor, *Test Engineering* (Chichester, UK: John Wiley, 2000).

5. H. Anthony Chan and Paul J. Englert, eds., *Accelerated Stress Testing Handbook: Guide for Achieving Quality Products* (New York: IEEE Press, 2001).

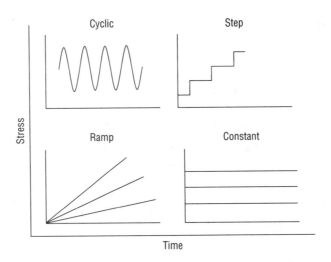

Figure 8.3   Examples of stress applications for HALT testing.

Candidate Stresses

Many different stresses can be applied in HALT testing. Usually these are external factors over which design teams have little or no control. Examples of just a few are extreme levels or changes in

- Ambient temperature, humidity, and atmospheric pressure
- Line voltage and AC frequency
- Vibration and shock in handling during shipping, storage, and installation

The goal of HALT is to precipitate failures, detected by monitoring the important functional performance responses of the system. A hard failure occurs when performance is far off target and cannot be returned to its specification by reducing the stress. It can be corrected only by adjustment or repair.

Highly Accelerated Stress Screening (HASS)

The purpose of a HASS test is to identify products that have weaknesses that might lead to failures during customers' use. In a HASS test, the applied stresses are higher than normal but lower than the stresses used in a HALT test. The goal is to make weak products fail, but not to damage the strong products. Generally the weak products are from a population different from the strong products. They are not products from the weak tail of the strong population but actually have different characteristics. The difference in strength is usually caused by errors in the manufacturing process, so these products can be expected to fail under the nominal stresses of customers' applications.

Figure 8.4 shows the strength distributions for the weak and strong populations as well as the bathtub curve. It is an oversimplification of what really happens to a population of products in the market. In fact, some weak products will not fail because they are never exposed to stresses severe enough to precipitate failure. There is an excellent discussion of this in Chan and

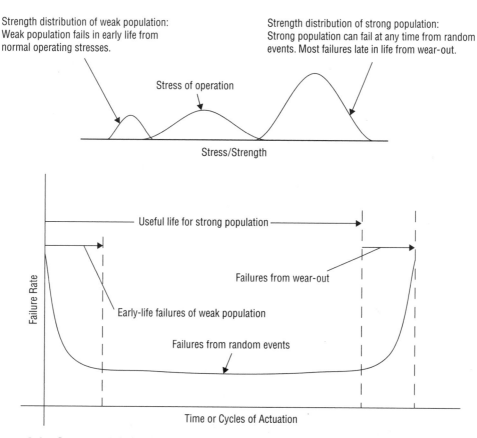

Figure 8.4   Strong population, weak population, and bathtub curve.

Englert.[6] We summarize several of the key points in the following section about the stress-strength distribution.

Understanding Stress versus Strength

Effect of Increasing Stressful Test Conditions

Figure 8.5 is a plot of strength versus applied stress, illustrating the stress levels in HALT and HASS test strategies. It describes at a high level what happens as you gradually increase the stress applied to a population of components or complex product systems. A good way to understand the relative stress levels and strength is to consider the consequences of gradually increasing the applied stress. As with all mental models, it's a framework for thought and not a flat assertion of the way things are. Real life is often complex, requiring some simplification to gain insight.

Stress level 1. This is the level of nominal operating stress for the population of products, with no overstress due to random events. When elevated stresses are applied, failures in the weaker population can be expected to occur sooner.

---

6. Chan and Englert, eds., *Accelerated Stress Testing Handbook.*

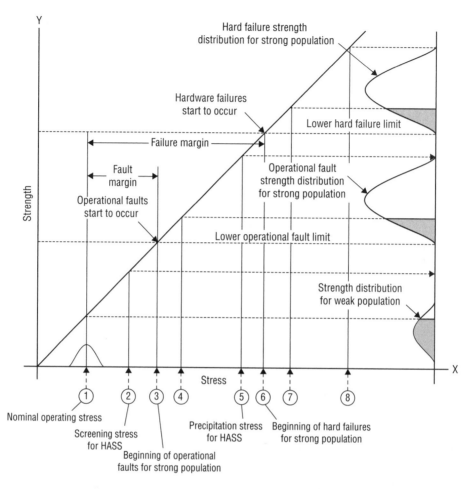

Figure 8.5   Operational fault and failure stress limits for HASS and HALT testing.

Stress level 2. This is a good level for a screening test. It should cause all of the weak products to fail but not cause any failures in the stronger population.

Stress level 3. This level of stress is where operational faults are expected in the stronger population of products. When an operational fault occurs, the product should return to fault-free operation when the stress is relaxed. There should be no permanent damage.

Stress level 4. At this stress level some portion of the stronger population should experience operational faults.

Stress level 5. At this level of stress most of the strong population should experience operational faults but not permanent damage. This is a stress level that could be used for a precipitation screen, causing strong products to fail but without permanent damage.

Stress level 6. This stress level should cause weaker products in the strong population to have permanent damage. Failures can be either unrecoverable faults or hard failures. In either case the problem is correctable only by a service intervention. Reducing the applied stress does not correct the failure.

Stress level 7. At this level of stress, a fraction of the strong products should experience hard failures.

Stress level 8. At this level of stress, most of the strong products should experience hard failures.

The Stress-Strength Distribution

Some statistics can help you understand a conceptual model relating the interactions of stress and strength and why some products fail while others survive. The discussion assumes some understanding of probability distributions, which we cover later in Chapters 21 and 22.

Think about two independent distributions. The first distribution describes the value of the maximum stress that each member of the product population sees during its lifetime. While some population members will be exposed to high stress levels, others will not. The second distribution represents the strength of the population members. When stress exceeds strength, a failure occurs. The stress distribution can be expressed as $f(x)$.

The population strength distribution is a bit more complicated since it is a mixture of the strength distributions for the strong and weak populations. You can express the strength distribution as

$$f(y) = Pf_1(y) + (1 - P)f_2(y) \tag{8.20}$$

where

P = fraction of the total population from the strong population

$(1 - P)$ = fraction of the total population from the weak population

$f_1(y)$ = distribution describing the strength of the strong population

$f_2(y)$ = distribution describing the strength of the weak population

The stress-strength distribution is a joint probability distribution, the product of the stress and strength distributions. It is expressed as

$$f(x,y) = f(x)f(y) = f(x)\left[Pf_1(y) + (1 - P)f_2(y)\right] \tag{8.21}$$

By definition, the total volume under the bivariate surface is unity, accounting for 100% of the population.

Figure 8.6 shows the stress-strength distribution for a population composed of members from a strong population and a weak population. Depending on the manufacturing processes used for the product, it is possible to have a population strength distribution with more than two subgroups, but two is enough to illustrate what goes on when products operate in their market.

The weak population includes all products produced with substandard quality, resulting in lower strength. The strong population includes all products made to specification for which the manufacturing processes are under control. During production of the strong population, there were no quality lapses that might yield products significantly different from the strong population.

The stress-strength distribution in Figure 8.6 has two peaks. In this example, 80% of the products are from the strong population, 20% from the weak population. The probability distributions for the strong and weak populations could be different. At the very least they could be the

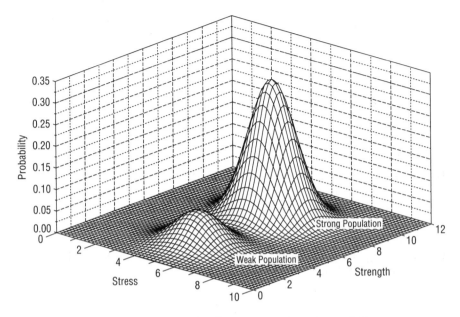

Figure 8.6   Stress-strength distribution showing strong and weak populations.

same distribution but with different distribution parameters. The stress and strength distributions are also likely to be different.

An interesting plot of contour lines from the stress-strength distribution is shown in Figure 8.7. Like a topographical map, the contour lines are closer together where the terrain is steep and farther apart elsewhere. The diagonal line dividing the plane in half is the locus of all points where stress equals strength. The area above the line, where strength is greater than stress, is the region of survival, and the area below the line, where strength is less than stress, is the region of failure.

A conclusion is that some strong products can fail, while some weak products survive. This happens because the stress and strength distributions are independent. A weak product may never be exposed to high stress during its service life. Strong products can fail if the mean strength of the strong population is too close to the stress-equals-strength line or the variability of the strong population is too large, causing the strength distribution for the strong population to overlap the strength-equals-stress line.

The plots can indicate ways to make products less vulnerable:

1. Reduce the size of the weak population or eliminate it by improving the quality and consistency of the manufacturing processes. Failures of the weak population usually occur early in life and do not require high levels of applied stress. A statistician would say that the mean strength of the strong population is significantly different from that of the weak population. The strong and weak products are produced by different processes, one good and one flawed.
2. Make the strong population stronger. Increasing the safety factor or design margin increases the difference between strength and stress. That makes the product more likely to survive random events when higher stress levels are applied.

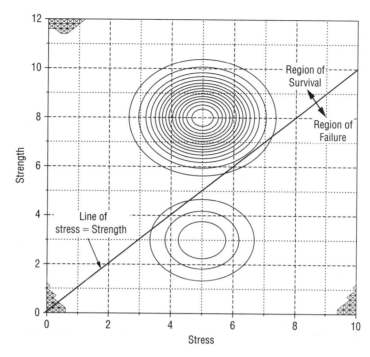

Figure 8.7   Contour lines for the stress-strength distribution.

3. Reduce the variation in strength for the strong population by reducing manufacturing process variability. There will be fewer low-strength members of the strong population that are vulnerable to high-stress events.
4. Use accelerated stress testing, such has HALT and HASS, to identify and eliminate weaknesses in the product design.

How High Should Stress Levels Be Set?

HALT identifies design and manufacturing weaknesses during product development. HASS screens out weak products during production. The statistical discussion is a caution against dismissing the results of stress testing just because the applied stresses are much higher than the nominal stresses expected in service. To reiterate, many of the failure modes observed in accelerated stress testing are the same ones that cause field failures at lower stress levels. They just occur more quickly.

HALT

Fundamentally the question of how high stresses in HALT should be is more of a business question than a technical one. It requires you to understand the range of stresses likely to occur in the field under customers' usage and their economic impacts. As you increase the applied stress and find a weakness, you have to decide whether or not to fix the problem. By correcting the problem, you effectively increase the stress threshold for failure. If you make a large enough improvement in a weak link, it is expected that a different problem will then become the limiting factor. And so it goes, until you conclude that the design is good enough.

HASS

The purpose of HASS is to screen out products from the weaker population without damaging the stronger products. This means that the maximum applied stress must be less than the stress threshold for a hard failure. The only way to establish those limits is first to perform HALT. You can then understand the stress limits that must be observed when running HASS and be able to screen out weak products. The other constraint on HASS stresses is not to unacceptably shorten the life of good products. So the severity of the test has to be below the stress threshold for hard failure in order to limit cumulative damage to an acceptable level. These limits are usually best established by experiment.

Tools and Systems That Support Testing

DMAIC Six Sigma Method: A Tool for Efficient Problem Solving

DMAIC is a data-driven process for improving the quality of products and processes. With methods derived from the early work of Deming, Shewhart, Juran, and others, DMAIC is the basis of the Six Sigma strategy developed by Motorola for improving business processes. While DFSS is focused on problem prevention, DMAIC[7] is focused on problem correction. The basic functions of performance, $y = f(x)$, have characteristic responses that can be measured, analyzed, and improved to achieve requirements. The functions have critical design parameters that can be refined, specified by design, and controlled by stable processes in manufacturing and service. As with any improvement methodology, the achievement of sustained quality improvements is enabled by management, with commitments of resources, time, and organizational competencies.

This problem-solving approach has several advantages. When institutionalized, it is a consistent approach that can be taught, practiced, tracked, and managed across a wide variety of problem types. It emphasizes the need to correct the actual consequences to customers. By being a sequential process, it ensures that the problem solutions are based on sufficient understanding of requirements, misbehaviors, and root causes. It resists the urge to jump directly to quick fixes, particularly those that treat the symptoms rather than the causes. To enhance the basic scientific and engineering capabilities of your organization, it brings the vast collection of Six Sigma tools to the problem-solving process, with measurement and analysis that provide data with higher integrity.

The DMAIC methodology involves five sequential steps, shown in Figure 8.8. The DMAIC acronym is derived from those steps. They are applied to each problem chosen to be solved. As the process proceeds, it may be necessary to cycle back to earlier steps if, for example, the chosen solution is found not to correct the real problem.

Figure 8.8   DMAIC Six Sigma is a systematic five-step process of problem solving.

---

7. Michael L. George, *Lean Six Sigma: Combining Six Sigma Quality with Lean Speed* (New York: McGraw-Hill, 2002).

Here are the steps:

DMAIC Activities

Define: What is the problem and what are its consequences to customers?

- Identify the problem to be corrected and quantify its performance-requirement gap.
- Describe the problem in terms of the level of severity of its consequences to customers and its probability of occurrence.
- Justify the expected improvement effort with an acceptable business case.
- Clarify the basic function behind the problem and its requirements, as derived from the needs of the customers for the product or process.
- Evaluate cause-and-effect relationships to identify potential root causes.
- Identify the boundary conditions of the problem's solution set.

Measure: How is the problem characterized?

- Conduct experiments to gather data that can characterize the behavior of the problem.
- Verify the integrity of the data.
- Validate your measurement and data collection system.
- Describe the problem with verifiable data and performance metrics.
- Determine whether or not the existing product or process design is under control.
- Determine whether or not the problem is stable or changing over time or under specific conditions.

Analyze: What are the root causes of the problem?

- Integrate collective wisdom and experimental conclusions to identify the potential cause(s) of the problem and their relative importance.
- Conduct experiments to verify the hypotheses that identify the primary root causes.
- Verify the statistical significance of the relationships.
- Separate product design causes from those inherent in manufacturing or service processes.

Improve: How can the product or process design be improved to satisfy the requirements?

- Identify criteria for evaluating alternative solutions.
  Note: Problem correction can be in the domain of the product designs, in their manufacturing and service processes, or both.
- Evaluate alternative solutions in the product design or in the manufacturing and service processes.
- Determine whether or not the capabilities of the design can satisfy the requirements.
- Select the best approach to improve the design.
- Develop the solutions, aimed at the root causes, to achieve the required performance and to be robust under stressful conditions.
- Specify the improved design and place it under the control of your product data management system.
- Verify that the improved design achieves the required performance and is superior to alternative approaches that are available.

Control: Sustain the improvement in production.

- Implement the changes in the product design or in its manufacturing and service processes.
- Implement product data management and production process controls to ensure that the solution to the problem will be permanent and stable over time.
- Identify remaining risks; establish preventive actions or contingency plans.
- Validate the measurement and data collection systems.
- Record the lessons learned to enable future problem avoidance.

More details about the DMAIC Six Sigma methodology can be found in many books, including the references here.[8]

Failure Reporting and Corrective Action System (FRACAS)

You may think of a failure-reporting system as just a database of problem reports (PRs) and their status. Okay, but the database is a knowledge management system with critical information about the relationships among problems, their causes, and their proven solutions. That information is the basis of diagnostics for product assembly and services repairs. It contributes to problem avoidance in the designs of future products. In order for this value to be realized, the information has to be in a concise, understandable, and searchable format with easy access by everyone on the project.

Our experience leads us to recommend that your Failure Reporting and Corrective Action System (FRACAS) be managed by an organization independent of product design, with editing rights to ensure the completeness of descriptive details, the assignment of responsibilities, and the accurate tracking of status. It then falls to management to set clear expectations for the routine and early use of this information. What a waste it is for organizations to solve the same problems over and over! The database can be even more valuable if it includes trouble-remedy information from product maintenance and repairs. It's understandable that organizations face barriers to using a single database for both pre- and post-launch problems, but think about how valuable this would be.

The FRACAS can use the DMAIC phases or a modification that fits your process model. The following are examples of status categories:

Define: PR has been created by test personnel to document the symptoms and circumstances observed during testing. Responsibility has been assigned to a subsystem team leader for analysis and corrective action.

Measure: Subsystem team is involved in the test to characterize the problem.

Analyze: Subsystem team is working to determine root causes and possible corrective actions.

Improve: Subsystem team is developing a solution for the problem. Prototype samples are on order to demonstrate the effectiveness of the solution and to enable initial implementation.

In verification: Samples of the problem correction have been provided for an independent test to prove that the solution is effective.

---

8. Lynne Hambleton, *Treasure Chest of Six Sigma Growth Methods, Tools, and Best Practices* (Upper Saddle River, NJ: Prentice Hall, 2008); Peter S. Pande, Robert P. Neuman, and Roland R. Cavanagh, *The Six Sigma Way Team Fieldbook* (New York: McGraw-Hill, 2002).

Closed: Experience with the problem's solution has demonstrated that the problem has been solved. Judgment should be by an independent group, such as systems engineering or quality assurance.

Information: The problem is determined to be a repeat of a previously documented PR that has not yet been solved. The PR is not counted in quality metrics.

Above all, the definitions of status levels should be clear for management and actionable for design teams. When project management discipline is imposed, the process can establish the predictability of improvements. When it can be forecast which problems will not be resolved by market entry, a corresponding quality defect level (QDL), equation (8.22), can be predicted, with its implications for the production and service business functions. It can be the basis for communications with customers about the capabilities they can expect from the new product.

Levels of Problem Severity

Not all failures have the same consequences for customers. The principles of failure reporting assign severity levels to problems as a way to prioritize the needs for corrective action. Some may be observable deviations in performance that are noticeable to customers or to a development test operator but may not detract from the product's usefulness. Others may interfere dramatically with the product's usefulness. Certain failures may prevent customers from using the product at all. Here are some useful definitions of severity levels that combine the concepts of probability of occurrence with the consequences of the failure. The concept is broadly applicable, being analogous to the expected value of the impact on customers.

Low level (1): This is a very minor problem, generally cosmetic in nature and not affecting the product's use. It usually has a low to medium impact on the product's users, is rare in frequency, and has an easy work-around available. It would be a suggestion for a future enhancement of the product's design.

Medium level (2): This is a minor problem that should be fixed. It has low to medium impact on the product's users. In general, the product can still be used for its intended purpose, although some deterioration in its functionality may be noticeable. The frequency of occurrence may be low and the consequences may not be apparent with general use. The problem may be in a feature rarely used by the majority of customers. An acceptable work-around is available.

High level (3): This is a fairly significant problem that must be fixed. The problem seriously interferes with reasonable use of the product and occurs with noticeable frequency, based on common customer usage. Acceptable work-arounds may be available, although customers' dissatisfaction may be evident. It may involve requirements that are not met and that affect a significant population of users. The problem may prevent shipments to customers if judged to be intolerable.

Critical level (4): This is a most severe problem that must be fixed quickly. The problem dramatically interferes with reasonable and safe use of the product and occurs frequently under normal use conditions. The impact affects a broad population of users. It would contribute to significant dissatisfaction among customers and may make the product unacceptable to them. There is no acceptable work-around. The problem is considered to be a "showstopper" and will prevent shipment to customers until it is corrected.

Quality Defect Level (QDL)

An essential measurement of the overall design progress is a count of the open problem reports. Since not all problems have the same consequences for customers, your counting of problem reports should reflect severity levels. A method that we like is adapted from the counting of software bugs. It biases the counting by the square of the level of severity for each open problem report. So a severity 3 problem would have 9/4 more impact in the ranking of problem priorities than a severity 2 problem.

The QDL is calculated by summing the products of the number of PRs at each severity level multiplied by the square of the severity level, as shown in equation (8.22):

$$QDL \;=\; \sum_{i=1}^{4} N_i i^2 \tag{8.22}$$

where

Ni = number of problems at severity level i, for i = 1 to 4

This is a valuable metric at the full product level, particularly since the consequences to customers can be more directly understood. By analogy it can be applied also to processes. Acceptance criteria for the total score and for the number of severity level 1 and 2 problems should be established for the project milestones, along with their implications for market entry.

In our experience, products were not allowed to enter their markets with unresolved problems of severity level 4. A problem with a severity level of 3 was allowed only by exception. Last, there was a limit to the total QDL at market entry. Throughout the development process there were expectations for the QDL to decrease substantially as testing found fewer problems and corrective actions were implemented. It was a topic for gate reviews, subject to management intervention when progress was not acceptable.

Tracking Reports

The tasks of identifying problems and correcting them quickly often are the focus of work during those later development phases that are expected to have more predictable timelines. It is essential that the reporting of status has details that are easily interpreted and compared to acceptance criteria. If management intervention is called for, you'd like to trigger that early so that its benefits have time to take effect. Among the metrics that we have found to be useful are

- Cumulative number of PRs categorized as either "open" or "closed"
- Number of open PRs categorized by their status in the DMAIC process
- Number of open PRs categorized by severity level
- DMAIC status of open PRs by severity level
- QDL metrics for open PRs
- Number of open software bugs categorized by severity level
- Number of open PRs by subsystem, assembly, or function within a subsystem
- List of PRs with a severity level of "high" or "critical"

Figure 8.9 is an example of a "dashboard" that you can adapt to the particular needs of your business. Keep it up-to-date and visible to development teams and management. Another plot that is valuable is the QDL projected into the future, using the target dates for implementation of the corrective action for each problem.

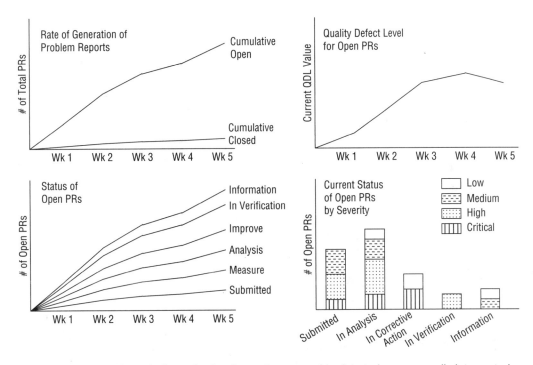

Figure 8.9 A dashboard of standardized reports can enable data to be more easily interpreted. The data can represent a subsystem or full system.

Table 8.2 Problem reports with high or critical severity

PR #	Date	Test Unit	Problem Description	Status	Severity Level	Subsystem	Responsible Team Leader	Target Date	Corrective Action

Table 8.2 shows an example of a format for listing those open PRs that represent the greater risks to the development project. The template enables the details to be understood easily. It is very important that these reports be updated frequently, such as weekly, and used in routine project meetings. They provide data that directly reflect progress.

The next section describes many types of tests that can be considered for product development projects. This range of test plans challenges management to understand the alternative strategies for experiments and the valid interpretations of the test results. The selected strategies and plans then have to be properly supported in project management plans.

Overview of Various Types of Tests

Specific test designs have different purposes. Some have the objective to learn about the product's vulnerabilities, while others are designed to demonstrate its performance, reliability, or usage life. Brief definitions can help you understand these options. Although ALT, HALT, and HASS are described earlier, they are included to complete the glossary. These descriptions, all focused on testing, are included in the glossary at the end of the book, but for readers' convenience we have included them here as well.

Accelerated life testing (ALT) is a method of testing used to develop a predictive model for reliability. ALT uses higher-than-normal stresses to force expected failures to occur sooner. It is more productive than testing at nominal stress levels.

Customer acceptance tests are designed by customers to provide unbiased external assessments of the product design by real users in real applications. They are performed with engineering models of the product. Because these tests are planned prior to the completion of the design phase, feedback from them can benefit the released designs.

Design verification tests are designed to demonstrate whether or not the technical product requirements have been satisfied. For customer-driven requirements, the test scripts are derived from customers' usage profiles. The tests are performed with engineering models of the production-intent design, with the configuration frozen.

DOE-based tolerance stress tests are designed to evaluate system performance as manufacturing tolerances are varied systematically within the framework of a statistically designed experiment (DOE). Often environmental conditions are stressed to force additional variation that might occur naturally during normal customer usage.

Environmental stress screening (ESS) uses extremes in environmental stresses to find latent defects or weaknesses that eventually might cause failures in more normal operation.

Fixed-length qualification tests are designed for a specified test length. The acceptance criterion is having fewer than a certain number of failures with a test of a given duration.

Highly accelerated life tests (HALT) push stress levels well beyond expected conditions to find the operational and destruction limits of the product's design. Weaknesses found under high stress levels have the potential to be sources of deterioration or failure under normal applications.

Highly accelerated stress screen (HASS) tests are used in manufacturing to filter out systems with latent defects. The goal is to make defective units fail without too large a reduction in the life of quality products. Clearly it is important to understand the allowable stresses for HASS tests. Minimizing damage to quality products requires that HASS stresses be no higher than what's required to make most of the weak products fail.

Highly accelerated stress tests (HAST) are designed to accelerate failures due to corrosion in seals, materials, joints, and the like. They have important applications for electrical and electronic devices. Generally these tests apply severe levels of temperature and humidity at elevated pressures to force moisture into electronic packages more quickly.

Life tests, under normal use conditions, can have two basic forms. Their duration and acceptance criteria can be based on a fixed duration of acceptable use or on the usage duration prior to a predetermined number of failures. Preventive maintenance must be incorporated if it is to be part of the product's reliability design.

Manufacturing burn-in tests operate products at the end of the assembly line to find manufacturing defects prior to shipment.

Reliability demonstration tests are designed to provide a measure of reliability that customers can expect under realistic applications and environments. These tests also enable teams to develop plans for preventive maintenance and for the replenishment of consumable materials, as well as for the availability of replacement parts or components. They are performed with preproduction units with their configuration frozen.

Reliability growth tests have the objective of not only finding problems under usage conditions but also demonstrating the benefits of the corrective actions that are implemented. The configuration of the units evolves during the testing with design upgrades.

Robustness development experiments are aimed at forcing variations to occur deliberately. Their objective is to identify controllable parameters and to optimize their set points to reduce the variations in performance and any shifting of its mean.

Sequential qualification tests[9] are designed to determine whether or not reliability has met its acceptance criteria. The system is run under conditions simulating customers' use, as represented, for example, by environments, duty cycles, applications, run lengths, or rough handling. If the system runs without too many failures, the system is judged to be acceptable and the test can be terminated early. Otherwise the tests continue until more conclusive data determine the acceptability of the design or judge that the design is not reliable enough.

Key Points

1. Reliable products start with good design practices that yield robust and durable products.
2. Tests must be designed to identify the errors and weaknesses in product designs, expected in any process managed by people.
3. Reliability growth testing of product systems must be preceded by subsystem development and system integration, with control parameters optimized at the system level.
4. HALT is aimed at identifying weaknesses in the product where the application of high stress can cause failure. It is an economic decision whether the strength of the product should be increased.
5. The purpose of HASS is to prevent defective products, having very low strength, from escaping into the market and causing failures in the hands of customers.
6. Developing reliable products and growing reliability requires problem-solving processes, tools, and skills.
7. Efficient development of products requires a good process for capturing and tracking problems and their resolution.

Discussion Questions

1. Why is testing a necessary tool in developing reliable products?
2. Do you think your organization is too reliant on testing or not reliant enough?
3. During a product development project, when is it time to start reliability growth testing?
4. What can happen if full system testing starts too early?
5. Have you ever had experience with unproductive system testing? If so, what was the reason?

---

9. A good source for guidance on fixed-length and sequential testing for reliability development is *Reliability Test Methods, Plans and Environments for Engineering, Development, Qualification, and Production*, MIL-HDBK-781A (Washington, DC: Department of Defense, 1996).

6. At times, management can be reluctant to act on opportunities to correct problems that are observed in HALT testing. They can question the need to fix problems caused by stress levels much higher than the stresses of normal operation. What facts and experience can be used to make a convincing argument in favor of stressful testing and related corrective actions? Have you ever had that discussion within your organization?

7. Is the decision whether or not to fix a problem observed during HALT testing a business decision or a technical decision? What factors should be considered when making the decision?

8. What must be done to establish stress limits for HASS testing?

9. If HASS testing identifies many weak products produced by the manufacturing process, what are some conclusions that you can draw?

10. Is HASS used within your organization? If so, how successfully?

11. What statistical tool can be used in production to minimize failures in HASS testing?

12. How will having good SOPs in manufacturing reduce HASS failures?

13. What are some of the benefits to having a well-designed FRACAS process?

14. How are problems and corrective actions managed in your organization?

SECTION III

Tools and Methods Supporting Reliability Development

The chapters in Section III provide descriptions of three fundamental building blocks toward much-improved reliability:

- Team-based practices that contribute to reliability development as an integrated tool set:

  Chapter 9: Understanding and Managing Customer-Driven Requirements

  Chapter 10: Quality Function Deployment

  Chapter 11: Concept Generation and Selection

  Chapter 12: Axiomatic Design

  Chapter 13: Tools to Reduce Risks

- Statistical principles of analytical and empirical modeling:

  Chapter 14: Descriptive Statistics

  Chapter 15: Inferential Statistics

  Chapter 16: Building Models Using ANOVA and Regression

  Chapter 17: Building Empirical Models Using DOE

  Chapter 18: Developing System Robustness Using DOE

- The application of reliability principles to the specification of set points that achieve more stable performance with the best design concepts available:

  Chapter 19: Robustness Optimization Using Taguchi's Methods

  Chapter 20: Tolerance Optimization

  Chapter 21: Reliability Functions and Life Distributions

  Chapter 22: Life Data and Their Implications

Understanding and Managing Customer-Driven Requirements

New products and services are developed to satisfy requirements. Okay, but what is the origin of those requirements? We see three fundamental categories:

 1. *Those driven by the needs and preferences of target customers*
 2. *Those driven by external standards and regulations*
 3. *Those driven by internal company standards and project objectives*

There is a fourth category that we discuss in Chapter 12, requirements that are derived internally from the selection of design concepts or manufacturing processes. It is expected that the value proposition for the product and services is focused on the customer-driven requirements, recognized by engaging customers or from speculating on future needs of which customers are not yet aware.

Customers have problems on which they are willing to spend money for solutions. So a critical challenge in understanding the needs of your customers is to identify how new solutions can respond to their perceptions of higher value. For a particular development project, is it through the benefits of superior features and functionality, or is it through lower costs either in price or in life-cycle costs, or is it both?

Do you have good ways to listen to your customers? Do your customers perceive your company to be one that truly hears what it is told or sees what they observe? Are your projects responsive to those inputs, or do they believe that they know more than their customers about their needs?

Our objective for this chapter is to encourage those who are developing improved solutions to engage their customers directly and often. We strongly encourage treating your customers as partners in your process, so please do not outsource the gathering of their needs. Involve those who need to use the information, not reporters.

We'll never claim to be authorities on "voice of the customer" (VOC) processes, but we'll share insights from our experiences. We ask that your development teams be students of the methods for engaging customers productively so that both they and your customers will benefit. You should expect these capabilities to enable product development to be faster, with higher quality.

There's a large body of literature and wisdom. At the end of the chapter we identify several authors whose work and writings should be followed. They do not always agree. None of them provides complete, stand-alone guidance, and some of their advice may be irrelevant to your business. However, their work is current and their intentions are well aligned with your intentions. The tactics for understanding customers' preferences, the relative importance of their needs, and the perspectives behind trade-offs that they make can be very complex. Our advice is to avoid junk science, so please pay attention to these wise people. They are "trailheads." Going down their paths will expose you to others whose advice may also be valuable. It's an adventure.

Objectives for Product Development

Your customers have objectives when considering new products. For example, they may want to avoid expected problems, gain new benefits, seek desired outcomes, enable new activities, or reduce their costs or hassles. Their perspective is in the context of the processes in their business or activities for the product's application. In general, customers' needs and preferences do not change quickly. However, as your competitors constantly provide newer features and functionalities, your customers' purchasing criteria and their sense of superior value can change rapidly.

Among the objectives for your company is the identification of opportunities for competitive advantages, to deliver superior value through a stream of winning products and services. Because product development takes some time, the requirements that are derived from your processes are expectations for your customers' acceptance criteria in the future. What will your competitors be offering after your product enters the market? Will your customers' purchasing criteria be the same as you understand them to be today, or are your markets changing rapidly? Your product requirements, particularly those that are customer-driven, have to reflect forecasts of those markets better than do those of your competitors.[1] That's a major contributor to the value in your future products being perceived to be better than their competition.

The product development process leads project teams through a logical sequence of information gathering and decision making:

1. Studies of markets and customers provide a broad understanding of customers' needs and of the value drivers for segments of customers. Market segments are defined by common value drivers.
2. A differentiating value proposition is selected for the product. It is derived from comparisons of the competitors' capabilities to deliver better solutions to the needs of target customers.
3. Detailed customer needs are gathered, analyzed, validated, and translated into product requirements that implement the chosen value proposition.
4. The baseline product architecture and design concepts are selected based on comparisons of the abilities of alternative design concepts to satisfy the product requirements.
5. Those baseline concepts are developed and optimized into designs for production-intent systems.

1. Hamel and Prahalad, "Seeing the Future First."

6. Customer acceptance tests and in-house verification tests determine whether or not the designs will satisfy the intended customers and their technical requirements respectively.

Understanding the needs of customers and translating them into good requirements are the basis for everything that follows. Knowing the priorities enables development teams to avoid spending too much time on capabilities that have low marginal value. If project teams don't know what measures of performance are important to their customers, their attempt to deliver a reliable product will be flawed.

After customers' needs are gathered, understood, and translated, a requirements management process maintains configuration control. In this chapter we discuss best practices for gathering and managing requirements, with attention to requirements that support reliability.

Types of Requirements

The technical requirements for new products and services cover a broad range of concerns. The requirements documents from previous projects can be guides to those requirements categories that are most relevant to the type of product being developed. The three fundamental categories mentioned in the chapter introduction may be useful for challenging creative thinking about the topics. Within each category there can be requirements for the functional operation of the product; for its features or nonfunctional attributes; for the ease of its use, production, or service; and for its compliance with standards.

Validation versus Verification

These two terms may seem to be synonymous. In practice they are quite different. *Validation* is a process of ensuring that requirements and their implementations agree with the intent of the source of the requirements. Do your customers agree that your project teams understand accurately and completely the needs that they expressed? Do the design concepts that your teams have chosen represent acceptable solutions to the problems that your customers have identified? Validation is a process of detailed feedback from representative customers.

Verification processes are internal tests to ensure that the technical requirements, derived from customers' needs, are satisfied by the production-intent designs. These are tests of compliance with requirements. They can be applied in Phase 4 to the product design. With some adaptation, they can become acceptance criteria for production processes, such as end-of-line system burn-in tests. Inherently, there is an assumption that the technical requirements are accurate and complete surrogates for the needs of target customers.

Customer-Driven Requirements

Requirements that are derived from the needs, preferences, and perceptions of quality of customers are the most challenging to gather and analyze. However, they are the primary source of your opportunities for competitive advantage. We've read the viewpoint[2] that customers hire a product to do a job for them. They buy the opportunity for its experience or to gain access to its benefits.

Not all needs can be addressed by your new product, or are even relevant to it. Our concern here is that your teams identify those needs that are essential to implementing the value

2. Clayton M. Christensen, "Finding the Right Job for Your Product," *MIT Sloan Management Review* (Spring 2007).

proposition chosen for the product. That value proposition is the strategy that differentiates the new product from existing products and from expected future products.

*A **value proposition** is an intention to deliver specific benefits relative to their costs. Whether or not it is better than alternatives available from competitors is a subject for competitive benchmarking.*

To be useful to product development, the needs must be translated from the language of customers (mostly qualitative) into the technical language of the engineered solutions (quantitative). Constrained investments of time, funding, resources, and technical capabilities limit those requirements that can be addressed adequately. So the requirements that the project team chooses to accept result from negotiations between the source of needs and the ability to achieve competitive solutions within the timelines and capabilities of the project.

Examples of customer-driven requirements can include the range of concerns for the product's physical form, its fit within work flows or activities, features that provide capabilities, usage scenarios, functionality that determines how well the product performs its various jobs over time and under stress, and the costs to the customers. Customers may also impose requirements for services such as training for users or for key operators. Their need for the availability of a repairable product applies to the designs' reliability as well as to service response and repair times.

Externally Driven Requirements

Markets impose requirements on products, many of which can be barriers to entry. These may be derived from industry standards or from import restrictions of various countries and localities. Regulations from government or other authoritative agencies provide requirements. They may reflect the environments within which the product must operate or be transported. They may vary over time as governments impose new laws or agencies upgrade regulations. Stressful conditions not controllable in the product design can be identified from the environments of the distribution systems and market applications.

Examples of externally driven requirements can include regulations for product safety, vulnerability to or generation of electrical interference, and ruggedness to environments encountered in shipping, storage, and rough handling. Import and export laws can impose requirements. Food and medical products have an enormous body of regulatory requirements to satisfy. Globalization strategies can be a source of requirements such as to accommodate a range of electrical power systems and space constraints. The countries around the world have many languages, cultures, and conventions to be embraced in the product's design.

Internally Driven Requirements

Your company imposes requirements on your new products. They may be found in internal mandates, policies, and standards. Many of the economic parameters that comprise an acceptable business case for the development project are sources of requirements.

Examples of internally driven requirements can range from compliance with trade dress and ergonomic standards to the costs for manufacturing, maintenance, and repair. Compliance with standards for manufacturability or serviceability can be derived from their related costs as well as from the processes chosen for those functions. For example, "design for assembly" requirements would be "derived" from a decision to use automated assembly. Distribution plans may impose requirements for packaging, labeling, or handling devices, for example.

Requirements for Reliability and Durability

The requirements for product reliability, usage life, and serviceability, for example, can be derived from any or all three of these categories. For example, higher reliability may not provide a competitive advantage in certain markets, but it could reduce service and warranty costs. In other markets it could be the dominant requirement driven by customers. We talk about reliability requirements in Chapter 2. The key point here is that these requirements reflect the ways in which customers perceive product reliability and life and the ways by which improvements to those characteristics can increase the value delivered to customers in chosen markets.

Criteria for Good Requirements

There are several guidelines for the writing of requirements, as translated from the language and context of their sources. These criteria are important to keep in mind as we talk about the gathering and analysis of customer-driven requirements. By reflecting on those criteria, development teams may realize that they do not have sufficient information or do not understand the information that they have well enough. What are the acceptance criteria against which to test your requirements? Here are some useful sets of criteria, some of which are similar but with particular descriptions that may resonate with your development teams.

"SMART"[3]

- **Specific:** clear, concise, significant, with supporting details recorded separately
- **Measurable:** quantifiable by practical methods, meaningful
- **Actionable:** achievable by a product, acceptable
- **Relevant:** focused on desired outcomes and the value proposition
- **Time-bound:** achievable within the time constraints of the development project

IEEE[4]

- **Abstract:** independent of the solution or technical target
- **Unambiguous:** stated clearly so that it can be interpreted in only one way
- **Traceable:** having the ability to be traced to a specific customer need or other valid source
- **Verifiable:** having the means to prove that the system satisfies the requirement

"4 Cs"[5]

- **Customers' language:** words and images expressed by customers, without technical jargon
- **Clear:** easily understood by everyone needing to use the information
- **Concise:** use of words that form necessary and sufficient statements
- **Contextually specific:** include the context and details of the application

---

3. Hambleton, *Treasure Chest of Six Sigma Growth Methods, Tools, and Best Practices.*

4. *IEEE Standard P1233* (New York: Institute of Electrical and Electronics Engineers, 1993).

5. Abbie Griffin, "Obtaining Customer Needs for Product Development," *The PDMA Handbook of New Product Development* (Hoboken, NJ: John Wiley, 2005).

Also, requirement statements should

- Describe a necessary, complete, and sufficient set of expectations for the product, without redundancy
- Begin statements with verbs
- Avoid compound sentences, which would indicate more than one requirement
- Avoid conjunctions, such as *and*, *but*, *or* …
- Describe units of measure and directions of improvement, such as "Accelerate faster than …"
- Characterize a constraint on the design
- Avoid non-positive formats, such as "Does not …"
- Avoid abstract words subject to varied interpretation, such as *reliable*, *attractive*, *easy*, *rapid*, or *etc*.

How Requirements Affect Project Success

Incomplete or Incorrect Requirements Can Jeopardize a Project

When requirements are incomplete or wrong, a number of bad things can happen to a development project. The impact can range from schedule delays caused by scope creep or scope gaps, to the failure of the product to its business goals after launch. Scope creep and scope gaps have been blamed for a significant fraction of projects experiencing delays in market entry.

In his book on project risk management, Kendrick[6] shows historical data on the causes of schedule slips covering over 200 projects. Here *scope creep* is defined as requirements continuing to evolve gradually after the product or system requirements are "finalized" and the project plan is in place. There is no dramatic "aha" moment when the magnitude of a large change causes project management to say, "Wait a minute! If we do this, we are going to miss our deadlines." The schedule threats are driven by a series of small, apparently innocent changes.

Scope gaps, on the other hand, are defined as new requirements that emerge quickly, creating differences between the revised customer requirements and the intended product capabilities. An example would be the launch of a competitive product that is clearly superior in some important performance dimension. A decision must then be made whether or not to change the design and plan a new schedule. Defects and dependency changes are arguably scope-related and often are caused by incomplete requirements.

There are two approaches to reducing problems from scope creep and scope gaps. Of course, the first strategy is to do an excellent job understanding customers' needs in the first place. After the requirements are "frozen," further refinements may cause only small consequences, and a disciplined change management process can protect the project plan. It's almost always more efficient if requirements are agreed to earlier in the process. The selection of design concepts and the development of systems can then proceed deliberately with the progressive freezing of their specifications.

However, in rapidly changing markets, that may not be possible. The second approach is to accept that requirements changes are inevitable and to adopt a strategy to respond to them gracefully. In that case the design architecture and development strategy should be more flexible, continuing to gather inputs and responding to them with little impact. When the

---

6. Tom Kendrick, *Identifying and Managing Project Risk* (New York: AMACOM, 2003).

design space is well understood, alternative designs are ready to go, and product architectures have robust interfaces, the consequences of late changes can be reduced. When those early requirements that have the most impact on the architecture and baseline design concepts are solidified, the expected later changes in requirements can be refinements that can be accommodated easily by adjustments or software revisions. The strategy of flexible product development,[7] embracing set-based design and agile development, is designed to assist projects developing products for rapidly changing markets. Change management will still demand an honest appraisal of the impact of the changing scope, including the risks posed to project schedules, resources, and costs.

In his book on Critical Chain Project Management, Leach[8] makes the point that the best way to manage scope creep is to do an excellent job of impact assessment and to communicate it clearly to the project customers. Often, if the changes are not truly essential, the response can be "Oh, never mind, I didn't realize what the impact of this change would be." Sometimes project leaders, concerned about being viewed as not team players or not having a "can-do" attitude, may be reluctant to block a change. In the short term, "going along to get along" makes life easier, but in the end, if the launch date is missed and project stakeholders are unhappy, it's best to have made the right decision. One advantage of a planned stream of new products is that there is the option of deferring the proposed new requirements to a follow-on product.

After a schedule slip has occurred, the last thing a project leader wants to hear is "Why didn't you communicate clearly the impact that this would have?" A way to manage this without leaving the project leader "alone in the boat" is a formal process for managing changes to requirements. It might require approval by a requirements change control board. This should be a non-bureaucratic process in which judgments are based on analysis of the expected benefits, costs, and risks.

The cost of a delayed product launch can have broad consequences, such as

- Reduced revenues because competitors have entered the market with products already meeting customers' needs, or because part of the product's life cycle has been "spent" on a schedule delay
- Project cancellation in anticipation of a late launch because the product has been superseded and recovery is unlikely
- Extended development costs to correct the design
- Scrapped tooling and parts inventories made obsolete by late design changes
- Lost or postponed orders to suppliers who depend on the flow of business expected from the new product
- Delays in follow-on projects due to delays in the reassignment of development resources

It Starts with a Good Idea

The gathering of customers' needs usually follows the genesis of an idea for a new product or service. Someone, somewhere in the organization, has identified needs that could be fulfilled by

---

7. Smith, *Flexible Product Development.*

8. Leach, *Critical Chain Project Management.*

developing a new product. Ideally this is derived from advanced product planning, which creates a portfolio of development projects based on studies of customers and markets, on competitive benchmarking, on the capabilities of technology development, and on the capabilities and capacities for developing new products.

However, opportunities aren't always identified in such an orderly fashion. There can be a chicken-or-egg problem of what comes first, the identification of needs or the creative concept for the product. "Technology push" often is the scenario, with project leadership claiming to know more than customers. Great products have emerged as the result of bright ideas driven by technologies, sometimes because of the good fortune of being in the right place at the right time, sometimes because of the force of a visionary champion. Although "technology push" may be successful at times, the risks of failure can be higher than a "market pull" scenario that chooses solution concepts in response to requirements. It may be that the scenario for your projects is a combination. In any case, those requirements with major design consequences should be solidified in the early phases, with conflicts between "push" and "pull" being resolved so that development can proceed deliberately. Either way, the requirements must reflect how your customers decide whether or not to purchase your product. How do they choose among alternative products?

Decisions Based on Emotions or Misperceptions

- **Emotion:** Sometimes, for reasons not well understood, our subconscious strongly favors one product over another. This has been studied by many marketing professionals and psychologists. To the extent that you can understand the underlying reasons for emotional decisions, you may be able to address them in your product offerings. Once understood, these decisions can be moved from the category of "emotion" to the category of "mechanism understood." For a fascinating discussion about why perfectly good products end up in landfills and how in other cases people can form emotional bonds with their purchases, see Jonathan Chapman's book.[9]
- **Flawed logic and incomplete or wrong information:** Customers can make bad buying decisions. Perhaps they lack an understanding of the value proposition or are biased by misinformation from competitors' marketing. Customers purchase the product that they expect to be the best in satisfying their needs. To the extent that your new product claims that advantage, your marketing message must help customers to understand that proposition

Decisions Based on Correct Understanding of the Facts

- **Product not seen to meet customer needs:** Your product may not have the features, functions, or level of quality desired. Its requirements may have missed the mark in capturing customers' needs. It may have been designed for a different market segment. The requirements may be on target while the execution of the design solution may be flawed. It may be that your competitors have beaten you with their own product being judged to be better. Customers will have made the correct decision.

---

9. Jonathan Chapman, *Emotionally Durable Design: Objects, Experiences and Empathy* (London, UK: Earthscan, 2005).

- **Perceived superior value:** If your teams have done their job well, understanding customers' needs and developing better solutions for those needs, at the right price and the right time, your customers should recognize that and reward you. Perceived value is the key concept. How do customers understand the benefits of the product in relationship to the purchase price or the cost of ownership? The product requirements must reflect those criteria.

Product Differentiation

Read the business press and you can see that escaping from "commodity hell" is a hot topic. This concern for the loss of competitive advantages means that companies are stuck using pricing tactics to eke out margins while selling products that are not differentiated. They don't offer a superior value proposition. We expect that the growing concern about the need for better differentiation stems from the increased range of choices in the market and the attendant pressure on providers of goods and services to be viewed as being different and better. Whether or not you're in commodity hell is a function of your product's value and how well it's positioned.

Take chicken, for example. People buy and sell frozen broiler futures by the boxcar load. It is a true commodity, traded daily on the Chicago Mercantile Exchange. However, does that mean that everyone in the chicken business is doomed to compete with price? Frank Perdue had a better value proposition. By emphasizing "consistently more tender-tasting" chicken,[10] he managed to position his product cleverly to move chicken from a commodity to a profit heaven. Over time, he conveyed the message that his product was different because *he* was different: "It takes a tough man to make a tender chicken."

The advantages of a differentiated product with a superior value proposition have been recognized for a long time.[11] Regardless of the product and its market, whether considering personal care products or farm equipment, companies are worried about how their product can stand out from its competition. The logical idea is that you should differentiate your products by offering greater value. By giving serious attention to opportunities for differentiation, your development teams are faced with challenges to gather requirements that represent those opportunities. If project teams see little difference between their requirements and existing competitive benchmarks, they should align their value proposition with a more valid opportunity. Differentiation has a dramatic effect on perceived product performance, the rate of customers' acceptance, the growth in market share, and the returns on the investments in product development.[12]

Why Is Differentiation Alone Not Enough?

Different is not necessarily better. Making your product offering different from its competition is one thing, but getting customers to recognize its superiority and buy it is another. Everett

---

10. Michael J. Lanning and Edward G. Michaels, "A Business Is a Value Delivery System," Lanning, Phillips & Associates, 1987.

11. Jack Trout and Steve Rivkin, *Differentiate or Die* (New York: John Wiley, 2000).

12. Cooper, *Winning at New Products.*

Rogers[13] outlined what's required for an innovation to be embraced and adopted. Some of the key questions you may consider include these:

1. What benefits does your new product offer that are not currently available from existing products? Differentiate your product in a positive way and offer your customers more value that they can recognize. If your product offers benefits not supplied by competitors' products, and/or it does so at a lower cost, it has a competitive advantage.

2. Are the advantages of the new product understood and believed? It's important to be better than the competition, but if those advantages are not made clear to your customers, your new product will not reach its market potential. Your marketing communications, developed in parallel with your product, have to convey the value proposition and details of its implementation that speak directly to your customers' purchasing criteria. That can be evident in labels, packaging, press releases, advertising, promotional displays, and other ways in which you communicate with your customers.

3. How difficult will it be for people to learn to use your new product? Are the procedures transferable from existing products? Customers can be frustrated easily with products that are confusing or difficult to use, and they factor concerns for their learning curve into their decision making. This is especially the case where the product will be applied to an important part of a business process or production operation. Those difficulties are additional costs to be compared to the anticipated benefits.

4. Is your new offering compatible with customers' work flow, activities, usage environments, or elements of infrastructure? For example, a new production process that will be embedded in a larger system has stakeholders that include the end users, the business of departments that may interact with the new system, the purchasing department concerned mostly with price, and possibly regulatory agencies that govern occupational health and safety. This example represents more complex concerns than for a consumer product, which might interact only with those who purchase it.

5. Can customers "test-drive" your product before they are asked to commit to its purchase? If so, that would tend to reduce the risks in customers' decisions. Examples of "try before you buy" are abundant in the world of software, where many suppliers aggressively encourage free downloads and 30-day trials. In the case of consumer products, such as the latest MP3 player, you can visit one of the electronics superstores for a hands-on tryout. However, packaging often prevents anything but a visual assessment. In certain cases it's impractical to deliver the product or system for a tryout, such as the design of a new production line. However, virtual prototypes can enable customers to visualize how things can be expected to work.

6. What are the impacts of failures in your product? Is your new product or system considered to be "mission-critical" or does it provide health and safety functions? If so, a failure can have serious consequences. It may be more difficult to convince customers to adopt your new solutions, and the barriers to innovation can be higher. This is especially true if the solutions currently being used are already acceptable. Your new innovation must then be perceived to deliver substantial improvements without increased risks.

---

13. Everett M. Rogers, *Diffusion of Innovations,* 5th ed. (New York: Free Press, 2003).

Value Drivers

One of the goals of development projects is to "deliver value to target customers that is superior to alternatives available to them after market entry." That concept has great power because it forces development teams to think like customers and to make decisions so their customers win. Similarly, "value" is a more useful attribute than "features," since it introduces incremental costs as an element of the requirements.

Your customers choose what to purchase by comparing the perceived value among available alternatives, including the option of not to purchase anything at the moment. So the task is to deliver a new product that customers will see as being better in the many elements by which a product's value is judged. That challenge demands that teams understand how customers perceive a product's value and how they make trade-offs among alternative bundles of benefits and costs. It also demands that teams understand how competitive products deliver value to those customers. These concerns emphasize the objectives of drawing valid conclusions from the analysis of customers' needs and preferences and of the value drivers for market segments, and of placing that information in the hands of decision makers.

The common perception of "value" is the comparison of the benefits to customers relative to their costs. A product feature that provides significant incremental benefits for the same or lower incremental cost would have higher value. Higher value could come from increased benefits or lower costs independently. That may be critical to customers' purchasing decisions. On the other hand, new and necessary benefits that have a higher cost than previous solutions could still be perceived as having higher value, in spite of their costs.

In general, products are developed for a population of customers who have similar needs and cost sensitivities. Value drivers may be specific benefits, for example, to enable an activity, to increase the quality of deliverables to customers of your customers, to improve efficiency, to reduce sensitivities to stresses, or to establish competitive advantages. They may reflect needs to reduce the costs of purchase or of the long-term ownership of the product. Given that customers vary in their value drivers and in the relative importance of them in purchasing decisions, market analyses are challenged to determine the value drivers that differentiate significant populations of customers. Market segments can be defined by those customers with similar bundles of value drivers.

You might think of a value driver as the title given to an affinity group of customer needs that are derived from VOC studies. It would be at the primary level of customer needs in QFD's House of Quality (see Chapter 10). The process for developing requirements then needs to decompose those value drivers into more detailed and specific needs so that their translation into technical requirements and cost constraints can be accurate and traceable.

The Kano Model

The principles embraced by the Kano Model[14] categorize customers' perception of value drivers into three groups. The model, illustrated in Figure 9.1, is derived from the work of Dr. Noriaki Kano, a Japanese consultant for Total Quality Management (TQM).

1. **Basic needs:** Certain value drivers are assumed to be prerequisites for the product's application. Customers may take these "threshold" attributes for granted and not mention them. However, failure to deliver these "Basic needs" will dissatisfy customers,

---

14. Lou Cohen, *Quality Function Deployment: How to Make QFD Work for You* (Englewood Cliffs, NJ: Prentice Hall, 1995).

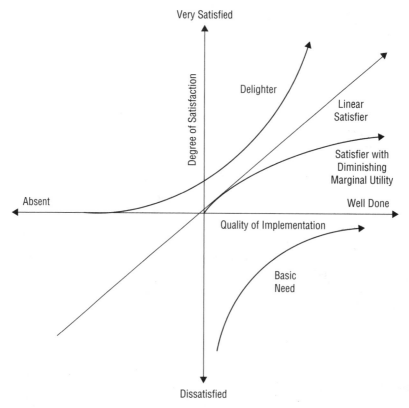

Figure 9.1   Kano Model

while excellent fulfillment of them will only avoid dissatisfaction. Their absence may make the product useless.

2. **Satisfier needs:** Value drivers on which customers tend to focus are those with which they have difficulties with current products and therefore want increased levels of benefit. The conclusions from market analysis need to be clear about the details of these "Satisfier needs," the competitive gaps among current products, the slope of the value/fulfillment curve (Figure 9.1), and whether or not there are diminishing returns. These attributes tend to be tied closely to price. They reflect the value proposition and are the domain of competition in the market. The slope of the curve is analogous to customers' rating of importance.

   For example, for passenger cars the efficiency of fuel consumption may be a strong value driver for those who drive long distances and who currently suffer low mileage. Rising fuel prices will tend to drive up the importance. However, the importance of that characteristic may diminish as the fuel efficiency increases or as the purchase price of more efficient cars increases. These preferences may differ greatly across countries and localities. In a population strongly worried about the environment, customers may not lose interest in fuel economy until it reaches much higher levels. They may be willing to pay more to get that benefit. Country settings, where gas stations are few and far between, may generate stronger interest in fuel economy. Or is the range of driving with a full tank the real value driver, as would be the case where higher horsepower is also a performance attribute? These "more-is-better" value drivers may be significant to customers' decisions to purchase additional products.

3. **Delighters:** Customers may benefit from new and unique attributes of which they are unaware. These capabilities surprise and excite them with clear benefits that they did not expect. They are "wow factors." However, since customers are unfamiliar with the opportunity for those benefits, they tend not to be able to tell you about their need for them. Product developers who work with customers may recognize those opportunities by observation or by being customers themselves. Creative features that provide unexpected benefits may be a significant source of product differentiation and thereby play a major role in customers' purchase decisions.

What is your strategy for winning customers? You can use the Kano Model to understand how Delighter, Satisfier, and Basic attributes are implemented in your strategy. Identifying the Delighters is an activity that will pay great dividends. Satisfier attributes will tend to be "linear differentiators." The more you deliver a Satisfier attribute, relative to competitors, the greater your product differentiation will be. The diminishing returns can be encountered when incremental units of performance come at increasing marginal costs.

Consider the "horsepower race" for automobiles or the resolution race for nonimpact printers, where competitors engage in a leapfrogging strategy with specifications. When you find a Delighter, it's probably unique relative to competitive offerings. If the other products have it, arguably it's no longer a Delighter. Being first with a Delighter also can have great value, since today's Delighter is tomorrow's Satisfier and, later, today's Satisfier is tomorrow's Basic need. The "wow factor" for various features and functions changes over time, as a result of competition. Devaluation is expected.

How Requirements Flow Down

At the top of the hierarchy you have the value proposition that summarizes why customers should want to buy your product. Depending on the product, the customer needs are captured in the language and context of customers, independent of the solutions. These statements tend to be in softer, nontechnical terms that require translation into the technical, explicit language and context of the solution. Those needs that are chosen to be addressed are embraced by the value proposition, supported by technical system requirements, and deployed into the requirements for the subsystems, assemblies, and components, and for the processes of manufacturing, service, and customer support. A given system requirement may be accomplished by one or more subsystems with their component functions and sub-functions. When deploying requirements throughout the system, development teams should evaluate the consequences for the production processes. They begin with the manufacturing of critical components and materials, through the assembly and installation, including the maintenance and repair of the system.

Figure 9.2 illustrates the flow-down of requirements. Chapter 10 on Quality Function Deployment describes this useful methodology in detail. Shown above the first dashed line are the activities that are usually included in the domain of requirements gathering and translation. Between the first and second dashed lines are the activities that lead to the selection of the design concepts. This deployment depends on the baseline system architecture and those subsystem concepts chosen to implement the product. These dependencies allow development teams to

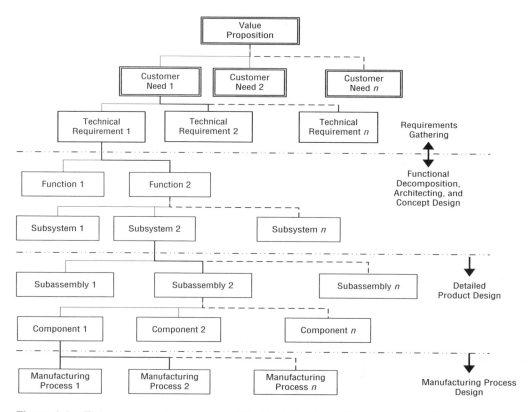

Figure 9.2   The product and process "tree" that is derived from the value proposition.

differentiate between those customers needs that have to be understood early and those that can be resolved later in the development project.

VOC Activities Vary Throughout the Development Process

Is the interest in customer-driven requirements limited to Phases 1 and 2 of product development? Our experience is that the more successful development projects are well supported by at least four levels of VOC throughout the development process, all of which are opportunities to exercise partnership with your customers.

Level 1

This type of customer engagement is characterized as ongoing research on the value drivers for broad populations of customers. The activities identify fundamental customers' needs, their perceptions of quality, preferences that affect trade-offs, the stressful conditions expected in the market environments, and their trends. They enable portfolios of development projects to be defined, based on value propositions that recognize future business opportunities. The proposed solution concepts in the portfolio help to drive technology development. For a specific development project, the customer-driven requirements have to be defined well enough to contribute to the charter for the project in Phase 0.

Level 2

In Phases 1 and 2, the engagement of customers is focused on those customer-driven requirements and stresses that are specific to the product under development. Customers' needs are gathered, analyzed, translated, and frozen progressively as decisions commit the project, downward through the architecture's hierarchy, as illustrated by Figure 9.2. Development teams engage their customers directly, not only to understand their latent needs and nuances but also to gain reasonable validation of the teams' understanding of the requirements and of the alignment of the selected product concepts with the intended applications.

Level 3

In Phases 3 and 5, customers are given the opportunity to provide feedback based on hands-on experience with engineering prototypes, preproduction units, and the service and customer support processes. Ideally, the feedback affecting the product design is needed in Phase 3, while in Phase 5 the feedback may be more relevant to support processes, since it is too late to change the design without potentially serious consequences. It's in Phase 5 that strategies for agile development can provide advantages when markets are changing rapidly.

Level 4

After product launch, customers provide feedback on the new product and its supporting services. The information may foster design changes for that product or specific requirements for follow-on products. It may stimulate revisions to user training or to preventive maintenance procedures and policies. Possibly customer support may need to be upgraded. The detailed feedback is essential for ongoing initiatives to improve customers' satisfaction.

However, it may not be easy to capture feedback that has sufficient detail and context to directly assist product development. The information systems may be designed for the needs of the product service and customer support organizations. So considerable effort may be needed to add details of interest to engineering and to ensure the timeliness of reports without frustrating those organizations. The best feedback in our experience has come from development engineers making customer visits to observe problems directly. Much can be understood by responsible people engaging unhappy customers in their own environments.

We have often found conflicts among development budgets, project schedules, and the fundamental activities of VOC. An overview of VOC steps may help you to plan and defend not only the detailed logistics but also the resources and funding necessary to engage your customers in a manner appropriate to your project.

The Major Steps in Requirements Gathering

Planning and Preparation

The activities of gathering and analyzing requirements have to be incorporated and defended in the project management plan. Many tasks depend on the resulting information. Often, this work is approached informally, not necessarily treated with the same process rigor that is reserved for developing the technical elements of the product. This can be a mistake. Development work is built upon sufficient and accurate sets of requirements. Otherwise, what are engineering teams suppose to develop? To do what? Delays at the front end can ripple throughout the project, jeopardizing those activities that depend on requirements. Activities that guess at requirements may have to start over if wrong.

VOC critics may argue that engaging customers can be a large effort with little return. They may feel that key project wizards already know more than current customers about future needs, or that customers tend to be poor reporters of their own needs. That viewpoint may satisfy the budget hawks but later put marketing in the position of having to convince customers to accept solutions that are looking for problems. It may be an unacceptable risk for a significantly new product, while possibly being reasonable for an incremental extension of an existing product.

The project leadership team then has to judge the expected value of a rigorous and ongoing engagement of target customers and be advocates for the strategies represented in their project management plans. A detailed plan that shows a no-frills approach to getting the required information will help build credibility and make it more likely to get management buy-in. For an in-depth treatment of gathering and processing customers' needs, see Burchill and Brodie.[15]

There are many elements to consider, such as whom to talk to, where to talk to them, how many interviews are necessary, who should engage the customers, how the conversations will be designed, what questions should be asked, how to observe customers without imposing a bias, and how to record the findings.

What strategies are to be implemented? Are group interviews, such as focus groups or customer councils, to be favored to get the benefits of interactive conversations among the participants and the economies of talking with several people at once? Or are one-on-one interviews to be favored to enable the probing for more depth in detail and to avoid the influence of power voices in the room? Are you to rely on indirect methods, such as surveys and telephone interviews, or on direct methods, such as observing customers in their own environment, inviting them to your customer innovation center, or, better yet, being a customer? You can watch your products being installed or answer customers' help calls, treating each complaint as a treasure. There are many trade-offs to consider, emphasizing the need for project teams to be students of VOC processes.[16]

Selection of Interviewees

Which customers will you interview? Customers can be classified in a couple of ways. One is by market segment; another is by whether they are leading-edge "early adopters" or conservative late adopters who wait until a new product is tried and proven by others. An interesting aspect of human behavior, which Rogers[17] saw in his fieldwork, is that even with highly beneficial innovations, such as ones affecting human health, there are no guarantees of wide acceptance. Rogers divided people into five distinct groups. The first group ("innovators") to adopt a new product was composed of those with advanced needs. Although small in number, they represented a leading indicator of where future markets were headed. They were followed by "early adopters," the "early majority," and the "late majority." The last group to adopt a product was called the "laggards." Those classified as innovators in one product area might have been laggards in another.

15. Gary Burchill and Christina Hepner Brodie, *Voices into Choices: Acting on the Voice of the Customer* (Madison, WI: Joiner Associates, 1997).

16. Abbie Griffin and John R. Hauser, "The Voice of the Customer," *Marketing Science* 12, no. 1 (Winter 1993); Francis J. Gouillart and Frederick D. Sturdivant, "Spend a Day in the Life of Your Customers," *Harvard Business Review*, January–February 1994.

17. Rogers, *Diffusion of Innovations*.

Some of the characteristics that can make a customer a good candidate for an interview include

- Enthusiasm for new ways of doing things
- Willingness to learn
- Likeliness to derive tangible benefits from the new innovation
- Experienced with similar products or systems
- Able to integrate a new product gracefully into their work environment or lifestyle
- Not risk-averse
- Cognizant of the value of a partnership with their supplier of products

Leading-edge customers may have idealized visions of future products and reflect that in their inputs. Whether or not their ideas are practical or achievable within the project capabilities is for project teams to judge.

Be certain to engage a wide range of people. The needs expressed by one group of customers may not comprise all of the needs of all customers. A conscious effort is critical to interview a range of customers representing the broad characteristics of the target market segment. Consider engaging people who are

- Currently satisfied users of existing products
  ("What do you like about our products or service?")
- Currently dissatisfied customers
  ("What problems are you having that we might correct in a future product?")
- Former customers
  ("What could have been done to change your mind?")
- Customers of your customers
  ("What attributes of the product outputs are most important to you?")
- Customers of your competitors
  ("What do you like about the competitive solutions?")
- People who have never been a customer
  ("What factors shape your buying decisions?")
- Customers' managers who approve purchase requests
  ("How does the product need to fit within your business infrastructure?")
- Customers' purchasing agents and those who advise them
  ("What product and price positions are appealing?")

The interviewing process is part of an overarching partnership with your customers. Engaging them constructively is a process itself to be experienced and improved over time.

Customer Engagement

An effective partnership with your customers is much more than a set of interviews. It may be an ongoing process with benefits to many development projects, and to your customers as well. Fundamentally, all of your organizations that touch your customers should be involved in some tangible and appropriate manner. Your teams need to be culturally competent, that is, speak the language of their customers and know their businesses or activities.

If you have read about "lean product development," you may know the anecdotal story of the Toyota chief engineer who spent substantial time in the United States with families who could benefit from vans and SUVs.[18] Similar stories are about Toyota development teams who drove existing vehicles for the same activities as did families, their customers. By submerging themselves in their customers' world, the development teams understood the nuances and reasoning of design details that would deliver clear and differentiated value, and the project leader became a visionary for future products. How else could they learn such lessons so well?

Be a customer. Use your competitors' products. Live with your customers in their environment. Watch what they do and how they have difficulties. Develop expertise in listening and observing.[19] You might even be able to videorecord customers using your product or those of your competitors. By being aware of needs that they tell you, the needs that they don't tell you, and the needs that they don't know to tell you, you can compete more effectively.

We remember asking a customer what he thought of our product's reliability. He said that there was no problem with its reliability. Our service representative was there every morning to ensure that the product worked well. Oh, boy, that was a message with a great many implications that we would not have understood without being in a one-on-one conversation.

Establish rapport with your customers and practice humility. Think like your customers. Experience what they experience. Feel their pain. The ideal is that your customers will become extensions of your product development teams.[20] Involve them in your creative processes. They may even have good ideas for new products.

Customer engagement is a cross-functional activity. It's a great opportunity for marketing and engineering to work together, to develop a common understanding of how customers make trade-offs and of the conflicts in designing better solutions.

Customer Interview Guide

Customer interviews require thought and preparation. Since you are asking people to give up a couple of hours of their time, it's helpful to ensure that the process is productive and yields useful information. There are mutual benefits. Be certain that the people who are interviewed understand that they have the opportunity to make contributions to the engineering of better solutions that they themselves may use in the future.

An interview guide is just a crutch. It gets you started and reminds you of the questions that you wanted to ask. We expect that you'll diverge from it in an actual interview. When developing an interview guide, there are some logical steps to include:

1. Identify topics that you want to explore during the interview.
2. Develop a number of high-level key questions for each topic.
3. Identify detailed questions to probe the depths of answers given to high-level questions.

---

18. Morgan and Liker, *The Toyota Product Development System*.

19. Sheila Mello, *Customer-centric Product Development: The Key to Great Product Development* (New York: American Management Association, 2002).

20. Jeff Jarvis, *What Would Google Do?* (New York: HarperCollins, 2009); Edward F. McQuarrie, *Customer Visits: Building a Better Market Focus* (Thousand Oaks, CA: Sage Publications, 2008).

4. Organize the results into a guide that can be used during the interviews.
5. Be selective and efficient in your choice of questions. Plan the interviews for durations of 60 to 90 minutes, allowing substantial contingency time for elaborate answers and unexpected topics that customers may introduce.
6. Select different lines of questions to be relevant to different interviewees. Not everyone will have inputs on the same topics. It's the summation of inputs that you're after.

There are many potential topics to explore, such as

- Elements of satisfaction or dissatisfaction with existing products
- Consequences of poor reliability in product systems
- Expected benefits from higher reliability and availability
- How costs influence purchasing decisions
- Desired enhancements to products or services
- Future activities or business processes that create new demands for products
- Customers' stories and visual impressions that illustrate ranges of applications, desired outcomes, and stressful conditions

Customer Interview Process

You are a guest and your customers' time is valuable, so please respect their schedule. Again, it's not wise to be overly directive about the path that the interview takes. Your interview guide is just a starting format, not a script. What might seem like a digression can often lead to valuable insights. Sometimes what started out as a concise, structured interview can evolve into a lengthy and detailed discussion of one or more "hot button" topics that you could not have anticipated. That can be gold! Sometimes, the interviewee has never been asked to provide his/her thoughts on the subject, so a lot of pent-up feelings and opinions can emerge.

Be certain that customers understand that the interview is not a sales call, nor is it meant as a service call to correct current problems they may be having. If major concerns for customer satisfaction arise, we suggest that you record the information and tell the customer that you will pass these concerns on to a responsive organization. However, don't spend interview time trying to solve problems. At a suitable time, follow up to ensure that the problems have been addressed. The ideal number of persons participating on the interviewer side of the table is three:

1. The interviewer is responsible for controlling the specific questions and the direction of the interview.
2. The scribe is responsible for capturing a transcript of the interview with as much verbatim content as possible. An audio recorder can assist. Some customers may be uncomfortable with being recorded, so always ask permission.
3. The observer is responsible for recording the context or images in the customer's descriptions. The visual image can be based on both real-time observations of the customer's environment and product usage as well as on the customer's "stories" about experiences using existing products or processes.

In our experience, alternating roles during an interview can add freshness to the discussion and more insight into questions.

Ask Better Questions

Not every question is a good question. In fact, not only should the questions be preplanned, but the way questions are asked can make a big difference in the value of their answers. It's not only the answers that you hear but also how the other person reacts to what you ask. Learn to ask better questions, demonstrating your relevant knowledge and an understanding of applications. Here are some ground rules that can help:

1. Keep questions open-ended, leading to explanations rather than "yes/no" answers. It's easy to filter the superfluous statements after the interview but impossible to respond to what was not said.
2. Customers may say, "I want ...," "Just give me ...," "It'd be great if ...," and express their unfulfilled needs in other ways. Among these statements are both needs and wants, with a range of priorities and contexts. Ask probing questions to understand the consequences of their needs not being fulfilled.
3. Probe for their desired outcomes, rather than for features that they need. Avoid talking about specific solutions.
4. Ask how they make trade-offs. What would they give up to get a more valued benefit? How do costs factor into their purchase and use decisions?
5. Keep questions nonjudgmental. There are no "bad answers." When asking "why" questions to probe for underlying reasoning or motivation, be careful not to give the impression that you think that the customer is doing something wrong. The idea of the "five whys" is to dig down for fertile ground until your customer tells you that he/she has already answered that question.
6. If the interview drifts away from your agenda, you'll need to decide what to do. As long as the discussion is relevant, it may be better to let the information flow. Often you can gain important insights into behaviors, values, and decision criteria from spontaneous tangents.
7. Ask the customer to elaborate or clarify when you have doubts about the meaning, context, or stressful conditions in their descriptions. Ask for scenarios to establish context, even demonstrations. Try "Show me" or "Can I try it?"
8. Listen reflectively. Tell interviewees what you heard them say, without interpretation. Be thoughtful about why your customers are expressing their viewpoints or feelings.
9. Keep your eyes wide open. Much can be learned by observing what customers do as well as by listening to what they say. These are valuable opportunities to see how your future product can be used and how competitive products are being used.
10. The work flow is an important dimension of the customer experience. Is it frantic, leisurely, or somewhere in between? Is work a "team sport" or does it depend more upon individuals? How does your product need to interact or interface with other products?
11. Don't be defensive. That will throttle the information flow. If your customers have negative feelings about current or past experiences with your product, listen and learn. If they are angry, allow time for that to be expressed in order for the discussion to shift to their actual needs. What gets in the way of their desired outcomes? Stories that customers tell about related experiences, frustrations, or even barriers to use can be very valuable. Negative feedback, if based on facts, can offer you an opportunity to improve.

12. Avoid giving any hints about what you think is the "right answer" to your question. It may inject bias into the interview and result in your customers telling you what they think you want to hear. Ask the same question in different ways to test the validity of their responses.

13. Before closing the interview, ensure that the key points have been covered to the extent that time permits. Give the customer a last chance to talk about important concerns that may not have been discussed.

14. Ask for your customers' help at a later time to validate the team's conclusions after the interview data have been processed. Be certain that your customers understand that you are there to understand their needs in order to develop the next product better. Their information can return benefits to them.

Asking better questions can be an art form. Listen carefully to see how productive your questions are and to identify ways to probe for insight more effectively with less potential for bias. Your development project teams depend on the knowledge that you bring to them.

It's very important to keep in mind the difference between "stated" requirements and "real" requirements. In his book on requirements engineering, Ralph Young[21] doesn't waste any time in making this distinction. Stated requirements are what customers tell you they need. Real requirements are what customers actually need. If unrecognized real requirements exist, so does the potential for not satisfying them. The worst scenario would be a "Basic need" not being recognized until late in a development project. That could have serious consequences for budgets and schedules, and it could be disastrous if not seen until after product launch. You can't assume that customers will necessarily identify all of their real requirements. The burden is on the development teams to tease out the latent and unspoken needs. If they get it wrong, the market will penalize your business.

Processing and Analyzing Interview Data

Customers are expected to give you plenty of data. Most will be in transcript form, ranging from relatively structured information to "stream of consciousness" statements. To organize the information, we suggest using a technique called the KJ Method, named for Jiro Kawakita (Japanese names have the last name first). It is an easy way to structure language data containing facts, opinions, and degrees of vagueness. The outputs from the KJ are inputs to the translation of customers' needs into the technical requirements that lead to concept selections.

"KJ" Organization of Requirements

The KJ Method organizes the customers' needs into affinity diagrams, helping you to organize disparate and fuzzy data. A "requirements KJ" starts with the many raw statements from customers, yet ends with a concise diagram that has requirements in prioritized groups with a hierarchy of layers displaying increasing details.

Who should participate? It's preferable that the team be cross-functional, including the perspectives of people who interviewed the customers as well as those who need to use the information. This may include marketing, engineering, sales, manufacturing, R&D, and quality assurance,

---

21. Ralph R. Young, *The Requirements Engineering Handbook* (Norwood, MA: Artech House, 2004).

and possibly even certain suppliers. How about including selected customers? This involvement develops a common understanding of the needs with the different functional viewpoints integrated.

What are some of the steps of a KJ transformation of customer-driven requirements?

Divide and Conquer (Work as Individuals)

Have each team member review the interview transcripts and write each statement on a three-by-five sticky note. Don't attempt to edit or combine statements; that will come later. This is an opportunity to edit the inputs into meaningful statements that are not redundant. Keep in mind that you are looking for customers' inputs that can lead to requirements. The specific details are not as important as the intuition behind them. Look for statements having specifics about

- Explicitly stated needs
- Customers' views about desired outcomes
- Needs that were observed, although not necessarily stated
- Problems that need to be resolved

If duplication exists in the interview transcript, it's not necessary to transcribe all of those statements. Select the statements that are clearer and more concrete. The best inputs may be based on personal experience.

Throw It at the Wall to See if It Sticks (Work as Individuals)

Place all of the sticky notes on a wall or whiteboard without trying to organize them. If a team member observes that a key input has been missed, this is a good time to add it. Redundant statements can be identified more easily with this display.

Identify the Uncertainties (Teamwork)

You have the opportunity to reduce the vagueness by going back to customers to ask for additional information or clarification. This step improves the overall quality and completeness of the data and prevents misunderstandings that could send the team off on low-value tangents. It's much better to ask your customer, "Please help me understand" than later to speculate about what he/she really meant.

Identify the Strong Voices (Teamwork)

You'll need to filter out the low-value statements as a team. An easy tactic is to have team members place a "sticky dot" on those statements that they feel meet the test of value: Do they know what is meant? Will this statement contribute to the requirements for the product? How well do the statements comply with the criteria for good requirements described earlier? Team members are asked to mark the statements that they feel should be retained for future use. Statements without a dot can be moved to a holding area. If another team member has already marked a statement with a dot, it's not necessary for it to be marked again since you are not yet voting on the level of importance.

Develop the Affinity Diagram (Teamwork)

The purpose of the affinity diagram is to identify groupings of statements that have something in common. As inputs to the QFD methodology you can also identify their relative importance. The importance levels attributed to groupings are derived from the impressions you form from interviews and observations while engaging customers in their own environment.

Time to Organize. Next, team members reorganize the customers' statements. Ask one person at a time to reposition the statements into groups with similar themes. If there are duplicates, place them in a holding area while the team considers other requirements. When team members find additional notes that are good fits with an already existing group, they should add them to the collection. If several groups have the same theme, they can be combined. For this task, it's best for participants to work silently, not discussing the details or their reasoning. There may be some outliers that don't appear to fit nicely into any of the groups. Do they represent a new affinity group or are they actually part of an existing group?

Naming the Groups. As a team, assign a name to each affinity group. The names should represent the underlying themes for each collection of needs statements. To capture a diversity of perspectives, ask team members to suggest names that they feel best reflect the theme, although not everyone has to agree on the name for each group. What is important is a common understanding of the implications of the statements. Group names should be written on sticky notes of a different color from those used for the customers' statements.

Organize the Hierarchy. Which statements are subsets or lower levels of detail to other statements? By placing subordinate statements into a tree format, your teams develop a hierarchy of details. Some statements may be seen as too deep in detail while others may beg for additional clarity. Consider an example of interview results for a clothes washer, illustrated in Table 9.1. The first column contains the "primary-level" strategic names for the affinity groups. The "secondary-level" tactical statements are the subordinate details for each affinity group. "Tertiary-level" operational details (not shown) may be necessary to get to the essence of customers' needs.

Select the Important Groups (Teamwork)

Not all requirements have the same level of importance for customers. With affinity groups, your teams can rank the groups into a useful order. This requires a good understanding of how customers might use the product and the stresses that the product might encounter. It's not necessary to be exactly correct in the initial ranking, since teams can later ask their customers for validation of the levels of importance. It can be expected that engineers and customers will organize and prioritize information differently.

There are several useful ways to assign relative importance. Here's one of them:

1. Give participants six votes each and ask them to allocate their votes as follows:
   a. Three votes for the most important group
   b. Two votes for the second most important
   c. One vote for the third most important
2. The vote allocation of each team member should be done privately, without discussion.
3. When all participants have decided on their rankings, they place a mark on the group names with their vote allocation. A convenient tactic is to mark each group's sticky note using pens or sticky circles of colors representing the importance weights. This makes it easy to quickly see how many first-, second-, and third-place votes each group receives.

Table 9.1 Hierarchy of customers' needs for a clothes washer

Primary	Secondary
Cleans clothes well	Clothes are clean Clothes are soft Removes excess soap, bleach, or softener Clothes not wrinkled
Full range of capacities	Accepts small loads Accepts large loads
Right height	Low enough for a shelf above control panel Washer height and tub depth allow easily reaching to bottom of tub
Easy to use	Controls are easy to use, easy to understand Guide to temperature for various fabrics and colors is displayed Conveniently measures and dispenses detergent, bleach, and fabric softener
Full range of operation	Handles white and colored, delicate and tough fabrics; hand wash; permanent press Cleans very dirty clothes Cleans "fine delicates"
Flexible operation	Able to manually change operating mode after start Able to add to load after start of cycle without spills or overflows
Green	Uses less energy Uses less detergent Conserves water
Modern or new capabilities	Filters lint from drained water Features for unusual items to be cleaned such as easily tangled clothes and sneakers Easily measure soap, bleach, and fabric softener Can turn on automatically at night when electricity costs are lower Notifies you with a pleasant audible sound when cycle is complete Can check wash cycle status remotely by phone or Internet
Quiet operation	Low vibration Won't move out of position due to vibrations Noisy enough so you can hear it Quiet enough to listen to TV or radio in next room
Attractive design	Range of optional color choices including chrome Something that looks more attractive than a cube Looks like a furnishing, not basement appliance Standard size
Easy to clean	Easily cleaned with water/sponge Surface not stained by detergents, softeners, or hard water

Validating Results of Requirements KJ with Customers

Process a Customer-Rated Importance Survey

After the team completes the processing and ranking of needs statements, they should validate their conclusions with customers. A survey can be useful for this step, either mailed or Web-based. It can be very simple, yet return valuable information. Ask customers to assign an importance rating to each of the requirements statements that have been extracted from the interview transcripts. For example, ask them to use a scale of 1 to 5 (1 = unimportant, 5 = extremely important).

Prior to this step, the requirements team has assigned a level of importance to needs statements. Possibly a key input was missed or its value misunderstood. The survey gives customers the opportunity to weight each of the requirements and thereby validate or correct the team conclusions. With the survey additional inputs can be requested in case the opportunity generates new ideas. Your teams will have taken a major step toward reducing their project risks.

While a customer importance survey helps you to understand the customers' feelings about each requirement, it doesn't help much to identify the opportunities for differentiation. The Kano Model described earlier has that purpose.

Classify Requirements Using the Kano Model

The Kano Model requires the development of a questionnaire that is focused on understanding customers' feelings for both positive and negative outcomes in meeting requirements. The CQM publication referenced[22] describes the use of the Kano Model and its variants in great detail. Since customers usually won't be aware of opportunities for "Delighters," development teams will have to infer them from what they hear and observe when engaging customers. Therefore the Kano questionnaire should be submitted to customers after the needs have been gathered and analyzed, as part of the requirements prioritization effort.

The Kano questionnaire asks customers two "what-if" questions. One is phrased positively, the other negatively. Positively phrased questions ask how the customers will feel if the need is satisfied and the solution implemented well. The negative questions ask how the customers will feel if the need is not satisfied or its solution implemented poorly. The answers are mapped to a classification matrix that determines the nature of the requirement. The Kano classification matrix, shown in Table 9.2, uses this terminology:

Basic need (B): Customers expect these features and are unhappy if they are not present.

Questionable (Q): Usually involves a logical inconsistency, for example, responding, "I like it that way" to both the positive and negative questions.

Delighter (D): Not expected by customers, but pleasing when present.

Reverse (R): We have it backwards. What we believe will be appealing to customers is not.

Indifferent (I): Customers do not care if the feature is present or not.

Satisfier (S): Increasing levels increase satisfaction; decreasing levels reduce satisfaction.

---

22. Center for Quality of Management, *Center for Quality of Management Journal* 2, no. 4 (1993).

A couple of possibilities from the clothes washer example may clarify this method.

Example 1

Positive Question	Negative Question
If the controls for the washer are very easy to use, how do you feel?	If the controls for the washer are not very easy to use, how do you feel?
Possible Answers	**Possible Answers**
(I like it that way.)	I like it that way.
I expect it.	I expect it.
I am neutral.	I am neutral.
I can live with it.	I can live with it.
I dislike it that way.	(I dislike it that way.)

By locating the customers' responses in the Kano classification matrix, you see that, in this example, "easy-to-use controls" would be a Satisfier. The easier the product is to use, the more the customers like it, while "harder to use" would drive satisfaction in the negative direction.

Example 2

Positive Question	Negative Question
If you are able to check the wash cycle status remotely using your phone or Internet, how do you feel?	If you are not able to check the wash cycle status remotely using your phone or Internet, how do you feel?
Possible Answers	**Possible Answers**
(I like it that way.)	I like it that way.
I expect it.	(I expect it.)
I am neutral.	I am neutral.
I can live with it.	I can live with it.
I dislike it that way.	I dislike it that way.

In this example, "remote status checking" would be a Delighter. If the capability is included, they would like it, but if not, they would not have expected it anyway.

The value in the Kano Model is the insight into the nature of the requirements. Understanding whether a requirement is a Basic need, a Satisfier need, or a Delighter can guide your decisions about how much emphasis to place on certain requirements. The Satisfiers and Delighters represent the opportunities for differentiation.

Table 9.2 Kano Classification Matrix

Customer Requirements Performance	Answer to Negative Question				
	Like	Expect it	Neutral	Live with	Dislike
Like	Q	D	D	D	S
Expect it	R	I	I	I	B
Neutral	R	I	I	I	B
Live with	R	I	I	I	B
Dislike	R	R	R	R	Q

(Left margin label: Answer to Positive Question)

Key Elements of Requirements Management

Approval of Requirements

After the customer-driven requirements have been validated, differentiated from expected competitive products, and evaluated for their contribution to the project's business case, key stakeholding managers should approve them. With their signatures, stakeholders indicate their agreement that

- The product is expected to deliver superior value to target customers.
- Satisfying the customer-driven requirements will be critical to the return of sufficient value to the business.
- The functional organizations are capable of implementing solutions to those requirements within the resources and constraints of the project.

Management of Creep and Gaps in Scope

You can reduce the risks of scope creep and gaps by doing thorough work with requirements at the beginning of the project. However, it can be expected that requests for new requirements will emerge. If a competitor has introduced a new product that is better than your planned product, the product's concept may need to be reevaluated. Possibly the challenge can be met by adding easily implemented features or functions. On the other hand, it may not be that simple.

Proposed changes to requirements should be evaluated carefully for their consequences to revenues, costs, schedules, and risks. What is the incremental value that they propose? How many more units would have to be sold so that the accumulated margin will cover the additional development costs? Often these assessments justify postponing a new requirement to a follow-on product.

Change Control

In our experience, a formal process for reviewing and authorizing requirements changes contributed well to reducing risks, particularly after the production-intent design is released. Every proposed requirements change was submitted for formal analysis. Depending on the nature of the change, evaluations included the consequences for resources, implementation timelines, inventories, schedules, sales forcasts, costs, service, and the many other parameters of the project. An approved change was thereby supported by a reasonable and acceptable business case.

Electronic Requirements Management

Requirements should be stored electronically, edited easily, and accessible to everyone on the project. Typically this means an online database accessible over the company intranet. There is a range of options for software applications that will make the job easier, establishing linkage among the requirements, verification test plans, test reports, and any other relevant information that can enhance traceability.

The function of a "requirements analyst" may be valuable, particularly to be a specialist in the software application for requirements management. That person should have editing rights, acting with the authority of the change control governing body. In organizations without much requirements infrastructure it can be essential for the requirements management process to receive visible recognition with reinforcement of the expected discipline. You do not want requirement change to be a casual process subject to knee-jerk reactions.

How Can Your Requirements Efforts Fail?

Bad Attitude

Have you heard this? "We know what customers need! We don't need to talk to them." In many organizations we can find people who presume to speak on behalf of customers. Having insiders acting as surrogates for customers can work if they have more than occasional informal conversations with customers. We've experienced situations where the inputs from surrogates were extremely flawed with disastrous consequences.

A variant on this concern is the assumption that there will be no competition. Customers will be expected to purchase the product because they have the need with no alternative solution. "What you see is what you get" (WYSIWYG)! In one of our experiences, that assumption was badly wrong.

Failure to Employ a Cross-Functional Team

People from different organizations see and observe customers with different viewpoints and interpretations. Salespeople think differently from in-bound marketing specialists. Design engineers think differently from manufacturing engineers. A business person thinks differently from a technical development person. Your project can benefit greatly from those varying insights being integrated. Support from functional managers may depend upon people from their organizations being thoroughly involved, increasing the credibility of the conclusions.

Lack of Management Support

An activity trap may confuse motion with progress. An example would be the criticism that if the product isn't being designed, with prototypes being fabricated and tested, nothing is happening and progress is stalled. In that scenario, the requirements-gathering activities may not be valued and are, at best, grudgingly allowed. Support can be shallow: "Okay, you can do it if you must, but don't let it affect the schedules or budgets!" Most grassroots initiatives requiring cross-functional support can have many obstacles, with success being difficult to win. In the strategic activity of requirements gathering, the lack of management support can be fatal.

Insufficient or No Budget

The project team may have failed to plan for sufficient time and resources. Either the activity was omitted from the project management plan, or the magnitude of the task was underestimated. Experience with previous projects can contribute to more realistic estimates for labor, timelines, and costs. There is plentiful evidence that a poor job with requirements is likely to have serious consequences for the project and its business plan. On the other hand, if the activities to engage customers well are viewed as investments with expected returns that are valued, their funding can be more easily approved.

Wrong or Incomplete Requirements

Among the most prevalent causes are

- A sample size for customer interviews that is too small, leading to missed inputs from important customers or market applications
- Failure to close the loop with customers to validate the requirements and their importance ratings

Requirements That Are Not Sufficiently Differentiated

Common root causes include

- A lack of benchmarking of competitive products
- A lack of innovators or early adopters represented in the sampling of interviews

Poor or No Process for Disciplined Requirements Management

A lack of assertive requirements management can open the door for scope creep with its consequences for the project schedules and costs. Design verification may fail to determine if all requirements have been satisfied, leaving the product inadequate and putting the value delivery objective at risk.

Key Points

1. Everything your projects do in developing a new product must be built upon a foundation of requirements. Good requirements enable successful outcomes.
2. The purpose of requirements gathering is to understand "real" requirements versus "stated" requirements. They identify the many elements of value that your customers expect your product to deliver, what your customers pay for. At times they can be unstated and must be inferred from observations of customers or from personal experiences.
3. The process of gathering requirements begins when you have an idea about how to satisfy the needs of customers. It is done best with direct customer contact. While in-house customer surrogates can be useful, your teams must understand the sources of their insight and the potential for bias.
4. You have three powerful tools for gathering and analyzing requirements:
   a. **Customer interviews:** Customer interviews should be well planned. Whom you talk to and the tactics in the interview have significant influences on the information extracted. There are best practices that can be followed in both selecting interview candidates and in conducting the interviews.
   b. **KJ Method for processing interview data:** The KJ Method helps to turn the raw data into useful information by translating the often qualitative, fuzzy statements into sharply focused customer-driven requirements. Teams then decide on the

relative importance of the requirements, using what they have both seen and heard while interacting with their customers. After completing the translation and prioritization of requirements, teams need to validate their findings with feedback from customers. An effective tactic is the customer-rated importance survey, combined with the Kano Model.

c. **The Kano Model for requirements classification:** The Kano Model helps teams understand and classify requirements. Its power is in how it can identify features and functions that will be Satisfiers and Delighters. Needs in those categories are opportunities for clear differentiation of your product and the potential for competitive advantages in the market.

Discussion Questions

1. By what processes do your development projects obtain their requirements? How do project teams know that the requirements that they strive to satisfy reflect the needs of their customers?
2. What barriers do your project teams face in engaging their customers directly, engineer to customer? How often can your engineering teams spend time in your customers' environment and how effectively can they learn from those experiences?
3. How well can your engineering teams use your products and those of your competitors?
4. When your engineering teams engage their customers, what questions do they ask? How do they ask those questions? How are the responses recorded? Are video recordings practical?
5. How well and how often do your project teams engage their customers in the early phases of development?
6. How representative are your customers' inputs of the broad needs of the selected market segment(s) across the countries and cultures in your target markets?
7. How vulnerable is the voice of your customers to being trumped by the voice of your management?
8. How well do marketing and engineering work together?

Additional Reading

Hooks, Ivy F., and Kristin A. Farry. *Customer-Centered Products.* New York: AMACOM, 2001.
Ries, Al, and Jack Trout. *Positioning: The Battle for Your Mind.* New York: McGraw-Hill, 2001.
Young, Ralph R. *Effective Requirements Practices.* Boston: Addison-Wesley, 2001.

Additional Authors Worth Following

Christina Hepner Brodie   John R. Hauser   Sheila Mello
Gary Burchill   Pamela Hunter   J. S. Pinegar
Clayton M. Christensen   Jeff Jarvis   Stefan H. Thomke
Lou Cohen   Gerald M. Katz   Eric von Hippel
Abbie Griffin   Edward F. McQuarrie   Anthony W. Ulwick

Quality Function Deployment

Quality Function Deployment (QFD) is a methodology to translate the needs of your customers into the requirements for your solutions. Through a number of workshops early in a development project, QFD integrates the voice of the customer (VOC) with the "voices" of your business and technologies. If carried to its fullest extent, it can trace requirements and align decisions for manufacturing, service, and customer support back to customers' needs. The primary purpose of QFD is to enable the selection of your baseline product concepts and the approaches for service and customer support to be based in part on criteria that drive the most value to your customers.

Since QFD was introduced into the United States from Japan in the mid-1980s,[1] it has been adapted by many companies to be a valuable tool for product development. Some of the early articulations of QFD were very complex, and without skilled facilitation experiences were disheartening. However, many companies have been encouraged and have invested in learning how to adapt QFD to the character of their projects and to use it effectively.

With efficient practice and sufficient management support, QFD can provide major benefits to your product development. An important principle is that this early investment in analysis, planning, and requirements can save substantial time later in your process. In this chapter we describe QFD, offer some useful insights, and refer to sources for you to explore more details.

What Is Quality Function Deployment?

"Quality function" does not refer to your organization for quality assurance. Instead, it is focused on the functions of your products and services that deliver quality in their benefits to your customers. The term *deployment* refers to the cascading of requirements from the outside in, from the level of the full product system to the subsystems, modules, components, parts, software, and materials, and to their manufacturing and assembly processes in production. Similar deployment applies to the requirements that affect repair and maintenance functions and those for customer support. All contribute to the value intended to be delivered to your customers.

---

1. John R. Hauser and Don Clausing, "The House of Quality," *Harvard Business Review,* no. 3 (May–June 1988).

Customers' needs can be vague and qualitative, not easily measured. Technical requirements, on the other hand, have to be specific and measurable. QFD translates those requirements from the language of customers, in the context of their applications, into the language of the engineers who are challenged to develop solutions. It's a methodology that leads development teams through focused data gathering, probing questions, and discussions. With a large visual format it records the decisions that prioritize requirements and sets their targets. That enables the selection of solution concepts that can implement their value proposition, increase the probability of higher customer acceptance, and support your pricing strategy.

QFD brings cross-functional participants together to organize information that is necessary to enable development activities. Without requirements, creative and differentiated solutions cannot be selected. Without system-level requirements being allocated, that is, deployed throughout the system's architecture, the design concepts for subsystems cannot be chosen. With information from competitive assessments, QFD guides teams to establish requirements that incorporate those attributes of the product and services that can make them competitively superior in the market. It can also be used to assist in the allocation of company requirements, such as for manufacturing or service costs or for compliance with user interface standards. By supporting both benefit increases and cost reductions, it is a value improvement process.

The format of QFD is a set of linked matrices that map the relationships between requirements and their sources. In practice it can be a large wall display onto which QFD teams record their data. That can provide an element of discipline to the process as teams work their way systematically through its various elements, gathering information, asking questions, discussing alternatives, and making decisions. The large display enables teams to see the big picture, to evaluate the patterns, and to focus on those requirements that are judged to be the most important to customers. Conflicts among requirements can be identified so that they can be resolved early in the process. Software applications are available to help the information be recorded, distributed, and preserved, and to reduce the work of computations.

The matrix format of the display is particularly useful when there are "one-to-many" relationships to be mapped. By that we mean that the satisfaction of a particular customer need involves compliance with requirements for more than one product function or feature. Similarly, the achievement of the solution to a particular product or service requirement contributes to more than one customer need. The same concept applies to the deployment of requirements throughout the product system, as top-level requirements are allocated to the various system elements. If the requirements need only one-to-one maps, or if each requirement is deployed directly to one subsystem function, a simple table can be more useful than a matrix. Much flexibility in the application is intended, and many companies stop at the highest-level matrix. Those who have mastered the mapping of requirements into manufacturing and service have "deployed the needs of their customers to the factory floor."

Whether at the full system level or for lower levels in the system, the requirements contribute criteria for the selection of the baseline architectures, design concepts, enabling technologies, and supporting processes. The strategy uses QFD to link the concept generation and selection process (Chapter 11) to the VOC. This integration has been called "Enhanced Quality Function Deployment."[2] The linkage of tools focuses on the selection of the available best solutions, enabling the trade-offs and consequences for the project management plan and the project's business plan to be evaluated and justified.

2. Clausing and Stuart Pugh, "Enhanced Quality Function Deployment," Design and Productivity International Conference, Honolulu, HI (February 6–8, 1991).

Expected Benefits of QFD

Requirements must be developed and agreed upon early, prior to the selection of their solution approaches. How can it be wise to choose the product concept without knowing its requirements? The model is that the product concept is chosen to satisfy requirements that have been selected to achieve the value proposition and thereby implement the portfolio strategies for product and price positioning. It's a "market pull" strategy. With more completeness, integrity, and accuracy in the early requirements, the probability of late, disruptive design changes is reduced. Fewer mistakes provide clear benefits to time to market and development costs. With a shorter time to market, or at least market entry that is not delayed, revenue streams start earlier and development costs ramp down sooner. Prioritized requirements lead to properly focused resources. The linkage of information in the QFD format makes it easier to evaluate the consequences of proposed changes to requirements and to decide how to react. You can see that it can be a valuable tool for the development teams.

More value is derived from application in the front-end phases of projects. During advanced product planning, competitive assessments and portfolio analyses lead to a value proposition for the new product or service. In this phase QFD enables planning teams to establish high-level requirements that can lead to the initial proposals for product concepts and their enabling technologies and, subsequently, to initial development plans and business plans. In the initial phases of product development, teams can use QFD to define more extensive and detailed requirements and to identify elements of the VOC that are not understood well enough. This dialog may stimulate additional data gathering and decisions as a proactive initiative.

The first step with these applications is to establish those system-level requirements that drive the selection of the product's architecture. That then enables the high-level requirements to be deployed to functions and subsystems so that lower-level concepts and technologies can be selected. Feedback from customers may validate the requirements or may add or modify certain ones. Teams can easily evaluate the impacts and make the decisions that lead to completion and approval of the concept for the product system. Your requirements management application can then provide easy access and change control with traceability to the sources of requirements. The objective is for the product concept to be reasonably stable and achievable so that the acceptability of the resulting project plans and business plan can be judged.

In most cases, the development and delivery of new products and services involve cross-functional team activities. Some of your organizational functions may not see how or why they need to participate in early collaboration, and others may perceive that there are cultural barriers among them or at least language barriers. QFD teams are established to include broad representation of those functions that either touch the customers in some way or develop solutions for those customers. By working through the methods, team members develop a common understanding of their customers' business or activities, of how the new products and services have to be developed, and of how the selected value proposition must be delivered. A shared understanding of the requirements and their rationale is developed as well as an appreciation for the range of viewpoints represented. Participants learn to ask better questions and to think with an improved integration of business and technical strategies. The improved collaboration that results increases the probability that the new market entries will have the quality and cost structures expected, will be launched smoothly and on time, and will be well differentiated from their competition.

The Structure of QFD

The highest-level structure of QFD is often called the "House of Quality." The name reflects its layout. You can see in Figure 10.1 that the linked portions of this matrix resemble the shape of a house with rooms, a roof, and a basement. Some companies have developed creative adaptations of the layout to suit their particular needs, so consider the format and computations to be flexible as long as the objectives of the method are satisfied. The process uses the rows in the relationship matrix as inputs, and the columns as outputs.

In the rows, the needs derived from the VOC and the voice of your business (VOB) are placed on the left side and the competitive assessments and project goals are recorded on the right. The column headings contain the technical requirements for the product and/or its services. The required features and functions are described at the top as categories for the requirements, and their priorities, competitive assessments, targets, and tolerances are documented at the bottom. It is expected that certain needs of customers can create conflicts among the technical requirements. The matrix in the "roof" of the house leads teams to identify those conflicts so that they can be resolved early in development.

Project Charter

The best starting point is to clarify the team's understanding of the objectives, constraints, and assumptions of their development project. These basic details should be written to ensure

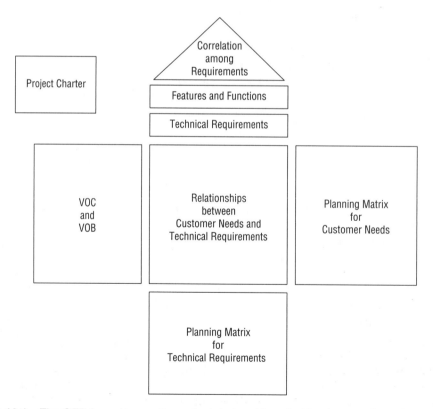

Figure 10.1   The QFD format is a pattern of related matrices that lead teams through data analyses and decisions in a disciplined manner.

that there is a common understanding of their implications. Figure 10.2 gives an example, which includes

- Definition of the target market segment and the value drivers that characterize it
- Description of typical customers and the range of their product applications
- Value proposition for the new product and/or service
- Portfolio objectives and constraints for the project, such as the date for market entry
- Key business and technical assumptions

You may be amazed that there can be wide misunderstanding of the markets to be served and of those attributes of products and services that customers value most. This is an initial agenda for marketing and engineering to get on the same page.

Customers have many needs that are either not fulfilled or only partially fulfilled by current products. However, it may not be realistic for a new product to satisfy all of them. The value proposition states the strategy for competitive differentiation in the benefits that will be delivered, compared to their costs. It creates a clear focus to select those needs of target customers that are to be addressed and to prioritize them. Naturally you want the more important requirements to be aligned to that value proposition.

Who Are Customers?

For many products, a range of "customers" must be considered. For products aimed at global markets, ranges of customers, characterized by geography, infrastructures, standards, cultures, and languages, have to be considered. Grouped as a market segment, they have common value drivers. Your product has to appeal to all of them or be able to be adapted to subsegments of them.

"Customers" are not necessarily just those who use or operate your product. There are those whose role is to purchase the product or to authorize its purchase as being suitable within their business infrastructure. Retailers have a lot to contribute, not only because they touch

Project Charter		
Project Name	Development Phase	Revision Date
Target Market and Value Drivers		
Characterization of Customers and Range of Product Applications		
Value Proposition for New Product/Service		
Key Business and Technical Assumptions, Objectives, Constraints Target Date for Market Entry		

Figure 10.2   The project charter should be kept visible and verified.

your customers directly, but also because they have to fit your product strategically within their display space.

Voice of Your Customers

In Chapter 9 we talked about gathering the voice of the customer. That's the source of information placed in the first room, on the left side of the House of Quality. Its articulation is in the language of your customers with the context of their applications and other imagery that can enhance the team's understanding.[3] If the product is to be new to the world, customers' inputs may be in terms of benefits that are needed, independent of the concepts for the solutions. If the product is to be an incremental design in the context of an existing product line and competitive products, customers' inputs may refer more to current capabilities that do not yet satisfy them.

Remember, from our discussions of the Kano Model in Chapter 9, that customers tend not to talk about all of their needs. They assume that you already know their "Basic" ones. They don't know enough to ask for "Delighter" features, so they will talk mostly about capabilities for which more functionality is better, the "Satisfiers" in the Kano Model. Your requirements team has to fill in the rest with their knowledge of the intended product and its applications and with their intuition about how to differentiate the product and delight their customers. A "Delighter" need can be treated as a "voice of your business" that reflects opportunities for differentiation. A capability that "delights" customers can be critical to gaining initial sales, while a "Satisfier" can be critical to repeated sales.

It is expected that expressed needs can have variations in levels of detail. In the "VOC room," the higher-level, somewhat vague statements of needs are placed in the first column. These are often called the primary or strategic needs, the ones that you would expect to reflect the value proposition or the value drivers for the market segment. If the VOC gathering has organized the many needs with the KJ Method, the "primary level" would be the headings for the affinity groups.

What do these high-level statements actually mean? For example, do you know what your customers mean by "high reliability"? Certainly you can expect a range of interpretations. Working with the VOC inputs, or with customers themselves, the next step is to decompose the high-level needs into lower-level statements, at the secondary or tactical level, and subsequently to the tertiary level, if necessary. Figure 10.3 illustrates a template for this tree diagram, usually with three levels being sufficient and manageable. Some companies find that more levels can be useful. The lower levels express the needs more clearly, with specific details that capture what is actually meant by the primary-level affinity categories. These verb-object statements must be unambiguous so that their translation into engineering language can be accurate. As a validation step, you want your customers to say, "Now you understand us."

In the example in Figure 10.4, a needs statement such as "high reliability" may have several different interpretations and implications, depending on customers' applications and their experiences with existing products. For a repairable product such as a high-volume printer, reliability can be related to service call frequency or to how well customers can fix a problem themselves. Specific to a printer are viewpoints such as dependability for long print runs or whether pages from different print runs look the same. This process of VOC interpretation and decomposition should be facilitated skillfully so that those customers' needs that are documented create a complete description without redundancy or trivial differences. Each line in the matrix should be worthy of the time that the QFD team has to invest in its analysis.

---

3. Mello, *Customer-centric Product Definition.*

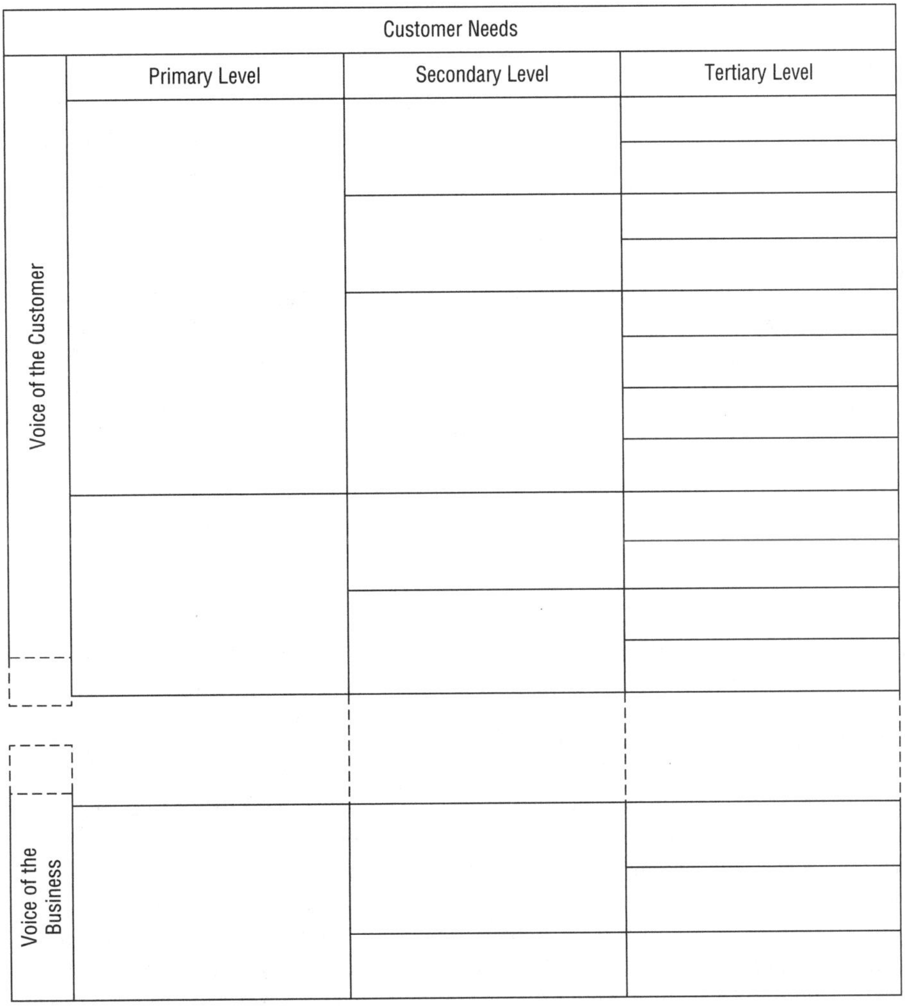
Customer Needs
Primary Level
Secondary Level
Tertiary Level
Voice of the Customer
Voice of the Business

Figure 10.3 The VOC room in the House of Quality documents the needs of customers, decomposed to be clear, accurate, and sufficient statements of their value drivers.

Planning Matrix for Customer Needs

Figure 10.5 illustrates a planning matrix with some entries to illustrate the process. The method assigns scores to represent the market segment's preferences for individual needs. They are not expected to be the same. The techniques provide ways to establish the relative importance among them. This is a numerical ranking, with a higher number representing a need that is more important.

Be careful; you're developing a product for a market segment, not just for a few vocal customers. We advise not using a large numerical range for ranking, since that may promote nit-picking about small differences that matter very little. It might create a false sense of precision in the data and force calculations within QFD to generate large numbers that can be confusing. How about "high," "medium," and "low" rankings, with associated numbers such as 5, 3, and 1 to use in calculations? The idea is to identify differences so that the team can choose where to focus

Customer Needs			
	Primary Level	Secondary Level	Tertiary Level
Voice of the Customer	High Reliability	Stable Image Quality	No variation during long print run
			No variation from run to run
		Trouble-Free Paper Handling	Very low jam frequency
			Easy jam clearance without bending
		Always Available	Very few service calls per month
			Easy access to phone support
			Maintenance and repair by key operator
			Off-hours preventive maintenance

Figure 10.4   This VOC example illustrates that a range of related needs may be satisfied by both the product design and the capabilities of service and customer support.

development efforts. This strategy can keep the process from getting too complex. Some companies have used scoring over a 100-point scale. That makes our brains hurt.

In Figure 10.5 we show the "importance" (aka desirability) aligned to the tertiary-level needs. The ranking of needs by customers should correlate with the value proposition and with the value drivers for the market segment. If not, something is wrong. Do you have a value proposition that customers will reward with revenues? How is it to be reflected in the capabilities of the product?

Opportunities for competitive differentiation can be found by benchmarking products currently in the market, or by forecasting the capabilities of those that you expect to be launched by your competitors. By looking at how well your current product(s) satisfy each need and how well you expect your competitors to do after your product enters the market, your QFD team can establish goals for each need that will enable the product and price positioning. They should use data as much as they can, rather than personal opinions. The goal setting is a numerical scoring, typically using a scale of 1 to 5 or 1 to 10, for example, with a resulting calculation of the relative improvement required for the new product, when compared to your current product(s).

Certain customers' needs, if met really well, can have major benefits to the processes of marketing, advertising, and selling. These can be scored also. Our advice here is to use a small scale—1 to 1.5, for example—to avoid an excessive influence over the priorities. You might expect these "benefit to sales" scores to be aligned to the value drivers for the market segment.

Now there can be a calculation of the overall importance of each customer need statement. In the QFD methodology, it is the product of the

- Importance of the need to target customers
- Degree of improvement set in the goal for the product
- Potential benefit to sales

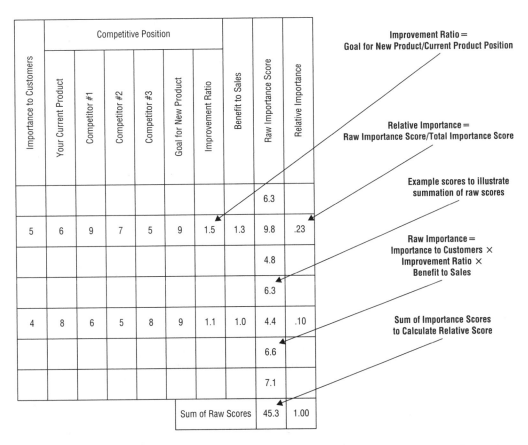

Figure 10.5 In the planning matrix the relative importance of customer needs can reflect not only their importance to customers, but also their development challenges and the leverage they can provide to sales.

For example, a high customer importance of 5, an improvement ratio of 1.5 and a sales benefit score of 1.3 would produce an overall importance rank of 5 × 1.5 × 1.3 = 9.8, while lower scores of 3, 1 and 1 would yield a rank of 3. The first is roughly three times more important than the second. The usual practice is to convert absolute importance ranks into relative or normalized ranking, by adding the raw scores and dividing the total into the individual rank. This % ranking is then used in the relationship matrix described next.

Requirements for Features and Functionality

Customers' needs may be satisfied partially by the capabilities of the new product, which can be characterized as features and functional capabilities for which measurable requirements can be assigned. They may also be contributed to by the capabilities of repair and maintenance services or of customer support. These requirements are often requirements for the entire project to satisfy via an integrated system of products and services. In our example for a repairable printer, both the reliability of the product and the preventive maintenance performed by either the customers or the service organization contribute to the customers' need for the printer's availability for use. What are the requirements for each? As with the customer needs rows, the attributes, features, and functions are arranged in columns as a tree diagram, with affinity headings that relate to the functional architecture of the product system.

Although the technical requirements are in the language of those who have to develop the solutions, they should not direct a specific solution. This independence enables the development teams to select the best product architectures, design concepts, and service approaches, for example, to achieve competitive superiority in the value delivered to their customers. In Chapter 9 we identified additional criteria for good statements of requirements.

As we mentioned in Chapter 9, products also have requirements that are internally driven, such as manufacturing and service costs, or the mandate to satisfy certain corporate standards. Externally driven requirements can also be critical. These can include regulations imposed by industry standards or government agencies or the need to accommodate a range of electrical power systems, space constraints, or foreign languages. QFD is most suited to translate those that are derived from stated customers' needs, but not necessarily the full range of sources of requirements that must be satisfied. It's a valuable tool, but it won't do everything that you will need. Other technical requirements can be defined directly from their sources or from previous products that are being leveraged.

Requirements Relationship Matrix

The cells at the intersections of the rows and columns map the relationships between the needs of customers and the corresponding technical requirements. For each cell, the QFD team is asked to judge how strongly the requirements will contribute to the satisfaction of the need. Usually this rating has just one of four levels: strong, moderate, weak, or none. Since the intention is just to understand the traceability and to establish priorities for development, additional scoring levels would add unnecessary details. Each evaluation can be recorded both as a symbol and as a numerical rating, with the rating scheme chosen to force differentiation. These decisions foster thoughtful considerations about how the needs of the customers can be satisfied most directly.

The objective is to establish strong differentiation among the cells by using a strategy for the contribution scores, with associated symbols, such as

Strong:	9	◎
Moderate:	3	○
Weak:	1	△
None:	0	Blank

The QFD team learns to ask better questions. It may find that the answers reveal unexpected relationships that have to be developed. It's good for teams to step back from the details. They often find that the pattern of symbols is easier to interpret. A glossary with clear definitions of terms, such as relationship strengths, can improve their discrimination.

Your team can benefit from a matrix that is not too large having not too many cells for which there are significant correlations. Otherwise the establishment of useful priorities can be diluted and the matrix can become overwhelmed by too many "trees to see the forest." For an entire product the matrix could be quite large, with hundreds of rows or columns resulting in thousands of cells. That's very difficult to manage. We prefer much smaller matrices that can be developed by working at a higher level, by merging some needs or requirements, or by subdividing a larger matrix in a logical way, such as by module or function. Your teams should identify significant rows and columns so that the cells with noteworthy correlations are no more than, say, 25% of them. Clever guidance by an experienced QFD facilitator can help your teams work more efficiently.

Each cell has two parts. One is the symbol previously mentioned, or its contribution score. The other is a priority rating, calculated by multiplying the contribution score by the overall percent importance that is calculated for the row in its planning matrix. A cell showing a significant contribution and relative importance—say, 9 and 22% respectively—would have a priority score of $9 \times 22 = 198$, while one at the opposite end of the spectrum might have a priority of $1 \times 10\% = 10$. You can see that using smaller numbers for the ratings still yields a significant differentiation in the priorities. Figure 10.6 illustrates a partial example for a relationship matrix.

After each cell is evaluated and scored, the priority scores of the cells in each column are added. This gives the overall priority for each technical requirement, representing the combined effects of the contribution to all of the customers' needs, the importance that customers place on those needs, the improvement necessary to meet the competitive position, and the potential benefit to sales.

If a row has no contribution from the columns, either an additional technical requirement should be identified or the need would be judged to be irrelevant, outside the ability of the product or service to satisfy. Similarly, if a column has no relationships with the rows, either the requirement is not necessary or a relevant customer need has not yet been articulated. A way to

| | Technical Requirements | | | | | | | | | | Planning Matrix | | | | |
| | New Product | | | | | | New Services | | | | | | | | |
VOC	Feature #1	Feature #2	Feature #3	Function #1	Function #2	Function #3	Repair/Maintenance Requirement #1	Repair/Maintenance Requirement #2	Customer Support Requirement #1	Customer Support Requirement #2	Importance to Customers	Improvement Ratio	Benefit to Sales	Raw Importance Score	Normalized Importance Score
	9/126					3/42				1/14				6.3	.14
		9/198						3/22		3/22	5	1.5	1.3	9.8	.22
	3/33		9/99			3/33				1/11				4.8	.11
				9/126					3/42					6.3	.14
		1/10			9/90					1/10	4	1.1	1	4.4	.10
						9/126								6.6	.14
							9/135	3/45	3/45	1/15				7.1	.15
												Totals		45.3	1.00

Figure 10.6 The requirements relationship matrix maps the technical requirements that apply to the product and/or services. The cells document the extent to which each requirement will contribute to the satisfaction of each need. The priority of each is that contribution multiplied by the normalized importance from the planning matrix.

assist this process is to cross-check the requirements against a list of generic requirements that have been applied to previous products. Your team may find some meaningful ones that have been overlooked.

Correlation of Requirements

The cells in the roof matrix map each technical requirement to each other one. The question asked here is whether the efforts to satisfy one requirement will support or inhibit the ability to satisfy the other. If one inhibits the other, you have a conflict that must be resolved. It may be that the source is a conflict between the intent of the design and the laws of physics. Physics will win unless the conflict is resolved. Ideally the requirements will be independent of each other, although in practice a few interactions may be expected. Resolving the conflicts may involve a change in the architecture of the product, a decision to disregard a low-priority customer need, a design compromise that balances the targets for these requirements, or some other creative solution. The objective is to identify those conflicts before commitments are made to difficult requirements or to flawed design concepts.

Many teams just note the positive or negative correlations with symbols such as ■ and *, as shown in Figure 10.7. Other teams assign numerical scores with positive or negative signs and even carry those indications down to the planning matrix in the "basement." We suggest just sticking with the symbols and focusing more on resolving the conflicts in the design requirements rather than making the calculations more complicated. An important outcome of this analysis is a plan for specific teams to work closely together where there is a conflict or to place the conflict within the responsibilities of one work group.

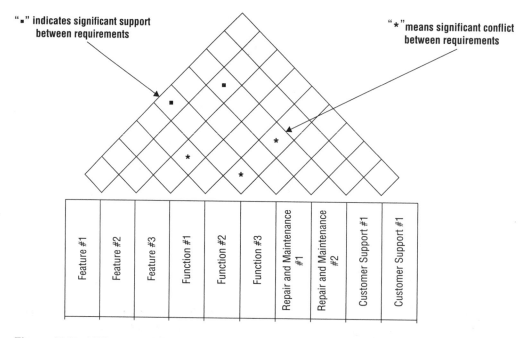

Figure 10.7   With the "roof" of the House of Quality, teams evaluate the interactions among the technical requirements. Those for which there are conflicts need to be resolved early.

Planning Matrix for Technical Requirements

All of the work in moving through the rooms in the House of Quality is aimed at establishing target values and tolerances for those technical requirements that are correlated with the customers' needs. The priorities are calculated to help the development teams focus on those requirements that have the largest impact on the project. Remember that the technical requirements also may be derived from sources other than customers, such as company standards and portfolio objectives, or external standards and regulations. Once identified, they may be included in the QFD process to identify potential conflicts with customers' requirements.

With the flexibility of QFD you can include other relevant information, as shown in Figure 10.8. This extension can enhance the selection criteria for the baseline designs. The additional details may contribute also to the selection of the approaches for product service and customer support to the extent that they contribute to satisfying customers' needs. The overall

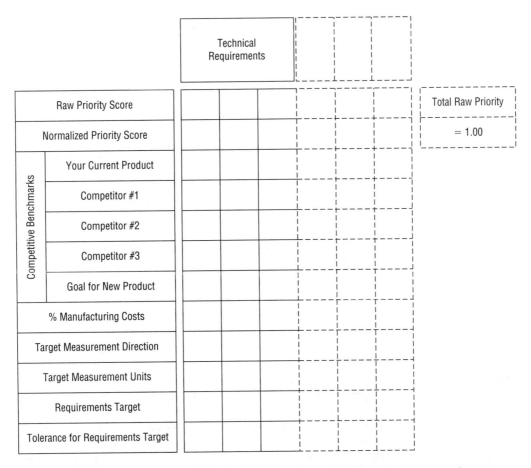

Figure 10.8   In the planning matrix for technical requirements, the evaluation expands upon that used for customers' needs. This is a flexible template to capture details leading to the setting of measurable targets and their tolerances.

priority of a requirement indicates how strongly your QFD team has judged the requirement to contribute to overall customers' satisfaction. The relationship matrix documents the rationale. The paradigm is to "make decisions so that your customers win" by translating their needs into the requirements for your products and services. Those target parameters become the criteria by which decisions are made for the verification of the product design and its quality assurance in production.

Data from your competitive products lab (you do have one, don't you?) may identify stretch targets for selected requirements in order for your new product to be "superior" in your market segment. This would be a side-by-side comparison of performance under the same conditions with the same measurement process. In the example, the unit of measure for each requirement is shown as the direction of its optimization, such as "Nominal is best," "The smaller the better," or "Larger is better." Teams can then judge the risks or difficulties in implementation toward satisfying each technical requirement.

Another modification is to estimate the development costs or manufacturing costs to achieve a requirement. This can focus the efforts of value engineering or the planning for lean product development. A high-risk requirement that has high priority toward satisfying customers' needs, a significant competitive challenge, and a distorted cost structure should receive particular emphasis in the project management plans. That is, the allocation of skills, funding, prototypes, and schedule must represent an achievable plan to develop an efficient solution to that requirement.

Sequence of Matrices

The first matrix of QFD applies to the level of the full system product and its services. The "deployment" of requirements means that the top-level requirements are decomposed and allocated throughout the elements of the design architecture to the level of the specifications for the manufacturing or procurement of components, parts, materials, and assemblies. In the case of services, the deployment is to policies, procedures, and service quality measures. Figure 10.9 illustrates the methodology as a series of linked matrices.

Linkage is accomplished by the output columns of the first matrix becoming the input rows of the next matrix. Not every column needs to be transferred, just those most critical to success. The columns of the next matrix are the technical requirements at that level. For each, imagine the product architecture that is chosen and the several subsystems that are defined to achieve the functions of the product. Each subsystem will have a QFD matrix. Likewise, customer support and product repair/maintenance will have specific matrices. The calculations and decisions in the columns will lead to the requirements for that subsystem or process. From that matrix your teams can generate subsequent matrices for assemblies, components, control software, consumable materials, or processes. It depends on how deeply the one-to-many relationships continue. Often the matrix approach is no longer useful beyond the subsystem level, and the specifications for the designs of parts and components can be linked one-to-one with the subsystem requirements. This depends on the architecture of the subsystems. Analogous advice applies to service and customer support. Although early descriptions of QFD identified upward of 30 matrices, many companies limit their practice to only a few, just those that add value and are worth the investment in the time of the QFD team.

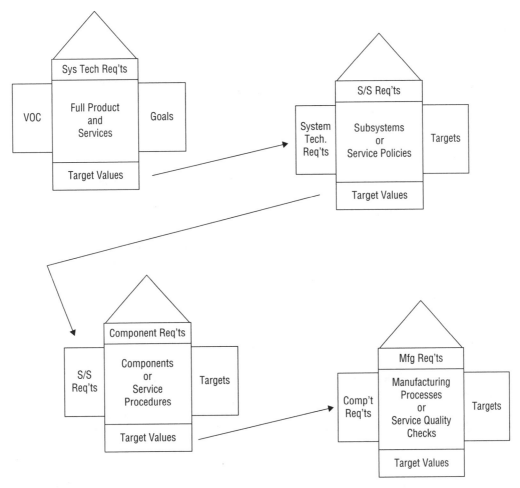

Figure 10.9 Through linked matrices, the VOC is deployed to the factory floor. As a consequence, the production requirements are traceable to their source, customers.

The QFD Team

Team Members

Most of the methods described in this book are for development teams. The teams are expected to be reasonably small working groups (fewer than a dozen people) of cross-functional representatives bringing a range of experiences and perspectives to the process. A QFD team should represent those organization functions that touch customers in some way, such as marketing, sales, service, customer support, manufacturing, design engineering, and partner development companies. Visionary or leading-edge customers can be helpful to validate the fidelity of the VOC interpretations and their translations. Representative suppliers, distributors, or retailers may add value, depending on your business model. Global regions should be considered also since their customers and requirements can vary greatly.

Team members must be committed to the project, staying together over several development phases to complete the process. For QFD, the team members have to be skillful in talking with customers and, more important, in listening to them. They may need to be trained in QFD, although their facilitator can do that within the process. The underlying principles are not difficult. Respect among the participants enables their insights, contributions, and creativity to flow into your new products and services. Power-voiced bullies can sabotage the process, so they may be a challenge to your facilitator. After all, the project objective is the delivery of a new product or service that will represent the best efforts of your entire enterprise and that will be better than your competitors' offerings.

Facilitator

Your teams will benefit greatly if they have a skilled QFD facilitator to drive the planning process with a spirit of clarity and urgency. You might consider this as a valuable internal resource to your development projects. He or she must be trained and experienced with the intent, structure, nuances, flexibilities, pitfalls, shortcuts, and documentation of QFD. In turn, this person can train the team, step by step, in the context of their location in the process. Familiarity with the technical details of the product is not particularly needed, although it might help.

Facilitators should be independent of the project team and neutral in their guidance. Their responsibility is to be efficient and persevering in leading the team through the process. Their insight into the key enablers of the business model can guide their thinking.

Recorder

The role of the facilitator is greatly enhanced by a person assigned to document the data and decisions in the format of QFD. This recorder will need to be proficient with your chosen QFD software application. He or she should also understand how to use facilitation materials and facilities to enable the workshops to be efficient. The job of the recorder is to enable the team to understand the insights and decisions revealed in the matrices and to preserve the conclusions for future use.

Facility

QFD is a workshop process taking many sessions over an extended period of time. The facility can assist by providing the space for breakout teamwork, substantial wall space for the large QFD charts, and the capabilities for computers and projectors to display the matrices. This is a project war room. The team must work; so must the room.

Suggestions to Improve the Facilitation of QFD

It is the job of the facilitator and recorder to keep the process manageable and timely to derive the benefits of the methodology. With repeated practice, facilitators can become aware of pitfalls to avoid and guidance to keep the process focused, adaptable, and value-adding.

Here are a few suggestions:

- Start your practice of QFD with a small product or pilot project using a deliberate learning process. An accessory for a product system is a good example. Select people to learn to be QFD facilitators. Their experience and insight will be critical to teams being efficient in using the methodology.
- Keep it simple. Focus on the critical few differentiators. Use scoring schemes with small ranges and coarse increments to avoid needless hairsplitting.

- Focus on what customers prefer, not what they dislike. The latter is to fix problems that should not have escaped to the market. Reuse data from leveraged products. Focus on the new and differentiating challenges. The list of complaints from customers can be used as a sanity check.
- Keep each matrix at a manageable size by avoiding customers' needs or technical requirements that are close to saying the same thing or have a minor level of difference. Use the importance and priority rankings to focus on those concerns that really matter to customers. An initial list of needs might be very large, so look for ways to group or filter them without losing the essence of the requests. The value proposition not only represents the opportunity for differentiated benefits but also enables decisions to delete irrelevant needs. The cost side of the "value" definition may be another filter. Consider situations for which breaking the QFD team into work groups can subdivide a large matrix into several more manageable sub-matrices.
- Impose time discipline on the process. Use reasonable workshop durations and frequencies to keep the energy level up. Allow time between meetings to gather new information. The matrices preserve the decisions and their rationale, so the teams can always revisit their work to challenge conclusions or to evaluate new information.
- Establish a project plan with milestones for the achievement of specific progress. Ensure that the timing of the deliverables is integrated into the dependencies and timelines of the product development project. The QFD team should be expected to deliver partial but firm information on a schedule that enables development work to begin while the QFD process continues to establish more mature information.
- Ensure that management understands the process and its benefits. You do not want the "voice of management" to override the conclusions of the QFD team. Likewise, you do not want a rush to design something just to show progress. Design what? A management steering committee may be useful toward this end. Ensure management's commitment to the timeline and to the assignment of participants to the QFD team.
- Keep the QFD conclusions as the visible rationale behind the resulting requirements document, its progressive freezing, approval, and change management.
- Be flexible and adaptive with the process. It is not a religion. There's no "silver bullet" here. It's just a planning tool to enable your teams to work their way through complex decisions in a reasonably efficient and disciplined manner.
- Consensus is not necessary. Reasonable agreement is. Resolve disagreements by investing the team's time in understanding the various explanations and viewpoints and in making decisions that people can live with. Caving in to powerful voices is not acceptable.
- If engineers disagree with customers, the customers rule.
- Keep the team from obsessing over scores. The overall impact of the priorities is of more concern. Avoid low-value tangents and dilution by too many details. Keep the attention on the big picture, on the higher-priority needs of your customers. If the tertiary-level needs are too many, work at the secondary level.

Key Points

1. QFD is a team sport. It's a methodology to be learned, practiced, and institutionalized.
2. It determines those technical requirements that are more strongly correlated with the objective of delivering superior value to customers in the context of competitive markets.

3. The requirements lead to the selection and development of superior design concepts.
4. QFD provides the traceability that fosters decisions so that your customers win.
5. QFD can be hard work, but it is sound engineering. When practiced early and well, it can provide substantial benefits to your projects.
6. This structured planning tool, in advance of costly development work, is a sound investment in the strategy to "do it right the first time."

Discussion Questions

1. How well are the consequences of changing customer needs evaluated throughout your product system?
2. How well are internal biases avoided in the setting of requirements?
3. How well are technical requirements traceable to their sources?
4. How well do development teams use common methods to establish product requirements?
5. How well do requirements teams avoid complexity that can bog them down in endless, trivial debates?

Further Reading

In addition to those references identified by footnotes, the following may be helpful sources:

Clausing, Don. *Total Quality Development: A Step-by-Step Guide to World-Class Concurrent Engineering.* New York: ASME Press, 1994.

Cohen, Lou. "Quality Function Deployment: An Application Perspective from Digital Equipment Corporation." *National Productivity Review,* Summer 1986.

———. *Quality Function Deployment: How to Make QFD Work for You.* Englewood Cliffs, NJ: Prentice Hall, 1995.

Hambleton, Lynne. *Treasure Chest of Six Sigma Growth Methods, Tools, and Best Practices.* Upper Saddle River, NJ: Prentice Hall, 2008.

Shillito, M. Larry, and David J. DeMarle. *Value: Its Measurement, Design, and Management.* New York: John Wiley, 1992.

Sullivan, Lawrence P. "Quality Function Deployment: A System to Assure That Customer Needs Drive the Product Design and Production Process." *Quality Progress,* June 1986.

Concept Generation and Selection

The process of product development involves many choices. The development project itself is selected from a portfolio of candidate projects. The value proposition for the product is chosen from alternatives. In the early development phases, teams select the architecture for the product and the design concepts for the subsystems, components, and parts. Much depends on decisions being made well since they strongly influence the eventual quality, reliability, and costs of the product and have significant consequences for the schedule of the development project, its risks, and its costs.

The process for selecting the best approach is a decision process that can be applied to situations beyond the product designs. Decisions are made about the structure of work teams, the selection of team leaders and key enablers, or the development strategies that become the bases of the project management plans. Gate reviews are decision processes.

Standardized decision processes, discussed in Chapter 6, have common elements. The focus of this chapter is on four of those elements: the requirements that must be satisfied, the generation of alternatives to be compared, the process for comparing them with relevant information and perspectives, and the selection of the best outcome.

Decisions in Product Development

Product development teams have the responsibility to make the right decisions at the right time to ensure beneficial consequences for their project. The consequences include commercialized designs that perform better than their competition or that can be manufactured and serviced with higher quality at lower costs. A superior design concept may require less development effort with lower risks and thereby enable a shorter, more predictable time to market.

Common to all decisions is the selection of the best option among available alternatives. It is based on knowledge that is gathered or developed to be necessary and sufficient to compare the alternatives and to be confident in the outcome. A standardized, systematic process for doing so is a best practice for product development. It enables project teams to make decisions efficiently, to accept the conclusions as being the best for the project, and to participate in the implementation with commitment. Most important, the process can be simple.

The consequences of not selecting the best design options can be the worst fate for a development team. A product architecture that depends entirely upon a risky new technical concept can jeopardize the ability of the project to solve fundamental performance problems or to achieve acceptable reliability. A design concept that depends on an immature manufacturing process may be plagued forever by higher costs and lower yields. So an effective decision process is essential for development teams.

In Chapter 3 we focused on the need for development teams to select the best strategies for reliability development. Those strategies are translated into plans that have dependencies, timelines, resources, and costs. A superior development strategy may enable higher levels of reliability to be achieved in a shorter time and with fewer disruptive design changes. However, good plans without sufficient resources are not plans at all. They are just demands.

These steps work together as a concept generation and selection process. It generates new solution concepts, eliminates alternatives that are weaker, strengthens those that have higher potential, generates additional options, and converges on the best available solution. As a facilitated team process, it must be driven by the best available knowledge, without the biases that often can plague group interactions. Stuart Pugh[1] introduced the process to emphasize the need for evaluating alternative concepts. Some people refer to it just as the "Pugh Selection Process." We have found it to be easily learned and broadly applicable to almost any decision.

Concept Generation and Selection

The structure of the method is a decision matrix, shown in Figure 11.1. It serves as a handy format for the expression of unambiguous selection criteria and the alternative design concepts. Acting as

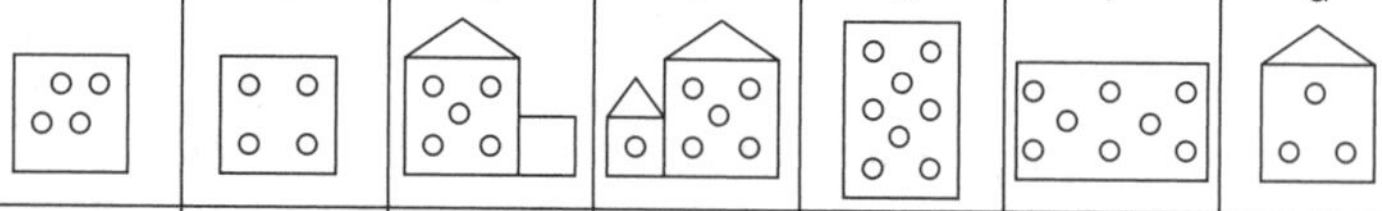

Selection Requirements	Concept A	Concept B	Concept C	Concept D	Concept E	Concept F	Concept G
Summary: $\Sigma + =$ $\Sigma S =$ $\Sigma - =$							

Rating: Better Than = +; Worse Than = −; Same As = S

Figure 11.1 The decision matrix provides a visual guide to the generation and selection process.

1. Pugh, *Total Design.*

a visual guide, it facilitates the decision and the documentation of the conclusions from team discussions. The matrix has the selection requirements organized in rows. The columns in the matrix represent the available alternative design concepts. One concept is selected as the reference, or "datum," against which the others are compared.

The cell at each row/column intersection represents the evaluation of the particular concept against one of the requirements. The question is not how well the design concept can be expected to satisfy the requirement. The objective is to select the best concept, so the process asks whether or not the particular concept is expected to be better than the referenced concept at achieving the requirement. The method strives to select the concept that is as good as or better than the other concepts for all of the most important selection criteria.

The methods of QFD link the requirements to the needs of customers and facilitate the allocation of system requirements to subsystems, assemblies, components, parts, and manufacturing processes.[2] At the level of the concept selection, the requirements become the selection criteria.

After the selection requirements have been defined, alternative design concepts are identified as potential solutions to those requirements. An initial comparison may identify concepts that are inherently stronger and those that are too weak and should be eliminated. Creative team discussions may identify additional concepts for evaluation. Possibly features of one concept can be combined with features of another to create a hybrid concept that is stronger than either of the original ones. So the process both reduces the number of concepts and expands it. That is, the process "generates" new alternatives and "selects" those to be improved or eliminated. That's why Stuart Pugh called it a "concept generation and selection process."

The process is an iterative one. It refines stronger concepts, eliminates weaker ones, and adds new ones. The overall choices are narrowed by repeated evaluations and decisions until one concept is judged to be the winner. Based on the information gathered and analyzed, the selected concept is expected to be superior to the other available design concepts at satisfying all of the selection requirements. Pugh termed this iterative process of drawing out new and improved concept ideas and then eliminating weak ones "controlled convergence." The idea, illustrated by Figure 11.2, is not a smooth funnel but one that diverges with newly generated concepts and converges as weaker concepts are eliminated.

The selection of the best concept for a subsystem is enabled by a product architecture that easily accepts alternative designs. For example, if the product architecture is not dependent on one subsystem concept and if the interfaces among subsystems are simple and robust (no degrading interactions), a better design concept can replace a flawed one later in the development process without major disruptive consequences. In this case, competing design concepts could be developed in parallel and the selection of the best one delayed until empirical and analytical evidence shows which is clearly the best for the application. This flexibility in design enables the project leadership to correct flawed decisions made too early or with inadequate information, or to react to late-changing market information. On the other hand, if the product architecture is dependent on the selected concept, development difficulties with that design can hold the project hostage.

The disciplined comparison of alternatives avoids false confidence in favored designs by subjecting each concept proposal to objective scrutiny. The rational process of criticizing concepts, improving them, generating new ones, and eliminating weaker ones avoids the tendency to select the first concept suggested or to promote the one being developed by the person with the

---

2. Clausing, *Total Quality Development.*

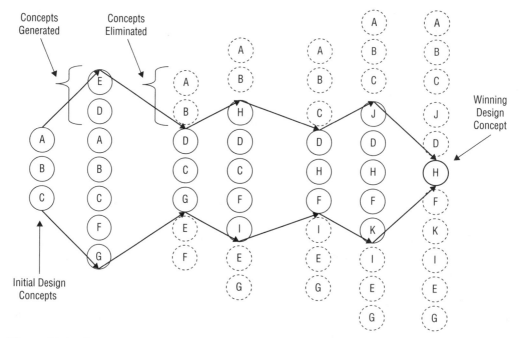

Figure 11.2   Through creative concept generation and disciplined elimination of weak concepts, the selection process converges on the best available design concept.

strongest voice in the room. The process reinforces the team's confidence in their selection by making the decision participatory, the rationale understandable, and the perspectives of all stakeholders heard. Fundamental to the process is the assumption that the selection requirements are necessary and sufficient, and that the basic quality of each concept is a summation of the evaluations across the selection requirements.

Identifying Selection Requirements

In general, the selection requirements are derived from the full system product requirements. They represent the expectations of customers, such as those for the product's features, functionality, quality, reliability, usability, and costs. Requirements may reflect problems with current products. They may represent your company's expectations for the design's robustness and for the quality and costs of manufacturing and service. Industry standards and government regulations may need to be included since they can be barriers to market entry. We find it very useful to articulate requirements at their secondary or tertiary levels of detail so that the selection team can have a clear and common understanding of their intent. Although some requirements may be truly subjective, objective metrics can reduce their ambiguity.

For this selection process teams should be interested only in those requirements that are critical to the choice of the best design concept. That should be a relatively small number, ten or fewer. This is different from a design verification process, which must ensure that all requirements have been satisfied. To decide which design concept is to be the baseline for further development, the development teams have to identify those criteria that will differentiate the alternatives and characterize the best one. Examples can include

- Demonstrated performance and robustness (a surrogate for reliability)
- Absence of stresses imposed on other subsystems (noise propagation)
- Design complexity (a surrogate for reliability) or technical risks
- Expected part count (a surrogate for both manufacturing costs and reliability)
- Power demands
- Flexibility in sourcing for components and materials (supply chain options)
- Number and ease of adjustments
- Expected manufacturing costs or manufacturing process capabilities (Cpk)
- Expected capabilities to repair or maintain the design
- Ease of system integration, for example, physical size or shape, simplicity of interfaces
- Ease of compliance with regulations or industry standards, such as environment, EMI, safety

Weighting of Selection Requirements

We like the idea of grouping requirements into prioritized categories such as "must have" and "nice to have." Those requirements that are most critical to the selection of a concept will be from the "must have" group. However, the ranking of the selection requirements and the assignment of weighting factors have difficulties that jeopardize the objectivity of the process.

In most cases the judgment of "relative importance" is subjective, representing just the opinions of team members. Given that the weights are intended to represent the opinions of customers, it is difficult to imagine that the collective wisdom of a small group of internal people can assign preferences that represent the many customers in a market segment.

Second, the assignment of importance factors is vulnerable to the biases of individual team members as they favor particular concepts. Weighting implies too much precision and may inhibit the qualitative judgments of the team. In our opinion it is better not to use weighting factors. We are more comfortable ensuring that all of the selection requirements are critical to the decision and that the selected concept is better than other concepts for all of those requirements. In that case, the weighting of requirements would not add value.

Generation of Alternative Concepts

Not all candidate concepts may be identified at the beginning of the selection process. Often it is the case that the winning concept is either a combination of elements from other concepts or is a new concept that emerges from the creative deliberations of the team.[3] The consideration of solutions to analogous problems in other industries may be a source of ideas. Likewise, the application of the TRIZ tool[4] (beyond the scope of this book) can identify ideas evident in patent searches. The initial concept comparisons and the reflections on the requirements serve to build insights into the characteristics of a superior design that can have a greater probability of being achievable and that can stand up to the expected competition in the marketplace.

Alternative design concepts may include the best design in a current product, the benchmark concept in a competitor's product, a new design concept emerging from technology development,

---

3. Ullman, *Making Robust Decisions.*

4. Clausing and Fey, *Effective Innovation.*

or a new idea for a concept that could be developed. It could be that several design concepts may be reasonable candidates for the product's application. Parallel development can then lead to more valid demonstrations of their comparisons later in the process. When the thought process is open to concepts that can be acquired or adapted, the "not invented here" pitfall can be avoided.

Most important, the options must be "available" to the project. What does that mean? If a design is already incorporated in one of your company's products, certainly it's available. If a concept used by another company can be purchased, and that process is acceptable to your business, it may be considered to be available. If a new concept under technology development is too far from being ready for transfer into product development, it is not available. If the time or cost to commercialize a concept is beyond the constraints of the project, the concept is not available to that project, although it might be to a follow-on project.

The process asks that each option be described in an abstract fashion but with enough detail to enable meaningful, unbiased comparisons. It may be useful to identify the technologies that enable the design concept so that their implementation risks can be evaluated. The descriptions can take whatever form the team needs to ensure a common and sufficient understanding.[5] Written descriptions may be supplemented by graphical or pictorial illustrations, schematic drawings, flowcharts, functional diagrams, sketches, or CAD layouts, as examples. They may be supported by empirical or analytical evidence. The concepts may have physical models that can be used for comparison tests. The objective is for each concept to be well understood so that the comparisons can be informed, not vulnerable to personal biases or uninformed opinions.

Tools described later in Chapter 13 can contribute well to comparisons among alternative concepts. A recent experience is a good example. Our pickup truck has a radiator to cool the transmission fluid. Its architecture has an internal wall shared with the radiator for the engine coolant. That wall failed recently, allowing antifreeze and water to mix with the transmission fluid. Water in the transmission is not good! Fortunately the transmission was not ruined this time. However, a simple FMECA could have identified that failure mode and its consequences, causing a different architecture to be selected without that failure mode.

The Selection Process

Remember that the question asked is whether the particular design concept is expected to be the same as, better than, or worse than the referenced concept at satisfying each selection requirement. If the concept is judged to be better, a + is entered in the cell. "Better than" might mean that the concept might be easier to commercialize, thereby having a higher probability of achieving the requirement. It might mean that the concept is less vulnerable to sources of stress (higher robustness) and therefore will contribute to higher reliability. The question is specific to the particular requirement. If the design concept is expected to be worse than the referenced concept, a − is entered in the cell. An *S*, for "same as," is entered if no substantial difference can be identified.

It's best if the differences are truly significant, not trivial or just a gut feeling. These judgments may be based on demonstrated evidence or on forecasts of expectations. The process does not mathematically sum the evaluations. For example, four + and one − ratings do not add to a three + rating. The intent is to be "same as" or "better than" for all of the selection requirements. One negative could be the downfall of a concept, depending on the severity of its handicap. The objective for now

5. Karl T. Ulrich and Steven D. Eppinger, *Product Design and Development* (New York: McGraw-Hill, 1995).

is to understand why a concept is not superior and to determine how it can be improved for those particular selection requirements that are rated negatively.

There may be a tendency for the team to assign extra emphasis to an assessment, such as ++ for a concept thought to be far superior to the reference. We suggest that this is a distraction and may encourage the team to dwell on small differences that are not significant in the end. It would be better to eliminate weaker concepts and then rerun the comparison with a different reference. A concept that is truly superior will become evident, particularly if the reference is one of the concepts in close contention for the selection.

Figure 11.3 illustrates a selection matrix after an initial round of concept generation and evaluation. A strong rating on one requirement does not offset a weak rating on another. The process is to strengthen the weaker attributes with creativity, new ideas from brainstorming, and insights gained from the evaluations of other concepts. Then eliminate the obviously inferior concepts and move forward to generate additional concepts that are new or hybrid ones for the next round of evaluation. Over a couple of iterations of generating and eliminating concepts, the superior concept should emerge with overwhelming strengths for the selection requirements.

It may be that the team does not have enough information to compare the alternatives thoroughly. That would be the time to place the selection process on hold while further development, testing, analysis, research, and other information gathering enable better judgments to be made later.

The discussions around the selection process enable the team to know the reason for the choice. All should feel comfortable with the decision and be able to live with the consequences,

Selection Requirements	Concept A	Concept B	Concept C	Concept D	Concept E	Concept F	Concept G
Through speed $>= (x)$	−	+		+	S	−	−
Robust for the application	+	−		+	−	−	−
Low technical risk	S	S		S	−	−	+
Function independent of manufacturing process	S	+	Referenced Concept	−	−	−	+
Simple interfaces with other subsystems	−	−		+	−	S	−
Manufacturing cost $<= (y)$	−	−		−	S	−	+
Easy customer maintenance	−	S		S	−	+	+
Summary:	$\Sigma + = 1$ $\Sigma S = 2$ $\Sigma - = 4$	$\Sigma + = 2$ $\Sigma S = 1$ $\Sigma - = 3$		$\Sigma + = 3$ $\Sigma S = 2$ $\Sigma - = 2$	$\Sigma + = 0$ $\Sigma S = 2$ $\Sigma - = 5$	$\Sigma + = 1$ $\Sigma S = 1$ $\Sigma - = 5$	$\Sigma + = 4$ $\Sigma S = 0$ $\Sigma - = 3$

Rating: Better Than = +; Worse Than = −; Same As = S

Figure 11.3 The objective of the process is to select the design concept that is superior to the other available alternatives for all requirements.

even though certain team members may see the selection as a compromise to their personal or organization agenda. A consensus decision, that is, a selection that everyone favors, may not be achievable. It's acceptable that team members agree that the decision is the best among its alternatives and that they commit to its implementation.

However, if a concept has a negative assessment for a requirement but is judged to be superior anyway, that particular requirement may not actually be a critical selection requirement. The selected design concept needs to make sense to the team. So it is important for the team to reflect on the insights they've gained from their evaluations, ensure that important selection requirements have not been missed, and strive to identify a design concept that has overwhelming strengths when compared to other alternatives. If members are not comfortable with the decision, it could be that critical selection requirements are missing or that the search for a better concept should be continued.

Exception for Requirements Weighting and Concept Scoring

Although we advise against this next step, in some cases it may be necessary to use a weighting and scoring process. That would involve ranking the importance of the selection requirements and assigning a score to the comparison of each concept for each requirement. This method may help to resolve close differences among competitive designs. However, there is a risk of the process being vulnerable to the manipulation of numbers to force the selection of a preferred concept with bias.

Our concerns about the assignment of importance factors may be overcome if the relative values of product attributes are firmly established. The needs and preferences of customers would have to be understood very well and demonstrated by valid feedback from them. An extreme example is when there is only one customer and that customer is very involved with the development process, even participating with the concept selection team. In that case the process is not so vulnerable to the personal opinions of the team members.

In the facilitation of this process, the scales for scoring and weighting should not be too large. For example, a numerical range of 1 to 5 or 1 to 10, with the higher score being better, could be manageable. The value of each cell would be the product of the concept score and the requirement's weight.

As before, we remain concerned about the tendency to sum the scores across the requirements. A design concept that is very strong for several requirements but noticeably weak for one critical requirement could have a high summary score and be thought to be the best. However, that one significant shortcoming could be the source of major development difficulties and hold the project hostage. As we advised earlier, a better solution might not have extreme strengths but instead would be reasonably strong on all selection requirements.

Our concerns remain about being falsely confident about the conclusions. The scores should provide guidance that helps the selection team understand the differences among concepts. The insights then should lead to the strengthening of concepts and the convergence to the best one.

Process Guidance

Table 11.1 summarizes the process of concept generation and selection in sequential steps that may be helpful to its practice and facilitation.

Table 11.1   Concept generation and selection is a process that compares available alternatives and converges on the best solution to critical requirements.

Steps	Guidance
Identify Requirements	
Identify critical selection requirements that the chosen design concept must satisfy.	Requirements must be independent of the design, unambiguous, traceable, and testable. Critical selection requirements can be derived from customers' needs, as well as from company standards and mandates, or from industry standards and government regulations. Manufacturing costs or project risks may be critical.
Decompose the selection criteria to be clearly understood and most relevant to the function.	The requirements for a design concept are derived from system-level requirements and their value proposition through a structured process such as QFD.
Ensure the selection criteria to be those that characterize the attributes of the best concept and that are the most effective at drawing out the differences among alternatives.	Select those critical few requirements that must be satisfied. Try to limit the requirements to a reasonable number, such as fewer than ten. They are equally important. That is, there are no implied weighting factors assigned to requirements.
Ensure that the requirements are both necessary and sufficient.	If the team wants to select a concept with a negative evaluation for a criterion, that criterion may not be critical to the selection. If several concepts cannot be differentiated clearly, some requirements are missing.
Describe Alternative Design Concepts	
Identify available design concepts that are reasonable candidates for a solution to the requirements.	An alternative concept must be real and achievable, not hypothetical. It may include a design currently in production, a new concept coming out of technology development, a backup concept being developed in parallel, a mature concept leveraged from previous products or analogous applications, or a design that could be acquired from other companies. It is unlikely that a "brainstorm" or "eureka" concept involving major technical hurdles or inventions could be "available" within the timeline.
Ensure that all concepts being evaluated are available to the business and can be commercialized within the time frame of the project.	If a new concept under development is not yet ready (robust) for transfer into a development project, it is not yet "available." Likewise, a candidate concept may be in current use by a competitor. If there are barriers due to intellectual property rights, the business partnering strategy, or its sourcing, it is not an "available" alternative.
Describe each alternative concept at the same level of detail in ways that enable clear comparisons.	Useful descriptions can be written, graphical (sketches, CAD illustrations, flowcharts, functional diagrams), physical, and/or analytical.

(continues)

Table 11.1 Concept generation and selection is a process that compares available alternatives and converges on the best solution to critical requirements. (*continued*)

Steps	Guidance
Prepare a Selection Matrix	
Prepare a decision matrix to guide the selection process.	Arrange the selection requirements in rows with the alternative design concepts in columns. The matrix is a visual guide to comparing the alternative concepts and a place to record the conclusions.
Select one of the alternative concepts to be a reference for the comparisons.	Usually the team should select the concept that appears to be in the middle of the rankings. Scale compression can result if an obviously strong or weak concept is used as the reference.
Compare Alternative Design Concepts	
For each requirement, compare each design concept to the referenced concept.	The task is not to demonstrate how good a concept is or to prove that a concept satisfies all of its requirements. That is the job of design verification. The task here is to select the concept that has the highest probability of being superior to available alternatives for those requirements that are critical to the design baseline.
Based on information available to the team, decide if the specific concept will be clearly better than the referenced concept (+), clearly worse than it (−), or about the same (S).	The phases of product development provide opportunities to improve the design concept and to integrate it into the full product system. So the question is not whether or not the concept is already the best, but rather whether it has a reasonable probability of being the best after further development. This judgment is the opinion of the development team, based on available information, their experience, and an achievable project management plan.
	Use the selection matrix to record the judgments in the cells that are the intersections of the rows and columns.
Identify the need for more information.	The team must decide whether or not it has sufficient information to make valid comparisons. It may be necessary for the process to be placed on hold while additional information is gathered, tests or analyses are run, or a specialist or customer is asked to join the team.
Review the selection matrix to determine if there are concepts that are overwhelmingly strong or weak. Delete weak concepts and those that cannot be improved enough to make them the best.	The objective is to select the concept that is clearly the best for all of the selection requirements. In the first evaluation, certain concepts may be superior for particular requirements, while others may be strong for other requirements. A negative score that can be accepted indicates that the requirement is not critical to the selection.

Steps	Guidance

Iterate Concept Generation and Elimination

If there is no obvious winner in the initial screening, generate additional alternative design concepts.	Identify any strong attributes that can be integrated to make a strong concept stronger. Hybrid concepts can be generated by combining features of various concepts to change a weak assessment into a stronger one. Possibly new concepts can be envisioned by seeing analogous applications that can be adapted. Be careful not to strengthen a negative feature at the expense of a positive one. Be aware that derivative concepts may have very close comparisons.
Select a new reference if the assessment looks to have a compressed scale or if a different concept appears to be superior.	Scale compression can be reduced by selecting a reference concept that is neither exceptionally strong nor weak.
Reevaluate the alternative design concepts, comparing their capabilities to the referenced concept for each requirement.	Often it is the case that the superior design concept is not among the original group to be evaluated, but rather it is an entirely new one or is one generated from the better attributes of competing concepts.

Select the Best Available Design Concept

Iteratively evaluate concepts, eliminate obviously weak ones, generate better ones, and reevaluate them until a clear, compelling choice with overwhelming strengths is identified.	Stuart Pugh termed this iterative process of expansion and contraction of choices "controlled convergence."
Reflect on the information and evaluations that converge to the winning concept. Clarify the team's description of the winning concept.	Ensure that the winning design concept is the best available solution for the requirements, that it can be developed to be better than its expected competition, that the judgments that led to these conclusions are agreeable to all team members, and that the consequences of the decision can be lived with.
Verify that the concept is superior to available alternatives.	Select a different reference concept and rerun the evaluation to see if the different perspective confirms the choice. A superior concept will remain evident when the reference is changed.

Teamwork

The concept generation and selection process is a cross-functional team activity, led by a process facilitator. As members of a small empowered team, the participants represent the various constituent organizations or disciplines and bring unique perspectives. They have responsibilities to implement the selected concept, so they have a stake in the outcome of the decision.

More efficient teams tend to be relatively small in order to be truly participatory. Selected team members should have a broad range of knowledge and experience. For example, people who are strong in advanced development should be balanced by people who are strong in manufacturing, assembly, or service. Current practitioners can provide practical inputs from their experiences with existing products and possibly even with competitive products. Suppliers may be asked to join the selection team, and customers may also be able to contribute to concepts that affect attributes such as usability and aesthetics.

Participants must be productive in the process by being able to contribute relevant information, data, and experience. They have to be creative in generating better concepts and in comparing them objectively. People who are prolific at suggesting ideas and who think "outside the box" can make the process much more fertile, as long as they remain practical.

As with QFD, facilitation of teamwork can play a valuable role in keeping the process moving forward efficiently, while maintaining its creativity and objectivity. Certainly the facilitator must be experienced with the nuances of requirements analysis, concept generation, and decision processes. This process leader keeps the focus on choosing the best concept and guides the team to avoid dwelling on distracting opinions or arguing about small quantitative differences. More important, the facilitator has the job of encouraging each person to participate creatively and diffusing rigid biases that can be introduced by influential, strong-voiced people. A useful tactic is to ask each participant to contribute several design concepts that they have conceived outside the group discussions. That will demonstrate respect for their independent contributions. Since the process is iterative and potentially time-consuming, the team may grow impatient and want to make a premature decision. The facilitator then has to encourage patience, with the clear objective of identifying the concept that will be proven to be the best in the competitive marketplace.

Key Points

1. The process of selecting the best design concept from available alternatives is a best practice for product development.
2. The process is facilitated to be disciplined and objective, with creative participation by people from a range of technical and business disciplines. The process is easy to learn.
3. For a new product to be competitively superior in the market, its design concepts must be developed to have higher performance and robustness, with lower costs, and an on-time market entry.
4. Excellence in product design cannot compensate for the selection of inherently weak design concepts.

Discussion Questions

1. How well does the selection of your design concepts follow a structured process?
2. What criteria are used to select the baseline design concepts?
3. How well do project teams consider alternative approaches?
4. How creative are the discussions that identify alternative design concepts?
5. How much development work precedes the selection of the baseline design concepts?
6. How well do project management plans enable designs to be delayed until the necessary information can be gathered or developed?

C H A P T E R 1 2

Axiomatic Design

Product development is a set of activities and decisions by which teams create new or improved solutions to problems, defined by requirements, while complying with relevant constraints. Those solutions are embodied in the controllable parameters in the product designs and in the production processes for replicating and maintaining those parameters. However, the resulting designs can become complex systems of interrelated attributes often determined by ad hoc processes.

Axiomatic design (AD)[1] is a methodology that improves the capabilities to model and evaluate the concepts for solutions. It provides criteria that lead to the selection of the best concept among alternatives, as well as to the focused improvement of the chosen concept. By concentrating first on the functions of the solution, prior to its physical embodiment in prototypes, AD can contribute to reduced cycle times, costs, and risks in development. It can improve the probability that the solution will satisfy those requirements that are important to your customers. It is a systematic process that can be applied to the development of both products[2] and processes,[3] without depending on extensive repetitions of "design-build-test-redesign-build-retest."

During the initial phase of product development, customers' needs are translated into functional requirements for the product. At the same time, development teams identify constraints that must be respected, some of which are derived from customers' needs. Alternative concepts for solutions to those requirements and constraints are identified and evaluated, and the best is selected to be the baseline for the product. Subsequent development may be needed to improve those concepts and make them more robust in the context of the product system and its applications. During the second phase, subsystem designs are integrated into systems and optimized, enabling their functional parameters to be specified. During the design phase those functional parameters are transformed into production designs, with critical design parameters either fixed or adjustable.

---

1. This chapter was developed in collaboration with Matt Pallaver and Julie Carignan, Axiomatic Design Solutions, Inc.

2. Basem Said El-Haik, *Axiomatic Quality: Integrating Axiomatic Design with Six-Sigma, Reliability, and Quality Engineering* (Hoboken, NJ: John Wiley, 2005).

3. Basem El-Haik and David M. Roy, *Service Design for Six Sigma: A Road Map for Excellence* (Hoboken, NJ: John Wiley, 2005).

One strategy for navigating this route is to build physical models in the early phases and manage the "design-build-test-redesign" process as well as possible. The consequences can be less than satisfactory. Flawed concepts can be chosen, with subsequent development attempting to compensate for bad decisions. The system-level architecture may be dependent on complex subsystem concepts. Coupling in the concept may force compromises among requirements by not being able to satisfy all requirements with the same parameter settings. Resulting development iterations with physical prototypes can be time-consuming and expensive, and ad hoc, trial-and-error approaches have a low probability of identifying how a concept can be improved, let alone optimized. They can be more focused on making the most of a bad concept than on recognizing that the concept is fundamentally flawed and needs to be replaced. The competitive business pressures that most of us face place an increased emphasis on effective analysis prior to developing the physical form of the product.

Axiomatic design contributes valuable methods and acceptance criteria to an engineer's DFSS tool kit. It guides development teams to think about functions first. Before attention can be given to the details of subsystem development, AD emphasizes a clear understanding of customer needs and their translation into high-level functional requirements, as well as the implications of high-level constraints. Only then can the baseline product concept be chosen to meet those requirements. The product architecture guides the decomposition of the system functions into the functions of subsystems, enabling the system-level requirements to be decomposed to the levels of subsystems, components, parts, and eventually their production processes. This structured approach avoids the premature development of subsystems and components without an understanding of the complete set of requirements or of the consequences of those concepts being selected.

The expectation of the process is that better designs can be developed in less time with fewer risks. In the view of Professor Nam P. Suh of MIT, where axiomatic design was developed, "Doing it right the first time, rather than spending most resources to correct mistakes made at the design stage through testing of prototypes, will pay off handsomely in terms of profit, technology innovation, efficiency and reputation."[4]

Axiomatic Design

Axiomatic design provides decision criteria based on axioms and design guidance based on theorems and corollaries. It provides a systems engineering process involving linked domains and hierarchies of information, a strategy for decomposing requirements and solutions throughout the architecture of the product and its production processes, and a method for the modeling of relationships among requirements and solutions.

Design Axioms

With AD, good designs are governed by axioms, theorems, and corollaries. These are clear, simple rules for understanding design problems and reducing the complexities in their solutions.

> An **axiom** is a proposition of fundamental knowledge that is generally accepted as being self-evident based on its merits without the need for proof.[5] Axioms are high-level criteria guiding the selection of the best solution concepts.

---

4. Basem Said El-Haik, *Axiomatic Quality.*

5. As a matter of interest, a *hypothesis* is a proposition of fundamental knowledge that needs to be proven.

> A **theorem** is a proposition that is not self-evident but can be demonstrated to be true by logical argument. A theorem is analogous to a guiding principle for design.

> A **corollary** is a proposition derived from a proven proposition with little or no additional proof needed. A corollary is analogous to a rule of good design.

Axioms are the foundation of axiomatic design. We talk about them first, deferring our discussion of theorems and corollaries to a later point in the chapter. The reference by Nam Suh[6] provides elaborations on these axioms accompanied by 28 theorems and 8 corollaries. We identify some of them as we describe AD.

Two axioms are basic to the methodology. A good design needs first to satisfy Axiom 1, then Axiom 2.

Axiom 1, the Independence Axiom: Maximize the independence of functional elements.

In an acceptable concept, the design parameters and their functional requirements are related in such a way that a specific parameter can be adjusted to satisfy its corresponding requirement without affecting other functional requirements. If more than one requirement is satisfied by a particular design parameter, the setting of that parameter forces compromises among those requirements. In that situation, not all requirements can be satisfied simultaneously.

Axiom 2, the Information Axiom: Minimize the information content.

Among those alternative designs that satisfy Axiom 1, the best concept is the one with the fewest design parameters and more relaxed tolerances. It is less complex, easier to reproduce, and less costly. For example, excessive part counts and design complexities can compromise reliability in use. Designs that are more robust increase reliability.

The prerequisite that the Independence Axiom be satisfied is important. The best engineering efforts to improve a bad concept may just result in a bad design that is improved. Design vulnerabilities are often the result of early decisions that selected poor conceptual designs.

The axioms supply criteria to understand how well a design will be able to achieve its intended functions and required performance. There are clear advantages to complying with them. The Independence Axiom assures that a design will be controllable, without iterations of adjustments and unintended compromises. The Information Axiom assures that a design can be developed to be robust, or that its production process can be developed to have a high-quality yield. Minimizing the information required to reproduce the design is equivalent to maximizing its probability of success in production.

An exact compliance with the design axioms might not always be achievable, for example, because of technical constraints or cost limitations. Various degrees of vulnerability may result, imposing risks for system integration and optimization, or for production implementation. However, by understanding the inherent vulnerabilities in the selection of the baseline concept, the project teams can focus their attention on reducing those vulnerabilities during development.

> **Conceptual vulnerabilities** are the result of violations of design principles and guidelines. They handicap the activities to satisfy requirements and optimize the designs.

---

6. Nam P. Suh, *Axiomatic Design: Advances and Applications* (New York: Oxford University Press, 2001).

Operational vulnerabilities are the result of a lack of robustness in production or customers' applications. They cause variations in performance due to factors beyond the control of the design or its processes.

Domains

Think of domains as logical groupings of captured information, recorded in the AD software application. They separate the functional and physical parts of the system design. Conceptually there are four basic domains, as illustrated in Figure 12.1. The mapping is from left to right, one domain to the next. The domain on the left represents *what* the development teams want to achieve, while the domain on the right represents *how* they choose to do this. Each can have useful sub-domains.

1. **Customer Domain:** This is the collection of benefits that customers desire, the customers' needs. This domain is one of the sources from which technical requirements are derived with details that are relevant to the value proposition for the product under development. In the methods of QFD, they are described in affinity groups and decomposed into more detailed and actionable statements in the first room of the House of Quality.
   Customers' needs (CNs): The unfulfilled needs and preferences of customers representing target markets are captured and analyzed by VOC methods. This VOC may be in the form of statements with customers' terminology and imagery, usage scenarios for the product, and use cases, for example. In the context of each application, "customers' needs" (CNs) define what they need the product to do for them, their desired outcomes. To the extent that development teams understand these details, the CNs are the functional needs that are implied by information gathered from customers. CNs are recorded and linked in the Customer Domain.

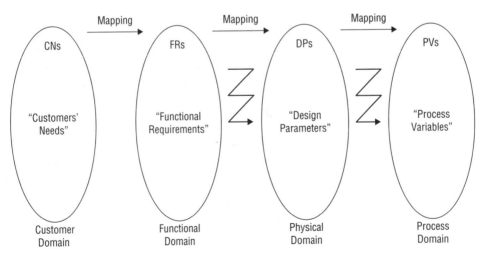

Figure 12.1 Domains of linked information.

2. **Functional Domain:** Technical requirements for the product are translated from the language of customers into the functional language of the solutions to be developed, although not specific to any preconceived solution. In QFD this is described in the requirements relationship matrix. When the system architecture of the product is decomposed, functional requirements can be deployed to subsystems, modules, and components down the product hierarchy. These data are recorded with parent-child linkage in the Functional Domain. Certain of these requirements may actually be viewed as constraints on the design, recorded separately.

Functional requirements (FRs): Functional requirements are derived from the needs of customers as well as from other sources. At the highest level, they represent the minimum number of requirements necessary to characterize the product's objectives. As an agreed-upon set, decomposed to be applicable throughout the design hierarchy, they characterize the functional needs of the solutions to be developed. They are subject to validation by customers and to verification by internal quality tests. Requirements management capabilities maintain this information and control relevant changes. FRs are recorded and linked in the Functional Domain.

Corollary 2 (Minimization of FRs): Minimize the number of FRs and constraints.

3. **Physical Domain:** The physical embodiments of the product functions are the product designs, as represented first in conceptual models, then in CAD, and later in physical breadboards, prototypes, and preproduction units. At the system level, their designs are integrated and optimized to achieve the required performance and robustness.

Design parameters (DPs): In the language of axiomatic design, design parameters are the elements of the design solution in the Physical Domain. They are selected and specified to control the outputs of product functions in manners that satisfy their requirements. At high levels in the product, DPs are the product concept, its system architecture, and subsystem concepts. At lower levels in the architecture, they are the critical control parameters that enable the functional responses to achieve their performance targets and to be robust against stressful conditions. For example, these DPs may be fixed dimensions of parts, adjustments in assemblies, critical properties of materials, code in software, or specified properties of purchased components. Systems for critical parameter management document them, manage their change implementation, and ensure that everyone works with valid set points. DPs are recorded and linked in the Physical Domain.

As we have mentioned often, requirements should be stated in ways that are independent of a particular solution. However, it may be that customers state their requirements by presupposing the solution that they think will meet their implied needs, thereby requiring a DP. If this type of requirement is nonnegotiable, it becomes a limit to the range of solutions that can be considered, beyond which might be more value-adding alternatives not known to customers. Depending on their relationship with the customers, development teams may have an advantage if they can have a conversation with their customers. Collectively they can explore creative alternatives to the underlying needs, rather than designing the specific concepts that were mentioned.

4. **Process Domain:** Manufacturing and assembly processes have the tasks to replicate the specified design configurations and parameters. Service processes maintain those parameters via preventive maintenance and repairs. Those processes themselves have controllable parameters that are developed and specified to establish and maintain the design parameters to be on target.

Process variables (PVs): Process variables are the selected processes and critical control parameters for fabrication, assembly, procurement, and service processes. Recorded and linked in the Process Domain, they are process solutions that are selected, developed, and specified to control the replication and maintenance of the product's design parameters.

The methods of axiomatic design trace the requirements to solutions across these domains and down through the architectures of the product and its production processes. The details of the CNs are independent of the architecture of the solutions. The documentation of these linkages is a major contribution to knowledge management, enabling the consequences of decisions in one domain to be mapped to their effects in other domains.

Constraints (CONs)

Often customers' needs may limit the design options. Those that are significant may affect large parts of the system and must be considered at every step of the decomposition. Some may not be functional in nature but may describe a limiting characteristic. Other constraints may influence those functional requirements derived for the lower levels in the design.

Constraints have several characteristics. They are emergent system properties that may affect one or more functional requirements or selected design concepts. They are not functions themselves but may be created by or directly linked to design concepts (DPs). They create limits rather than tolerance ranges, with values whose precision is not very important. For example, customers may need to impose limits on the design, such as its weight, shape, stability, or ease of use. The applications may limit acoustical noise or stray light. The business applications of the product may impose cost limitations. There may be other CONs that are derived from the intended applications or value propositions, such as reliability or specific appearance characteristics.

Often constraints are derived from regulations, industry standards, and corporate mandates and are subject to compliance approvals. Examples include product safety, electromagnetic interference, human interfaces and their variations among markets in different countries and cultures, or specifications for interfaces to other products.

Using budgeting and allocation methods, hierarchies of constraints can be generated that parallel the hierarchies of functional requirements in the system architecture. With decomposition, the constraints at lower levels are solution-specific.

In implementation, constraints have two types of impacts on designs. They can serve as filters, creating limits to the range of acceptable solutions (DPs). In such a case, there is no direct link between a DP and a constraint. It simply has to be verified that the solution does not violate the constraint. The second case is when the constraint serves as a source of a "derived functional requirement." In this case the CON can be linked to an FR. As with other requirements, derived FRs must be collectively exhaustive and not redundant at their level in the design hierarchy. CONs are recorded and linked to the Functional Domain.

Differentiate Constraints from Functional Requirements

At times this can be a difficult task. The basic difference is that an FR requires a specific DP, while a CON is not met by an individual DP. If a constraint can be satisfied completely by a specific design concept, the constraint is actually a functional requirement. For example, product safety regulations, or those for electromagnetic interference, may seem to be constraints since their compliance is a pass-or-fail assessment for the entire product. However, if a specific detail of the regulation leads to a specific design choice, that regulation statement should be converted to a functional requirement, subject to verification.

There can be "global constraints" that remain binding even as the FRs and DPs are decomposed into lower levels of the design hierarchy. On the other hand, there can be constraints that map directly as a child of an FR at some level within the decomposition. By recognizing this difference, development teams can employ the constraints in their selection of the design concepts for the specific location in the system design.

Design Hierarchies

The system architecture of a product design has functional layers, subdivisions, and branches, often mapped as a tree diagram, such as a bill of materials. At a high level these are the systems of the product. Decomposition of a system, through the layers of its architecture, breaks it down to subsystems, modules, assemblies, components, parts, materials, software, and procedures. In the terminology of axiomatic design, these layers represent the "design hierarchy." At the lowest level are the basic scientific functions that we described with the parameter diagram in earlier chapters. The functions have inputs and outputs, stresses imposed on them, and the critical functional parameters that can be specified to control their performance and robustness. These are logical decompositions of the system-level functions. They may also represent how development is to be allocated to work teams. Functional development may involve single technical disciplines, such as software or materials development, or be inherently cross-discipline integrations of hardware, software, materials, and human processes. Collectively they must be developed and integrated to achieve the required performance and stability of the full system product. Optimizing system performance does not necessarily mean optimizing the performance of each subsystem. It may be necessary to compromise certain subsystems in order to compensate for shortcomings elsewhere and to optimize full system performance and robustness.

The Process Domain also has hierarchies, logically decomposing production processes. The parameters that control those processes, the process variables (PVs), must be specified and maintained as are those for the product design. The axiomatic design methods map clear and deliberate linkages between the PVs and their corresponding DPs, providing traceability among these parameters and their requirements across the layers of the system hierarchy. The capabilities support and maintain the evaluations, trade-offs, and rationale behind decisions.

Zigzagging

To get it right, the development process tends to iterate from higher levels in the architecture to lower levels, from one domain to the next. A high-level set of FRs enables the DPs at that level to be selected. Those DPs enable the higher-level FRs to be decomposed to the next-lower level in the hierarchy. Decisions made at one level in the hierarchy have consequences for the decisions at lower levels.

The decomposition of the full system functions into sub-functions enables the best architecture to be selected. The system architecture then is the framework that enables the FRs to be decomposed and assigned to the sub-functions. Those deployed requirements, layer by layer through the hierarchy, are acceptance criteria for the selection and development of solution concepts (DPs) for each layer.

Not all requirements are achievable, so there's a logical negotiation back and forth among requirements and achievable solutions, between the "what" and the "how" of the solutions. Axiomatic design calls these movements "zigzagging." Figure 12.2 illustrates this as an iterative process of movements among domains and hierarchy levels. It represents paths to be taken to develop design parameters, evaluate trade-offs, improve interfaces, and other steps that are then integrated to achieve the system-level design. Note that the FRs, DPs, and PVs are decomposed throughout their architecture, while the CNs are not. The end result is the specification of those parameters that have been selected to control the performance of the product and its variability.

These development activities can be complex. Axiomatic design provides systematic methods to avoid many pitfalls and can lead to lower development costs and risks, shorter time to market, higher levels of quality and reliability, and increased competitive advantages in the market.

Design Matrix

In general, a subsystem concept has the responsibility to deliver several functions. Each function has to be developed and integrated with the others to comprise the design of the subsystem. Consider a basic technical function, modeled as a parameter diagram in Figure 12.3.

Each function has an output response for which there is a functional requirement (FR), decomposed through the product architecture to the level of the function. The function has controllable design parameters (DPs) that are developed and specified to cause the output of the function to satisfy its requirement. So each functional requirement is dependent on the relevant design parameters. In the ideal case, termed an "uncoupled design," the FR is controlled by one DP.

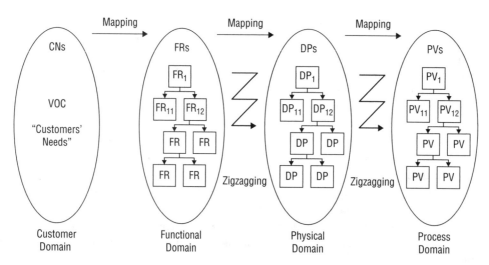

Figure 12.2 Mapping across domains and zigzagging through hierarchies.

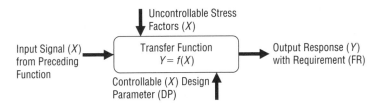

Figure 12.3 Diagram of a basic function, for which $Y = f(X)$.

System Functional Requirements (FRs)	Design Parameters (DPs)		
	DP_1	DP_2	DP_3
FR_A	A_{A1}	A_{A2}	A_{A3}
FR_B	A_{B1}	A_{B2}	A_{B3}
FR_C	A_{C1}	A_{C2}	A_{C3}

Figure 12.4 A design matrix is a visual representation of the mapping of DPs to FRs.

In cause-and-effect analyses or in FMECA, certain failures of the function can be traced to deviations of its design parameter.

Within a system design having many functional requirements and many lower-level design parameters, there could be multiple relationships among FRs and DPs. It might take several design parameters to achieve a functional requirement, and several of those parameters may be depended upon to achieve other functional requirements at the same time. This can become quite complex, adding risks to the implementation of the concept. In matrix algebra this many-to-many relationship can be expressed by $[FR] = [A][DP]$, where the A_{ij} terms are the sensitivities of the FR_i to changes in the DP_j. These sensitivities can range from being significant to trivial to nonexistent. They are essential elements of the transfer functions represented in matrix form in equation (12.1):

$$
\begin{bmatrix} FR_a \\ FR_b \\ \vdots \\ FR_m \end{bmatrix} = \begin{bmatrix} A_{a1} & A_{a2} & \cdots & A_{an} \\ A_{b1} & A_{b2} & \cdots & A_{bn} \\ \vdots & \vdots & \cdots & \vdots \\ A_{m1} & A_{m2} & \cdots & A_{mn} \end{bmatrix} \begin{bmatrix} DP_1 \\ DP_2 \\ \vdots \\ DP_m \end{bmatrix} \tag{12.1}
$$

Matrix $[A]$ is called the "design matrix." It defines the relationships among the FRs and DPs in a design concept, with m FRs in rows and n DPs in columns. The sequence of the FRs in the matrix represents the sequence of setting the functional adjustments, while the sequence of DPs represents the sequence of adjusting the design parameters to achieve those requirements. Visually the matrix equation looks like Figure 12.4, an easy-to-read map of the more significant dependencies. For each cell in the matrix, the question is whether or not the functional require-ment in the row is affected in a major way by the design parameter in the column. If so, an X is placed in the cell. If not, the cell is left blank or assigned 0. At this conceptual stage in develop-ment this "relationship" may be based on an intuitive understanding of the concept rather than a mathematical transfer function. Later experimentation with the concept may define the transfer functions and be able to assign a sensitivity coefficient.

Populate the Design Matrix

At each level in a decomposition process, the axiomatic design methods recommend analyzing the design matrix that relates the relevant functional requirements to the design parameters. The analysis of the matrix is critical to developing and selecting the best design concepts for the appli-cation. With this visible mapping, development teams have the ability to see the effects of chang-ing one design parameter on all of the dependent functional requirements.

To build a complete matrix of the dependencies, the team should evaluate each FR-DP intersection to determine when varying the particular solution DP would have a significant impact on the FR in question. This impact, or dependency, should be performance-specific, con-sidering the actual operation of the system, and be independent of the relationships with other DPs. Dependency relationships that may exist in another context, such as manufacturing, can be ignored. The X placed in the cells represents a significant "nonzero" coefficient for the relation-ship between each FR and the DP chosen to satisfy that FR. This relationship must exist for the functional requirement to be satisfied.

How might a design matrix appear for an actual concept? There are three fundamental pat-terns to recognize, illustrated in Figure 12.5.

In the "uncoupled design," all of the dependencies are mapped along the diagonal. There are no off-diagonal relationships. This pattern is interpreted thus:

Use DP_1 to set FR_A; then use DP_2 to set FR_B, and so forth.

The diagonal pattern indicates that each FR is controlled by only one DP. No specific sequence is required since the FRs are independent. This is an ideal concept, satisfying the Independence Axiom. When development improves the robustness of one function, no other functions are affected.

We mentioned theorems earlier. A few summarize the comparison of design matrices:

Theorem 4 (Ideal Design): In an ideal (uncoupled) design, the number of DPs equals the number of FRs, and each FR has a unique DP. In the design matrix there will be no off-diagonal relationships.

Theorem 6 (Path Independence of Uncoupled Design): The information content of an uncoupled design is independent of the sequence by which the DPs are changed to satisfy the given set of FRs.

In the "coupled design," three of the four FRs are controlled by the DPs not along the diag-onal in the matrix, and the off-diagonal relationships are on both sides of the diagonal. This is complex. The pattern is interpreted thus:

Use DP_1 to achieve FR_A. It will also influence FR_B. Then use DP_2 to adjust FR_B to its requirement. So far, so good. Now use DP_3 to set FR_C and contribute to FR_D. However,

System Functional Requirements (FRs)	Design Parameters (DPs)			
	DP$_1$	DP$_2$	DP$_3$	DP$_4$
FR$_A$	X			
FR$_B$		X		
FR$_C$			X	
FR$_D$				X

"Uncoupled design": All FR-DP dependencies are along the matrix diagonal.

System Functional Requirements (FRs)	DP$_1$	DP$_2$	DP$_3$	DP$_4$
FR$_A$	X		X	
FR$_B$	X	X		
FR$_C$			X	
FR$_D$			X	X

"Coupled design": There are FR-DP relationships on both sides of the diagonal that are circular dependencies.

System Functional Requirements (FRs)	DP$_1$	DP$_2$	DP$_3$	DP$_4$
FR$_A$	X			
FR$_B$	X	X		
FR$_C$		X	X	
FR$_D$	X			X

"Decoupled design": There are FR-DP relationships that are off-diagonal, but they are in a triangular pattern that is not circular.

Figure 12.5   Analysis of design matrices can reveal complicating dependencies.

DP$_3$ also influences FR$_1$, which had previously been set. Now you have to iterate back to DP$_1$ and correct FR$_A$. However, FR$_B$ is now disturbed.

You can see that you now have an iterative mess. This "coupled design" violates the Independence Axiom. It indicates inherent weaknesses in the concept, creating risks for the product's ability to satisfy its requirements. A coupled design is complex and can be troublesome to implement, so it should be replaced by a much better concept, one that does satisfy the Independence Axiom.

In the "decoupled design," there are off-diagonal relationships, but they are located below the diagonal, forming a triangular grouping. The pattern is interpreted this way:

Use DP$_1$ to achieve FR$_A$. It will also influence FR$_B$ and FR$_D$. Use DP$_2$ to refine FR$_B$ and contribute to FR$_C$. Use DP$_3$ to refine FR$_C$ and DP$_4$ to refine FR$_D$. Job done!

You can see that there are no circular iterations. However, there is a sequence of addressing the FRs. If that sequence is implemented, the setting of one FR does not disturb another FR that had been set previously. Unlike the coupled example, there is no rework. The FRs remain independent and the Independence Axiom is satisfied. Although more vulnerable than an uncoupled design, it still has less risk than a coupled design.

> *Theorem 7 (Path Dependency of Coupled and Decoupled Designs): The information contents of coupled and decoupled designs depend on the sequence by which the DPs are changed to satisfy the given set of FRs.*

So uncoupled and decoupled designs satisfy the Independence Axiom, while a coupled design does not.

Example

Historical examples of coupled and uncoupled designs are the ironclad warships of the American Civil War. Some ship designs had fixed gun mountings. The direction in which their guns pointed was determined by the direction of the ship. The FR for the shell's trajectory was coupled with the FR for the ship's direction through its steering mechanism. However, the USS *Monitor* introduced a rotating gun turret, among other innovations, which uncoupled the design. The FR for the gun direction could be satisfied independently of the FR for the ship's direction.

A word of caution: These matrix mappings can be very enlightening, but they can become very large. We advise keeping them small and focused on areas where you suspect coupling. That can avoid them becoming too overwhelming in detail and too burdensome to be useful.

Improvement of Design Concepts

Axiomatic design assists development teams either in decoupling a coupled design concept or in creating alternative concepts that are inherently uncoupled or decoupled.

Decouple the Design

There are several useful approaches to reducing the coupling in a design concept. The ideal of an uncoupled design may not always be achievable, so a decoupled design may be a reasonable compromise.

Suppose you are faced with the task of improving a concept that appears to be coupled. What can you do?

An initial approach is to determine if changing the sequence of FR and DP settings can decouple the design. Figure 12.6 illustrates a hypothetical transformation of the design matrix by rearranging the rows and columns, the sequence of adjustments.

Several other matrix patterns provide insight and illustrate theorems.

> *Theorem 1 (Coupled Design): When the number of DPs is less than the number of FRs, either a coupled design results or the FRs cannot be satisfied. Figure 12.7 illustrates this.*

	Design Parameters (DPs)			
System Functional Requirements (FRs)	DP$_1$	DP$_2$	DP$_3$	DP$_4$
FR$_A$	X		X	
FR$_B$	X	X		
FR$_C$			X	
FR$_D$			X	X

This design is "coupled" in its initial description. It shows that DP$_1$ and DP$_3$ are both required for FR$_A$, and that DP$_3$ is also required for FR$_C$ and FR$_D$. Setting DP$_3$ for FR$_C$ and FR$_D$ would cause a change to FR$_A$ and the need for iterative adjustments, a circular dependency.

	Design Parameters (DPs)			
System Functional Requirements (FRs)	DP$_3$	DP$_1$	DP$_2$	DP$_4$
FR$_C$	X			
FR$_A$	X	X		
FR$_B$		X	X	
FR$_D$	X			X

"Decoupled design": By rearranging the sequence of FRs and placing the setting of DP$_3$ ahead of DP$_1$ in the adjustment procedure, a triangular pattern is created without circular dependencies. The design is "decoupled."

Figure 12.6 Rearranging the sequence of FRs and DPs allows the design to be decoupled.

Theorem 2 (Decoupled Design): When a design is coupled because of the greater number of FRs than DPs, it may be decoupled by adding DPs to make the number of FRs and DPs equal. In the design matrix, a subset of n × n elements can be arranged in a triangle. Figure 12.8 shows this pattern.

The coupled design in Figure 12.7 has been decoupled by adding a new DP.

Robustness development may be useful to decouple a design. It takes advantage of nonlinear relationships between DPs and FRs, which can be identified with designed experiments. By specifying a set point for a design parameter where the influence on an FR is much reduced, a relationship above the diagonal, which would cause a circular dependency, can be reduced to being trivial, that is, $\Delta FR \sim 0$. This ensures that the variations in the FR due to variations in the DP are small compared to the tolerance for variations in the FR. Another tactic with the same effect is to assume a fixed value for that DP, that is, $\Delta DP \sim 0$.

Theorem 3 (Redundant Design): When there are more DPs than FRs, the design is either a redundant design or a coupled design. Figure 12.9 illustrates a redundant design.

The selection of the best solution concepts depends on criteria by which alternatives can be compared. Once a concept is selected, its improvement depends on understanding its weaknesses and how to recognize an improved functional design.

System Functional Requirements (FRs)	Design Parameters (DPs)	
	DP$_1$	DP$_2$
FR$_A$	X	
FR$_B$		X
FR$_C$	X	

Figure 12.7   DP$_1$ is required to satisfy both FR$_A$ and FR$_C$. The FRs are not independent.

System Functional Requirements (FRs)	Design Parameters (DPs)		
	DP$_1$	DP$_2$	DP$_3$
FR$_A$	X		
FR$_B$		X	
FR$_C$	X		X

Figure 12.8   The addition of another DP enables FR$_C$ to be refined without affecting FR$_A$.

System Functional Requirements (FRs)	Design Parameters (DPs)			
	DP$_1$	DP$_2$	DP$_3$	DP$_4$
FR$_A$	X			
FR$_B$		X		
FR$_C$			X	X

Figure 12.9   FR$_C$ is controlled by both DP$_3$ and DP$_4$. The design is redundant.

System Decomposition

Each high-level requirement is the source of requirements allocated to lower levels in the design hierarchy. Through linked matrices, the allocation of requirements drives the ability of one or more subsystems to perform as necessary in order for the full system design to satisfy its requirements. These decomposed requirements contribute to the selection criteria used to choose the best design concept for each level in the system hierarchy. By mapping requirements to solutions, teams may identify the need to change the allocation of requirements in order to make the full system performance more achievable. That would reduce product risks.

As illustrated in Figure 12.2, the decomposition process "zigzags" between the functional and physical domains and up and down in the design's hierarchy. For each FR, development teams evaluate alternative "design parameters" and select the best solution concept. Each selection of a concept has the consequence of generating a new level of derived requirements.

For example, consider an illumination system that must have "on" and "off" modes. Electrical lighting and flame lighting are considered as alternative design concepts for a camp lantern. Suppose that the flame lighting concept is selected. This decision, in turn, creates a functional requirement for the source of the flame, as well as requirements for a way to ignite the flame and later to extinguish it. These are "derived requirements."

By analogy, the selection of a DP can generate a new constraint. In the illumination example, the flame concept may create a constraint for minimum distance from flammable materials, a "derived constraint." Development teams should analyze the consequences of their FR-DP decisions at each level in the decomposition to identify unintended conflicts and dependencies. There may be better choices that can be made early, before committing to the development of the physical embodiment of the concepts.

Set Expectations for the Decomposition Process

Axiomatic design quality insists on a thorough analysis of needs, constraints, and high-level functional requirements. Consequently, development teams should expect to stay high in the design hierarchy until this prerequisite work is finished. Often teams are anxious to create parts, material batches, and code. However, because this work takes place early in the development stage, it is critical that alternative design concepts not be constrained and that teams have the most freedom to select the best available concepts.

> *Theorem 17 (Importance of High-Level Decisions): The quality of designs depends on the selection of FRs and the mapping from domain to domain. Wrong decisions made at the highest levels of design hierarchy cannot be rectified through the lower-level design decisions.*

Often design teams can be frustrated with this early exploratory work. They may have a "technology push" view of their work, believing that they already know what their customers need. They may have been working on an innovative design concept and feel that the new product is the best opportunity to commercialize their work. They may not feel the need or have the time to seek feedback from customers with conceptual prototypes. What is missed? Often it can be small elements of the system design that cause the lost development time, cost overruns, and missed requirements, along with the emotional stresses that often plague projects. So development teams should understand the purpose of the process and be tolerant of the early decomposition work. It is critical to consider and analyze the underlying framework of the requirements and constraints as well as the conceptual design solutions that they propose to meet those requirements. Delays in market entry can be extremely costly in lost revenues and extended development efforts. So thorough analysis and planning prior to deliberate design should be expected as being efficient, to avoid mistakes that can be difficult to correct later.

Define the Strategy for High-Level Decomposition

Consider specifically what is to be designed. Is it the product, the process for using the product, or the process for producing it? Often development teams tend to lump these together. For example, a strategy for automated assembly may dominate the thinking of development teams. Does it conflict with attributes intended for customers' use? It is better to consider each separately. A requirement for automated assembly does not apply to the selection of the concepts for achieving the functional requirements derived from how the product is to be used. However, it may very well apply to the configuration of the product's architecture, the assembly strategy, and the fasteners. To avoid this confusion, the development leader should keep the team clearly focused on the product domain being decomposed.

Develop a First-Pass Constraints Domain

After careful analysis, teams often find that many expressed customer needs are actually constraints, with only a limited number of CNs surviving to drive functional requirements. CONs are not inputs to the design. Instead, they are limits on the potential set of solutions that can be considered. They have to be revisited constantly at all levels in the decomposition to ensure that design decisions do not violate them. To facilitate the axiomatic design process, the development teams should populate a designated "Constraint Domain." This helps to avoid the need to search for constraints in the much larger CN list.

Design decisions that select the solution concepts may create additional constraints that apply to the balance of the design. For example, the selection of color and texture for external surfaces can impose constraints on other parts of the design that must be compatible. These are the "derived constraints" mentioned before, the consequences of the selections of design concepts.

Traceability links are then established from the selected DPs to the derived constraints. If teams identify constraints that cannot be traced back to the CN analysis, the CN analysis should be updated to include the actual source of the constraints. As constraints influence the design further, they can be linked to affected functional requirements. Such a populated Constraint Domain will meet the traceability and constraint-reporting requirements of certain industries.

Work with Derived Constraints

Derived constraints can enable the development process to converge on a baseline design, or later on a specified production-intent design. However, if certain decisions are made too early, they can prematurely prevent creative solutions to problems not yet identified. An example might be the scenario of late-arriving changes in requirements that might be expected in rapidly changing markets. Feedback from prototype tests by customers might clarify or change requirements for the better but may contradict decisions made early in the process. An early decision on an appearance parameter, such as a paint color, may later be contradicted by the selection of a material that cannot withstand the paint's curing temperature.

Derived constraints must be captured in order to represent the complete system. It may be helpful to distinguish them from customer-sourced constraints since they are created by the decisions of the development teams. It is not that decisions cannot be overturned. It's the awareness of these problems that's important. In the preceding example, concurrently developing the material and its surface finish means that those decisions can be mutually agreeable.

In axiomatic design, an important procedural step and a goal of decomposition is a check on the consistency in mapping constraints from the parent level to the level of the children. Proper flow-down requires that each constraint at the parent level apply to at least one child FR in the decomposition.

Decompose Functional Requirements and Map Them to Design Parameters

Usually these high-level decisions are focused on systems and subsystems since the level of abstraction is too high for specific components down at the level of engineering science or physics. However, as the decomposition process continues to map requirements and solutions through the design's hierarchy, the level of the functional "parameter diagram" is reached, as described in the chapters on robustness development.

The children FRs that are necessary to implement the functionality of the parent FR have to provide every requirement without duplication. However, no extra FRs should be present that are not required in order to specifically meet the needs of the parent FR. Sometimes, extra FR-DP pairs appear in decomposition levels because the development teams know from experience that a given component or subsystem has been called for in the past for the type of system being designed. FR-DP pairs can be inserted anywhere in the tree, particularly to ensure that the system is complete.

FR-DP pairs that are not necessary to meet the parent DP are said to be "dangling." Identifying dangling pairs may help to identify missed FRs elsewhere. Once identified, a dangling FR-DP pair should be moved to position it properly in the hierarchy and to link it to the parent of its derivation. This is important. In response to a design change later in the product's life cycle, a change impact analysis should be able to identify all design elements that are affected. For example, an innovative solution (DP) at a higher decomposition level could eliminate the derived need for the dangling FR-DP pair. If the dangling pair was not properly linked in the hierarchy, it would be missed in the design change, throwing off system analyses such as manufacturing cost calculations.

At every level in the design hierarchy, the DP decisions have to be tested against their constraints. In axiomatic design, this is a manual process of reviewing the design parameters against the Constraint Domain. Can a specified critical design parameter be replicated by efficient manufacturing processes? Does the selection of a particular material for the fabrication of a part comply with environmental regulations?

Ensure Conformance to the Rules of Decomposition

Throughout this discussion, rules for decomposition have been introduced, or at least implied. Here is a useful summary:

1. **There should be one DP for each FR.** This is a characteristic of an ideal, uncoupled design concept. Alternative DPs can be considered, but in the final design only one should exist to fulfill its relevant FR. See Theorem 4.
2. **FRs should generate two or more derived FRs.** If a parent FR has only one child FR, the child FR is a restatement of the parent and one or the other should be eliminated.
3. **Consider more than one design concept.** Usually there are multiple solution concepts and decompositions for a design problem. The challenge is to select the best available concept for the application. It is fundamental to DFSS that alternative

concepts be identified and compared, and the best available concept selected to be baseline.

Corollary 7 (Uncoupled Design with Less Information): Seek an uncoupled design that requires less information than coupled designs in satisfying a set of FRs.

4. **The hierarchy of FR statements should remain necessary and sufficient require-ments for the full system.**
   a. During decomposition, sets of child FRs should be
      i. Independent of each other
      ii. Complete descriptions of the needs of their respective parent FRs
   b. The sum of FRs at a given decomposition level should describe what the design must do to meet the requirements of the parent FR.
5. **The requirements statements should comply with acceptance criteria** (see Chapter 9). As development teams contend with initial design concepts, they may have to accept some vague requirements, serving as placeholders until development processes can provide more clarity.
6. **Prematurely working low in the decomposition hierarchy constrains higher-level options.** Recognize that jumping to the bottom of a decomposition hierarchy has the effect of limiting the options for higher-level DPs. For example, the premature selection of a subsystem concept may imply a specific system architecture and force interface requirements on other subsystems that have yet to be chosen. When the drive to develop the details of a design, or to jump directly to implementation ideas, is delayed, the ability to consider creative solutions at the upper levels is retained.
7. **If there are no relationships among parent FRs and DPs, there should be no rela-tionships among their children FRs and DPs.** Likewise, if two FRs are dependent on the same DP, the source of the relationship can be tracked to the corresponding child FR-DP pairs. This can be very useful in identifying and removing sources of coupling.

Consequences for Development Teams

The division of work is a critical element of a project. How should the scope of responsibilities for development teams be defined? Is it by a logical grouping of design parameters (DPs) chal-lenged to satisfy a variety of requirements? This is a somewhat traditional view, dividing the physical embodiment into systems and subsystems. Work groups are then assigned to develop those systems and sub-functions. In the situation where more than one DP influences an FR, the DP-oriented groups will have to cooperate.

Defining systems will also expose the need for interfaces among systems, some of which may not have been anticipated. It's best that a complex interface be within the responsibilities of a work group, not between two work groups. If certain parts of this system design are to be developed by suppliers or partner companies, the interfaces among these separable elements should be precisely defined so that these separate organizations can act as if they are integral to your parent company.

Another strategy to think about is to define the development teams by functional requirements (FRs). Although the strategy would be more cross-disciplinary, it would place the efforts to satisfy a set of requirements within the responsibilities of a team. In our experience that has worked very well. It does require cross-discipline cooperation, a challenge for team leaders and functional managers. The team structure then is oriented more toward system performance than functional excellence.

A Familiar Example

Consider a simple example of a coupled design, the dual-knob faucet design illustrated in Figure 12.10. It is used frequently in the AD literature. Technically it is a coupled process design, embodied physically in hardware. The adjustment of the left knob affects the flow of hot water and thereby the water temperature. Similarly, the adjustment of the right, cold-water knob affects both system flow rate and temperature. Iterative adjustments of both knobs are required to satisfy the functional requirements for temperature and flow rate.

Notice the X above the diagonal in the design matrix. This indicates the circular dependency inherent in a coupled design. You have to adjust DP_1, then adjust DP_2, and then DP_1 again, and so forth, to get both temperature and flow rate to the levels desired.

Identifying and resolving coupling in a design is often an opportunity to add value or to reduce risks. It can be a source of performance improvements and competitive advantages and an object of new intellectual property. Design departments wanting to implement a high-quality design process should be very sensitive to tracking and dealing with functional coupling. If no decoupled design can be found within the constraints of the system, and an uncoupled design cannot be substituted, the design might proceed as long as the project leadership understands the nature and consequences of the coupling and compensates for it during implementation.

Remove Coupling in a Design

Often development teams may feel that their design concepts are generally decoupled. However, without the analysis of functional dependencies, there is no mechanism to visualize the coupling. When it does appear, development teams can apply useful approaches that can remove that coupling. One approach is to modify the set of DP solutions. This is a creative process that

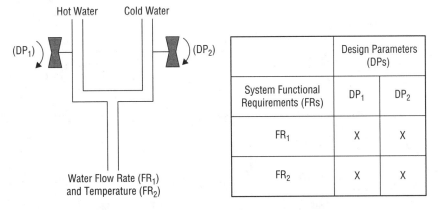

Figure 12.10 With a coupled design, both requirements are dependent on both parameters.

benefits from an understanding of how particular FR-DP relationships create coupled designs in the first place. Consider how this might be done.

One approach is to add another design parameter (Theorem 2), as shown in Figure 12.8. In the faucet illustration, suppose that the hot and cold valves are adjusted to achieve the required water temperature (FR_2) while nearly at full flow. Then a third valve located downstream is used to reduce the flow rate to its requirement (FR_1). That design matrix would look like Figure 12.11. It would depend on a human procedure to open valves DP_1 and DP_2 to set the temperature before valve DP_3 was adjusted. You can see that the circular dependency is gone and the design is now "decoupled." The decoupling included rearranging the sequence of the FR settings, FR_2 before FR_1. You can see that the decoupled design works, but it remains complicated.

Constraints on the project may prevent a major concept change, but it could be that the best decoupled design might not be good enough. Now you have a tough decision. Do you scrap the project or force a delay so that a new, uncoupled design can be developed and substituted?

The opportunity that remains is to separate the functional relationships of the DPs from the FRs. That's the strategy of the single-lever faucet design. Consider Figure 12.12. In this example,

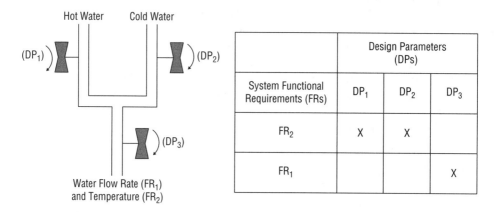

	Design Parameters (DPs)		
System Functional Requirements (FRs)	DP_1	DP_2	DP_3
FR_2	X	X	
FR_1			X

Figure 12.11   Adding a third valve and a procedure can decouple the design.

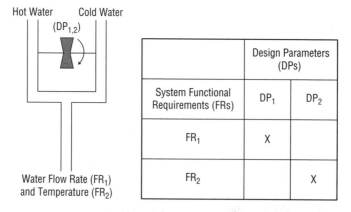

	Design Parameters (DPs)	
System Functional Requirements (FRs)	DP_1	DP_2
FR_1	X	
FR_2		X

Figure 12.12   The bidirectional lever concept for a faucet is an uncoupled design.

the difference is in the chosen architecture, enabled by both the technology in the valve concept and the project decision to adapt the bidirectional lever as the user interface. The side-to-side movement (DP_2) changes the water temperature (FR_2), and the up-down movement (DP_1) changes the water flow rate (FR_1). This concept is uncoupled because each requirement is affected by only one parameter. Since there are many examples of both types of faucets in the market, there must be differences in the perceptions of customers, that is, the needs and preferences among the various applications for water faucets. Otherwise, why would there not be a dominant design?

An interesting scenario is that of a redundant design, illustrated by Figure 12.13, where there are more DPs than FRs. With a redundant system, FRs can be satisfied independently by a subset of DPs, so the design can be considered to be uncoupled. An example of this situation might be that of an appliance that uses facility water but also has internal controls (DP_4, DP_5) of temperature and water volume (FR_3). Each FR can be satisfied by a subset of DPs, so the design is uncoupled. Redundant designs can have higher costs and lower reliability, although they can be appropriate for particular applications. These familiar illustrations, although a subset of the many examples in the axiomatic literature, show the value of the visible relationships displayed by the design matrix.

Another approach to resolve coupling is to impose a constraint on the DPs. Figure 12.14 is an illustration. When coupling arises because of a DP impacting an off-diagonal FR over some subset of its system range, that off-diagonal relationship can be removed by constraining the system to

	Design Parameters (DPs)				
System Functional Requirements (FRs)	DP_1	DP_2	DP_3	DP_4	DP_5
FR_1	X				
FR_2		X		X	
FR_3			X		X

"Redundant design": The FRs can be satisfied by a subset of the DPs.

Figure 12.13   An analysis of the design structure would reveal the redundancy in the concept.

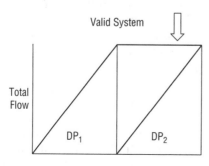

Valid System

FR_A = Control Flow
FR_B = Control Temperature

Total Flow

DP_1

DP_2

	Design Parameters (DPs)	
System Functional Requirements (FRs)	DP_1	DP_2
FR_A	X	
FR_B	X	X

Figure 12.14   Setting limits on DPs can effectively decouple a design.

operate over only that portion of the design range for which the DP does not impact the off-diagonal FR. That effectively decouples the design.

Consider the example of the bi-valve water faucet. If the maximum valve aperture of each hot- and cold-water port equals or exceeds the aperture of the spigot, the temperature of the water outlet can be determined by maximizing the opening of one of the valves (DP_1) and displacing some of its flow by manipulating the other valve (DP_2). This may be acceptable for a shower, but probably not for a sink.

Key Points

1. The goals of axiomatic design are the same as those of product development, namely, higher customer satisfaction, higher performance and robustness, reduced development time and costs, and reduced technical and business risks.
2. AD emphasizes early conceptual work with functions, enhancing problem prevention.
3. AD provides useful tools to model and analyze solution concepts to detect fundamental flaws and risks that can be corrected early. The conclusions can lead to the selection of better concepts and to the identification of improvement opportunities for development.
4. By developing a structure for functional requirements in a hierarchical decomposition, AD ensures the efficient traceability of solutions back through higher-level functions to the needs of customers.
5. The capabilities of the AD application include the documentation of developed knowledge in a manner that is both traceable and manageable. This linkage enables the consequences of changes to be understood throughout the full product system.
6. AD can provide significant value to product development by reducing several categories of risks:
   a. The risks of incomplete performance requirements. Missing requirements will become evident during the construction of the decomposition tree.
   b. The risks of redesign to achieve the required product reliability. The nodes of the functional decomposition are potential sources of performance failures. They enable the risks in robustness and reliability to be analyzed during development, prior to the design phase.
   c. The risks of excessive problem solving in the design phase. AD can contribute substantially to problem prevention. It thereby reduces the threat to schedules due to unanticipated problems late in the design process.
   d. The risks of failing to satisfy the needs of all stakeholders. The models of functional solutions can be validated systematically against the assessment of stakeholder needs before the design work is completed.

Discussion Questions

1. Is the use of quality methods by your development teams ad hoc or systemic?
2. How do your teams manage the requirements of their product and ensure their accuracy and completeness in representing your customers?

3. How are the requirements for assemblies and components derived?
4. How well do your development teams think through their functional problems before jumping to "build-test-fix" iterations?
5. How is the knowledge developed during the process maintained for use by current and future development teams?
6. How vulnerable are your project schedules to quality problems that persist into later development phases?

Tools to Reduce Risks

What do your project teams talk about in their routine coordination meetings? We expect that they evaluate the progress made over the past week and look forward to the work of the next week or so. When a major review is on the horizon, they focus on getting deliverables completed and preparing reports for management. Okay, so what's missing?

Among the disappointments in our experiences with product development is a lack of attention to risks in the design of the product and in its manufacturing and support processes. Similarly, there's a lack of investment in reducing risks in the practice of project management.

We've often seen project teams focus on selecting baseline technical concepts but not on understanding the development risks in commercializing them. They agree to reliability requirements but do not know how to achieve those requirements. They focus on solving current problems without understanding the risks in problems that they have not yet encountered.

Analyses of current problems and of risks for future problems can lead to better solutions, fewer difficulties later, and more predictable schedules. Analyses? By that we mean the use of tools and methods to understand the functional details of a system, how vulnerable the system is to misbehaviors of its component parts and their interactions, and how to prioritize efforts to solve current problems and prevent future ones from occurring.

Project Team Meetings for Risk Reduction

A best practice for team meetings is a routine agenda with time allocated to identifying risks in the project's implementation as well as those risks inherent in the design of the product and of its manufacturing, service, and customer support processes. Those risks are prioritized, and plans are implemented either to reduce the risks or to compensate for the consequences if the risk events occur.

A focus on risks asks what can go wrong and how it can be prevented. If a risk cannot be prevented, what can be done to reduce the consequences, to reduce the probability of the risk event occurring, or to recover easily the quality or safety of the product's or system's performance? For a non-repairable system, how can you extend the product's useful life?

The reduction of risks doesn't just happen with good engineering practice. It takes management attention and a prudent allocation of labor, prototypes, test equipment, facilities, and, most of all, time. Those allocations may compete with other demands for those same resources. Certainly, the meetings of cross-functional project teams are appropriate forums to ensure that the application of resources is the best for the overall project. For example, an attack on an element of risk may require the project to disregard the organization structure and apply an "all hands on deck" strategy to focus on problem-solving efforts for one subsystem.

In our experience, project teams that do well with these challenges not only devote routine agenda time to risks but also employ useful tools to harness the ideas and solutions of the many people in the project. The information derived from these tools enables teams to make better decisions and to solve problems more quickly. The use of efficient tools is not just an activity to be noted on a checklist of deliverables or on a project scorecard. The objective is to get the benefits of the information generated and the actions taken based on that information.

Useful Tools Provide Valuable Benefits

In this chapter we look at several tools that can facilitate the work of the cross-functional project teams. They are representative of a wide range of tools that are broadly applicable and easily adapted to situations. Our intent is to illustrate certain tool concepts and to encourage your teams to adapt them into common practice. Usually guidelines describe them as being applicable to the design of the product. However, with a little modification they can also be applied to the manufacturing, service, and customer support processes, to project management, and even to the development of the project's business plan. The information that one tool generates can be an input for the use of other tools. Most of the tools and methods described are part of an integrated tool set. Our recommendation is to reinforce an expectation that teams have effective tool kits that they use early and often in ways that add value. Often the methods that generate data help to justify the claim for the integrity of the data.

We expect that you may be familiar with some of these tools and even use them at times. There are many published sources that describe a wide range of useful methods and explain how to use them in detail. We'll give you some references that we know about. They will lead you to additional sources that you may also find useful.

Usually teams find that creating the information necessary to use these methods is an education itself. The tools provide handy ways to document the collective wisdom, inspiration, and ideas of participants. Lessons learned from previous projects can be captured in a concise manner, or referred to if previously recorded. Early analyses can identify fundamental flaws in the product architecture, subsystem concepts, or manufacturing processes when changes can be made more easily without major consequences.

In general, these are brainstorming and system integration tools for cross-functional teams, often assisted by a facilitator. It is best if participants include people with unbiased perspectives who can bring lessons learned from other products and development projects. They tend to provide fresh perspectives that can identify gaps in the team's collective understanding. Even persons outside the design project may be valuable contributors, particularly if they have "street smarts" from the field or are veterans of design projects with analogous problems.

There are four fundamental tool concepts that we address: risk analyses, matrix analyses, flowcharts, and tree analyses. You can find many other tools that are derived from these concepts.

Risk Analyses

The failure modes, effects, and criticality analysis (FMECA) assists teams in identifying potential product failures and prioritizing them based on their probability of occurrence and their potential consequences to customers. The priorities lead to targeted preventive actions or to plans for contingency actions. As the name implies, preventive actions reduce the expected consequences of a failure or reduce its probability of occurrence, while contingency actions correct the consequences after the failure has actually occurred. By repeating the FMECA periodically throughout the project, teams can monitor their progress in reducing the product's risk of failure.

The FMECA model is directly applicable to risks in other elements of a project. A good example is the evaluation of the risks associated with project management.[1] There may be concerns about the potential shortage of the right resources, the absence of time buffers, or the lack of predictability in the work that has been outsourced. The project's business plan is another "deliverable" for which a FMECA application can provide substantial insight and, if acted upon, improve the predictability of the business outcome.

Matrix Analyses

Think of a matrix as a two-dimensional list that enables you to map the relationships between one set of parameters and another. The cell at the intersection of a row and a column represents the relationship between the row element and the column element. The relationship may be described as a positive one that needs to be developed well. It can have a level of importance that can drive the allocation of resources. The relationship can also be negative, representing a conflict that may need to be resolved in a requirement or in a solution concept.

A matrix is particularly useful when the relationships are one-to-many. As an example, in the translation of customers' needs into product requirements, several technical requirements may have to be satisfied in order to achieve a particular need. Likewise, the achievement of a technical requirement may contribute to several customer needs. This is a one-to-many mapping. If you had just a one-to-one situation, the matrix would be reduced to a list.

An adaptation of the matrix tool is the linking of several matrices to map the deployment of objectives and plans from high levels to lower, actionable levels. QFD is an example. The columns of the first matrix become the rows of the next matrix in the sequence.

A matrix can be used for many applications, such as

- How organizational goals map into project objectives and subsequently into the implementation strategies, plans, and specific activities
- How the value proposition reflects the business opportunity or customers' value drivers for target market segments
- How the selected design concepts differentiate the product from its expected competition

QFD (Chapter 10) and axiomatic design (Chapter 12) apply a matrix tool. We illustrate two other applications later in this chapter.

---

1. Smith and Merritt, *Proactive Risk Management.*

Flowcharts

A flowchart is a mapping of system functions, activities, or information flows that have dependencies. For functions and activities, the map uses verb-noun pairs to describe the objective of a flow element. The flowchart describes the dependencies of elements on inputs and the flow of their output to other elements in the flow.

Systems Engineering

For systems engineering, these dependencies may indicate that a particular architectural function needs to receive an input signal or information from an "upstream" function. The resulting functional decomposition is necessary to allocate the system-level requirements down to subsystems and to individual technical functions, such as were illustrated in the parameter diagram described in Chapter 7. It also is needed to integrate these sub-functions back into full system performance and reliability.

Project Management

In a project management plan, the starting of an activity has requirements for resources and information that are outputs from the completion of earlier activities. Likewise, the completion of part of an activity provides information that enables another activity to begin. When the needs for partial information are mapped, dependent activities can work more in parallel. This is the essence of concurrent engineering and collaborative development, where product design and manufacturing or internal and outsourced functions work in parallel. To be useful to cross-functional project teams, integrated project management plans must map these dependencies to enable, for example, engineering and marketing or other organization functions to work together efficiently.

Parameter Mapping

A flowchart is very useful for mapping the dependencies among information in a project. Complex systems have many parameters in sub-functions that control a system-level function. It may be that certain parameters cannot be determined until a prerequisite parameter is selected and frozen. It's logical, then, that these dependencies and the work to freeze their set points progressively should be reflected in the project management plan.

Information Mapping

Another example is that of the dependencies among project documents, illustrated in Figure 4.11. The financial analyses in a business case use parameters such as the product's market price and its manufacturing and service costs. A flowchart can help the project team understand the relationships among, for example,

- Reliability requirements, market value drivers, and service costs
- A product's requirements, its potential market price, and sales volume
- Sales volume, unit manufacturing costs, and supply chain capacities

These relationships can enable people working on technical solutions to understand the business consequences of their actions and decisions as well as the business imperatives that influence their daily work.

Tree Analyses

A tree diagram is a mapping with many branches, some linked and others independent. It can employ one of two strategies: Start with a particular outcome of interest and map its contributors, or start with an event and map its consequences. It may be a functional response for which the contributing design concepts, control parameters, or external stresses need to be understood. It may be a system problem for which possible root causes need to be identified. A tree analysis maps these contributors or causes in ways similar to branches of a tree or, in the case of the cause-and-effect diagram, as the bones of a fish, an analogous format. Major branches represent causal themes, or affinity groups, and smaller branches represent root causes at more specific levels of detail.

Tree analyses are useful for both system designs and problem diagnosis. They can illustrate how system elements are related and how the functions of the product are to be achieved. They can show the decomposition of a product design into its component subsystems, assemblies, and components. Conceptually, a bill of materials is a tree diagram, although its form is usually an indented list. Organization structures are often illustrated as tree diagrams. These last two are examples of formats oriented vertically; earlier examples are often arranged horizontally.

A tool like fault tree analysis (FTA) includes symbols to illustrate the logic of the relationships, while the cause-and-effect diagram is focused more on capturing the results of creative brainstorming without the logic. A decision tree (Figure 6.2), like a fault tree, can employ probabilities to estimate the expected value of outcomes.

Think of how much more value-adding your collaborations can be as you employ these tools. When used in a manner appropriate to the project, they can facilitate cross-functional teamwork, integrate business and technical initiatives, and enable systematic risk reduction. People will not want to miss the project team meetings. Gate reviews will no longer be crises of last-minute preparations, but rather summaries of work that is ongoing.

Risk Analyses

Failure Modes, Effects, and Criticality Analysis (FMECA)

A "risk" is something that could go wrong in the future, a failure whose occurrence has some uncertainty. Analyses of risks lead to priorities for potential problems, which can then be subject to preventive actions. Each potential failure mode that can be identified is analyzed to define

- The system-level effects of the failure, if it does occur
- Its probable causes
- The probability of occurrence for each of its causes
- The consequences of the specific failure for your customers or your business

Many companies include in the analysis the ability to detect a pending failure before it occurs so that actions can be taken to mitigate the consequences. If a failure has already occurred, a Failure Reporting and Corrective Action System (FRACAS), described in Chapter 8, is more effective.

FMECA[2] is a method of risk assessment and mitigation. Figure 13.1 illustrates a simple format for FMECA that leads teams through the analysis for each potential failure mode.

---

2. Hambleton, *Treasure Chest of Six Sigma Growth Methods, Tools, and Best Practices*; U.S. Department of Defense, *Procedure for Performing a Failure Mode, Effects and Criticality Analysis*, MIL-STD-1629A (Washington, DC: Department of Defense, 1980); J. D. Andrews and T. R. Moss, *Reliability and Risk Assessment, 2nd ed.* (New York: ASME Press, 2002).

Component or Function	Description of Potential Failure Mode	Probability of Occurrence — Probable Causes of Each Potential Failure	Severity of Effect — Effect of Failure	Detection Rating — Concept for Failure Detection	RPN	Preventive Action Plans — With responsibilities, dates, deliverables, verification	Contingency Action Plans — With trigger event

Risk Priority Number (RPN) = Probability of Occurrence × Severity of Effect × Detection Rating

Figure 13.1 The FMECA table can be easily adapted to the character of your project and to your focus on risks.

It contributes to risk reduction through preventive actions taken in anticipation of potential problems. It can also stimulate contingency actions that are planned to be enacted if the anticipated problem does occur. When the tool is applied at the level of a component or assembly, outside the context of the full system, the "criticality" of the consequences to customers cannot be determined, so the method is then called FMEA, with *criticality* deleted.

Failure modes can be identified in any relevant part of a system. Potential failures can be identified for system-level functions as well as for any level in the design hierarchy, such as for hardware, software, consumable materials, packaging, and handling. Failure can be anticipated in human interactions with the product and for customers' applications, power sources, and usage environments. The consequences may be performance loss or annoyance that can reduce customers' satisfaction, or even safety problems or risks to regulatory compliance. They may be unacceptable drivers for the costs of manufacturing, service, or customer support, or serious threats to project schedules. With slight modifications, FMECA is applied to technical processes such as manufacturing, service, or distribution, or to business processes such as project management, procurement, sales, and business case development. For example, the focus on manufacturing could be on operational steps, the process controls, and the consequences for manufacturing costs, quality variations, or yields.

In some cases a failure mode may be anticipated, but the possible consequences for the full system may not be understood. In that case stressful testing may be necessary to force the problem to occur artificially and then to determine the actual consequences. The method treats each cause of a failure mode separately in the context of the system architecture and the principles of its operation.

The FMECA table documents the conclusions from the analysis. Many companies have modified and expanded it to suit their particular needs, such as to include design or process controls, implementation plans for mitigation steps, and follow-on assessments of risks after preventive actions. It's a very flexible process. Regardless of its adaptation, the FMECA method is basically the same. Companies just adapt the details of the analysis to their application.

The information contained in the FMECA table is a somewhat quantitative assessment of each root cause for each potential failure mode. The assessment includes the collective wisdom of team participants who have an insightful understanding of the architecture of the product or process, design concepts, theories of operation, market applications, and the inherent vulnerabilities to stresses. Brainstorming with tools such as the cause-and-effect diagram can be a creative way to identify potential problems. The assessment may be greatly enabled by illustrations of the product or process concept, CAD layouts, function analysis system technique (FAST) diagrams, physical prototypes, and other aids to understanding the system. Lessons learned from similar or analogous products and from earlier development projects may contribute realistic risks for consideration. It is an easy method for the anticipation of problems.

We've been very pleased with the improvement of decisions at project gate reviews when FMECA is used to assess project-level risks and to identify appropriate actions, responsibilities, and implementation target dates. Across the several phases of a development project, repeated use of FMECA can demonstrate the reduction of project-level risks, assuming that preventive actions have been implemented. It can also highlight increases in risks due to newly identified failure modes, or due to the lack of progress if the project resources have not yet acted upon previous risk analyses. There may be circumstances in which the capabilities to reduce a major risk are outside the control of the project team. That would be a good situation for action to be assigned to the project's governance committee. Monitoring risks can trigger management intervention when risk reduction has not been achieved. That can provide major benefits to a project, particularly if the actions are not too late. We particularly like the ability of the method to stop the talk about ambiguous "issues and concerns" and focus instead on consequences and probabilities of occurrence.

The FMECA methodology determines a priority for each failure mode, a "risk priority number" (RPN). The RPN is an index calculated from the combined effects of the probability of occurrence of the failure mode, the severity of the consequences, and how difficult it could be for the failure to be detected so that the consequences can be prevented.

$$RPN = \text{Probability} \times \text{Severity} \times \text{Detection} \qquad (13.1)$$

where

> Probability = probability of occurrence of the failure mode
> Severity = severity level of the consequence
> Detection = ability to detect failure mode

The assessment team assigns a risk rating for each failure cause.[3] Participants estimate the three factors in the RPN calculation and rank the "ability to detect" so that a higher rating means more difficulty in detecting the failure mode before it occurs. The RPN can then be a quantitative evaluation of risk that can be repeated during development to monitor risk reduction. Companies often extend the FMECA format to document these repeated assessments.

Some companies apply a ten-point rating system for all three metrics. Our worry is that so many rating options can imply more precision than is realistic. Often a simpler three- or four-point rating system can be just as useful and is less vulnerable to arguments over small differences that may not matter much. Table 13.1 is an example of a ten-point rating scheme.

Another scheme, shown in Table 13.2; is the four-point severity level ranking similar to one that is often used for software bugs. Not only are the severity levels clearly defined and well

---

3. Leonard A. Doty, *Reliability for the Technologies* (New York: Industrial Press, 1989).

Table 13.1   Many references for FMECA describe a ten-point rating scheme for the probability of occurrence, the severity of the effect, and the inability to detect the approaching failure

Rating	Probability of Occurrence	Severity of Effect	Detection
10	Very high: Failure is almost inevitable	Hazardous without warming	Cannot detect
9		Hazardous with warming	Very remote chance of detection
8	High: Repeated failures	Loss of primary function	Remote chance of detection
7		Reduced primary performance	Very low chance of detection
6	Moderate: Occasional failures	Loss of secondary function	Low chance of detection
5		Reduced secondary performance	Moderate chance of detection
4		Minor defect noticeable by most customers	Moderately high chance of detection
3	Low: Relatively few failures	Minor defect noticeable by some customers	High chance of detection
2		Minor defect noticeable by discriminating customers	Very high chance of detection
1	Remote: Failure is unlikely	No effect	Almost certain chance of detection

Table 13.2   A four-point rating scheme, similar to this one adapted from one used for software bugs, may be easier to use and avoid nit-picking over trivial differences

Severity Rating	Consequences of Failure Effect
4	**Critical:** The effects of the failure mode are most severe. The problem dramatically interferes with reasonable use of the product and must be resolved quickly. The impact affects a broad population of users. There is no acceptable work-around. It is considered to be a "showstopper" that will prevent shipments to customers.
3	**High:** The problem is fairly significant and must be resolved prior to market entry. The problem seriously interferes with reasonable and common use of the product. Acceptable work-arounds may be available. It may involve a requirement not met which affects a significant population of users and common applications. The problem may prevent shipments to customers if judged to be intolerable.
2	**Medium:** Minor problem that should be resolved. It has a low to medium impact on users with common applications, but it may not be apparent for many applications. In general the product can still be used for its intended purpose, but with functionality that is compromised. It may be a feature rarely used by a majority of customers. An acceptable work-around is available.
1	**Low:** A very minor problem, generally cosmetic in nature and not affecting the intended use of the product. Its impact on the use is low to medium, and an easy work-around is available.

differentiated, but their scores are squared to magnify the differentiation. Another way is to use a three-point scale, such as "high," "medium," and "low," and assign the numbers 9, 3, and 1 to the ratings, as is done in the QFD methodology. With whatever scheme you choose to use, the objective is to differentiate potential failure modes well enough to guide smart decisions about preventive actions.

There are three basic solution strategies for anticipated failures:

1. **Develop preventive actions.** Examples of preventive actions include the selection of baseline design concepts that are less vulnerable and the development to increase their robustness. This is a good reason to use FMECA early to compare alternative solution concepts. A preventive action could be to add instrumentation to enable an approaching failure to be headed off by maintenance or by an operator procedure. Integrated process control could be a strategy, if the benefit/cost estimates are favorable. The objective is to reduce or eliminate the probability of occurrence or to reduce the severity of the consequences to your customers or to your business.
2. **Develop contingency action plans.** For failures with low severity, teams can plan actions to be taken by customers or customer support people when the failure actually occurs. For higher severity levels, a contingency action might be enacted by manufacturing or service. Contingency plans may be enabled by instrumentation to monitor performance variations or shifts in functional responses before their consequences become severe.
3. **Do nothing.** This approach assumes that the judgment of a low-risk priority is a reasonable reflection of the overall impact on customers, and that customers will accept that risk.

An important result of using FMECA is the collection of the action plans so that their implementation can be managed and verified. Suitably, the FMECA table includes, for each potential failure cause, documentation of the

- Action to be taken
- Responsibility for the action
- Target date for implementation
- Implementation status

Keys to any risk reduction initiative are the prioritizing of development efforts, the allocation of the right resources to accomplish the solutions, and the verification that the solutions work and are implemented prior to market entry. The alternative is to tolerate the risk. That may be acceptable if the consequences are acceptable or if the approaching failure can be detected and the contingency plans are acceptable.

Many companies modify and expand the tool to suit their own needs. Here are two adaptations that you may find useful:

- A failure mode may have more than one effect on the system. These may be treated together or separately, depending on the effect on the action plans.
- The effect of a failure mode may not have 100% probability of occurring. That is, it may not be certain that the failure mode will always cause the consequence, particularly

if the failure is at a lower level in the system hierarchy. In that case, the probability of the consequence becomes the product of the probability of the failure mode and the probability of its effect.[4] In this manner the additional uncertainty is included in the RPN calculation.

The information from FMECA analysis can have significant benefits, such as

- Risk mitigation plans whose implementation can be prioritized, managed, and verified
- Improved allocation of resources to obtain the most benefit for the project
- More knowledgeable project decisions, such as those influencing the system architecture, design concepts, business processes, and project management
- Improved plans for designed experiments to excite failure modes
- Design changes made earlier with more informed priorities
- An efficient documentation of the vulnerability of the product, process, or development project
- An identification of the need for diagnostic procedures
- Plans for feedback from customers' usage in the market to determine if the actions were effective in preventing failures or if the contingency plans were acceptable

Matrix Analyses

Many different matrix designs are used in business to map or translate one set of parameters to another. QFD (Chapter 10) translates needs to requirements. The design matrix (Chapter 12) maps requirements to design parameters. For product development, two other matrix applications may be useful: Customer-Oriented Product Concepting (COPC), as an alternative to QFD, and GOSPA, standing for Goals–Objectives–Strategies–Plans–Actions.

Customer-Oriented Product Concepting (COPC)

COPC[5] is often useful when the relationship between a customer need and a product requirement is one-to-one rather than one-to-many. The familiar matrix in QFD then is reduced to a table. Similarly, the mapping to the solution concept may be one-to-one, enabling the solution concept to be included in the table. Figure 13.2 illustrates the form that documents the information. Is COPC suitable for your project?

You can see that it has many of the elements used in QFD, such as customers' ratings of the importance of a function, the assessment of relative capabilities among competitors, and the importance of a function to sales. It also develops a technology path by identifying criteria that are important to manufacturing. Alternative design concepts or technologies are rated against each manufacturing criterion and each customer requirement, as is done with the Pugh concept selection matrix. You may find this tool easier to use than QFD and, if appropriate, just as valuable.

---

4. Smith and Merritt, *Proactive Risk Management.*

5. M. Larry Shillito and David J. DeMarle, *Value: Its Measurement, Design, and Management* (New York: John Wiley, 1992).

Product	Customer Requirements	Competitive Analysis				Planning			Manufacturing	Design/Technologies		
		Market Segments and Customer Needs							Product Manufacturing & Design			
	Importance	Current			Improve-ment Ratio		% Score		% Wgt	Customer Value 1	Customer Value 2	Customer Value 3
Function	Feature	Us	O M 1	O M 2	O M 3	Our Goal	Sales Point	Score	Mfg Criteria	Tech 1 Manuf. Assmnt	Tech 2 Manuf. Assmnt	Tech 3 Manuf. Assmnt

Figure 13.2   The COPC matrix is a shortcut alternative to QFD.

Table 13.3   A GOSPA assessment can ensure that the planned activities are those most important to the project and its objectives

Goal	Objective	Strategies	Plans	Actions			
				Who	**When**	**Why**	**How Much Cost**

Goals, Objectives, Strategies, Plans, and Actions (GOSPA)

The mapping of goals to actions is an important team exercise that can be facilitated by the GOSPA[6] tool. Your project is an investment in a new product that is expected to create a new business or extend an existing one. It has project-specific objectives that implement the goals of the portfolio of new product development projects. So it's essential that your project team understand clearly how well their strategies are aligned to those goals and objectives. The analysis can be simple, such as a one-to-one map that can be documented by a table, like the one illustrated in Table 13.3.

For a more complex one-to-many mapping, a sequence of linked matrices will enable more insight to be represented. Figure 13.3 illustrates the concept. Each objective has several implementation strategies, and each strategy has several implementation plans defining several actions to be taken. The objective is to ensure that the right resources are applied at the right time to actions that are well aligned to the objectives of the project and the overall business goals.

---

6. Hambleton, *Treasure Chest of Six Sigma Growth Methods, Tools, and Best Practices.*

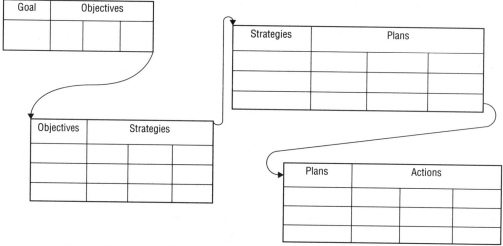

Decomposition is mapped by the columns of a matrix becoming the rows of the next matrix.

Figure 13.3 If there are one-to-many relationships among the strategies, plans and activities, a sequence of linked matrices may be very useful to the GOSPA analysis.

Flowcharts

A flowchart is very effective at illustrating how the elements of a system or process are intended to work together. The three applications that we describe here are adaptable. For example, they can be used for a process, for a product's architecture, and for a reliability model.

Precedence Diagramming Method

The Precedence Diagramming Method[7] maps activities and their dependencies. You may think of it as a PERT chart, although that is erroneous. PERT is oriented to events rather than to the activities leading to those milestone events. In a precedence diagram (Figure 13.4), the relationships among dependent activities can be of four types:

- Finish-to-start: One activity must be completed before another one can begin.
- Finish-to-finish: One activity must finish before a second one can also finish.
- Start-to-start: One activity must begin before a second one can also begin.
- Start-to-finish: One activity must begin before a second one can finish.

The finish-to-start relationship is very common. Often in the case of concurrent work, planning teams must subdivide an activity to show the availability of partial information that may be sufficient to enable a parallel task to begin. It may not be necessary for the entire activity to be completed. It may also be that the leading activity may deliver a sequence of progressively maturing information that is needed by the parallel activity on a predictable schedule. Both activities will need to be subdivided in order for the interfaces to be understood. This is

7. Project Management Institute Standards Committee, *A Guide to the Project Management Body of Knowledge* (Upper Darby, PA: Project Management Institute, 1996).

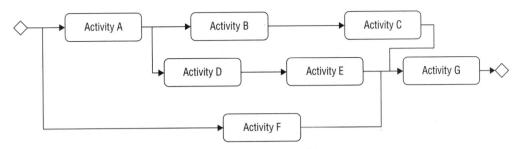

Figure 13.4 A precedence diagram maps dependencies and sequences. It can apply to the flow of activities, information, critical parameters, and other information that is linked.

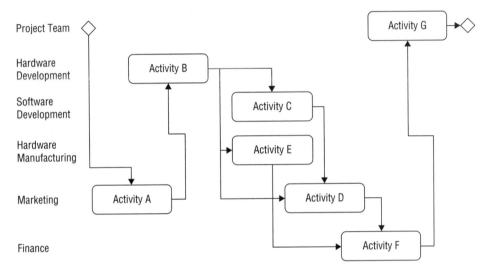

Figure 13.5 A "swim lane" diagram maps dependent and parallel activities as they are aligned with their responsible organization.

particularly useful to ensure that tasks start on time, a tactic that is very productive toward shorter time to market.

An easily understood example is the release of a mechanical design for tooling. Purchasing the block of steel for the tool may require knowing only the exterior dimensions of the part. Later steps in the tool design require defining part features, while the parameter dimensions are needed even later in the tool design sequence. Planning this progressive freezing and release of information enables product design and tool design to work in parallel with a shorter cycle time.

When a precedence diagram is distributed across organization functions, it is often called a "swim lane" diagram (Figure 13.5) to illustrate the parallel work and dependencies. It's a very useful tool for cross-functional teams to plan their interactions. It can be essential for collaborative work that requires internal activities to be integrated with those that are outsourced to partners or suppliers.

People tend to know the activities for which they are responsible, but not how their work fits with the work of others. Even just the visual representation can be enlightening. In one case a flowchart that we developed provided so much clarity for a sales process that it was deemed to be confidential intellectual property. In another case the attention was focused on the deliverables at each output from one function having acceptance criteria to satisfy as inputs to the next function. It was not a complicated process, but it added tremendous value.

Functional Analysis System Technique (FAST) Diagram

The FAST diagram,[8] shown in Figure 13.6, maps how the sub-functions connect the designed solutions (the *hows*) to the basic functions (the *whys*). It represents the structure and logic of the dependencies among the sub-functions and is useful for prioritizing the functions that are designed into a product. The FAST diagram was derived from the work of Charles Bytheway at Univac in the early 1960s, and several variations have been developed as companies adapt the methods to their particular needs.

The diagram defines each basic function of the product that customers expect, those that represent the highly valued needs. It then maps those sub-functions to be developed to enable the system functions to be achieved. The sub-functions are arranged into a hierarchical order that describes their cause-and-effect relationships. Those sub-functions that are essential to delivering value to customers represent the "critical functional path." The method also identifies supporting functions that are not essential to the basic function. For example, they may contribute to the product's productivity or its manufacturability, or deliver secondary customer "wants."

For an office printer, a basic function can be the delivery of a printed sheet of paper. Many sub-functions must work well to satisfy that need. A supporting function may be the attenuation

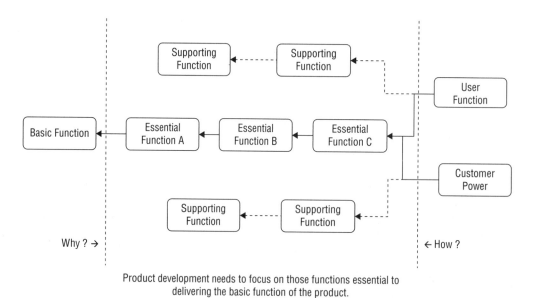

Product development needs to focus on those functions essential to delivering the basic function of the product.

Figure 13.6   A FAST diagram maps those sub-functions that are most essential to the delivery of the product's basic function to its customers.

8. Shillito and DeMarle, *Value.*

of acoustical noise. This differentiation helps teams to focus their attention on those functions that are more critical to accomplishing the basic function of the product, instead of viewing the product just as a collection of features. It may help to eliminate or reduce requirements that detract from the basic functions and identify opportunities to simplify the design, reduce its risks, or reduce its costs in areas that are not essential.

As for any functional diagram, the functions are not actions, but rather the objectives of actions described with verb-noun pairs, such as "apply force," "pump liquid," or "fill bin." At the boundary on the left are the basic functions to be delivered to customers. Beyond the boundary on the right are the inputs from the product's user, such as electrical power. With the essential functions mapped along the middle as the "critical functional path," those sub-functions off the critical path are the ones subject to simplification and cost reduction, while those along the critical path should be the focus of robustness development and reliability growth. Certainly the technique can also be applied to product service and customer support, which touch customers directly. Variations of the FAST diagram can include cost and impor-tance estimates as a way to ensure that the allocation of production costs is correlated with the value of the function to customers, as is the focus of value engineering methods.

Reliability Functional Block Diagram

The reliability of a product system can be modeled at a high level by mapping those major contributors to reliability. Each "block" is a function, component, or interface that is sub-ject to potential failure. The resulting block diagram[9] illustrates the product system as a network of functions related to reliability. It is not a schematic block diagram of the system but rather a functional layout that reflects how the operation of the product is partitioned. The consequence of each subunit's local reliability depends on how it's arranged by the sys-tem architecture.

An analytical model of the product system's reliability can be derived from this func-tional diagram. The model can identify areas in the product design for which additional investments in reliability development can yield necessary improvements in full system relia-bility. "What-if" studies with the model can enable alternative approaches to be evaluated for their benefit/cost trade-offs. Asking each subsystem to improve by the same percentage is the right answer only if the reliability of each is the same as is each of their opportunities for improvement. Otherwise equally distributed resources will be wasted. A more productive approach is to focus on achievable improvements that will have the most benefits at the full system level. For a design being leveraged from an existing product, for which valid reliabil-ity data already exist, a reliability model can enable a reasonable prediction of the reliability for the new product. A model with substantial details can assign a unique population distribu-tion to each contributor that recognizes the failure characteristics of that element. We discuss reliability modeling in Chapters 21 and 22.

With a simple architecture, the more important elements of reliability are arranged in series, as illustrated by Figure 13.7. This means that if any of the elements deteriorates or fails, the consequences will be suffered at the full system level. Certainly this is the case for

---

9. Andrews and Moss, *Reliability and Risk Assessment.*

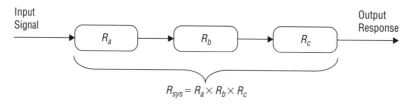

Figure 13.7   Many products are configured with components that act in series.

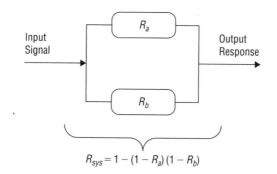

Figure 13.8   A parallel design configuration has two identical components, one acting as a backup for the other.

many familiar complex systems such as your car's engine or your office printer. If units act in series, their reliabilities (probability of working) are multiplied to estimate their combined effect:

$$R_{sys} = R_a \times R_b \times R_c \qquad (13.2)$$

and their failure rates are added:

$$\lambda_{sys} = \lambda_a + \lambda_b + \lambda_c \qquad (13.3)$$

In certain circumstances teams may choose to incorporate redundant features in the product design. If one feature fails, the product continues to operate using the backup element. The product will be inoperable only if both components fail. This architecture is modeled as a parallel system, as in Figure 13.8. Its reliability can be expressed by

$$R_{sys} = 1 - (1 - R_a)(1 - R_b) = R_a + R_b - R_a R_b \qquad (13.4)$$

The development of reliability block diagrams can be complex mapping exercises. For example, there can be mostly series elements, but with redundant functionality such as a major standby system. The standby system may be at the component level, such as a backup power source. Figure 13.9 is a simple example of a system with series and parallel elements. You can see that the reliability model can grow in complexity quickly.

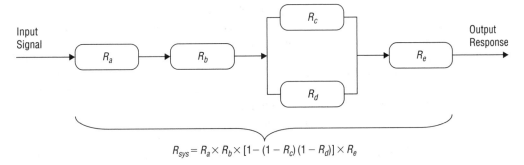

$$R_{sys} = R_a \times R_b \times [1 - (1 - R_c)(1 - R_d)] \times R_e$$

Figure 13.9 A product system can be a complex architecture of series and parallel components and subsystems.

Tree Analyses

Cause-and-Effect Diagram

Development teams can focus their work more deliberately by identifying those parameters that may have the most significant influences over a design. As illustrated by the parameter diagram (Figure 7.2) and its place within a functional flow, teams can identify the performance effects and their causes, the controllable and noncontrollable parameters that affect the functional response. The analysis can be set up to look at either improving performance or diagnosing problems with performance, either demonstrated or anticipated. Work groups can then

- Select better design concepts by comparing the cause-and-effect dependencies of alternatives
- Reduce the complexity of a design by analyzing its cause-and-effect dependencies and simplifying its configuration
- Improve a design's performance by optimizing the set points of controllable parameters
- Design stressful experiments in ways that are expected to force performance problems to occur deliberately
- Correct problems by eliminating their causes rather than by treating their symptoms
- Prevent problems by anticipating causes that can be eliminated in design, manufacturing, service, or customers' use

The scientific and engineering understanding of the basic functions is the starting point, enriched by the history of trouble/remedy experience with prototypes and with past and current products. It's a brainstorming process to gather collective wisdom about the several ways that effects can be caused.

A cause-and-effect diagram is a relationship network, a horizontal tree diagram, arranged to look like the skeleton of a fish, giving it the name "fishbone diagram." The tool was developed by Dr. Kaoru Ishikawa, which leads to it also being referred to as an "Ishikawa diagram."[10]

---

10. Hambleton, *Treasure Chest of Six Sigma Growth Methods, Tools, and Best Practices.*

The head of the fish is the effect being analyzed, either a desired response or an unwanted problem. The major bones are causal themes, or affinity groups of possible causes, and the smaller bones capture the potential root causes in increasing detail.

It is common for cause-and-effect analyses to be applied to a problem, a negative effect. There are many methods that teams can use for root cause analysis, several of which may be familiar in the context of the technology or problem consequence. The diagram itself is a way for teams to map the potential causes and their relationships. It can be applied at the system level, where the "effect" can impose consequences on customers, or at lower levels in an architecture, where the effects are stresses imposed on other elements of the system. Cause-and-effect analyses do not include logic, although the chosen affinity groups can stimulate creative thinking. At first the information mapped can be speculative, potential causes without certainty about their validity or the magnitude of their contributions to the effect. Diagnostic experiments can then identify those drivers of failure, among the many proposed, that are the actual causes of the specific problem. The collective wisdom of knowledgeable people contributes greatly to the planning of experiments that can prioritize the parameters to be examined. However, teams may have to design additional experiments to explore the potential that parameters not yet understood are actually more influential. The development of robustness depends on this.

A key to the process is the clear articulation of the outcome being evaluated. Both the effect and the proposed root causes must have modifiers that describe what responses or parameters might have changed from their design intent and in which direction would be the deviation that correlates with the effect.

Record what your team knows as part of the design documentation. Figure 13.10 is an example for our friend the football place kicker. Think about the parameter diagram. With your understanding of the process, what changes in critical parameters can cause the deviation from target? Which ones can be controlled by your process? What measures do you have in place to

Figure 13.10   The ability to control the mean and variability of field goal kicking depends on both controllable and noncontrollable factors.

detect deviations of these parameters? What tactics can you devise to correct the function or to compensate for a misbehaving factor that you cannot control? With this understandable example you can develop diagrams for the particular function you are developing. Be certain to include the adjectives.

Companies have used several different categories for potential causes. They can stimulate thinking about how a problem could be caused in ways that are relevant to the applications. Examples of the categories used often include

- "6Ms": Machines, Methods, Materials, Measurements, Manpower, Mother Nature
- "4Ps": Policies, Procedures, People, Plants/Technologies
- Other useful categories: Facilities, Equipment, Users, Environments, Design, Manufacturing, Suppliers, Service, Customer Support, Consumable Materials, Management, Market Applications, Management Systems, Handling, Transportation

These are affinity groups, similar to those in the KJ analysis of requirements. As each category of potential causes is decomposed, smaller and smaller "bones" in the diagram can record the linkage and hierarchies among suggested causes.

This tool can add much value to development work:

- It can facilitate creative brainstorming by people who have knowledge about the system and instincts about the potential root causes of problems. Their suggestions can include theories and pet ideas since it's a brainstorming process without prejudgment about the validity, probability, or priority of the suggestions. It's not just a map of demonstrated root causes.
- The method is quite adaptable. It can be used for technical processes, such as manufacturing, or for business processes for which there is a system that delivers an outcome.
- It does have the context of the system architecture. The potential consequences of variations in low-level parameters or of component failures are as important as the team's understanding of the theory of operation for subsystem concepts and the integrated system.
- It is a useful tool to document the lessons learned from experience with similar or analogous products, as well as the current thinking about the problems at hand.
- Certain potential causes may be candidates for designed experiments or analytical modeling. If they are controllable parameters, they may be candidates to be designated as critical functional parameters. If they are not controllable, they may be useful as experimental stresses (aka "noises") or as surrogates for combinations of stresses.
- Once specific root causes are proven to have strong influences, they can be included in a FMECA process that leads to preventive actions or contingency plans.
- The suggestions can contribute to diagnostic procedures to assist the troubleshooting of problems.

Event Tree Analysis

An "event tree" and the more familiar "fault tree" are very similar. Both are graphical illustrations of potential deteriorating events within a product system using the logic of the system architecture. The event tree diagram is a cause-to-effect map, while the fault tree diagram is the reverse viewpoint.

With an event tree,[11] forward logic is introduced to map how an initiating event, such as a component failure or the deterioration of a design parameter, causes a sequence of unwanted consequences that result in an undesired outcome. Such an outcome may be the loss of a feature or its functionality or an increased risk to the safety of the customer. The initiating event may have several different consequences, so the process maps all of those possible outcomes, as shown conceptually in Figure 13.11.

A thoughtful design of a complex system may have one or more devices to prevent a local failure from affecting the whole system, or at least to reduce the consequences at the system level. Each of these has a probability of working. For example, a circuit breaker is a barrier to the consequences of an excessive current rise. Smoke detectors are warning devices to enable people to act appropriately.

The event tree is one source for maintenance and repair diagnostics. The process describes the event that starts the sequence of failure modes. Through the system, protective devices are mapped in a tree structure. Each protective device is described as a negative statement, such as "The smoke alarm fails to turn on." The device, acting as a barrier to the propagation of the failure, is treated as either working or not working. For example, the statement "The smoke alarm fails to turn on" can be either true or false. Additional events or aggravating factors can be mapped with their possible outcomes. The failure sequence leads to consequences at the full system level.

The event tree model can be used to evaluate risks. The event that initiates the failure sequence has a probability of occurrence. Each protective device has a probability of success. The severity of the system-level consequence has a cost. Modeling the probabilities enables the expected value of the consequence to be estimated as an assessment of the risk.

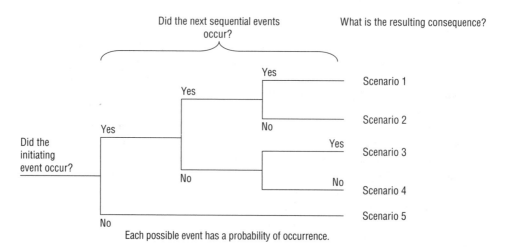

Figure 13.11   An event tree diagram maps the consequences of an initiating event, such as a component failure, through a series of sequential events to arrive at a range of scenarios with different outcomes and probabilities.

---

11. Andrews and Moss, *Reliability and Risk Assessment*; P. L. Clemens, "Event Tree Analysis" (Jacobs Sverdrup, 2002), available at www.fault-tree.net.

Fault Tree Analysis (FTA)

With an event tree, teams map their design by beginning at an initiating event, such as a component failure, and working outward to the system-level consequence. With an opposite strategy, a fault tree[12] begins at the system-level consequence. The process then works backward through the system architecture and design concepts to identify all possible root causes and their logical relationships. An event tree analyzes the alternative consequences of a single deteriorating event, while a fault tree analyzes a single system-level consequence to determine how it could have happened. The FTA incorporates lower-level root causes, their interactions, and other logical connections. Both use tree structures to create a graphical illustration that is easy to understand. Both can include probabilities of occurrence to enable estimates of system-level reliability. Both can provide insights that lead to improvements in the product design or in the diagnostic procedures used to drive repairs.

A fault tree diagram is constructed with a top-down viewpoint. It models the paths within a system by which component failures or deteriorations can lead to a loss in system-level performance, a failure, or a safety concern. Events or conditions that contribute to the fault are linked by the logic of their relationships in the system architecture, using standard logic symbols such as "AND" and "OR" gates and others noted in Figure 13.12.

	**AND Gate** If all input events occur simultaneously, the output event will occur		**Priority AND Gate** Output event occurs if all events occur in the right order
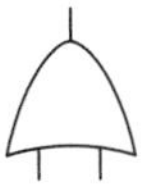	**OR Gate** If at least one input event occurs, the output event will occur		**Exclusive OR Gate** Output event occurs if one, but not both, of the two input events occurs
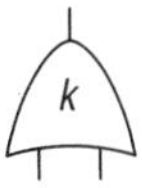	**Voting Gate** Output event occurs if at least a specified number (k) of possible events occur	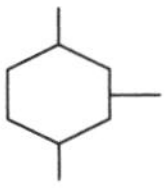	**Inhibit Gate** Input produces output when conditional event occurs
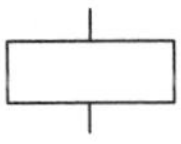	**Top Event** A top-level event that is a result of lower-level events		**House Event** Logic event which either occurs or does not occur with certainty
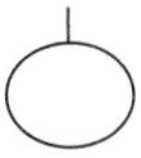	**Basic Event** The lowest-level event, the limiting resolution in our analysis	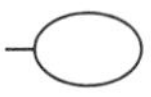	**Conditional Event** Used with inhibit gate

This is not an exclusive list. More symbols found in the references may be useful.

Figure 13.12 In a fault tree diagram, logic symbols are used describe failure events and their relationships that lead to a system-level consequence.

---

12. Hambleton, *Treasure Chest of Six Sigma Growth Methods, Tools, and Best Practices;* Andrews and Moss, *Reliability and Risk Assessment;* Patrick D. T. O'Connor, *Practical Reliability Engineering* (New York: John Wiley, 1991); P. L. Clemens, "Fault Tree Analysis" (Jacobs Sverdrup, 1993), available at www.fault-tree.net.

Suppose you have two failure events, *A* and *B*. With an AND gate, their probabilities of failure are multiplied since both conditions must exist:

$$P(A \text{ and } B) = P(A) \times P(B) \tag{13.5}$$

For an OR gate, the probabilities are added, since either condition will cause the consequence. If the failure events can interact, the general case is

$$P(A \text{ or } B) = P(A) + P(B) - P(AB) \tag{13.6}$$

If the failure events are independent, $P(AB) = 0$ and the probability reduces to

$$P(A \text{ or } B) = P(A) + P(B) \tag{13.7}$$

Figure 13.13 is an example of a fault tree diagram applied to model reliability. You can see that the development team would have to think through their conceptual design to describe its logic. That exercise alone could identify opportunities for improvement. It can be a useful application of reliability modeling to support concept comparisons and selection.

There are some rules that apply to both the event tree and fault tree models. For example:

- The deteriorating events must be anticipated. Preceding an FTA with a FMECA or a cause-and-effect analysis can provide this information.
- A fault tree analyzes a single outcome at a time, while the event tree analyzes a single initiating event.

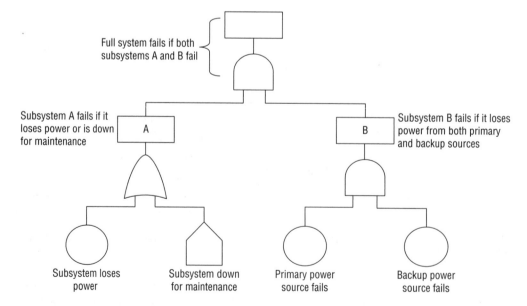

Figure 13.13   A fault tree diagram can show how a failure at a low level in a design hierarchy may contribute to the failure of a full system functional response.

- Each contributor to the failure must be anticipated.
- Each contributing component or parameter must be represented by two outcomes, such as "It worked properly" or "It did not work properly."
- Contributors must be independent of each other.

As is true for an event tree analysis, the information gained from a fault tree analysis enables

- The analysis of risks in the system design, ranging from human errors to annoyances, repairable problems, and safety concerns
- The identification of improvements required to the system architecture, to subsystem design concepts, and to their technologies
- The identification of improvements to service diagnostic procedures, or to the protective devices upon which the system depends
- The justification of resources required to be allocated for those improvements

Key Points

1. A major reason for using tools such as those described here is to prevent problems from occurring.
2. Most important is the reduction in those corrective actions late in the project that can have disastrous effects on project schedules and costs.
3. Efficient analyses can reduce the dependency on physical tests and their related demands on resources.
4. Clever analyses early in a development process can contribute to the reductions in the number of failure modes escaping to the market with the subsequent benefits to customers' satisfaction and to their costs of ownership.

Discussion Questions

1. How well trained are your development teams in practical engineering tools?
2. Do your teams have access to expert practitioners who can guide them in better use of sound tools?
3. How consistently are effective tools used to ensure the integrity of data and to increase the probability of success for design concepts?
4. To what degree do managers expect the use of standardized tools to support data, recommendations, and decisions?

Further Reading

In addition to those references identified in footnotes, the following may be helpful sources:

Clausing, Don. *Total Quality Development: A Step-by-Step Guide to World-Class Concurrent Engineering.* New York: ASME Press, 1994.

Damelio, Robert. *The Basics of Process Mapping.* Portland, OR: Productivity Press, 1996.

George, Michael L., David Rowlands, Mark Price, and John Maxey. *The Lean Six Sigma Pocket Toolbook.* New York: McGraw-Hill, 2005.

Ginn, Dana, Barbara Streibel, and Evelyn Varner. *The Design for Six Sigma Memory Jogger*. Salem, NH: Goal/QPC, 2004.

Levin, Mark A., and Ted T. Kalal. *Improving Product Reliability: Strategies and Implementation*. West Sussex, UK: John Wiley, 2003.

McDermott, Robin E., Raymond J. Mikulak, and Michael R. Beauregard. *The Basics of FMEA*. Portland, OR: Productivity Press, 1996.

Rausand, Marvin. *System Reliability Theory*. New York: John Wiley, 2004.

Sundararajan, C. (Raj). *Guide to Reliability Engineering: Data, Analysis, Applications, Implementation, and Management*. New York: Van Nostrand Reinhold, 1991.

Descriptive Statistics

The development of a new product is a journey punctuated by many changes to its design and to its production processes with the goal of improving the product delivered to customers. In some cases the changes are in a positive direction and life is good. At other times it seems that you can't win. The reality is that progress is usually forward, yet accompanied by the occasional step backward. If you are hard pressed to improve the performance of the original concept, it may be that your design concept is flawed. Statistics provides tools that help you to make decisions based on data rather than on feelings or intuition. Whether it's a consumer product, a transactional business process, or a new drug, you want to demonstrate that the new designs represent improvements over current and previous designs. You do this through purposeful changes followed by measurements and analyses of the resulting data using statistics. In addition to enabling data-driven decisions, statistics also allows you to model the functions of the product and its processes in order to predict future performance. We explore the statistical tools for modeling in Chapters 15, 16, and 17.

Why Do You Need Statistics?

Imagine a world with no variability, where all products of a given design have identical performance. If you produced a million copies of a device, each would have the same performance as the first. If every time you repeated an experiment you got the same result, you would not need replicates. Clinical trials for medical products and drugs would be simple, requiring a sample of one, since the benefits and side effects would be consistent. If there were no variability in the world, the science of statistics would not be necessary, and countless students would be spared "statistics anxiety." Of course, this is a fantasy, for we know that real life is otherwise. Phenomena such as the scattering of subatomic particles, the pressure of a gas, and quantum effects all invoke the laws of chance.

In physical systems, you know that no two objects are identical, particularly if you increase the magnification enough. Even if you could achieve manufacturing nirvana and produce "cloned" products, uncontrolled factors such as the use environment, customers' applications, and other external effects can cause performance to vary across a population of systems. On top of this, physical objects can deteriorate through mechanisms such as diffusion, migration, fatigue, creep, wear, and corrosion, causing performance to drift over time. In living organisms, population genetics, environmental effects, random events, and evolution have, over time, caused tremendous diversity, complexity, and variability. Human responses show great variability, as observed in behavioral experiments in laboratory settings. So our world is variable, whether you consider physical science, life science, or human behavior. It makes life interesting but also creates challenges when you try to develop something new, whether it's an automobile, an implanted defibrillator, a dishwasher, or a psychological assessment tool.

Statistics enables you to accomplish three fundamental tasks that are important in developing new technologies, products, and processes:

1. Detect significant changes in performance of a system due to purposeful design changes.
2. Develop empirical relationships or models to predict behaviors across the design space.
3. Make inferences about the performance of the large number (population) that will be created when manufacturing processes are scaled up to full production rates.

Each of these tasks requires development teams to be able to separate purposefully caused changes (the signal) from random variations (the noise). The tools of descriptive and inferential statistics enable that differentiation.

Descriptive statistics: Using statistics such as the mean and standard deviation, descriptive statistics helps you to characterize the location and spread of the distribution of product performance.

Inferential statistics: When you need to make comparisons between population parameters, you can rely on the tools of inferential statistics such as the t-test, chi-square test, and F-test. These probability tools help you to infer the likelihood that two samples came from the same population.

In Chapter 15 we discuss inferential statistics and show how it plays an essential role in developing new products and processes.

Graph the Data First

Before you spend much time generating descriptive statistics for a data set, it's a good idea to use your right-brain thinking skills and look at graphical displays of the data. A plot of the data can tell you much that cannot be seen by looking at the mean and standard deviation.

For example, Anscombe[1] presents four data sets that are quite different, although they have the same descriptive statistics, are fit with the same linear model, and have the same coefficient of determination.[2] For each data set, X is the independent variable and Y the dependent variable. Y1, Y2, and Y3 are all responses measured at the same values of X1.

---

1. Francis J. Anscombe, "Graphs in Statistical Analysis," *American Statistician*, February 27, 1973.

2. The coefficient of determination is a statistic that tells us the percent of total variation in the data explained by the model that's been fit to the data. For linear regression it's the square of the correlation coefficient.

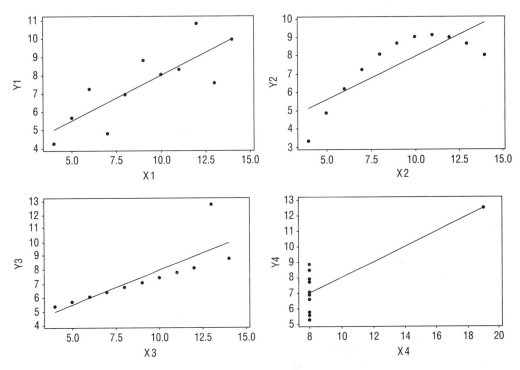

Figure 14.1 Plot of Anscombe's quartet showing why it's best to first plot the data.

$X1 = X2 = X3 = (10, 8, 13, 9, 11, 14, 6, 4, 12, 7, 5)$
$Y1 = (8.04, 6.95, 7.58, 8.81, 8.33, 9.96, 7.24, 4.26, 10.84, 4.82, 5.68)$
$Y2 = (9.14, 8.14, 8.74, 8.77, 9.26, 8.10, 6.13, 3.10, 9.13, 7.26, 4.74)$
$Y3 = (7.46, 6.77, 12.74, 7.11, 7.81, 8.84, 6.08, 5.39, 8.15, 6.42, 5.73)$
$X4 = (8, 8, 8, 8, 8, 8, 8, 8, 8, 8, 19)$
$Y4 = (6.58, 5.76, 7.71, 8.84, 8.47, 7.04, 5.25, 5.56, 7.91, 6.89, 12.50)$

The plots of the data sets, shown in Figure 14.1, illustrate the importance of looking at the data graphically.

Graphical Display Tools

The plots shown in this chapter were created using Minitab, a computer application for statistical analysis. We have not shown "how to" screen shots for using the Minitab graphical tools because we want to concentrate on the statistical elements. We describe the following useful graphical tools:

- Scatter plot
- Matrix plot
- Box plot
- Individual value plot

- Time series plot
- Histogram
- Dot plot
- Probability plot
- Control chart

They are a subset of the graphical tools available in Minitab, appropriate only for continuous valued data. We emphasize this advice: When confronted with a data set, resist the temptation to jump immediately into calculating the descriptive statistics. Instead, take a minute to plot some graphs. You might be surprised.

Scatter Plot

The scatter plot is the simplest type of x-y plot. It's a good starting point to analyze a data set and to uncover potential relationships among the variables. A strong relationship between two variables in a scatter plot is not necessarily evidence of cause and effect, and it doesn't describe the functional relationship that might appear to exist. It's strictly a tool to enable your eyes and brain to do some processing and conclude that possibly a deeper look is warranted. The scatter plot in Figure 14.2 indicates some relationship between X5 and X2. On the basis of the plot alone you cannot say if it's a cause-and-effect behavior, but you can see that further investigation is justified.

When to Use the Scatter Plot

Use the scatter plot to uncover possible relationships between variables. In the hierarchy of relationships, the scatter plot is at the initial level. A display such as Figure 14.2 should stimulate your interest in understanding further what's going on between X2 and X5.

Matrix Plot

The matrix plot is just a collection of scatter plots, useful for looking at several pairs of variables on one plot. If you have many variables, you can look at their pair-wise relationships at once.

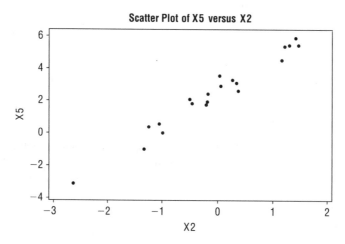

Figure 14.2   Scatter plot of two variables

Matrix plots look at patterns or correlations, but they do not necessarily imply causation. Looking at the matrix plot shown in Figure 14.3, you can quickly deduce the following:

X2-X1: No apparent relationship
X3-X1: Curvilinear relationship
X4-X1: Negative linear relationship
X5-X1: No apparent relationship
X3-X2: Possible curvilinear relationship
X4-X2: No apparent relationship
X5-X2: Positive linear relationship
X4-X3: Curvilinear relationship
X5-X3: Possible curvilinear relationship
X5-X4: No apparent relationship

Note that the matrix cells above and below the diagonal show both relationships between the variables. For example, cell 1 in column 2 shows $X1 = f(X2)$, while cell 2 in column 1 shows $X2 = f(X1)$. Consequently, it's really necessary to display only either the upper right or lower left part of the matrix. However, it is useful to display both since they look different and offer some additional insight into the relationships.

When to Use the Matrix Plot

Use a matrix plot when you have many variables and want to identify possible relationships between variables. It illustrates many pair-wise comparisons at once using the scatter plot. Think shotgun instead of rifle.

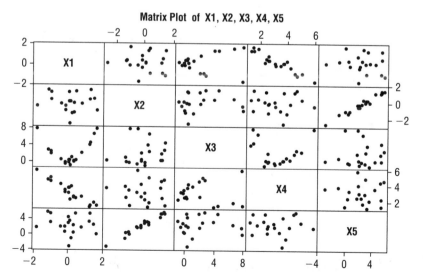

Figure 14.3 Matrix plot showing pair-wise scatter plots of five variables.

Box Plot

The box plot, in Figure 14.4, packs a lot of information into a small space. It's also known as the box-and-whisker plot, introduced by John Tukey.[3] It shows both the location and the spread of a data set. The box encompasses the middle 50% of the data, including the second and third quartiles.[4] The line within the box locates the median, and the whiskers extend 1.5 times the interquartile range. Some applications employ a slightly different definition. Points beyond the ends of the whiskers are outliers and are denoted by asterisks. By looking at each of the plots, you can tell how the data are distributed with respect to the median. Equal-length whiskers and equal-width quartiles on either side of the median indicate that you have a symmetric distribution of data with respect to the mean and median.

When to Use the Box Plot

Use the box plot to see how data are distributed across a range and whether or not the distribution is symmetrical. The box plot packs a wallop, giving you lots of information in a small package.

Individual Value Plot

Figure 14.5 shows the individual value plot. It is just a plot of the raw data for each level of the independent variable. While the box plot gives you the "big picture," the individual value plot shows more details about how the data are distributed for each sample.

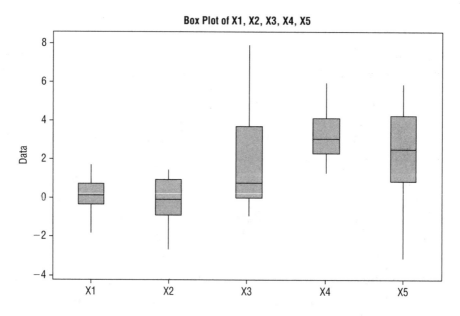

Figure 14.4   The box plot shows the distribution of a variable and more.

---

3. John W. Tukey, *Exploratory Data Analysis* (Reading, MA: Addison-Wesley, 1977).

4. For a data set arranged in order of increasing value there are four quartiles, each quartile encompassing 25% of the data points.

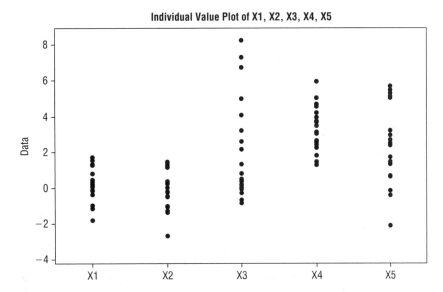

Figure 14.5   Individual value plot for the data shown in the box plot of Figure 14.4.

When to Use the Individual Value Plot

A useful approach in data visualization is to use the individual value plot together with the box plot. The box plot shows summarized data, while the individual value plot shows the raw data.

Time Series Plot

Often you are interested in determining whether there are time-related variations in the output of a process. You may be concerned for a manufacturing process that replicates components, a sales process that generates orders, or a natural process such as weather formation. In any case, variations that exhibit trends may be explained by temporal changes affecting the process. Questions such as "Does product quality change when the shift changes?" or "Is there a seasonal variation in sales volume?" can be answered by making a time series plot. The plot is very simple, being just a plot of observations in their order of occurrence. In Figure 14.6 you see the high and low temperatures at the Des Moines airport[5] for the first of the month, 1951–1952. The large cyclical variations are climate, while the smaller, higher-frequency variations are weather.

If you want to see more of the variability, you need more data. Figure 14.7 shows daily high and low temperatures for the month of July 2007. It's apparent that if you need to see variations on a smaller scale of time or distance, you need to increase your sampling frequency.

When to Use the Time Series Plot

The time series plot is great for uncovering periodicity in data caused by factors such as time of day, seasonality, and shift changes. It's simple; just plot the data in their order of occurrence.

---

5. All Des Moines weather data courtesy of Iowa State University Department of Agronomy and the National Weather Service Cooperative Observer Program.

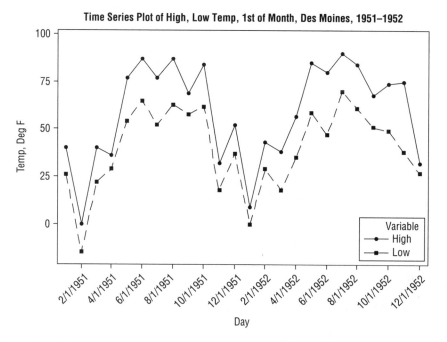

Figure 14.6   The time series plot can be used to see seasonal variation.

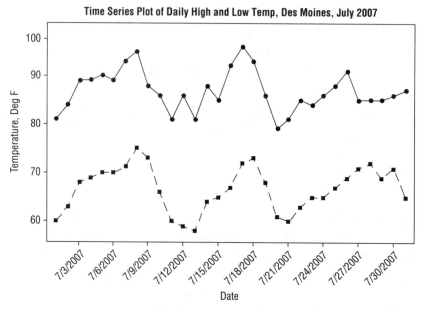

Figure 14.7   At a shorter time scale the data reveal the daily variability.

Histogram

The histogram is a cousin of the probability distribution plot. As a non-parametric plot, the histogram is helpful for seeing if a dependent variable is distributed in some interesting way. As a first try, analysts let the statistics application choose the number of bins. However, usually there is the option to override the automatic selection and specify the number. The histogram displays both location and spread, as illustrated in Figure 14.8, with a normal distribution fit as an overlay.

In order to plot a histogram, the data are binned. This means that points within a certain distance of each other are grouped for the purpose of counting the frequency of occurrence. There is an ideal range for the number of bins that is "just right," enabling you to obtain insights into the underlying distribution of the data. Using too few bins may obscure some interesting characteristics, while too many bins may introduce artifacts that are not real. You may lose the meaningful pattern in the data and not "see the forest for the trees."

Consider the problem of artifacts in data. Depending on the resolution of the measurement systems that generate the data and the associated quantization, you may see artifacts that are not descriptive of the behavior of the data that would occur in the real world. The histogram in Figure 14.9 shows large differences in frequency of occurrence for adjacent temperatures. With daily temperature being a continuous variable, that pattern of variation does not make sense. As a result of the binning process in this case, the histogram doesn't represent the way that nature behaves. Figure 14.10 shows that when fewer bins are selected, the artifacts are eliminated. Keep this problem in mind when you are faced with an odd-looking histogram that just doesn't seem right.

When to Use the Histogram

The value of the histogram is in its ability to show how data are distributed across a range. Are they clumped together or spread out? Does the histogram have a unique shape or does it appear to be noninformative? If the histogram has a unique shape, is there a probability distribution that's a

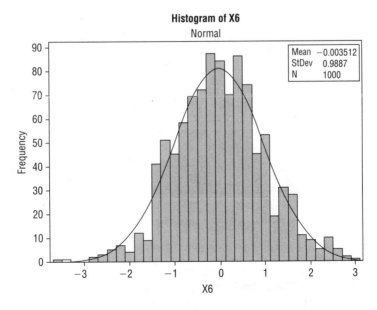

Figure 14.8   The histogram shows the distribution of a variable.

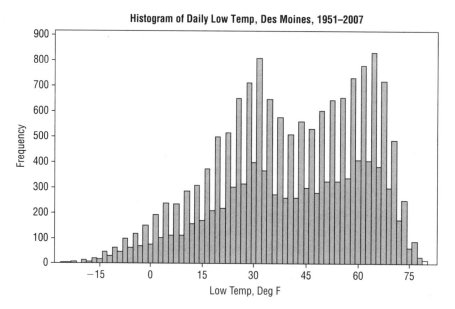

Figure 14.9   Daily temperatures for 57 years in 72 bins with automatic binning

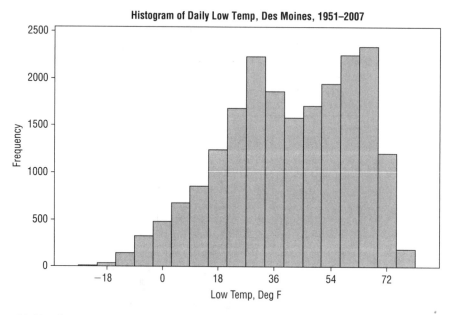

Figure 14.10   Same data as in Figure 14.9 with 20 bins instead of 72

good fit for it? For example, when analyzing the results of a reliability life test, you might expect the data to show a central tendency, possibly with a long tail going out in time or actuations. It is useful to fit a distribution to the data since that will enable you to estimate early failures, a concern for both customers' satisfaction and the producer's warranty expenses.

Dot Plot

The dot plot, Figure 14.11, is a legacy of the mainframe era when most plots were character graphics. Although at times they were the best thing available, they remain useful even though they are not pretty. The dot plot is similar to the histogram, showing how a variable is distributed. One of the useful attributes of the dot plot tool in Minitab is its ability to show the distribution of multiple variables across a range. A dot plot of multiple variables can provide insights not available with calculations of the descriptive statistics.

When to Use the Dot Plot

Use the dot plot when you want to display data from several sources to see how they compare. Suppose in Figure 14.11 that the values Y1, Y2, and Y3 are sales results for a particular month for persons in three different regions. How do they compare? Why are there differences? What would you have to do to increase the sales of Y1 and Y3 to the level achieved by Y2?

Probability Plot

The probability plot is used to look at the "goodness of fit" for a data set and a specific probability distribution. The scale for the y-axis has been transformed so that the probability plot is a straight line. So the y-axis scale will be distribution-dependent while the x-axis is linear. Figure 14.12 shows the fit of the temperature data to a normal distribution. The conclusion is that 57 years' worth of July 1 low temperatures are normally distributed. You can also look at the goodness of fit for multiple distributions by plotting them "four up," as shown in Figure 14.13.

You see in Figure 14.13 that, although the normal distribution has the best Anderson-Darling[6] goodness-of-fit statistic, there are three candidate distributions that have a decent fit. Which one should you choose? In general, simpler is better, so the normal distribution is a safe choice.

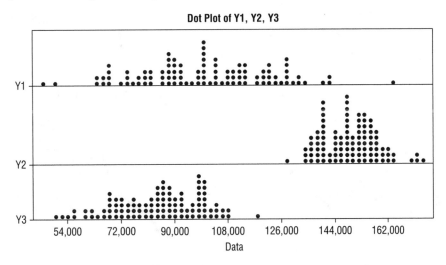

Figure 14.11   The dot plot can show how several variables are distributed.

---

6. Anderson-Darling is one of several widely used goodness-of-fit statistics, the others being Kolmogorov-Smirnov, Shapiro-Wilk, and Ryan-Joiner. These statistics are used to make relative comparisons of how candidate distributions fit a given data set.

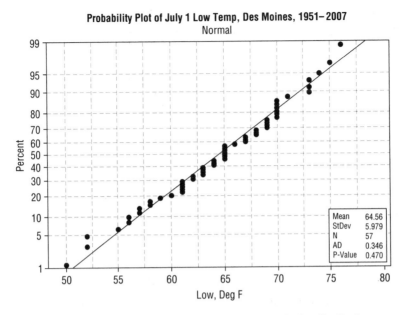

Figure 14.12   The probability plot shows how the data fit a particular distribution.

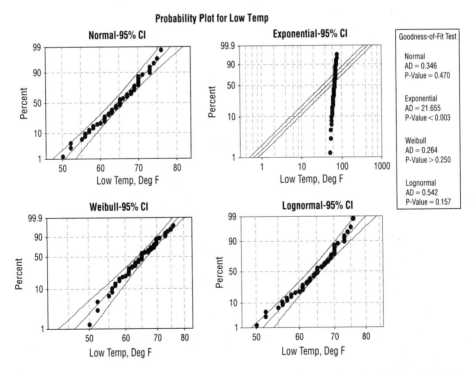

Figure 14.13   You can use multiple probability plots to see how various candidate distributions fit the data.

When to Use the Probability Plot

Use the probability plot when you want to see how a candidate distribution fits a data set. The test for normality in Minitab automatically produces a normal probability plot so you can see if the normal distribution is a good fit to your data. When you analyze life data, you use the distribution ID plot to generate probability plots to evaluate the goodness of fit for various candidate distributions.

Control Chart

The control chart is a special-purpose time series plot, the backbone of statistical process control (SPC). SPC monitors the output of a process to enable its variability to be limited. Control charts have been applied effectively to a wide range of processes, ranging from manufacturing to business transactions. The basis for SPC is the notion that the variability in a process has two very different causes. The first and more benign is "common cause variation." The roots of common cause variation are the random small forces impacting a process, causing variations that are in some ways analogous to Brownian motion. For example, humans do not accomplish repetitive tasks exactly the same way each time. At some level even robots have measurable variability in their performance. This variability can translate into measurable variations in the product. You should not try to "fix" common cause variation that is within acceptable control limits. Doing so can result in moving the process mean off target.

The second type of variation, which indicates process problems, is called "assignable cause variation." This can involve sudden shifts or trends in one or more important characteristics of process output. Things to look for are sudden shifts in the process mean, upward or downward trending of the mean, or increasing process variation. Assignable cause variation demands corrective action to bring the process back on target or to find the cause of the increasing process variation.

Common cause variation is usually Gaussian[7] in nature. If you generate a probability plot of the data, it will be a best fit for the normal distribution. The upper plot in Figure 14.14, labeled "Individual Value," shows the diameter of parts produced by a manufacturing process. The lines labeled "UCL" and "LCL" are the upper and lower control limits. These limits represent ± 3 standard deviations of the observed data. When reviewing the data, look for rare events such as the point labeled "1" that is below the LCL for the individual observations. Does the occurrence of a rare event mean that something has gone wrong with your process? Not necessarily, but it does mean that you need to investigate to determine if there is an assignable cause, or if it really is a rare event that should occur, on average, about three times in a thousand chances. So another clue that alerts you to trouble is the frequent occurrence of "rare" events. Statistical analysis applications that produce control charts for data generally have built-in checks for rare events, such as a certain number of consecutive points above or below the mean, or upward or downward trends exceeding a certain number of points. The basis for these tests is the likelihood of these patterns occurring in a stationary random process where the mean and standard deviation are constant rather than trending up or down or having abruptly shifted.

The lower plot in Figure 14.14 is the "moving range." It is the difference between the consecutive individual values shown in the upper plot. The moving range is used in the estimation of process variability and the calculation of the upper and lower process control limits.

---

7. Named in honor of Carl Friedrich Gauss, a Gaussian process has output that is normally distributed.

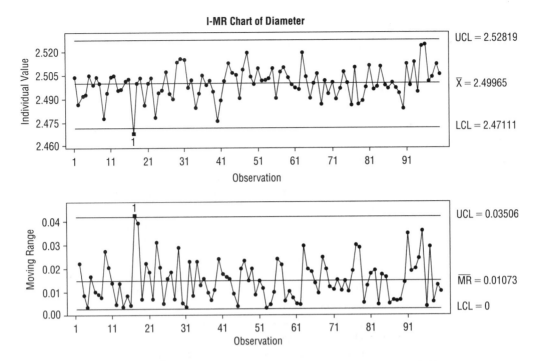

Figure 14.14   The control chart shows the random variation in a process variable.

When to Use the Control Chart

The control chart is a great diagnostic tool to evaluate process performance over time. It can indicate if the process has been stable over time (common cause variation), or if it has experienced sudden jumps or trends (assignable cause variation). It's applicable to any process that produces sequential results in an ordered way, ranging from mortgage approvals through products moving along a production line.

Descriptive Statistics

Measures of Location and Spread

The most basic characteristics of a data set are location and spread. If you are designing a product, it's likely that you want to put performance on target and minimize the spread of performance when measured over a number of systems being used in typical customer environments. To achieve on-target performance with minimal variability, intuition tells you that you probably need to manage the variations of those design parameters having the most pronounced effects on performance. Consider typical measures of location and spread. The mean and the median are two parameters that locate a distribution of data.

Mean

Usually you cannot measure the mean of an entire population. You have to estimate it from the mean of a sample. As a convention, the sample mean is designated by $\overline{X}$ and the population mean by μ. The calculation of each is straightforward:

$$\overline{X} = \frac{\sum\limits_{i=1}^{n} X_i}{n} \tag{14.1}$$

$$\mu = \frac{\sum\limits_{i=1}^{N} X_i}{N} \tag{14.2}$$

Here n is sample size and N is the number in the population.

Median

The median is the middle value if you arrange your data set in increasing or decreasing order. In the data set 1, 2, 4, 7, 10, 12, 15, the median is 7. It's the middle data point.

You don't have to calculate the median unless you have an even number of data points, in which case the median is calculated by averaging two neighbor points, as with

$$M = \frac{X_{n/2} + X_{n/2+1}}{2} \tag{14.3}$$

where n is the total number of points in the sample, arranged in either increasing or decreasing order.

For some data the median gives you a better sense of the central tendency of a population than does the mean. A good example is household income, for which a relatively small number of very high incomes can skew the distribution, making the mean less representative as a measure of central tendency.

The source of most statistical problems is in the spread or dispersion of the data, its variability. Variability is caused by the random factors that affect your process or product.

Standard Deviation

The standard deviation is the root-mean-square of the deviations from the mean. It can be calculated for any data set, regardless of whether or not the data follow a statistical distribution. For the normal distribution, the standard deviation happens to be the distance from the mean to the "point of inflection." Here the rate of change in the slope of the curve transitions from increasing to decreasing, or vice versa. Figure 14.15 shows how population data are distributed around the mean and the percentage of data within so many standard deviations of the mean.

The standard deviation of the population is

$$\sigma = \sqrt{\frac{1}{N} \sum\limits_{i=1}^{N} (X_i - \mu)^2} \tag{14.4}$$

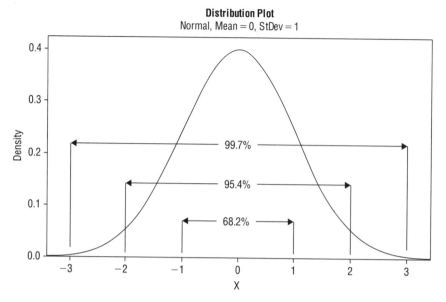

Figure 14.15   Plot of standard normal distribution.

Similarly, the standard deviation of a sample is

$$S_n = \sqrt{\frac{1}{n-1}\sum_{i=1}^{n}(X_i - \overline{X})^2}$$

(14.5)

The standard deviation has the advantage of being expressed in the same units as the data. A large standard deviation means that the data points have a large amount of variability.

Variance

The variance is the sum of the squared deviation of all the data points from their mean, which is the same as the square of the standard deviation. Its units are the square of the units for the data.

The variance of a population is defined as

$$\sigma^2 = \frac{1}{N}\sum_{i=1}^{N}(X_i - \mu)^2$$

(14.6)

For a sample, the variance is given by

$$S_n^2 = \frac{1}{n-1}\sum_{i=1}^{n}(X_i - \overline{X})^2$$

(14.7)

Here, n is the sample size, $\overline{X}$ is the sample mean, and the degrees of freedom in the divisor $(n-1)$ makes the sample variance an unbiased estimator of the population variance.

A small variance means that the data points are clustered close to their mean; they have low variability. The standard deviation is the positive square root of the variance.

Useful Properties of the Variance

In statistics, the variance is a valuable metric because it has some useful properties.

Variances are additive. If you have several uncorrelated random variables, the variance of their sum is the sum of their variances. If Y is equal to the sum of a number of independent random variables, X_i,

$$Y = X_1 + X_2 + X_3 + \cdots \qquad (14.8)$$

then the variance of Y is

$$\sigma_Y^2 = \sigma_{X_1}^2 + \sigma_{X_2}^2 + \sigma_{X_3}^2 + \cdots \qquad (14.9)$$

Don't be fooled by the term *additive property*. If Y is the difference between two random variables, the variance of Y is still the sum of the variances.

If Y is equal to the sum of a constant and a number of random variables,

$$Y = C + X_1 + X_2 + X_3 + \cdots \qquad (14.10)$$

the variance of Y is unchanged. Adding a constant just shifts the central location of the distribution, leaving the spread of the data unchanged.

Multiplying a random variable by a constant either compresses or stretches the distribution and changes the variance. If Y is equal to the sum of a number of random variables, each multiplied by a constant,

$$Y = C_1 X_1 + C_2 X_2 + C_3 X_3 + \cdots \qquad (14.11)$$

the variance of Y is

$$\sigma_Y^2 = C_1^2 \sigma_{X_1}^2 + C_2^2 \sigma_{X_2}^2 + C_3^2 \sigma_{X_3}^2 + \cdots \qquad (14.12)$$

Range

The range is another basic estimate of the spread. It is described by the lowest and highest values in the data set. For example, if your data set is

Y: 1, 2, 7, 6, 21, 25, 15, 2, 6

the range is minimum = 1, maximum = 25.

The Normal Distribution

Properties of the Normal Distribution

The normal distribution is bell-shaped and symmetrical. It is a two-parameter distribution, described by its mean and variance. Because data that describe so many different phenomena are well characterized by the normal distribution, it is useful to transform the normal distribution to standard form. If X is normally distributed with a mean μ and standard deviation σ, then

$$Z = \frac{X - \mu}{\sigma} \qquad (14.13)$$

is also normally distributed but with a mean of 0 and standard deviation of 1. Once you have the standard form for the normal distribution, you can easily draw some useful conclusions about distributions of normally distributed data. The spread of the data may be a focus of concern. For example, if you characterize a sample of a product that you are developing, it's very important to know how much variability in performance you can expect. This variability can be translated into customers' dissatisfaction and possibly warranty costs.

Figure 14.15 shows a plot of the standard normal distribution. The x-axis shows the number of standard deviations from the mean. The percentages correspond to the fraction of the total population that lies within ± 1, 2, and 3 standard deviations of the mean.

As you can see, only 0.3% of the population lies beyond $\pm 3\sigma$ Not shown, because of the scale, are the $\pm 6\sigma$ limits, which are of importance when discussing "Six Sigma" quality. Only about two parts per billion are farther than ± 6 standard deviations from the mean. This seems to be a negligibly small number until you think of high-volume manufacturing. If you make a million of something, about 3000 items will have characteristics with values beyond $\pm 3\sigma$ of the mean.

The statistical distribution that you encounter the most, by far, is the normal distribution. Here are some of the reasons for this:

1. **Many things in nature are normally distributed.**

   If you look at height data for humans, for example, you find important subgroups, most notably ethnicity and sex. Height data show that various ethnic groups have different mean heights, and that men are taller than women as a population. However, for the same sex and within an ethnic group, height is approximately normally distributed.

2. **Man-made processes often produce things having normally distributed characteristics.**

   There are many different processes for manufacturing, with a myriad of product characteristics that are of interest. Whether you are considering silicon wafers, stamped parts, injection molded parts, chemical products, or life sciences products, many important product characteristics are well described by the normal distribution.

3. **The central limit theorem drives many "end results" toward normality.**

   A reason that the normal distribution has such wide application is the central limit theorem (CLT). The CLT is a fundamental theorem in probability theory that states that the mean of a large number of independent and identically distributed (i.i.d.) random variables, each with a finite mean and standard deviation, will be normally distributed. The requirement of being i.i.d. can be relaxed under certain conditions.

   As an illustration, suppose that you have a manufacturing process shown as a block diagram in Figure 14.16. You are interested in the total time to build a widget that moves from one workstation to another where different processes are performed as part of the overall production process.

   Each process step requires a completion time, which you can assume to be described by some arbitrary statistical distribution. If you kept track of the times for each step as many widgets moved down the line, the cycle time for a given step would

Figure 14.16   Steps in a manufacturing process

not be a single value but would be distributed across a range. If you further assume a simple process with good flow, without bottlenecks, queues, or waiting times, the total time for completion is just the sum of all the times for the individual steps:

$$\text{Total Time} = \sum_{i=1}^{n} t_i \qquad (14.14)$$

It turns out, thanks to the CLT, that the distribution of Total Time will approach normal, and normality makes life easier for you because of the large statistical toolbox for analyzing these types of data. The CLT helps simplify the analysis of problems as diverse as tolerance stacks, process cycle time, time for the critical path through a project network, and both process and design capability.

4. **Data that are not normally distributed can be transformed to normal form.**

In some cases nature is more difficult because some phenomena have skewed distributions naturally. When this happens, transformations can be helpful in converting a data set that has a skewed distribution into a more symmetric distribution, at times even a normal distribution. Let's look at an example where the underlying engineering science results in a response that is not normally distributed, but where an appropriate transformation helps.

Suppose you have small cylindrical samples of the same material and you are interested in their force-deflection relationship. In the force-deflection test you impose a constant deflection on the tip of each sample and measure the resulting force, as shown in Figure 14.17. Assume that the diameter (d) of the samples is normally distributed and that the samples are all of the same length.

For a cantilever beam with a boundary condition of zero rotation, the force-deflection relationship at the tip is

$$y = \frac{PL^3}{3EI} \qquad (14.15)$$

where

$E =$ the modulus of elasticity, which is material-dependent

$I =$ the moment of inertia of the beam cross section.

For a circular section the moment of inertia is

$$I = \frac{\pi d^4}{64} \qquad (14.16)$$

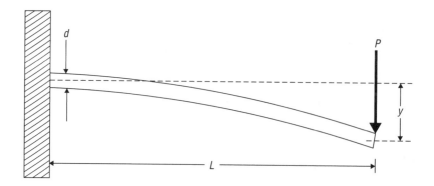

Figure 14.17   Cantilever beam deflection

So the relationship for the force, P, in terms of everything else is

$$P = \left(\frac{\pi E y}{18L^3}\right)d^4 \tag{14.17}$$

If the diameter, d, is normally distributed, your measured force, P, will not be. However, if you define a new variable,

$$F = P^{0.25} = \left(\frac{\pi E y}{18L^3}\right)^{0.25} d \tag{14.18}$$

you can see that F is normally distributed since your independent variable, d, is normally distributed and everything else is constant. This is an example of how your technical knowledge can justify the use of a data transformation. Of course you might ask, "If we're so smart and have a model, why bother with the test?" The point is that often a model is not a good enough predictor to be relied upon without supporting data. However, at times, a "back of the envelope" analysis can give you insight and enhance your understanding, even if it's not sufficiently accurate to replace experimental data.

A good rule is to use transformations when the underlying science can explain why the transformation makes sense. Be cautious when using transformations solely because they work.

5. **Many tools of inferential statistics are based on normal data.**

Fortunately, techniques for analyzing normal data are well developed, statistically powerful, widely used, and included in all statistical analysis applications. In the next chapter we discuss the tools of inferential statistics. A common theme is analyzing sample data to determine the location and spread of population statistics, such as the mean and variance. While it's possible to calculate the mean and variance for any data set, you need to rely on one of the sampling distributions to estimate confidence intervals for these statistics. The use of sampling distributions, such as the Student's t, chi-square, and F-distributions, assumes normality of the population data.

One exception, where you don't often encounter the normal distribution, is in survival data such as the distributions of time or cycles to failure. Other distributions, such as the Weibull, exponential, and lognormal, are frequently seen. We discuss these distributions in Chapter 21.

Key Points

1. Statistical tools are applied to understand how performance is affected by changes in design or process parameters.
2. Always plot the data before embarking on an analysis. A plot can quickly give you insights that might be missed if you look only at descriptive statistics.
3. The two most frequently used population statistics in product development are the mean and standard deviation.
4. The normal distribution shows up regularly in data analysis. The root cause of this is that many natural and man-made processes generate attributes that vary with similar frequencies in either direction about their mean.
5. Sometimes non-normal data can be transformed into normal form.
6. Some processes produce non-normal data and can be described by other probability distributions. One example is time between failures where often the Weibull, lognormal, or exponential distributions will fit the data.
7. If a distribution other than the normal fits the data well, it's preferable to use that distribution rather than transforming the data.

Discussion Questions

1. Why are some characteristics normal and others not?
2. If a manufacturing process generates non-normal data, is that a sign that something is wrong with the process?
3. What do the mean and standard deviation tell you about process quality?
4. Which is more difficult to change in a process, the process mean or the standard deviation? Why?
5. Where can you put the process control chart to good use?
6. Why would you want to plot a data set on a normal probability plot?
7. What does the histogram in Figure 14.18 tell you about the data?

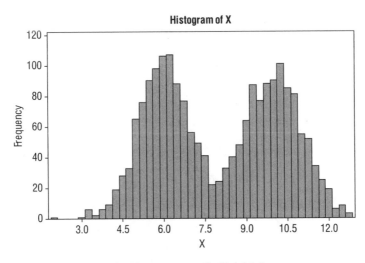

Figure 14.18   What's the cause of a histogram profile like this?

Inferential Statistics

In the design of a product or process, it is essential to understand the system behaviors in the presence of all the sources of variation that can affect performance. Performance variations can be caused by external factors, such as environmental stresses, or by internal factors in your system, such as the variations in the properties of parts, materials, or assemblies. A proper interpretation of the results of testing and experimentation demands that you separate the signal from the noise, requiring that you use the fundamental tools of inferential statistics. In this chapter we introduce some of those tools. Our goal is to explain enough that you can understand the implications for decisions, be encouraged to learn more about the methods, and be able to have value-adding conversations with practitioners.

Questions You Need Statistics to Help Answer

One of the major challenges in product development is to understand and control variability. Variability in your product's performance is almost always a bad thing. It may be caused by sources in your manufacturing and assembly processes or by an inherent vulnerability in the design to the stresses in the product's use. Variable processes lead to variable performance, which results in quality loss, as discussed in Chapter 2. Statistical tools are essential to help you answer questions such as these:

1. Does the performance of a product or manufacturing process satisfy its requirements and have acceptable variability?
2. What forecasts can you make about the performance of the future population of products, after market entry, based on the conclusions from experiments with preproduction samples?
3. Are the responses in a manufacturing process or a product's performance stable enough under the stresses of usage, operating modes, environments, and other external factors?
4. Has a change to the design of a product or manufacturing process actually changed the distribution of performance of the population of products?
5. Are there significant subgroups within the global population, such as differences within a manufactured batch or among batches?
6. Do samples of populations from different suppliers indicate significant differences in the capabilities of those suppliers?

Valid answers to questions such as these will help you to develop products that have superior performance with an acceptable level of variability.

Population versus Samples

You need statistics to develop an understanding of population characteristics based on samples of the population. A population of products includes every unit, while a sample is only a part of that population. Usually you study samples, rather than the population, for a couple of good reasons:

1. You may have access only to samples.
2. You may not have the time or money to obtain data from the entire population, even if they could be available.

You then have the responsibility to obtain a sample that represents the population well. In a representative sample, responses occur with about the same relative frequency as they do in the general population. When you have a representative sample, you can use the frequency of a result to make inferences about the likely frequency in the larger population.

One concern is the potential existence of subgroups in a population. In a subgroup a certain response may be highly represented but not characteristic of the rest of the population. To handle subgroups, investigators use techniques such as "stratified sampling" where samples are drawn from subgroups in proportion to the relative size of the subgroup. The process of sampling is not trivial, especially in fields such as medical science, opinion polls, and clinical trials. For guidance in these areas, there are focused publications and specialists that should be consulted.[1] The need for more sophisticated sampling techniques occurs when subgroups have been commingled and are presented as part of a larger population. It could be that products, parts, or materials manufactured a year after market entry perform differently from those produced at the beginning of production. In manufacturing you have control over this up to a point. It is far easier to understand subgroups before all the parts produced by a process have been dumped into one large parts bin. Factors such as material lot variations and tool differences should be understood early before all the parts and materials are in production.

During development, teams test a number of manufactured parts and have to make inferences about the population based on characteristics of the sample. They may measure the performance of a number of prototype systems and have to draw conclusions about the performance of the population of products that will be produced after market entry. This is not to say that subtleties in sampling, such as subgroups, can't make decisions complicated. Subgroups do exist, such as different lots of materials, parts from different mold cavities, and different suppliers. These require care in sampling and data interpretation. But in general, sampling concerns are easier to manage within the factory or the laboratory. When subgroups are found in a population of objects that ideally should be members of a homogeneous population, you have to understand the cause of those subgroups. Is the existence of subgroups benign and to be expected? Does it signal more serious problems with the consistency of products and processes? Consider subgroups to be an opportunity to learn more about your products or processes.

---

1. William G. Cochran, *Sampling Techniques* (New York: John Wiley, 1977); Steven Piantadosi, *Clinical Trials: A Methodologic Perspective, 2nd ed.* (New York: John Wiley, 2005).

Once your product is in the market, the task of gathering representative samples can become complicated quickly by factors such as variations in customers' applications, usage environments, performance expectations, cultures, and geographic distance. In the laboratory you have the luxury of measuring and characterizing the product, making the link between product configuration and performance much stronger. Whether you are in the laboratory or out in the field, keep in mind the need for representative samples.

What Is Random and What Is Not?

You use samples to support decisions with data. As the data go, so go your decisions. So it's important to be confident that your samples are truly representative. The risk increases when using samples from suppliers, since you don't have much control over the characteristics of the samples or of the sampling process itself.

Samples should have these important qualities:

1. Be randomly chosen
2. From a stable process
3. Not from a truncated distribution where out-of-specification parts are eliminated
4. Not from a process that's been tweaked into submission and is running under conditions that are not sustainable

If a supplier's process is not producing consistently good quality, the supplier may be tempted to include only the good parts in the evaluation samples. The supplier's strategy might be first to get the order and then to worry about stabilizing and aligning the processes.

This is a discussion about sampling distributions. While some sampling distributions are different from the familiar normal distribution, they are closely related in their dependence on the normality of the population data. In the next section you will see why sample size is a powerful tool that helps to separate the signal from the noise.

The Power of Averaging

At times you may have to recover a meaningful signal that's obscured by a high level of noise. For example, Figure 15.1 shows the first 50 of 10,000 points sampled from a normal distribution with a mean of 1 and standard deviation of 10.

If someone asked you to estimate the mean using the sample of the first 50 points, you would probably demur. Let's assume that the points represent a stream of data generated by a very noisy process. However, you still need a useful estimate of the mean. With such a poor signal/noise ratio[2] (S/N = 0.01), a small sample size to estimate the mean would not give a useful or valid result, unless you were lucky. While more elegant and efficient filtering schemes exist, in this case brute force can be brought to bear. You can make a good estimate of the mean of the underlying population by averaging a sample that is large enough. Figure 15.2 shows how the sample mean converges to the population mean as additional data points are added to the sample. If the data come from a stationary random process (the mean and standard deviation are constant), taking a

---

2. The signal-to-noise ratio, a term from electrical engineering, is defined as (power of the signal) divided by (power of the noise), where power is proportional to the square of the amplitude. In our example, the amplitude of the signal is 1 and the noise 10; therefore, S/N = 0.01.

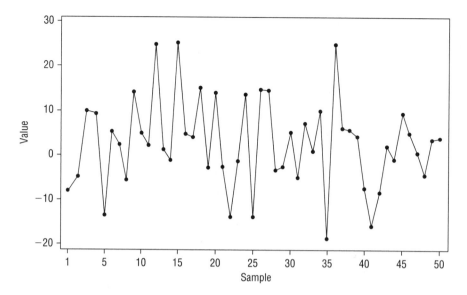

Figure 15.1   Time series plot of data

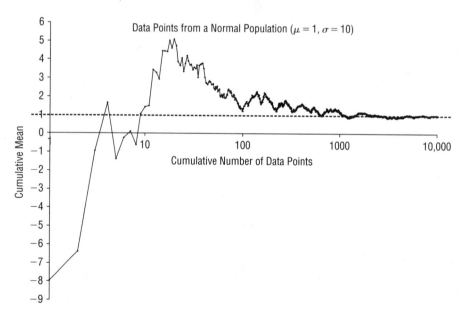

Figure 15.2   Cumulative average of points sampled from a normal distribution

large enough sample will work. As you add more points to the sample and continue updating the cumulative mean, the variability gradually washes away and the sample mean emerges from the noise, asymptotically approaching the underlying population mean.

The averaging process is one of low-pass filtering, eliminating the high-frequency variations, leaving you with the mean. This is a trick that has long been used in signal processing when searching for a small signal obscured by a large noise. In signal processing, a major concern is the processing time since many filters are implemented in software. However, if you're making physical products, very large samples can be cost-prohibitive.

Fortunately, with a manufactured product, you would not expect variability in the data to be so large. Usually the S/N is much larger, so good estimates of the mean can usually be made by averaging sample sizes of tens rather than tens of thousands. Since averaging will help you to improve your estimate of the population mean, sample size is an important consideration. In general, increasing sample size improves the accuracy of the estimate.

The example of a stream of many data points from a very noisy process shows the power of averaging to extract the process mean. The law of large numbers[3] shows that larger samples are an option for reducing uncertainty. Later in this chapter we describe some examples relevant to manufacturing processes. First we discuss high-level objectives for using inferential statistics.

Making Comparisons and Assessing Progress

During product development, teams are interested in both the statistical distribution of the characteristics of components and the performance of the product system. There are two parameters of main interest:

1. Location of the mean
2. Variation or spread of the data

Usually both of these descriptive statistics can be described by a statistical distribution. Often you are interested in comparing the mean and standard deviation to a reference, such as their requirements. For example, perhaps you are aiming at a certain value for a part dimension and want to know if your process is aligned with the target or is at least in the neighborhood. As another example, suppose you have samples of a part manufactured by two different suppliers and need to know if the processes are equivalent. Getting a good handle on the mean and spread of the data requires that you to learn how to use some common statistical distributions.

Useful Distributions for Making Comparisons

Table 15.1 shows five distributions that are frequently used for making comparisons using basic inferential statistics. Later we describe the use of these distributions in some detail. If you are interested in the functional form of any of these distributions, the book by Evans, Hastings, and Peacock[4] is a good reference. Unless you plan to study statistics further, it is unlikely that you will ever need to know the mathematical details for these distributions. They are complicated, and with the availability of good statistical analysis software, the functional form of these distributions is not required knowledge for the average person developing products and processes.

In the next section we discuss the advantage of having normal data. The normal distribution is used to introduce the concept of hypothesis testing. It is also the basis of one- and two-sample tests. In most cases, the t-distribution is used since you rarely know the population variance(s).

---

3. John E. Freund, *Mathematical Statistics, 5th ed.* (Englewood Cliffs, NJ: Prentice Hall, 1992).

4. Merran Evans, Nicholas Hastings, and Brian Peacock, *Statistical Distributions, 2nd ed.* (New York: John Wiley, 1993).

Table 15.1   Tests and distributions used when making comparisons

Comparison	Reference Distribution
Sample mean to a reference (σ known)	Standard normal (one-sample Z-test)
Sample mean to a reference (σ unknown)	Student's t (one-sample t-test)
Means of two samples (σ known)	Standard normal (two-sample Z-test)
Means of two samples (σ unknown)	Student's t (two-sample t-test)
Sample variance to a reference	Chi-square
Variances of two samples	F-distribution
Sample proportion to a reference	Hypergeometric (Fisher's exact method)
Two-sample proportions	Hypergeometric (Fisher's exact method)

Normality of Data

There is an important ground rule: Methods we describe that compare means and variances assume that the data come from a normal population. Usually you work with the "standard normal distribution." It has a mean of 0 and a standard deviation of 1. You can shift from raw data to the standard normal form by transforming the data, as shown in Figure 15.3. The analysis of the standard normal form has been tabulated, allowing you to apply the tabulated information to any problem involving normally distributed data. Of course, with the wide availability of statistical analysis programs for computers, it's not necessary to engage in the laborious process of converting your raw data to normal form.

However, when using any of these applications, it's nice to appreciate what's going on "under the hood." Also, if you know the background and are marooned on a desert island without your laptop, all will be well as long as you have the right book with you.

Since normality is a requirement, you should test the data for normality before going too far in its analysis. Statistics applications have convenient tools to test for normality. In Minitab, for example, either the normality test tool or the probability plot for the normal distribution can do it. While some tests are relatively robust to non-normality, it is still best to look at your data graphically first to determine how far the data depart. While data and associated specification limits often can be transformed into normal form, it's not recommended without the advice of a statistics professional. This guidance would be essential when a major decision is to be based on conclusions from hypothesis tests about differences in population statistics.

Statistical Inferences about Means

When the Normal Distribution Is the Sampling Distribution of the Mean

If you take many samples from a normal population and calculate the means of these samples, the distribution of the sample means is normal. The sampling distribution of the mean is centered on the population mean, although it has a smaller standard deviation. Just as you did with the normal distribution describing the population, you can also normalize the sampling distribution of the mean. If you know the population's mean to be μ and its standard deviation to be σ, the "z-statistic" is given by

$$z = \frac{\overline{X} - \mu}{\dfrac{\sigma}{\sqrt{n}}} \qquad (15.1)$$

For example, if you have historical data on the variability of a manufacturing process, you will know the process standard deviation, and if the parameter X is normally distributed, you will know that z is also normally distributed. In equation (15.1), n is the sample size and $\overline{X}$ is the sample mean. When you have normal population data, you can use the standard normal distribution for the distribution of sample means, as shown in Figure 15.4.

The "standard error of the mean" is the equivalent standard deviation for the sampling distribution of the mean. It is given by

$$\text{standard error of the mean} = \frac{\sigma}{\sqrt{n}} \qquad (15.2)$$

where

σ = the standard deviation of the population

n = sample size

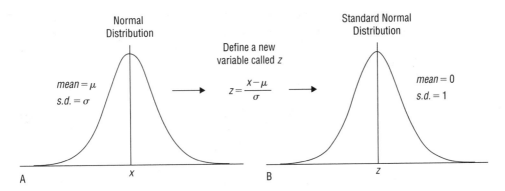

Figure 15.3 Getting from the normal to the standard normal distribution.

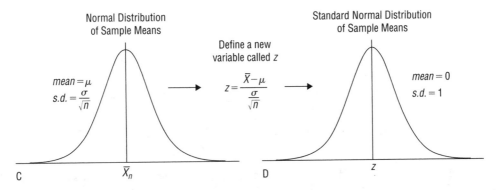

Figure 15.4 The normal and the standard normal distribution of sample means.

Intuitively, it makes sense that the standard deviation of the sampling distribution of the mean is less than the standard deviation of the population. Variability of the sample means is less than variability of the population because of the averaging process. Figure 15.5 illustrates the plot of the standard error of the mean versus sample size for a unit standard deviation. It is significant that the "knee" in the curve is at about 30 samples. Statisticians often use this rule of thumb to recommend the minimum sample size for a good estimate of the population mean. You can use smaller samples to reduce costs. However, keep in mind that it adds uncertainty to your inferences and risks to the resulting decisions. Later we discuss how to estimate the required sample size.

Caution on the Use of the Normal Distribution

In most situations the population variance is not known. The best strategy is to use the t-distribution as the sampling distribution of the mean. Use of the normal distribution is a holdover from pre-computer days when tools such as powerful statistics applications and desktop computing were not available. With the tools available today, the statistics calculations are easily done, so it is better to use the t-distribution.

Student's t-Distribution

The Student's t-distribution is the normalized sampling distribution of the mean. You use it when you don't know the variance of the population and have to estimate it using the standard deviation of the sample. When the population variance is unknown, rather than the z-statistic, you use the "t-statistic." It is determined by

$$t_{n-1} = \frac{\overline{X} - \mu}{\dfrac{s_n}{\sqrt{n}}} \qquad (15.3)$$

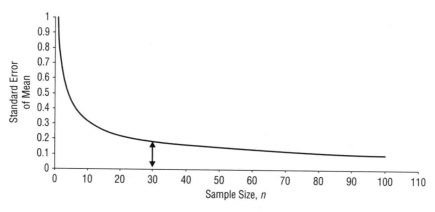

Figure 15.5   The plot of standard error of the mean shows the diminishing returns of sample size in improving the accuracy of estimates of the population mean.

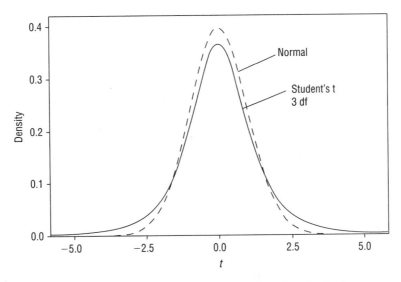

Figure 15.6   Distribution plots for standard normal and Student's t-distributions

where

n = sample size

$n - 1$ = number of degrees of freedom

s_n = standard deviation of the sample

There is a family of t-distributions with a different distribution for each value of degrees of freedom (df). The use of the sample's standard deviation when calculating the t-statistic introduces additional uncertainty into your estimate. Let's look at a plot of the Student's t-distribution, shown in Figure 15.6, and compare it to the normal distribution. You can see that the t-distribution is bell-shaped and symmetric. It is a one-parameter distribution, since all you need to know is the degrees of freedom. The Student's t-distribution looks like the normal distribution. In fact, by the time you reach a sample size of about 30, they are practically indistinguishable.

In the next section we introduce some important aspects of making comparisons and show how you can make good use of both the normal distribution and the Student's t-distribution.

Hypothesis Testing, Confidence Intervals, and p-Values

Making Comparisons

When you compare samples of components or prototypes of a product, usually you compare means and standard deviations of the data derived from measuring critical parameters or performance responses. Two types of comparisons are of interest:

1. Compare either the mean or the standard deviation of a sample to a hypothetical reference value, such as a requirement. In this case you have reason to believe that the

mean or standard deviation should be within some range, based on either design intent or historical data. There are three possibilities:

a. It's equal to some value.

b. It's less than some value.

c. It's greater than some value.

When we say equal to some value, the mean or standard deviation of the sample is within a required range of the reference value and given the variability in the data, it is not possible to make a statistical distinction between the mean of the sample statistic and the reference value.

2. Compare the mean or standard deviation of two samples to determine if they are statistically different.

Let's look at how you can use the Student's t-distribution to make inferences about the mean of a population. We assume here that the variance of the population is not known, requiring the use of the t-statistic.

Confidence Intervals for the Mean

You can use the Student's t-distribution to calculate confidence intervals for the population mean. The t-statistic is defined as

$$t_{1-\alpha/2,\,n-1} = \frac{\overline{X} - \mu}{\dfrac{S_n}{\sqrt{n}}} \tag{15.4}$$

In equation (15.4), note that we have added another subscript. The subscript α is the fraction of total area under the curve in the tail of the distribution, and $n-1$ is the number of degrees of freedom, as in equation (15.3). You can use this relationship between t and $\overline{X}$ to calculate a confidence interval for the mean. The value of the t-statistic for a given sample size and α level can be looked up in a table or calculated using most of the current statistical analysis software applications. If you know the value of the population mean μ, and you repeatedly draw samples of size n from the population, you know they will be distributed following the t-distribution. Figure 15.7 shows how you can get the t-distribution from the sampling process.

When you take samples from your normal population and compile your population of sample means, you will find that the mean of the distribution of sample means is the same as the distribution of the population mean. Also, the distribution of sample means is bell-shaped with many samples close to the mean. You will find fewer samples as you move farther away. Out in the tails of the distribution, shown in Figure 15.7, there are sample means that are seen infrequently. α is the probability of getting sample means at or farther away from the population mean than $\pm t_{1-\alpha/2,\,n-1}$. So small probabilities (p-values) indicate rare events where the hypothesized mean is far from where the data say it's expected to be.

The confidence interval for the mean is specified by the upper and lower limiting values. If you took repeated samples from a normal population with a specified mean, over the long run $(1-\alpha) \times 100\%$ of your sample means would be within this range. The more confident you are required to be, the smaller will be α and the broader will be the range. You have to decide how confident you need to be. For hypothesis testing, $\alpha = 0.05$ gives a 95% confidence interval,

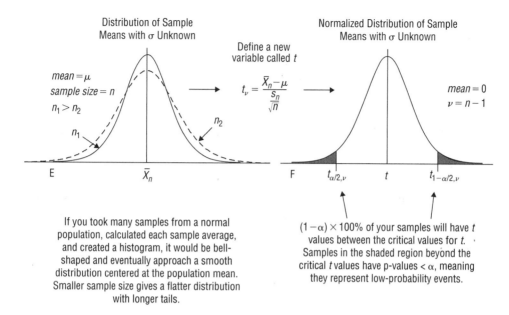

Figure 15.7   Getting from the distribution of sample means for unknown population standard deviation to the Student's t-distribution.

a level that is widely used. However, there is nothing sacred about 95%. Deciding what confidence interval to use is a business decision, not a statistical one. It depends on the risks associated with being wrong. Using equation (15.4), you can specify confidence limits for the population mean.

The two-sided confidence interval for the population mean is

$$\overline{X} - t_{\alpha/2,\,n-1}\frac{S_n}{\sqrt{n}} < \mu < \overline{X} + t_{1-\alpha/2,\,n-1}\frac{S_n}{\sqrt{n}} \tag{15.5}$$

Over the long run $(1 - \alpha) \times 100\%$ of your intervals will contain the true population mean. The one-sided upper confidence limit (UCL) is

$$\mu_{UCL} = \overline{X} + t_{1-\alpha,\,n-1}\frac{S_n}{\sqrt{n}} \tag{15.6}$$

Over the long run $(1 - \alpha) \times 100\%$ of your sample means will be less than the UCL. The one-sided lower confidence limit (LCL) is

$$\mu_{LCL} = \overline{X} - t_{1-\alpha,\,n-1}\frac{S_n}{\sqrt{n}} \tag{15.7}$$

Over the long run $(1 - \alpha) \times 100\%$ of your sample means will be greater than the LCL.

How can you use this? What is the appropriate interpretation? Suppose you have a population with a normal distribution, but its mean and standard deviation are not known. When you draw a sample of size n, there is a probability of $1 - \alpha$ that the population mean is between the upper and lower confidence limits, calculated using equation (15.5). For equation (15.5) you must look up in a table or calculate on the computer the critical values of the Student's t-distribution. The other factors in the equation are the summary statistics from your data set.

By examining equation (15.5), you can learn some important characteristics of confidence intervals for the mean:

1. With a smaller sample size your confidence limits are wider, since the t-distribution is broader. The standard error of the mean is larger, since it varies inversely as the square root of sample size.
2. With a smaller α error, your confidence limits are wider since you must include more of the distribution tails to increase the probability that your sample mean is within the limits.

Equation (15.5) shows that you can reduce risk by increasing sample size and narrowing the confidence interval. The risk here is not the α error we have been discussing but rather the risk of your product or system performance varying over a wider range.

Estimating Required Sample Size

Let's calculate a sample size based on your requirements for accuracy in estimating the mean. Suppose you want to have a 95% probability of having your sample's mean be within a range of $\mu \pm \delta$. This means that you choose $\alpha = 0.05$ and use the z-statistic rather than the t-statistic. If you tried to use the t-statistic, as in equation (15.4), you would find that picking the right value for t requires knowing sample size, the characteristic you're looking for to begin with. So use the z-statistic. We cover this later in more detail when we discuss the requirements for sample size in the context of comparing a sample mean to a reference value. Using the z-statistic, the value of δ is given by

$$\delta = z_{1-\alpha/2}\frac{s_n}{\sqrt{n}} \tag{15.8}$$

and the required sample size is

$$n = \left(z_{1-\alpha/2}\frac{s_n}{\delta}\right)^2 \tag{15.9}$$

Note that in applying equation (15.9), before you can calculate sample size you need to know the standard deviation or have an estimate of it. There's no way around it. Either you need prior knowledge about population variability or you have to take a sample to estimate it and then calculate the required sample size.

How the Hypothesis Testing Process Works

1. First decide on the allowable α error that you will accept.
2. Select a hypothetical value for the population mean. Your selected value could be a desired outcome, μ_0, or it could be based on historical experience.
3. Your "null hypothesis" is that the population mean is equal to μ_0.

4. If you have no reason to believe that the population mean is either greater than or less than μ_0 your alternate hypothesis is that the population mean $\neq \mu_0$.
5. Take a sample from the population and calculate its mean and standard deviation.
6. Calculate the upper and lower confidence limits for the population mean using equation (15.5).
7. If μ_0 falls outside the upper and lower confidence limits for μ, conclude that it's very unlikely that the population mean could be μ_0 and reject the null hypothesis.
8. If μ_0 falls within the calculated confidence limits, accept the null hypothesis since there is not enough evidence to conclude that $\mu \neq \mu_0$.

To understand these points better, consider some distribution plots and how to interpret the results of typical hypothesis tests. For each example we show a couple of alternatives for the null hypothesis.

Steps in Hypothesis Testing Using the One-Sample t-Test

Step 1: Formulate the null hypothesis.

Assume that the null hypothesis is $H_{01}(\mu = \mu_{01})$. This could be based on historical data or on a desired outcome such as the specification for a critical parameter or a response metric.

Step 2: Formulate the alternative hypothesis.

Alternate hypothesis 1: $\mu \neq \mu_{01}$

This alternate hypothesis is the first of three we describe. If you choose this alternate hypothesis, you have to calculate the upper and lower confidence intervals for the population mean, shown in Figure 15.8 as μ_{UCL} and μ_{LCL}.

Step 3: Draw a sample from the population.

For sample size, use equation (15.9), which is based on the maximum error you are willing to accept in the estimate of the mean.

Step 4: If the standard deviation of the population is not known, use equation (15.5) to calculate the sample mean, $\overline{X}$, and the upper and lower confidence limits for the population.

If the hypothesized mean is within the calculated confidence interval, conclude that you don't have enough data to reject the null hypothesis $H_{01}(\mu = \mu_{01})$.

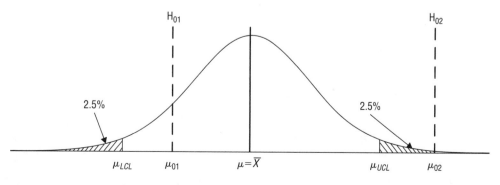

Figure 15.8   Confidence interval showing two different null hypotheses, $\alpha = 0.05$

If the null hypothesis had been H_{02} ($\mu = \mu_{02}$), you would conclude that, since μ_{02} is outside the confidence interval, it's unlikely to be the population mean. The conclusion would be that you reject the null hypothesis and accept the alternative, $\mu \neq \mu_{02}$.

Alternate hypothesis 2: $\mu < \mu_{01}$

If you choose this alternate hypothesis, you have to calculate the upper confidence limit for the population mean, shown in Figure 15.9 as μ_{UCL}.

If the hypothesized mean is to the left of the calculated UCL, conclude that you don't have enough data to reject the null hypothesis H_{01} ($\mu = \mu_{01}$).

If the null hypothesis had been H_{02} ($\mu = \mu_{02}$), conclude that, since μ_{02} is beyond the UCL, it's unlikely to be the population mean. Reject the null hypothesis and accept the alternate ($\mu < \mu_{02}$).

Alternate hypothesis 3: $\mu > \mu_{01}$

If you choose this alternate hypothesis, you must calculate the LCL for the population's mean, shown in Figure 15.10 as μ_{LCL}.

If the hypothesized mean is to the right of the calculated LCL, conclude that there is not enough data to reject the null hypothesis H_{01} ($\mu = \mu_{01}$).

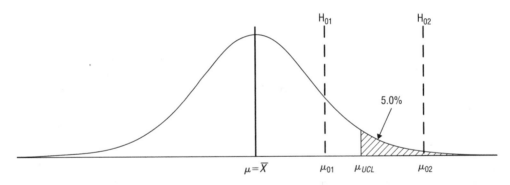

Figure 15.9   UCL with two different null hypotheses, $\alpha = 0.05$

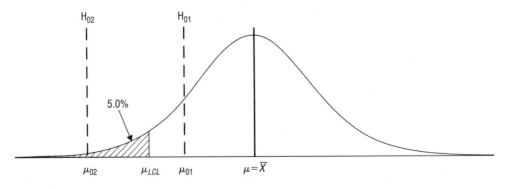

Figure 15.10   LCL with two different null hypotheses, $\alpha = 0.05$

If the null hypothesis had been H_{02} ($\mu = \mu_{02}$), conclude that, since μ_{02} is to the left of the LCL, it's unlikely to be the population mean. Reject the null hypothesis and accept the alternate ($\mu > \mu_{02}$).

How to Interpret p-Values

Today, most statistical analyses are done using one of the many computer applications that are available. Since any analysis of hypothesis tests involves the p-value, it's important to understand its meaning.

> The **p-value** is the probability that you would observe the data you have if the null hypothesis were true.

Figure 15.11A can help you understand this. Suppose you draw a sample from a population. You form a null hypothesis that the population mean is μ_0 and an alternate hypothesis that the population's mean $\neq \mu_0$. You then calculate the sample's mean, $\overline{X}$, and the sample's standard deviation, s. Using these sample statistics and equation (15.5), you calculate the confidence interval (CI) for the population mean (μ_{LCL}, μ_{UCL}). You can see that the null hypothesis is outside the confidence interval, so you reject the null and accept the alternate. If you were using a statistical analysis program, you would also get a p-value as part of the results. It so happens that the shaded area under the curve to the right of μ_0 is $p/2$.

Let's look at the problem another way and assume that the population's mean is actually μ_0. Consider the probability that you draw a sample from the population that is equal to or more extreme than your sample with a mean of $\overline{X}$. In Figure 15.11B you can see graphically what is happening by sliding the population distribution to the right and centering it on μ_0. The shaded

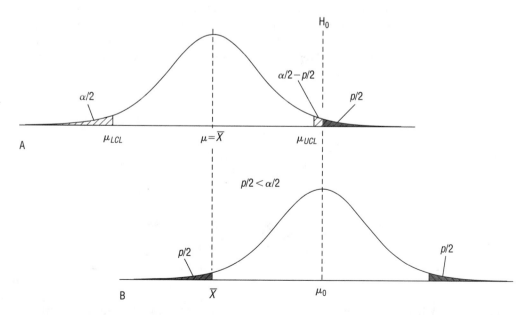

Figure 15.11 Looking at p-values in a couple of different ways

area under the curve to the left of $\overline{X}$ is labeled $p/2$. This is the probability of drawing a sample farther to the left. Since you are doing a two-tailed test, you also have to include the area in the right tail, also labeled $p/2$. The sum of the area in both tails is p, which is the probability of interest and the value calculated by your computer application. In a one-tailed test all the area associated with the p-value is in a single tail.

Sometimes You Will Be Wrong

Type I and Type II Errors

A confidence limit is a probable range of values for a population statistic. It's calculated using sample statistics and the appropriate normalized reference distribution. When you establish a confidence limit for a population statistic, you are specifying the probability that the population statistic of interest lies within those limits. There are two ways to be wrong when testing a hypothesis.

1. A "Type I" error (α error) is the rejection of the null hypothesis when actually it is true.
2. A "Type II" error (β error) is the failure to reject the null hypothesis when actually it is false.

The situation is illustrated graphically in Figure 15.12, which shows two overlapping distributions. The null hypothesis, H_0, is shown by the left-hand distribution, with its mean at μ_0. The alternate hypothesis, H_A, is shown to the right with its mean at $\mu_0 + d$. $\overline{X}_{crit}$ is the UCL for your sample means if the population's mean is at μ_0. The crosshatched area to the right of $\overline{X}_{crit}$ is the α error ($\alpha/2$ for a two-tailed test). The crosshatched area to the left is the β error, the probability of a Type II error.

When you reject the null hypothesis, it's possible that you are making a mistake. Suppose that, as shown in Figure 15.12, your population mean is actually at the hypothesized value of μ_0. If you drew a sample that had a mean just to the right of $\overline{X}_{crit}$, you would incorrectly reject the null hypothesis. This is a Type I error. Over the long run you will make Type I errors $\alpha \times 100\%$ of the time.

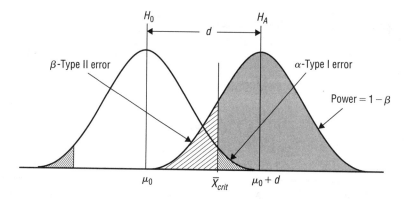

Figure 15.12   When making comparisons, you must consider both ways that you can be wrong.

The other type of error you have to consider is the Type II error. Refer again to Figure 15.12. Assume your null hypothesis to be $\mu = \mu_0$ and that you draw a sample with a mean slightly less than $\overline{X}_{crit}$. Suppose that the population mean is actually not μ_0 but is $\mu_0 + d$. In this case, the sample mean is within the calculated confidence interval. Based on the null hypothesis, you would incorrectly fail to reject the null hypothesis that the population mean is μ_0, when in fact it is $\mu_0 + d$. This is a Type II error with its associated β error. Note that the β error is a conditional probability. It's the probability of making a wrong decision, given that the mean is actually at $\mu_0 + d$. The implication is that to reduce the risks associated with incorrect decisions, sample size is an important business and technical consideration when planning a test.

Minimum Detectable Difference

Using equation (15.4), you can derive an expression for the minimum difference in population means that is detectable and the sample size that is required to detect that difference. Using the allowable α error, you have for the right-hand critical value of $\overline{X}$ for the H_0 distribution

$$\overline{X}_{crit} = \mu + t_{1-\alpha/2,\,n-1}\frac{s_n}{\sqrt{n}} \tag{15.10}$$

Use the allowable β error to determine the left-hand critical limit for the H_A distribution:

$$\overline{X}_{crit} = (\mu + d) - t_{1-\beta,\,n-1}\frac{s_n}{\sqrt{n}} \tag{15.11}$$

Using these two equations, you can solve for the minimum detectable difference for the test in terms of the standard deviation of the sample, the sample size n, and the α and β errors.

$$d = \frac{s_n}{\sqrt{n}}(t_{1-\alpha/2,\,n-1} + t_{1-\beta,\,n-1}) \tag{15.12}$$

So if you need to detect small shifts in the mean of a population with a given standard deviation, you may need to increase the sample size. Your other option is to reduce the process variability, which would be an objective for a development project. This value for d assumes a two-sided test. If you were doing a one-sided test, you would use $t_{1-\alpha,n-1}$ rather than $t_{1-\alpha/2,n-1}$ in equation (15.12).

Sample Size Requirements

Also of interest is the required sample size, given by rearranging equation (15.12).

$$n = \left(\frac{s_n}{d}\right)^2 (t_{1-\alpha/2,\,n-1} + t_{1-\beta,\,n-1})^2 \tag{15.13}$$

This equation is instructive. It tells you that for given α and β errors the required sample size varies directly with the sample variance and inversely with the square of the difference you are interested in detecting. The dilemma in using this equation is that you don't know n and, as a result, don't know how many degrees of freedom to assume for the t-statistic. The easy solution is to use the z-statistic instead of t.

The z-statistic is the limiting value for t as sample size is increased. If you consult a table of t-statistics for $\alpha/2 = 0.025$, the difference between the value of t for $n = 10$ and $n = \infty$ is about 15%. Using z instead of t, you would underestimate n. However, you can apply a safety factor to the sample size to compensate. So the additional precision in the estimate of n, by using a fancier calculation such as an iteration on t, is probably not worth it. On top of that, uncertainty in the population's standard deviation adds uncertainty to the calculation of sample size, making the more precise calculation additional work with little payoff. The expression in terms of the z-statistic and population standard deviation, estimated from the sample, is

$$n = \left(\frac{\sigma}{d}\right)^2 (z_{1-\alpha/2} + z_{1-\beta})^2 \tag{15.14}$$

If sample size is a given, you can calculate the minimum detectable difference:

$$d = \frac{\sigma}{\sqrt{n}}(z_{1-\alpha/2} + z_{1-\beta}) \tag{15.15}$$

Let's calculate a sample size based on your accuracy requirements for estimating the mean. Suppose that you want to specify the probability, $1 - \alpha$, of taking a sample and having the sample mean be within a range of $\mu \pm \delta$. This means that the maximum allowable difference between the sample mean and the actual population mean is half the width of the confidence interval, which is expressed by

$$\delta = \frac{\sigma}{\sqrt{n}} z_{1-\alpha/2} \tag{15.16}$$

The required sample size to stay within this limit of error is

$$n = \left(\frac{\sigma}{\delta} z_{1-\alpha/2}\right)^2 \tag{15.17}$$

Note that the "power of the test" (see below and Figure 15.12) and the associated β error are not factors in the confidence limit calculation. Realistically, you would look at situations where $d > \delta$. Assuming that you would first choose a sample size to pick the confidence limit width, you would calculate d in terms of δ for any sample size. It then depends only on the acceptable α and β errors that are chosen.

$$d = \delta\left(1 + \frac{z_{1-\beta}}{z_{1-\alpha/2}}\right) \tag{15.18}$$

Alternatively, if you choose the sample size based on the ability to detect a difference in means, you would express δ in terms of d.

$$\delta = \left(\frac{z_{1-\alpha/2}}{z_{1-\alpha/2} + z_{1-\beta}}\right)d \tag{15.19}$$

So there are a couple of useful results relating the minimum detectable difference in population means to the maximum error in the estimation of the mean for a specified α error.

Table 15.2 summarizes the errors that you can make when testing hypotheses. These errors are not mistakes that are made but rather are a result of the laws of probability. You have to choose the allowable probabilities for Type I and Type II errors, considering the consequences associated with being wrong.

Power of the Test

The power of the test is the probability of not making a Type II error. Figure 15.13 shows the effect of increasing sample size. It narrows the distribution of sample means, giving you some nice options. The standard error of the mean (equation (15.2)) varies inversely as the square root of sample size. So the distribution of sample means gets narrower by $1/\sqrt{n}$. Increasing sample size can help you increase the probability of correctly rejecting the null hypothesis if it is false and accepting it if it is true. You have some choices to make, such as these:

1. Reduce the α error.
2. Reduce the β error, increasing the power of the test.
3. Hold the test power constant and detect smaller differences.
4. Reduce the α and β errors and detect smaller differences.

So when you increase sample size, there is the opportunity to use the additional information to do a number of good things. All these benefits raise the question "What's not to like about a larger sample size?" The answer, of course, is that it will require more time and money. This cost has to be judged in the context of the reduced α and/or β error. Again, you have a business concern to grapple with, not a statistical one.

The results of your $\alpha - \beta$ error trade-off depend on the product and on how off-target performance or outright failure affects your customers. The consequences can be very serious for making a misstep in the design and for testing certain products, less so for others. Figure 15.13A shows the α and β errors for a test. Once you have set sample size and the difference that you want to detect, the only way to decrease the β error is to increase the α error. For some products, while you might be willing to assume an increased α in order to reduce β, the best available way to reduce the risk of making an incorrect decision is to test a larger sample size.

Table 15.2   Probability of Type I and Type II errors

		Reality	
		Null True	**Null False**
Conclusions of significance testing	Null true Fail to reject	$(1 - \alpha)$	β (Type II error)
	Null false Reject	α (Type I error)	$(1 - \beta)$

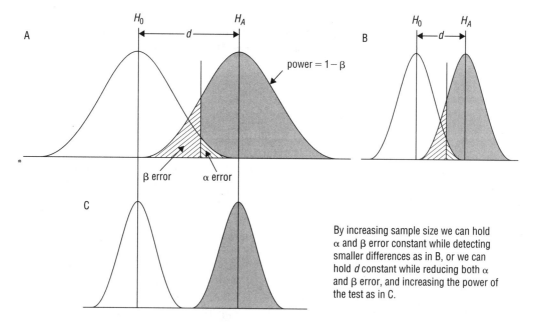

Figure 15.13   Increasing sample size can increase the power of the test and decrease minimum detectable difference.

Most applications for statistical analysis have calculators that make it easy to evaluate the trade-offs between power and sample size for various values of the population standard deviation. Shown here are results from Minitab:

Power and Sample Size

```
1-Sample t Test

Testing mean = null (versus not = null)
Calculating power for mean = null + difference
Alpha = 0.05 Assumed standard deviation = 1
```

| | Sample | |
Difference	Size	Power
1	5	0.401390
1	10	0.803097
1	15	0.949086
1	20	0.988591

Comparing Population Mean to a Standard—One-Sample t-Test

In the one-sample t-test, with a sample of size n from a large population, you would want to draw conclusions about the mean of the population. Could it be equal to a certain value, or is it more likely to be greater than or less than a specified value?

Here are examples of situations where you might be interested in comparing means:

1. After changing a manufacturing process, you have to determine if the mean of the process remained on target or has moved.
2. You switch suppliers and want to know if the new supplier is meeting your requirements.
3. After making a design change, you have to determine the effect of the change on some product function that's critical to quality. Has the functional output remained on target or has it shifted by a statistically significant amount?

Of course, the reason for needing statistical tools to gain insight is the variability in your data. Any physical process in the real world produces outputs with some level of variation. At some level, all processes are random.

Example 1: Understanding if a new process is on target

Suppose you have made a change to a manufacturing process to adjust the weight of a part to hit the specification of 10 grams. You want to show that you have met the specification. You know that not every part will weigh exactly 10 grams but are willing to accept a population mean of 10 grams.

Null hypothesis:	Mean weight = 10 grams
Alternate hypothesis:	Mean weight ≠ 10 grams

Note that your alternate hypothesis states that the mean part weight is not equal to 10 grams. This means that the part weight could be either higher or lower than the target of 10 grams. It is a two-sided test. If you have no reason to believe that the weight of the parts produced with the new process is biased in either the high or low direction, you should use a two-sided test.

For the study you use a sample size of 20 parts. You find that the sample's mean is 10.5 grams and its standard deviation is 1.1 grams.

The question to answer is "What's the likelihood that the sample, with a mean of 10.5 grams, could have come from a population with a mean of 10 grams, given the sample's standard deviation of 1.1 grams?" Note that you should always formulate the null and alternative hypotheses before seeing the data. Retrospective hypotheses aren't fair game.

Using equation (15.5), you can calculate the two-sided confidence interval for the mean. First either look it up or, using a tool like Minitab, calculate the critical value for the t-statistic. Since you are doing a two-sided test and using an α error of 0.05, you need the critical value of t for 19 degrees of freedom with 2.5% of the area in the right-hand distribution tail.

Looking up the critical value for t in a table, or using the computer, you get $t_{0.975,19} = 2.093$.

The critical value of the t-statistic is used to calculate the upper and lower confidence limits for the population mean:

$$10.5 - 2.093\left(\frac{1.1}{\sqrt{20}}\right) < \mu < 10.5 + 2.093\left(\frac{1.1}{\sqrt{20}}\right)$$

$$9.985 < \mu < 11.015$$

Since the hypothesized population mean is (just barely) within the confidence intervals calculated using the sample, you would choose not to reject the null hypothesis that the population mean = 10. Again, keep in mind that your conclusion is subject to change as more data accumulate.

If you have a tool like Minitab, it's not necessary to go through a multiple-step process to test your hypothesis. Using the one-sample t-test tool, you can input the sample either as raw data or as summarized data described by the mean, standard deviation, and sample size. The results from Minitab are displayed here:

One-Sample T

Test of mu = 10 vs. not = 10

N	Mean	StDev	SE Mean	95% CI	T	P
20	10.500	1.100	0.246	(9.985, 11.015)	2.03	0.056

The value for t that Minitab gives in this table is not the critical value but is calculated using the hypothesized mean and sample mean:

$$t = \frac{10.5 - 10}{\frac{1.1}{\sqrt{20}}} = 2.03$$

Since the calculated value for the t-statistic is less than the critical value (2.03 versus 2.093), you fail to reject the null, consistent with the null hypothesis mean being within the confidence interval for the population mean.

Example 2: Understanding if a design change has improved performance

Suppose that you have made a design change to increase the pullout force for a fitting. The goal is to be at least 95% confident that the average pullout force for the new design is greater than the pullout force of 50 pounds for the old design. You have a sample of 15 fittings made using the new design and have tested the pullout force for each. The summarized data are

$\overline{X} = 62.3$ pounds

$s = 15$ pounds

Null hypothesis: Mean pullout force for new design = 50 pounds

Alternate hypothesis: Mean pullout force for new design > 50 pounds

In this case, you hope that the design change will increase pullout force. So using a one-sided test, there is a 95% probability that the mean pullout force for the new fitting is greater than the calculated lower bound. If the lower bound is greater than 50 pounds, you can reject the null and be confident that you have made an improvement. The equation for the lower bound is

$$\mu > \overline{X} - t_{1-\alpha, n-1} \frac{s_n}{\sqrt{n}} \tag{15.20}$$

Looking up the value for t and substituting the values in equation (15.20) gives

$$\mu > 62.3 - 1.761 \frac{15}{\sqrt{15}}$$

$$\mu > 55.48 \text{ lb}$$

Minitab gives the same result:

One-Sample T

Test of mu = 50 vs. > 50

N	Mean	StDev	SE Mean	95% Lower Bound	T	P
15	62.30	15.00	3.87	55.48	3.18	0.003

Your interpretation of the result is that the best estimate of the mean pullout force for the redesigned fitting is 62.3 pounds, and there is a 95% probability that it is greater than 55.48 pounds. The "glass is half-empty" view is that there is a 5% probability that it is less than 55.48 pounds.

Statistical Inferences about Variances

Before we discuss the comparison of means of different populations, we will discuss the inferential statistics of the variance. When comparing population means, you need to understand whether the population variances are essentially the same or significantly different. Using the chi-square and F-distributions, you can make inferences about that.

Sampling Distribution of the Variance—the Chi-Square Distribution

Often you may be interested in knowing how much variability exists in the output of a process. To make inferences about this we turn to another sampling distribution, the chi-square distribution. Technically, the chi-square statistic can be defined as the sum of the squares of the values of the t-distribution:

$$\chi_v^2 = \sum_{i=1}^{n} t^2 = \sum_{i=1}^{n} \left(\frac{X_i - \overline{X}}{\sigma} \right)^2 \tag{15.21}$$

Substituting the definition of the sample variance

$$s^2 = \frac{\sum_{i=1}^{n} (X_i - \overline{X})^2}{n - 1} \tag{15.22}$$

in equation (15.21) gives

$$\chi^2 = \frac{(n - 1)s^2}{\sigma^2} \tag{15.23}$$

The one-sided UCL for the variance is

$$0 \leq \sigma^2 \leq \frac{(n - 1)s^2}{\chi_{1-\alpha,n-1}^2} \tag{15.24}$$

The one-sided LCL for the variance is

$$\sigma^2 \geq \frac{(n - 1)s^2}{\chi_{\alpha,n-1}^2} \tag{15.25}$$

The $(1 - \alpha) \times 100\%$ two-sided confidence limits are

$$\frac{(n-1)s^2}{\chi^2_{\alpha/2,n-1}} < \sigma^2 < \frac{(n-1)s^2}{\chi^2_{1-\alpha/2,n-1}} \tag{15.26}$$

This is the statistic of interest for testing a sample variance against a variance for a hypothetical population. The process is similar to testing a sample mean against a hypothetical population mean. First take the sample, calculate the sample variance, calculate the confidence limits, and finally see where the hypothetical population variance is relative to the confidence limits. It's the same process, but a different sampling distribution and test statistic. The chi-square distribution for a sample size of ten (nine degrees of freedom) is shown in Figure 15.14.

As for the Student's t-statistic, in order to get the critical values for the chi-square statistic you need either a table of calculated values, found in the back of most statistics texts, or statistical analysis software, such as Minitab.

Comparing a Population Variance to a Standard

Example 3: Understanding if a new process has reduced variability

Suppose that you have implemented a new laser cutting process for fabricating a part. You believe that the new process has less variability than the old process for a particular critical dimension. There are historical data for the variance of the old process.

Null hypothesis: $\sigma^2_{new} = \sigma^2_{old}$

Alternate hypothesis: $\sigma^2_{new} < \sigma^2_{old}$

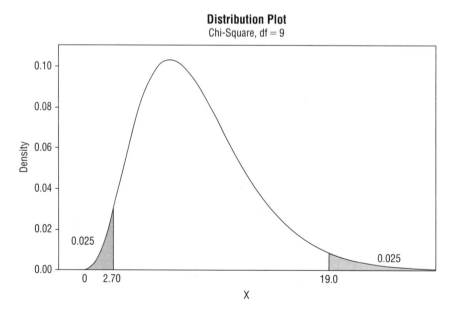

Figure 15.14   The chi-square distribution is skewed for small sample size.

Since you know the variance of the old process, you can use the chi-square distribution to compare the variability of a sample from the new process to that variance. You use a one-sided test, believing that the new process is better than the old, with less variability.

Historical variance for old process: $\sigma^2_{old} = 0.008$

Sample variance for new process: $\sigma^2_{new} = 0.005$

Sample size: $n = 30$

Degrees of freedom: $v = n - 1 = 29$

First calculate, or look up, the critical chi-square value, $\chi^2_{1-\alpha,n-1}$, with $n = 30$ and $\alpha = 0.05$.

$$\chi^2_{0.95,29} = 17.71$$

Calculate the UCL for population variance using the new process variance of $\sigma^2_{new} = 0.005$.

$$\sigma^2_{UCL} = \frac{(n-1)s^2}{\chi^2_{1-\alpha,\,n-1}} = \frac{(30-1)(0.005)}{17.71} = 0.00819$$

Since the UCL for the variance of the new process is greater than the old process variance, you conclude that there's not enough evidence to reject the null hypothesis on the basis of a sample of 30.

The statement "not enough evidence" is important to keep in mind. As more evidence accumulates, you will need to revisit this conclusion. For example, suppose you test 30 additional components and find the variance still at the same level. You would conclude that the new process is indeed less variable than the old. So the difference in variances need not decrease further for you to conclude that you have reduced it. Additional data just allow you to be more confident in that conclusion.

Comparing Two Population Variances—the F-Distribution

When comparing sample variances from two populations, you are trying to determine if the samples could have come from populations with similar variability. In fact, the F-distribution can be formed by repeatedly taking two samples from a large normally distributed population and calculating the F-statistic. By this method you can establish the probability of having two sample variances that differ by a certain amount. The confidence interval for the ratio of the population variances is

$$\frac{s_1^2/s_2^2}{F_{\alpha/2,n_1-1,n_2-1}} < \frac{\sigma_1^2}{\sigma_2^2} < \frac{s_1^2/s_2^2}{F_{1-\alpha/2,n_1-1,n_2-1}} \tag{15.27}$$

The smaller of the two variances is on top when forming the ratio. Also, most tables of the F-statistic are one-tailed. So if you are looking up the values of F in a table, you can use the fact that

$$F_{1-\alpha/2,n_1-1,n_2-1} = \frac{1}{F_{\alpha/2,n_2-1,n_1-1}} \tag{15.28}$$

This gives the following confidence interval:

$$\frac{s_1^2/s_2^2}{F_{\alpha/2,n_1-1,n_2-1}} < \frac{\sigma_1^2}{\sigma_2^2} < \frac{s_1^2}{s_2^2} F_{\alpha/2,n_2-1,n_1-1} \tag{15.29}$$

Of course, the easy way to perform the F-test is to use the computer.

Example 4: Understanding if two suppliers have similar quality processes

Suppose that you are considering two new suppliers to manufacture a critical component. You don't have any historical data for either, so you will have to rely on the samples submitted from their processes. A major concern at this stage of evaluation is the variability of their manufacturing processes. Over the long run, the supplier with less variability should be able to produce higher-quality parts, since it's usually easier to adjust the mean of a process than to reduce its variability. If you don't have a defensible opinion about which supplier's process has less variability, choose a two-sided test with 2.5% in each tail of the F-distribution. The alternate hypothesis asserts the inequality of the variances.

Null hypothesis: $\sigma_{vendor1}^2 = \sigma_{vendor2}^2$

Alternate hypothesis: $\sigma_{vendor1}^2 \neq \sigma_{vendor2}^2$

Each supplier submits 20 samples for evaluation, yielding the following variances:

$\sigma_{vendor1}^2 = 8.93$

$\sigma_{vendor2}^2 = 23.6$

The critical value for F which establishes the UCL for the ratio of sample variances is $F_{19,19,0.975} = 0.396$, as shown in Figure 15.15.

Using equation (15.29), you can calculate the 95% confidence interval for the ratio of the variances:

$$\frac{8.93/23.6}{2.526} < \frac{\sigma_1^2}{\sigma_2^2} < \left(\frac{8.93}{23.6}\right)\frac{1}{0.396}$$

$$0.15 < \frac{\sigma_1^2}{\sigma_2^2} < 0.96$$

The conclusion is that the two samples came from populations that are statistically different. If the confidence interval for the variance ratio had included a ratio of 1, you would conclude that there is not enough evidence to reject the null hypothesis.

The quicker way to reach this conclusion is to compare the F-statistic for the ratio of sample variances to the critical value. Our F-statistic is $F = 8.93/23.6 = 0.378$, which is less than the critical value of $F_{19,19,0.975} = 0.396$. Since the calculated value of F based on the samples is outside the confidence interval, you conclude that there is sufficient reason to reject the null hypothesis of equality of the variances.

Minitab gives the following results for the problem. The table shows the estimated confidence intervals for each sample standard deviation, as well as the results of the F-test. Note the

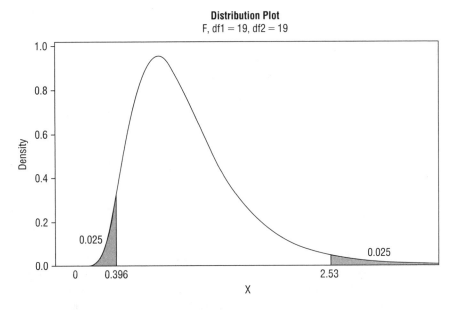

Figure 15.15   Confidence interval for the F-statistic

test statistic of 0.38. This compares to your result of 0.378, which is outside the 95% confidence interval for the F-statistic for this problem.

Test for Equal Variances

95% Bonferroni confidence intervals for standard deviations

```
Sample   N    Lower    StDev    Upper
     1  20  2.18932  2.98831  4.62978
     2  20  3.55910  4.85798  7.52645
```

```
F-Test (Normal Distribution)
Test statistic = 0.38, p-value = 0.040
```

Note that the p-value is slightly less than 0.05, which is consistent with our F-statistic being just outside the confidence interval.

Comparing Two Population Means

Using the Normal Distribution

When comparing the means of samples from two populations, you are usually interested in estimating the difference in the means of those populations. If you know the population standard deviations, you can use the z-statistic. The random variable used for this test is calculated by taking a sample from each population and defining a new variable that is the difference between the sample means. Remember from our discussion of descriptive statistics that the sum or difference of two normally distributed random variables is also a normally distributed random variable.

$$\overline{X} = \overline{X}_1 - \overline{X}_2 \tag{15.30}$$

The z-statistic of interest is

$$z = \frac{(\overline{X}_1 - \overline{X}_2) - (\mu_1 - \mu_2)}{\sigma_{\overline{X}_1 - \overline{X}_2}} \tag{15.31}$$

The confidence interval for $(\mu_1 - \mu_2)$ is

$$(\overline{X}_1 - \overline{X}_2) - z_{1-\alpha/2}\sigma_{\overline{X}_1 - \overline{X}_2} < (\mu_1 - \mu_2) < (\overline{X}_1 - \overline{X}_2) + z_{1-\alpha/2}\sigma_{\overline{X}_1 - \overline{X}_2} \tag{15.32}$$

The calculation of $\sigma_{\overline{X}_1 - \overline{X}_2}$ depends on the sample size from each population as well as whether or not the population standard deviations are equal. Consider two possibilities:

1. **Unequal standard deviations:** If you have reason to doubt that the population standard deviations are equal, first perform an F-test. If the conclusion of the F-test is that it's unlikely that the standard deviations are equal, combine the population standard deviations as shown here. In equation (15.33) we assume that sample sizes n_1 and n_2 could be different.

$$\sigma_{\overline{X}_1 - \overline{X}_2} = \sqrt{\frac{\sigma_1^2}{n_1} + \frac{\sigma_2^2}{n_2}} \tag{15.33}$$

   There are a couple of reasons that assuming unequal standard deviations should be the default unless proven otherwise. The first is that you usually don't know whether the standard deviations of the two things being compared are equal. The second is that by assuming inequality, you do not have to do the additional hypothesis test for equal variances, thus increasing your error rate, alpha.

2. **Equal standard deviations:** If you have reason to believe the population standard deviations are equal based on recent history or on the results of an F-test, you can combine the sample standard deviations to get an improved estimate of the population standard deviation. The sample standard deviations are combined by weighting them in proportion to the number of degrees of freedom for each sample and dividing that by the total degrees of freedom. This gives a pooled estimate of the population standard deviation:

$$\sigma_p = \sqrt{\frac{(n_1 - 1)\sigma_1^2 + (n_2 - 1)\sigma_2^2}{n_1 + n_2 - 2}} \tag{15.34}$$

The end result for the standard deviation of the differences is

$$\sigma_{\overline{X}_1 - \overline{X}_2} = \sigma_p \sqrt{\left(\frac{1}{n_1} + \frac{1}{n_2}\right)} \tag{15.35}$$

Caution

When specifying standard deviations for hypothesis tests, if you don't have historical knowledge, use the standard deviations of the samples as the best estimates. Be cautious when relying on past history. If the quality of a process has declined, probably the process variance has increased. You should check the sample variances against the historical values using the chi-square test. If the historical data for population variance are within the confidence interval estimated using current data, you are on safer ground using historical data.

Sample Size Required to Detect a Given Difference in Population Means

You need more samples to detect the difference in the means of two populations than to detect the same difference between a single population mean and a reference value. This is because having two populations with equal variances effectively doubles the variance of the random variable you define as the difference. Depending on your knowledge of the population variances, you would use either equation (15.33) or equation (15.35) to calculate the standard deviation for $\overline{X}_1 - \overline{X}_2$. For the case where the population variances are different, equation (15.36) can be used to estimate required sample size, assuming equal sample sizes from each population.

$$n = \left(\frac{\sigma_1^2 + \sigma_2^2}{d^2}\right)(z_{1-\alpha/2} + z_{1-\beta})^2 \qquad (15.36)$$

For equal population variances you can use them if you know them, or use a pooled estimate. As we discussed earlier when we developed equation (15.14), we use z instead of t to estimate sample size in order to avoid the need for an iterative calculation of t that would yield a more accurate result but not add much value.

Using the Student's t-Distribution

If using the Student's t-distribution when you have small samples and don't know the standard deviations of the populations, the confidence interval for the difference in population means is given by

$$(\overline{X}_1 - \overline{X}_2) - t_{1-\alpha/2,v} s_{\overline{X}_1-\overline{X}_2} < (\mu_1 - \mu_2) < (\overline{X}_1 - \overline{X}_2) + t_{1-\alpha/2,v} s_{\overline{X}_1-\overline{X}_2} \qquad (15.37)$$

If the population standard deviations are equal, use the following for the standard deviation of the difference in sample means:

$$s_{\overline{X}_1-\overline{X}_2} = \sqrt{\left(\frac{1}{n_1} + \frac{1}{n_2}\right)\left(\frac{(n_1 - 1)s_1^2 + (n_2 - 1)s_2^2}{n_1 + n_2 - 2}\right)} \qquad (15.38)$$

In equation (15.38), s_1 and s_2 are the standard deviations of two samples and n_1 and n_2 are the sample sizes. In equation (15.37) the number of degrees of freedom for t is $v = n_1 + n_2 - 2$. This is fundamentally the same calculation we did when we pooled the sample standard deviations using the normal distribution to estimate the confidence interval.

When population variances are unequal, use

$$s_{\overline{X}_1-\overline{X}_2} = \sqrt{\frac{s_1^2}{n_1} + \frac{s_2^2}{n_2}} \tag{15.39}$$

If you are working with small samples and the variances are unequal, it's necessary to calculate the adjusted degrees of freedom for the t-statistic[5] in order to estimate the confidence interval using the Student's t-distribution. The t-test using the adjusted degrees of freedom is known as Welch's t-test.

The adjusted degrees of freedom for the t-statistic is given by

$$v = \frac{\left(\dfrac{s_1^2}{n_1} + \dfrac{s_2^2}{n_2}\right)^2}{\dfrac{1}{n_1-1}\left(\dfrac{s_1^2}{n_1}\right)^2 + \dfrac{1}{n_2-1}\left(\dfrac{s_2^2}{n_2}\right)} \tag{15.40}$$

Example 5: Understanding if there is a statistically significant difference between two manufacturing processes

Suppose that two different manufacturing processes are used to produce rolls of a polymer. You want to know if both processes will produce materials within the thickness specification. Thickness is measured for a representative sample taken from each of 15 runs for each process. Based on experience with both processes, it is known that the process variances are equal. Summarized data for each process are as follows:

Process 1: $\overline{X}_1 = 0.035$ $s_1^2 = 0.004$ $n_1 = 15$

Process 2: $\overline{X}_2 = 0.045$ $s_2^2 = 0.009$ $n_2 = 15$

Null hypothesis: $H_0: \mu_1 - \mu_2 = 0$

Alternate hypothesis: $H_a: \mu_1 - \mu_2 \neq 0$

The first thing to do is validate your belief that the process variances are equal. If you use the F-test, you get the following result from Minitab:

```
Test for Equal Variances

95% Bonferroni confidence intervals for standard deviations

Sample   N     Lower      StDev      Upper
     1  15  0.0443882  0.0632456  0.107198
     2  15  0.0665822  0.0948683  0.160797


F-Test (Normal Distribution)
Test statistic = 0.44, p-value = 0.141
```

---

5. B. L. Welch, "The Generalization of Student's Problem When Several Different Population Variances Are Involved," *Biometrika* 34 (1947): 28–35.

You can conclude that the variances have overlapping confidence intervals and that the samples could have come from equivalent populations. This allows you to assume equal variances and use a pooled estimate for the population variance to calculate the standard deviation of the differences. Use equation (15.38) for this:

$$s_{\bar{X}_1 - \bar{X}_2} = \sqrt{\left(\frac{1}{15} + \frac{1}{15}\right)\left(\frac{(15-1)0.004 + (15-1)\,0.009}{15 + 15 - 2}\right)}$$

$$s_{\bar{X}_1 - \bar{X}_2} = 0.02944$$

Using equation (15.37), and your pooled estimate of the standard deviation, the confidence interval is

$$(0.035 - 0.045) - 2.048(0.02944) < (\mu_1 - \mu_2) < (0.035 - 0.045) + 2.048(0.02944)$$
$$-0.0703 < (\mu_1 - \mu_2) < 0.0503$$

You can use Minitab to get the same result:

Two-Sample T-Test and CI

```
Sample   N    Mean    StDev  SE Mean
1        15   0.0350  0.0632   0.016
2        15   0.0450  0.0949   0.025


Difference = mu (1) - mu (2)
Estimate for difference: -0.0100
95% CI for difference: (-0.0703, 0.0503)
T-Test of difference = 0 (vs. not =): T-Value = -0.34 P-Value = 0.737 DF = 28
Both use Pooled StDev = 0.0806
```

The conclusion of this analysis is that there is no statistical difference between the processes. The difference observed in the sample means is not statistically significant and is in the middle of the confidence interval for the difference.

This analysis concluded that there was not a statistically significant difference between the processes. However, you cannot conclude that the processes are equivalent, since you did not set up your hypothesis test to assess equivalence. We discuss equivalence testing later in the chapter.

Paired Comparisons

Sometimes you need to make comparisons of a new process to an old one, or one material to another. However, using the two-sample t-test would not be a good approach. Let's consider a situation where paired comparisons are the way to go.

Example 6: Paired comparisons of before and after production process improvements

A supplier is planning to modify its process for assembling a product. After a lean initiative involving multiple kaizen events in the production organization, potential improvements to the assembly process have been identified. A pilot study has been run to test the improvements to determine if both the process cycle time and its throughput rate have been improved. Table 15.3 shows the data from the study.

Table 15.3   Cycle time improvements in production process

Step	1	2	3	4	5	6	7	8	9	10	$\bar{X}$
before	89	52	78	111	119	61	47	110	48	50	76.5
after	84	46	75	95	110	49	33	102	38	38	67.0
d	5	6	3	16	9	12	14	8	10	12	9.50

Building the product has ten steps, and a different person is responsible for each step. This is typical of a situation for which you would analyze the results using a paired comparison approach. In Table 15.3 the "before" and "after" times are the times in seconds for each step in the assembly process. The "before" times precede the improvements from the lean initiative. The row labeled "d" is the difference between before and after for each step. We use a paired comparison rather than a two-sample t-test on the before and after data. The reason for this is that variability in the before and after rows of data includes the contribution of the effects of the ten different people and the fact that each step may differ in complexity and required completion time. The paired test eliminates the variability due to the people and task differences and focuses on the effect due to the change in method. So you calculate the confidence interval for the difference (d).

Our equation for the confidence interval for the mean difference is the same as equation (15.37), only we use the mean and standard error of the differences. In this case, for the analysis of the paired data, the degrees of freedom for the t-statistic is $\nu = n - 1$, rather than $2n - 2$ as it would be if we compared the before and after data.

$$\bar{X}_d - t_{1-\alpha/2, n-1}\left(\frac{\sigma_d}{\sqrt{n}}\right) < \mu_d < \bar{X}_d + t_{1-\alpha/2, n-1}\left(\frac{\sigma_d}{\sqrt{n}}\right) \qquad (15.41)$$

$$9.50 - 2.26\left(\frac{4.12}{\sqrt{10}}\right) < \mu_d < 9.50 + 2.26\left(\frac{4.12}{\sqrt{10}}\right)$$

$$6.56 < \mu_d < 12.44$$

This result shows that you have reduced the mean cycle time from 76.5 to 67.0 seconds (>10%) and that there is a 95% probability that the reduction is between 6.56 and 12.44 seconds. Had you done a two-sample t-test using the before and after data, you would not have seen a significant improvement because the variation among people and the tasks would have obscured the improvement. The following table from Minitab gives the summary statistics for your data and the calculated confidence interval for the improvement in the mean of the differences. The improvement is statistically significant, as evidenced by a p-value that's zero out to three decimals. Of course, whether the improvement is practically important from a business standpoint is an entirely different question. Again, you must keep in mind the difference between statistically significant and practically important.

Paired T-Test and CI: before, after

```
Paired T for before - after
            N   Mean  StDev  SE Mean
before     10  76.50  28.87     9.13
after      10  67.00  29.47     9.32
Difference 10   9.50   4.12     1.30
```

```
95% CI for mean difference: (6.56, 12.44)
T-Test of mean difference = 0 (vs not = 0): T-Value = 7.30 P-Value = 0.000
```

Comparing Proportions

Processes with a binary outcome, such as pass/fail, yes/no, or good/bad, are described by the binomial distribution. These binary outcomes will occur with a certain frequency or proportion in the population that you estimate using the sample. Generally, when you are working to improve product quality and robustness, you should avoid metrics such as these and instead seek continuous variables as measures of performance.

However, in some circumstances you have to deal with proportions. For those situations, the following describes a process for calculating confidence intervals and testing hypotheses. As we did with the mean and variance, we are interested in comparing a sample proportion to a reference and also comparing two sample proportions to see if they might have come from equivalent populations. An excellent reference that covers many aspects of dealing with proportions is Riffenburgh.[6]

There are two ways to test the hypothesis of the equality of proportions: a normal approximation and Fisher's exact test.

Normal Approximation to Binomial Distribution

A widely used rule of thumb for choosing a sample size large enough to permit the normal approximation is the following:

$$np > 5 \text{ and } n(1-p) > 5 \tag{15.42}$$

where

n = sample size

p = probability of an event occurring

$(1 - p)$ = probability of it not occurring

If you satisfy this rule of thumb, you can calculate confidence intervals for proportions using the normal distribution. You can also compare a proportion to a reference as well as compare two proportions to assess the likelihood that they came from the same population.

6. Robert H. Riffenburgh, *Statistics in Medicine* (Burlington, MA: Elsevier, 2006).

Using the normal approximation, the following equation gives the confidence interval for the event frequency in the population based on a sample size n:

$$p - z_{1-\alpha/2}\sqrt{\frac{p(1-p)}{n}} < \pi < p + z_{1-\alpha/2}\sqrt{\frac{p(1-p)}{n}} \qquad (15.43)$$

In equation (15.43) π is the unknown proportion in the population you are trying to estimate and p is the sample proportion. You can use variants of equation (15.43) to calculate upper and lower confidence intervals on population proportions based on your sample proportions. Do one-tailed and two-tailed tests.

Note the limitation of this approach. In equation (15.42), a sample size of 100 is required if the probability of the event occurring is 5% or 95%. So, because of the large sample sizes required, use of the normal approximation becomes impractical for very small probabilities or those close to 100%. A better way is Fisher's exact test, discussed in the next section.

Fisher's Exact Test

Fisher's exact test is used to calculate the probability of getting a given combination of categorical results and so is applicable to problems comparing event frequency (proportions) for two samples. The p-value for a given result is calculated exactly by adding the probabilities of all the less likely combinations having the same marginal totals in a two-by-two contingency table. It's a calculation-intensive process using the hypergeometric distribution and requiring evaluation of factorials that can be large depending on the sample size. For that reason, it's best done by statistical software. Fortunately, applications such as Minitab have that capability. That being the case, there is no need to use the normal approximation to the binomial distribution if the option exists for using a statistics application having one of the exact methods.

Equivalence Testing

Equivalence Test versus Difference Test

Suppose that you have developed a new process, with reduced costs, for manufacturing an electric motor. The motor is currently in production and its characteristics are well known. In this situation you don't have a process that's supposed to deliver an improved result. Instead you have to determine if the old and new processes provide results that are the same within some range of indifference. If you can show that they do, you can say that the processes are equivalent. In that case, you can implement the new cost-saving process, confident that you are producing acceptable products. Other examples of situations where you might be interested in demonstrating equivalence include the following:

- After a value engineering project on a medical device you feel that you have made significant cost reductions without reducing its quality and reliability. You have to show that the performance of the value-engineered device is equivalent to that of the existing product.
- Your supplier tells you that if they substitute a variant of an existing reagent in a diagnostic test kit, production costs will be reduced. This may enable higher margins and perhaps increased market share. You must show that the new test is equivalent to the old test.

- You are building a computer model to simulate a complex system. At what point in the model's refinement can you say that it's good enough? You could compare model results with actual performance and use equivalence testing to make your case.

Although equivalence testing looks at differences, it does so in a way that's different from the two-sample t-test. In a difference test there is a hypothesis pair that is generally of the form

$$H_0: \mu_1 - \mu_2 = \delta$$
$$H_a: \mu_1 - \mu_2 \neq \delta$$

In a difference test using the two-sample t-test, by setting the α risk to a small level you place the burden of proof on showing that any difference between the means is significant at that α level. If $p < \alpha$, you reject the null hypothesis and accept the alternate. However, you fail to reject the null hypothesis if you fall short with $p > \alpha$. So the evidence has to be pretty strong to reject the null and accept the alternate hypothesis. In the typical difference test, the objective is to understand if the difference between two treatments is significantly different. In the equivalence test, you must show that they are not different by more than a certain amount. Next we describe two approaches for testing equivalence: the confidence interval method and the method of "two one-sided tests" (TOST).

Confidence Interval Method for Showing Equivalence

The confidence interval method for showing equivalence is intuitive and straightforward. The steps in the process are as follows:

1. Decide how large a difference, δ, you can accept between means before you conclude that you do not have equivalence. This is the most difficult and important decision you will make in the process. It's not a statistical decision, but a decision requiring knowledge of how performance variations affect both the end user of your product and your business. As mentioned in earlier examples, it's a business decision. This is an opportunity to apply the concept of quality loss discussed in Chapter 2.
2. Take samples from both populations.
3. Calculate the confidence interval for the difference in population means. Note that this is a two-sided confidence interval with α in each tail. So if you want 95% confidence, use a 90% confidence interval with 5% in each tail of the distribution. You do it this way because you don't know in advance whether the difference will be biased toward $+\delta$ or $-\delta$.

   The confidence interval for the difference in means is

$$(\overline{X}_1 - \overline{X}_2) - t_{1-\alpha,n-1}s_{\overline{X}_1-\overline{X}_2} < (\mu_1 - \mu_2) < (\overline{X}_1 - \overline{X}_2) + t_{1-\alpha,n-1}s_{\overline{X}_1-\overline{X}_2} \qquad (15.44)$$

4. If the confidence interval for the difference in population means does not include either $\pm\delta$, reject the null and assume equivalence. This means that

$$(\overline{X}_1 - \overline{X}_2) - t_{1-\alpha,n-1}s_{\overline{X}_1-\overline{X}_2} > -\delta \text{ AND } (\overline{X}_1 - \overline{X}_2) + t_{1-\alpha,n-1}s_{\overline{X}_1-\overline{X}_2} < \delta \qquad (15.45)$$

Figure 15.16 shows the confidence interval for the difference in means with the possibility of the difference δ being either positive or negative.

Two One-Sided Tests (TOST)

Another procedure for equivalence testing is called "two one-sided tests" (TOST). In this type of test there are two null hypotheses asserting that there is a difference between treatments. The two null hypotheses would have the following form:

$$H_0: |\mu_1 - \mu_2| > \delta$$

This can be expressed as two one-sided hypotheses:

$$H_{01}: \mu_1 - \mu_2 > \delta$$
$$H_{02}: \mu_1 - \mu_2 < -\delta$$

If you fail to reject either or both, conclude that the treatments are not equivalent. If you reject both null hypotheses, accept the alternative hypothesis:

$$H_a: |\mu_1 - \mu_2| < \delta$$

This means that the two treatments satisfy the definition of equivalence. As you will see, you can derive the TOST from the confidence interval analysis. If either end of the confidence interval for $\mu_1 - \mu_2$ is outside the interval $\pm \delta$, you fail to reject the null and cannot assume equivalence.

$$IF\ (\overline{X}_1 - \overline{X}_2) - t_{1-\alpha,n-1}\sigma_{\overline{X}_1-\overline{X}_2} < -\delta \quad OR \quad (\overline{X}_1 - \overline{X}_2) + t_{1-\alpha,n-1}\sigma_{\overline{X}_1-\overline{X}_2} > \delta \quad (15.46)$$

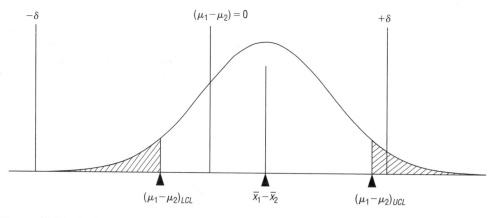

Figure 15.16   In this case we can assume equivalence since $\pm \delta$ are both outside the confidence interval for the difference in population means.

So if both of the following conditions are true, you must reject the null and assume equivalence.

Left-side test:

$$\frac{\delta + (\overline{X}_1 - \overline{X}_2)}{\sigma_{\overline{X}_1 - \overline{X}_2}} > t_{1-\alpha, n-1} \tag{15.47}$$

Right-side test:

$$\frac{\delta - (\overline{X}_1 - \overline{X}_2)}{\sigma_{\overline{X}_1 - \overline{X}_2}} > t_{1-\alpha, n-1} \tag{15.48}$$

Together these conditions comprise the two one-sided tests. If either condition is untrue, you fail to reject the null and must assume that they are not equivalent.

Example 7: Equivalence testing of cooling of two motor designs

The manufacturing process for an electric motor has been modified to reduce its cost. Samples of 30 original and 30 new process prototypes have been tested under stressful conditions to evaluate several performance parameters, including the temperature of the motor windings. Based on the expected warranty costs, a decision has been made that if, on average, the new motor windings run no more than 10°C hotter, the design changes will be approved. The summarized test data are as follows:

Old manufacturing process: Mean temp = 140.5°C   Standard deviation = 5.5°C
New manufacturing process: Mean temp = 145.2°C   Standard deviation = 11.2°C
Maximum allowable temperature increase: $\Delta T_{max} = 10°C$

For this decision, you are concerned only if the new process motor runs hotter than the old. It would be great if it ran cooler. To make the bookkeeping easier to follow, you define your difference in means as the mean temperature of the new design minus the mean temperature of the old design. Your right-hand test becomes

$$\frac{\Delta T_{max} - (\overline{T}_{new} - \overline{T}_{old})}{s_{\overline{T}_{new} - \overline{T}_{old}}} > t_{1-\alpha, n-1} \tag{15.49}$$

If you perform an F-test on the variances, you see that they are unlikely to be equal. So you assume unequal variances to calculate the standard deviation in the difference of the means. Using the standard deviations of your samples and equation (15.33) gives the following for the standard deviation of the difference in the means:

$$s_{\overline{T}_{new} - \overline{T}_{old}} = \sqrt{\frac{s_{new}^2}{n_1} + \frac{s_{old}^2}{n_2}}$$

$$= \sqrt{\frac{11.2^2}{30} + \frac{5.5^2}{30}} = 2.27$$

Now calculate the left-hand side of equation (15.49):

$$t = \frac{10 - (145.2 - 140.5)}{2.27} = 2.33$$

The critical t value for 0.05 in each tail is $t_{0.95,29} = 1.70$. Since your calculated value is 2.33, you can be confident that the two designs are equivalent for cooling. Therefore, the UCL for the difference in the population means should not exceed the maximum allowable temperature increase of 10°C.

Some Concerns with Significance Tests

We have devoted considerable time to hypothesis testing and the interpretation of the p-value. While the p-value is pervasive in statistics when you are making comparisons and evaluating differences, there has been considerable discussion about how the quest for statistical significance has occasionally led to unfortunate results. An interesting paper by Sterne and Smith[7] discusses some of the misconceptions about, and abuses of, significance tests.

For quite some time in medical science, there have been vigorous debates about significance tests and some of their negative effects on how science is done and what gets published.[8] This has resulted in proposals to place less emphasis on significance when judging the efficacy of new and improved products and treatment protocols. Editors of medical journals have been asking authors for confidence intervals in addition to p-values. The use of terms such as *significant* or *not significant* has been discouraged. While the debate outside of medical science has not been as intense, the ideas discussed in the medical science community are worth considering by anyone doing product development.

R. A. Fisher, the father of significance testing, viewed the p-value as evidence to be considered when making a decision to accept or reject the change being evaluated. He was not dogmatic about a particular cutoff threshold, but he did acknowledge that an error of $\alpha = 0.05$ was reasonable. Since that time, $\alpha = 0.05$ as the threshold for significance has taken root and resulted in some pathology, especially in research publications. The main concern is that real improvements can be overlooked because of the focus on statistical significance. The issue is, in some ways, similar to the "goalpost" view of meeting specifications, where just inside a specification limit is "acceptable" and just outside "not acceptable." The goalpost viewpoint of specifications and the idea that $p = 0.051$ is not acceptable while $p = 0.049$ is acceptable are both silly. Performance of most systems is a continuous variable. This is not meant to minimize the real dilemma of decision making in the presence of uncertainty, required when you have to determine whether or not the new change you are studying is justified. Figure 15.17 shows the dilemma caused by the abrupt transition from not significant to significant.

Some bad practices are caused by an overemphasis on significance testing. Examples include these:

- After the results of an experiment are in, switching from a two-sided test to a one-sided test, retrospectively, to change the conclusion from "not significant" to "significant" and, as a special bonus, to increase the power of the test.

7. Jonathan A. C. Sterne and George Davey Smith, "Sifting the Evidence—What's Wrong with Significance Tests?" *British Medical Journal* 322 (January 27, 2001).

8. Douglas G. Altman, *Practical Statistics for Medical Research* (London: Chapman & Hall, 1991).

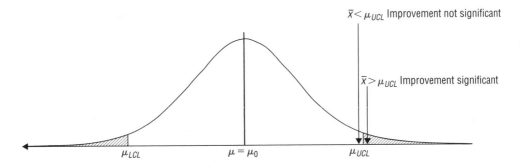

Figure 15.17   The binary outcome of significance testing may drive bad decisions.

- Investing too much importance in the p-value rather than treating it as another piece of evidence.
- Repeating experiments until you get a significant p-value for the result. With enough repetitions, the laws of probability will work their magic to provide significant results, even for a bad idea. It's just a matter of time and persistence.
- Meaningful results may not be published unless statistically significant.

While we hope that scenarios such as these occur infrequently, incentives can drive ill-advised behaviors. If you don't want the behavior, you should change the incentives.

Much of the overemphasis on p-values has occurred in the publication world. In practice in industry, p-values are not weighted as heavily as in publications. Generally, confidence intervals are presented along with p-values.

Interval Estimates or p-Values?

The problem with basing your decisions solely on p-values is that you will miss improvements that are real and possibly beneficial both to your business and to the end users of your product. This is bound to happen at times if you adhere to the policy that $p < \alpha$ is good and $p > \alpha$ is not. A small p-value, even if not significant, is evidence that you probably have moved the response toward a different and possibly better result. The confidence interval for the mean gives you a better perspective about what you have accomplished than does the binary result of significant versus not significant.

Key Points

1. The t-test is used to detect differences in means. How small a change is detectable depends on the level of noise in the data and the size of the samples.
2. With a large enough sample size you can detect arbitrarily small differences in sample means. However, significant differences are not necessarily important differences.
3. To perform the one-sample t-test, you take samples from a population and determine the probability of the population mean being in a certain range.
4. To perform the two-sample t-test, you take samples from two populations and determine the likelihood that the populations have the same mean.
5. The sampling distribution of the variance is the chi-square distribution.
6. A requirement of the t-test is that the data come from a normal distribution.

7. When comparing means of populations, you must determine if you can assume equality of the population variances.

8. Compare the variance of two samples using the F-distribution.

9. When you make comparisons and test hypotheses, there is a finite probability that you will make an error.

Discussion Questions

1. What is the main difference between comparing a sample mean to a reference value and comparing samples from two populations?

2. Describe α error and β error.

3. Why do you think α error is described as "producer's risk," while β error is described as "consumer's risk"?

4. What factors should be considered in choosing acceptable levels for α and β error?

5. What is the "power of the test"?

6. What is the sampling distribution of the mean?

7. What are some important conditions for being able to use the t-test?

8. When do we use the chi-square distribution?

9. Are there any conditions or limitations for using the chi-square distribution?

10. What sampling distribution is used for comparing the variance of samples drawn from two populations?

11. When you draw samples from two populations, what are the parameters of interest?

12. What factors should you consider when choosing the sample size to estimate the difference between the means of two populations?

13. What is a hypothesis?

14. What is the meaning of a p-value?

15. Which test has greater power, a one-sided or a two-sided t-test?

16. Which of the following distributions are symmetric and which are nonsymmetric: Student's t, chi-square, F-distribution?

17. What probability distribution describes the occurrence of binary events?

18. Under what conditions is the normal distribution a good approximation for the binomial distribution?

19. Describe equivalence testing and how it differs from difference testing.

Building Models Using ANOVA and Regression

Models[1] are essential tools in developing new products. With a good model we can predict performance as the important factors vary. We can also use a model to select operating points that are less sensitive to noise, including outside effects from the environment as well as variations internal to our system. There are two types of models: first-principles and empirical. In this chapter we discuss two of the tools that are important in building empirical models, ANOVA and regression.

Uncovering Relationships and Proving Cause and Effect

As developers of products and processes, we have a need to understand "cause-and-effect" relationships in order to understand the important factors that affect performance. This understanding is essential to the design process and, as you will see later, important for critical parameter management. The best way to show cause and effect is to run an experiment. If you can show that changing the level of a factor in an experiment causes a variation in the response, you have demonstrated causation.

Depending on what you are studying, there may be situations where the only available data are retrospective data, or there may be ethical and practical reasons preventing you from running an experiment. The problem is that retrospective data can be used to show association but not prove cause and effect, unless it can be shown that confounding factors are absent. Essentially a confounding factor is an important factor that affects the response that has not been explicitly included in the study. Retrospectively showing the absence of confounding factors, which is proving that something doesn't exist, can be a difficult challenge.

The Difficulties of Using Historical Data

One of the better-known examples of the difficulties that can arise in proving cause and effect was the trouble epidemiologists had early on in proving that cigarette smoking was a primary cause of lung cancer. For many years there was plenty of evidence of a statistical association

---

1. In this chapter we use the term *model* to signify a mathematical representation of a system or process as opposed to a physical prototype that represents the software and hardware of that intended design.

between tobacco usage and a number of pathologies, including lung cancer and cardiovascular disease. However, the weakness of early studies lay in the fact that they were based on retrospective data. In this type of case-control study, investigators studied a large group of smokers and compared the incidence of specific diseases with a control group of nonsmokers. The early retrospective studies were criticized and labeled inconclusive by some investigators and, not surprisingly, by the tobacco companies.

By definition, a retrospective study uses past history. As a result, unless the suspected causal factor is varied over the course of the study, it's not possible to conclusively show cause and effect.

One of the early and influential, forward-looking, or prospective studies[2] was by Doll and Hill, who evaluated a group of 40,000 medical doctors in the UK. The study results showed a highly significant difference in the incidence of lung cancer between smokers and nonsmokers drawn from the group. A key result was that the amount of tobacco consumed was shown to have a dramatic effect on the likelihood of disease, especially lung cancer. Following the publication of the early studies in the 1950s, the floodgates opened and many investigators launched new studies. Over time the evidence accumulated, and concerns about tobacco use grew into a major public health issue.

The tobacco industry fought back, making the statement that "association does not prove causation" their mantra. They argued that confounding factors or "lurking variables" could be the explanation and that perhaps tobacco usage wasn't the cause of disease, but rather that other behaviors or factors correlated with tobacco usage were the real problem. They even went as far as to form the Tobacco Industry Research Committee and co-opt some prominent scientists under the guise of fostering "good science." The real agenda, of course, was to try to offset the growing evidence of the dangers of smoking by creating doubt about the results of studies, as well as set standards of proof that epidemiological studies could never satisfy. This was combined with vigorous lobbying, public relations campaigns, and, where necessary, ad hominem attacks to discredit scientists who published results considered threatening to the business. This was especially the case for the issue of "secondhand smoke,"[3] which the tobacco industry correctly viewed as a serious threat.

Fortunately, much of the public in Europe and the United States did not buy the argument that the evidence was weak, and the percentage of the population that smoked started to shrink as governments responded to the public health threat with campaigns to discourage smoking. In 1964, based on the accumulated evidence, the advisory committee to the U.S. Surgeon General published the report *Smoking and Health*.[4] In 1965 the U.S. Congress forced the warning labels on tobacco products that have been modified over the years as more of the effects of tobacco were understood.

---

2. Richard A. Doll, and Bradford Hill, "The Mortality of Doctors in Relationship to Their Smoking Habits: A Preliminary Report," *British Medical Journal* ii (1954): 1451–5.

3. Elisa K. Ong and Stanton A. Glantz, "Constructing 'Sound Science' and 'Good Epidemiology': Tobacco, Lawyers and Public Relations Firms," *American Journal of Public Health* 91, no. 11 (November 2001); Pascal Diethelm and Martin McKee, "Lifting the Smokescreen: Tobacco Industry Strategy to Defeat Smoke Free Policies and Legislation" (European Respiratory Society and Institut National du Cancer, 2006).

4. Advisory Committee to the Surgeon General of the Public Health Service, *Smoking and Health* (1964).

After a long decline in the population of smokers, the result was conclusive, showing that as the number of smokers declined, the number of lung cancer cases declined also.[5] So, over many years, an "experiment" was run by an informed and concerned public; fewer people started to use tobacco, and many existing smokers quit. An excellent historical summary, with many references, is given by Doll.[6]

The landscape is dramatically different today compared to the 1950s. Most places of work and public assembly are smoke-free, and there are no more tobacco product ads or jingles on radio or TV. The tobacco industry has paid billions in judgments and settlements.

The historical record of the struggle to show that tobacco was a proximate cause of disease is a fascinating story. It shows how events can unfold and resistance can develop into a high-stakes game when scientific results are viewed as a threat by powerful and moneyed interests. It also has relevance to the statistical use of data in product and process development.

You have to be cautious when using retrospective data. In product development, with the exception of medical and pharmaceutical products, you aren't encumbered by some of the ethical, legal, and practical issues that constrained the people studying the link between tobacco and disease. As long as your experiment does not put people's health or safety at risk, if you want to run a study you can do it and not have to wait years for the results. In this chapter we discuss how you can show causation and ultimately develop useful models.

Identifying Cause-and-Effect Relationships

Understanding how things work is essential to achieving successful outcomes. You need to identify the important factors that affect performance. There are several situations in which you might find yourselves:

- An existing process or system has been functioning, but you have a limited understanding about what makes it work. If you continue with your understanding at the "black box" level, without knowledge of those factors that play significant roles in determining performance, you will be vulnerable to changes that could cause performance variations to fall outside specification limits. Possibilities could include unwitting changes to either the design or the manufacturing process. You may have access to retrospective data that describe system performance and possibly the settings of numerous candidate control factors that may affect system response.
- You are in the early stages of development and need to understand the relative importance of the factors that control system performance. Your objective is to understand the causes of performance variations that can reduce customers' satisfaction. You may have some theories about how things work, possibly supported by some first-principles analyses.
- A system or product has suffered deviations from performance requirements or outright failure or breakdown. You know that there are problems but don't understand which sources of variation are causing those problems.

---

5. Richard Peto, Sarah Darby, Harz Deo, Paul Silcocks, Elise Whitley, and Richard Doll, "Smoking, Smoking Cessation and Lung Cancer in the UK since 1950: Combination of National Statistics with Two Case-Control Studies," *British Medical Journal* 321, no. 5 (August 2000), 323–29.

6. Richard Doll, "Uncovering the Effects of Smoking: Historical Perspective," *Statistical Methods in Medical Research* 7 (1998): 87–117.

Building a model requires that you understand those cause-and-effect relationships that determine performance of the system. Sometimes this is straightforward. At times it can be a process of discovery. In the following sections we discuss a few of the tools people use to help with the process. The list is not all-inclusive but covers some of the more effective ways to approach the problem.

Correlation

Sometimes you have after-the-fact data and are interested in understanding whether there is a relationship among the factors contributing to the data. Looking at the correlation of factors can be a first step in uncovering relationships that may lead to understanding cause and effect. It's very important to understand that correlation does not necessarily establish cause and effect. An example is the correlation between children's reading ability and body weight. The causal factor determining reading ability is mental development, which correlates with age, which correlates with physical development. If two factors are correlated, it just means that when one changes, the other does also. However, the absence of correlation is evidence of no cause-and-effect relationship between two variables. Correlation is measured by a statistical metric, the Pearson correlation coefficient. The threshold for a correlation being statistically significant depends on the size of the data sets and the acceptable α error. Graphical analysis can be a powerful tool for looking at the correlation between one or more factors. As we advised in Chapter 14, "First graph the data."

Hypothesis

When you try to establish a cause-and-effect relationship, begin with a hypothesis about how you think things work. Your hypothesis might be based on past experience or on an understanding of first principles. When you form a hypothesis, try to test it by using your understanding of first principles.

Cause-and-Effect Diagrams

The cause-and-effect diagram, introduced in Chapter 13, is a tool to identify candidate factors that you think might be important to determine a response of interest. When applied to product or system performance, it can be used to map controllable parameters and noise factors that may cause the effect, such as an unacceptable variation in performance. Cause-and-effect diagrams are applicable to any system or process, displaying how the team thinks the system might work. Confidence in your beliefs will vary widely, depending on the system, your first-principles understanding, past experience with similar systems, and the breadth of technical expertise of the development team. The objective is to identify candidate factors and noises that might have a strong cause-and-effect relationship with an important functional response.

Brainstorming

A small group of people with technical competence and experience can be very helpful when creating a list of candidate factors for model building. Brainstorming in conjunction with cause-and-effect diagrams can be a powerful approach to capturing the core knowledge and first-principles understanding.

First-Principles Analysis

A first-principles analysis can greatly increase your confidence in identifying important factors that might affect a response. First principles are useful in situations where the device or system can be modeled using the laws of engineering physics. This would include most devices of an electromechanical, thermal, chemical, or fluidic nature. Modeling and simulation are being applied even in life sciences, which have systems with much higher levels of complexity than those in the physical sciences. As first-principles understanding grows in any field, and computers become more powerful, the use of models becomes more widespread. Even though a first-principles look at a system might not yield a model, first principles can help you get started in the right direction in planning your development of an empirical model. A good starting point for building any model, whether a first-principles or empirical model, is the P-diagram, described in Chapter 2.

Empirical Models

There are a couple of important tasks that you must do when building an empirical model, and there are different tools for each. The first is to identify the important functional responses or effects produced by the system. You have to understand customers' needs and link the responses of your system to the satisfaction of those requirements. By definition, a response of the system that affects an important customer need is an important response. A factor that controls a significant part of an important response is an important factor. Deciding the importance of a factor is one task for which statistics is a powerful tool. There may be a number of factors that can influence a response, but probably they are not all equally important. You need to keep in mind that statistical significance is all about your ability to measure the change in response as you change the factors. It depends on three elements:

1. The size of the sample
2. The strength of the effect
3. The amount of ambient noise

Recall that with a large enough sample size, when comparing samples you can detect arbitrarily small differences in a given response. So statistical significance is really about detecting differences in the presence of noise. It has nothing to do with practical importance.

Practical importance is focused on how large a difference needs to be in order to be important from an engineering or scientific standpoint. If you are trying to improve a product, process, or system, it's about changes that are important to customers, end users, or stakeholders, such as your business. When you design an experiment, you have to have a sample size that is large enough to detect practically important differences with the desired statistical power. We discuss practical importance in more detail in Chapter 17 when we cover design of experiments.

Once you identify possible relationships, the next step is to build a model. After gathering data, you test your model in a couple of ways. The first test, analysis of variance (ANOVA), is one of the fundamental statistical tools for model building. The second test, deciding on the practical importance of a factor, is in the domain of the subject matter experts. It must be informed by demonstrations of how changes in system performance affect the users of your product or process and how that ultimately affects the business.

In your quest to understand cause and effect you have to ensure that your experiments have, at the minimum, the factors of practical importance. Selecting factors to include in any experiment requires you to use a little bit of art, a fair amount of science, and as much relevant experience as possible. If you include too few, you will fall short of your goal and have to go "back to the drawing board." However, if you include some factors that are not important, you can rely on ANOVA to help you weed them out.

The power of ANOVA having been highlighted, it is important to understand some of its limitations. Just as the t-test becomes less effective when evaluating multiple paired comparisons, ANOVA has practical limitations when multiple factors and responses are considered. The standard tools of one-way and two-way ANOVA are limited by the size of the model. Larger models with multiple factors and responses can be analyzed using tools such as the General Linear Model (GLM)[7] and MANOVA, which are outside the scope of this book. In Chapter 17 we discuss the use of DOE for building empirical models. When there are multiple factors and responses, DOE will be your tool of choice for building empirical models. We discuss ANOVA before covering the basics of regression.

ANOVA

Understanding the Importance of Factors

In your first attempts to build empirical models you will generally want to err on the side of including a few too many factors, rather than too few. You do this to be certain that you don't overlook any important factors. The challenge then becomes one of weeding out the unimportant factors after you have some experimental data. ANOVA is a good tool to use.

ANOVA is an important statistical tool that can be used as a stand-alone tool. Along with regression analysis, it is built into most computer applications used for design of experiments. For discussion purposes we limit our focus to the simplest ANOVA, the one-way fixed effects model, where it is assumed that the underlying variables that determine the response are nonrandom. The assumption for this model is that the sample data come from normal populations that differ in their means. ANOVA breaks the variations down into the contribution of the error and the contribution of the variations due to the treatments.

Fundamental to ANOVA is the use of the F-statistic to compare variances. Recall from Chapter 15 the definition of the F-statistic:

$$F = \frac{s_1^2}{s_2^2} \tag{16.1}$$

where s_1^2 and s_2^2 are the variances of the data from two processes. In the case of the simple ANOVA model, the F-statistic is composed of the ratio of mean squares as unbiased estimators of variance and effects due to factor level differences. You then compare the portion of the total variance explained by the changes in factor level to that of the noise. If the factor level contribution is large enough compared to the noise, you conclude that it is significant.

---

7. The General Linear Model is a powerful statistical tool for building empirical models. It is available in statistical computer applications such as Minitab.

To illustrate, assume that a fixed effects model applies and you have one factor that determines the response. Assume further that the factor will be set at m different levels. At each factor level setting you have a sample size of n. This is called a balanced one-way (for one factor) fixed effects model.

The model for the one-way fixed effects case is

$$y_{ij} = \mu + \beta_i + \varepsilon_{ij} \qquad (16.2)$$

Equation (16.2) expresses the measured response as the sum of three terms. The first term on the right-hand side, μ is the overall grand mean of all observations. The second term, β_i, is the contribution of the factor set at level i. The last term is the error, which accounts for the distribution of observations around the population mean for level i. Assume that the errors are normally distributed with a mean of 0 and a common variance. Figure 16.1 shows the comparison of factor effects at multiple levels.

Figure 16.1 shows a general case where the response y depends on a single factor x. Factor x is set at various levels x_i, and several replicates of the response are measured for each setting of x.

We give the following result, without proof, for the decomposition of the sum of the squares. For the derivation, see Freund.[8]

$$\sum_{i=1}^{m} \sum_{j=1}^{n} (y_{ij} - \bar{y})^2 = n \sum_{i=1}^{m} (\bar{y}_i - \bar{y})^2 + \sum_{i=1}^{m} \sum_{j=1}^{n} (y_{ij} - \bar{y}_i)^2 \qquad (16.3)$$

On the left-hand side of equation (16.3), we show the sum of the squares of the difference between each observation y_{ij} and the grand mean $\bar{y}$. This is the total sum of squares, SS_{total}.

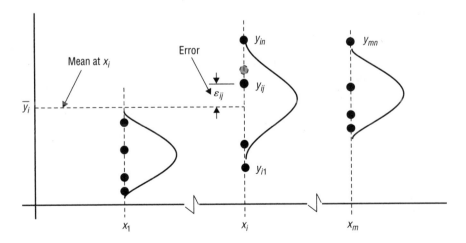

Figure 16.1 Comparing a factor effect at multiple levels using ANOVA

8. Freund, *Mathematical Statistics*.

The first term on the right-hand side is the sum of squares of the difference between each of the sample means, $\bar{y}_i$, and the grand mean, $\bar{y}$. Designated, SS_{factor}, it is the variation due to changing the level of the factors. The second term is the sum of squares of the error, the difference between each individual observation in a sample and the sample mean. This is called SS_{error}. Our assumption is that the errors follow a Gaussian or normal $(0, \sigma^2)$ distribution. Equation (16.3) can be expressed as

$$SS_{total} = SS_{factor} + SS_{error} \tag{16.4}$$

For calculation purposes, the following equations are useful:

$$SS_{total} = \sum_{i=1}^{m} \sum_{j=1}^{n} y_{ij}^2 - \frac{1}{mn} \left(\sum_{i=1}^{m} \sum_{j=1}^{n} y_{ij} \right)^2 \tag{16.5}$$

$$SS_{factor} = \frac{1}{n} \sum_{i=1}^{m} \left(\sum_{j=1}^{n} y_{ij} \right)^2 - \frac{1}{mn} \left(\sum_{i=1}^{m} \sum_{j=1}^{n} y_{ij} \right)^2 \tag{16.6}$$

You can use equations (16.5) and (16.6) to calculate SS_{error}:

$$SS_{error} = \sum_{i=1}^{m} \sum_{j=1}^{n} y_{ij}^2 - \frac{1}{n} \sum_{i=1}^{m} \left(\sum_{j=1}^{n} y_{ij} \right)^2 \tag{16.7}$$

This leads to the F-test, which we covered in Chapter 15 when we discussed comparing sample variances to see if they are likely to be from the same population. In this case we have partitioned the variance into two contributing factors, the effects of the factor level changes and the Gaussian component, which we call the error or noise. If the contribution of the factor level changes is large enough compared to the contribution of the noise, you would conclude that the factor level changes give you a statistically significant response. To determine statistical significance you first formulate the F-statistic. In this case the F-statistic is the contribution to the variance of the factor level changes divided by the contribution to the variance of the noise.

To form the F-statistic, first calculate the mean-square (MS) contributions of the factors and the error. This requires you to divide SS_{factor} and SS_{error} by their respective degrees of freedom.

$$MS_{factor} = \frac{SS_{factor}}{m - 1} \tag{16.8}$$

$$MS_{error} = \frac{SS_{error}}{m(n - 1)} \tag{16.9}$$

So now you can calculate the F-statistic used to assess model significance.

$$F = \frac{MS_{factor}}{MS_{error}} \tag{16.10}$$

Then compare your calculated value to the critical value of F, which depends on the acceptable α error and the degrees of freedom for both the ANOVA model and the noise. For details see the section on F-tests in Chapter 15. If your calculated value exceeds the critical value, you would reject the null hypothesis, that the factor level changes have no effect. You would accept the alternative hypothesis, that the response to at least one of the factor levels is significantly different from the noise.

Example 1: Evaluating the cutting performance of three different saw blade designs

A product development team has developed a new saw blade for cutting granite. They plan to test two new candidate designs against the existing design to evaluate cutting performance. The test is run with a constant cutting force, blade surface speed, and lubrication conditions. The response is the time to cut through a standard test sample. It is a smaller-is-better response. Reduced cut time is important to customers. If significant reductions are possible, while maintaining blade life, it's a competitive advantage. Market research has concluded that a reduction in cutting time of one minute in the standard test would indicate a performance improvement that would be important to customers. You are interested in how the changes in blade design compare to the natural variation of the cutting process. From past history, the variability in cutting time is known to be caused by both process variability in the blade's fabrication and the variability in the granite. ANOVA will tell you which blade design might be superior based on its reduced cutting time.

Using the current blade design and a range of granite samples, the following historical data exist for the cut time test:

Average cut time: $\mu = 13.5$ minutes

Standard deviation: $\sigma = 0.5$ minute

You use the one-way ANOVA power and sample size calculator in Minitab to determine the required sample size. Since you are evaluating three blade designs, there are three levels for the factor. The result of using the power and sample size calculator is that you need a sample size of six to detect a reduction of one minute in cut time with a power of 80%.

For each new blade design, six blade samples were used to cut six granite samples. Data from the cut time test are shown here. All times are in minutes.

Blade 1 cut times: 14.58, 13.49, 13.86, 13.29, 13.18, 13.55
Blade 2 cut times: 13.96, 13.68, 13.46, 13.77, 13.66, 14.03
Blade 3 cut times: 12.88, 13.09, 12.43, 12.51, 12.98, 12.38

The results of the one-way (single factor at multiple levels) ANOVA are shown below. The first conclusion is that blade design is a significant factor, but the ANOVA does not tell you which blade is the best. However, when you look below the ANOVA table, there are confidence intervals on cut time for each blade. This tells you that with overlapping confidence intervals, Blade 1 and Blade 2 are not significantly different. However, Blade 3 shows a reduction in cut time of nearly one minute. This is a practically important improvement that customers will prefer.

Results for: Cut time ANOVA

```
One-way ANOVA: Blade 1, Blade 2, Blade 3

Source  DF    SS     MS      F      P
Factor   2  4.014  2.007  15.17  0.000
Error   15  1.984  0.132
Total   17  5.998


S = 0.3637  R-Sq = 66.92%  R-Sq(adj) = 62.51%
                           Individual 95% CIs For Mean Based on
                           Pooled StDev
Level     N   Mean  StDev  --+---------+---------+---------+-------
Blade 1   6  13.658  0.508                         (-----*-----)
Blade 2   6  13.762  0.209                          (-----*------)
Blade 3   6  12.712  0.309  (-----*------)
                           --+---------+---------+---------+-------
                           12.50     13.00     13.50     14.00


Pooled StDev = 0.364
```

This example of one-way ANOVA shows the power and simplicity of the tool. Two-way ANOVA, used to assess the effects of two factors at two or more levels, is a bit more complicated, but with so many excellent applications available it is no more difficult to use than the one-way ANOVA. There are some limitations in the simple tools, such as the need for a balanced design where there are the same number of observations at each level of the factors. Beyond the simple tools are more powerful tools such as the GLM that can handle unbalanced designs (not necessarily the same number of observations at each factor level), covariates (an uncontrolled factor that can affect the results of an experiment), and regression.

Model Building Using Regression

Regression is a rich topic in statistics with tools for modeling processes and systems that include both continuous and discrete variables. Tools such as logistic regression are useful for modeling problems with binary outcomes. For example, you may want to understand how the probability of contracting a disease varies with exposure time. There are many excellent references that cover advanced topics in both ANOVA and regression.[9] We will not cover these topics in the following discussion but will focus on the linear regression of continuous variables. By concentrating on this area, we can address a couple of important issues that are relevant to the next topic of design of experiments.

Linear Regression

Suppose the model is a linear regression that can be written as an additive polynomial model. The limitation for linearity is that it must be linear in the coefficients, although not necessarily for the

---

9. John Neter, Michael Kutner, Christopher Nachtsheim, and William Wasserman, *Applied Linear Statistical Models* (New York: McGraw-Hill, 1996).

independent variables. This is not a major limitation for physical devices. An example of a linear model is

$$y = \beta_0 + \beta_1 x + \beta_2 x^2 + \varepsilon \qquad (16.11)$$

Nonlinear regression, which we will not cover, is a tool that is useful in modeling some biological and economic systems. An example of a nonlinear model is

$$y = \frac{1}{\beta_0 + \beta_1 x} \qquad (16.12)$$

This model is nonlinear because β_0 and β_1 appear in the denominator of the equation, and after long division is performed, equation (16.12) is not linear in the coefficients β_0 and β_1. We show these examples only to clarify the difference between a linear and a nonlinear model. Because of the x^2 term in the independent variable, confusion can be caused when a model such as equation (16.11) is described as a linear regression. In fact, any polynomial model can be easily transformed into a linear model. For example, if you have a quadratic model, such as equation (16.11), you can define new independent variables where $x_1 = x$ and $x_2 = x^2$.

How Do You Build Regression Models?

Simple Models with One Predictor Variable

Building a regression model with a single predictor variable (called a control factor in DOE) is easy. If you are trying to establish a $y = f(x)$ relationship, just take a number of evenly spaced data points and use the method of least squares to estimate the model coefficients. The good news is that if you are using a commercial statistics application, the analysis function is built in and all you have to do is run the experiment, take the data, and let the statistics program do the rest. Let's take a quick look at the method of least squares and how it works.

Suppose that you fit a simple linear model, as shown in Figure 16.2, to a set of observations. Assume that you have n observations of y at each of m values of x. If the model is a good one, the observations will be clustered around the fit line that's described by equation (16.13).

$$y = a + bx + \varepsilon \qquad (16.13)$$

The "residuals" are defined as the difference between the individual observations and the fitted line. The underlying model assumption is that the residuals are normally distributed with mean 0 and variance σ^2. If the model shows an appropriate fit to the data, there will be no evidence of major temporal or spatial trends in the residuals. Residuals are also called "error" and are given by

$$e_{ij} = (y_{ij} - a - bx_i) \qquad (16.14)$$

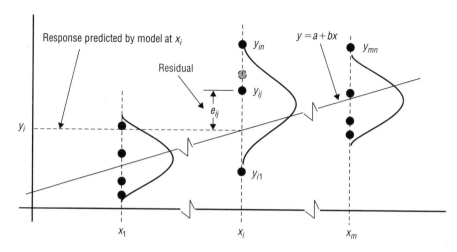

Figure 16.2   Simple linear regression

Our condition for the coefficients is that they minimize the sum of the squared error. The conditions for this are that the derivative of the squared error with respect to each model coefficient is 0, leading to the following requirements:

$$\frac{\partial}{\partial a} \sum_{i=1}^{m} \sum_{j=1}^{n} e_{ij}^2 = \frac{\partial}{\partial a} \sum_{i=1}^{m} \sum_{j=1}^{n} (y_{ij} - a - bx_i)^2 = 0 \qquad (16.15)$$

$$\frac{\partial}{\partial b} \sum_{i=1}^{m} \sum_{j=1}^{n} e_{ij}^2 = \frac{\partial}{\partial b} \sum_{i=1}^{m} \sum_{j=1}^{n} (y_{ij} - a - bx_i)^2 = 0 \qquad (16.16)$$

Using equations (16.15) and (16.16), you can solve for a in terms of b:

$$a = \frac{\displaystyle\sum_{i=1}^{m} \sum_{j=1}^{n} y_{ij}}{mn} - b\frac{\displaystyle\sum_{i=1}^{m} x_i}{m} \qquad (16.17)$$

You can use equations (16.16) and (16.17) to solve for b:

$$b = \frac{\dfrac{\displaystyle\sum_{i=1}^{m} \sum_{j=1}^{n} x_i y_{ij}}{n} - \dfrac{\displaystyle\sum_{i=1}^{m} x_i \sum_{j=1}^{n} y_{ij}}{mn}}{\displaystyle\sum_{i=1}^{m} x_i^2 - \dfrac{\left(\displaystyle\sum_{i=1}^{m} x_i\right)^2}{m}} \qquad (16.18)$$

This is the process to calculate the model coefficients for the case of simple linear regression. The result is that the coefficients a and b will place a straight line through the data set that minimizes the sum of the square of the residuals. Before using the model, you have to determine whether or not the model is adequate.

Is the Model "Good"?

You have to examine your model to see if it meets certain minimum criteria. These are some of the main criteria:

- Is the model statistically significant? This asks whether the model explains enough of the variability in the data to be useful. You can think of the data as being produced by two processes. The model is a mathematical description of some physical process that converts inputs to outputs. The other process is the random process that produces Gaussian variations about each sample mean. You would like the contribution of the model to total variability to be significantly larger than the contribution of the noise. Recall that in Chapter 15 we discussed the F-distribution that is used to compare the variance of two processes.
- Is the model a good fit to the data? A qualitative look at goodness of fit would probably conclude that the data set of Figure 16.3 is a good fit to the straight line passing through the samples, since the data samples all straddle the fit line. Our eyes and brains are pretty good in this case. What if the fit were not so good? You need an objective way to determine the goodness of fit that's based on statistics. You can fit a straight line to a couple of data points and a quadratic model to three data points, but is it a good idea? In each of these cases you have no way to test the model to see if it's a good fit. An important part of planning any experiment is to include enough data points to allow the estimation of error.
- Is the model an efficient one? If you can build a model that is statistically significant, has all the important factors, and fits the data without overfitting it, you are in good shape. You might be tempted to add higher-order terms, thinking that they will fit the data better. However, building a model is a situation where less is more. We discuss the concept of "parsimonious models" in the next chapter about designed experiments. In a nutshell, the idea is to make your model no more complex than is necessary.

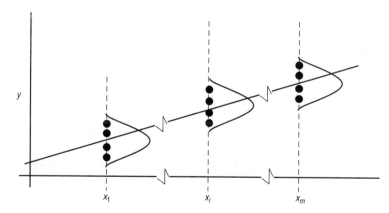

Figure 16.3 A good fit.

Statistical Significance of the Model

To answer the question of the "goodness" of the model, we use ANOVA. This is analogous to a single-factor ANOVA. In the case of simple regression you are interested in the statistical significance of the model instead of a single factor. The approach described here applies to data sets where you have multiple observations at each level of x_i. To start, express the total sum of the squares as the sum of two components:

$$\sum_{i=1}^{m} \sum_{j=1}^{n} (y_{ij} - \bar{y})^2 = n \sum_{i=1}^{m} (y_i - \bar{y})^2 + \sum_{i=1}^{m} \sum_{j=1}^{n} (y_{ij} - y_j)^2 \qquad (16.19)$$

On the left-hand side of equation (16.19) is the total sum of the squares of all observations relative to the grand mean. The first term on the right-hand side is the total sum of squares for the contribution of the model. The second term is the sum of squares for the total error.

You can express this equality as

$$SS_{total} = SS_{model} + SS_{residuals} \qquad (16.20)$$

The pure error is your reference for comparison when you consider both the statistical significance of the model and its lack of fit. For a given factor x, pure error is defined as the difference between the individual observations with $x = x_i$ in a sample and the mean at x_i.

$$\varepsilon_{ij} = y_{ij} - y_i \qquad (16.21)$$

Since pure error has the same definition as the error in our earlier discussion of ANOVA, you can use equation (16.7) to calculate the sum of squares of the pure error.

Applying what you learned in ANOVA, calculate both the mean-square contribution of the model and the pure error and compare them using the F-statistic.

The F-statistic is defined as

$$F = \frac{(MS)_{model}}{(MS)_{pure\ error}} \qquad (16.22)$$

The mean-square contribution of the model and the pure error are defined as

$$(MS)_{model} = \frac{(SS)_{model}}{(df)_{model}} \qquad (16.23)$$

$$(MS)_{pure\ error} = \frac{(SS)_{pure\ error}}{(df)_{pure\ error}} \qquad (16.24)$$

Using equations (16.23) and (16.24), you can form the F-statistic to assess the statistical significance of the model. If the mean-square contribution of the model is large enough compared to the contribution of the pure error, you conclude that the model is statistically significant. Details of performing the F-test are discussed in Chapter 14.

$$F = \frac{MS_{model}}{MS_{pure\ error}} \qquad (16.25)$$

Degrees of Freedom

This is a good time to review degrees of freedom (df). We make the following assumptions about your data set and model. It is a completely balanced case and thus the simplest:

Number of treatments = m. Treatments are the number of unique combinations of factors, or levels of a factor. For a simple linear model, the number of treatments is the number of levels of the predictor variable.

Number of replicates = n. Replicates are response data resulting from repeated runs of treatments.

Number of observations = mn

Number of predictor variables in the model = k

This results in the following degrees of freedom for each of the important sum of squares. The total number of observations is mn. We use one degree of freedom when we calculate the grand mean.

So the total degrees of freedom is

$$(df)_{total} = (df)_{model} + (df)_{residuals} = mn - 1 \tag{16.26}$$

There is one degree of freedom per predictor variable in the model, so

$$(df)_{model} = k \tag{16.27}$$

Calculate degrees of freedom for the residuals by subtracting degrees of freedom for the model from the total degrees of freedom.

$$(df)_{residuals} = (df)_{total} - (df)_{model} = mn - 1 - k \tag{16.28}$$

There are m levels of the independent variable and n replicates. For each set of replicates or samples, use one degree of freedom to calculate the sample mean.

$$(df)_{pure\ error} = m(n - 1) \tag{16.29}$$

In the next section we discuss how to decompose the residuals into two components: pure error and lack of fit. After decomposing the residuals, you can determine the statistical significance of lack of fit.

How Well Does the Model Fit the Data?

Fitting a model to a data set is an iterative process, where you look at the data and perhaps make an educated guess informed by your understanding of the science and your past experience. You have made an assumption about the nature of the model. Is it a straight line, a quadratic, or perhaps another shape? Having made a first cut, you need an objective and quantitative way to answer the question of whether or not the model is good. The way we do this is by assessing the statistical significance of lack of fit.

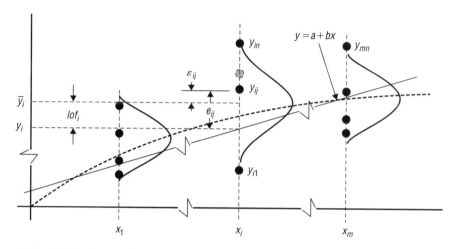

Figure 16.4   Evaluating goodness of fit

Figure 16.4 shows the situation. We have fit a model to the data given by equation (16.13). Now the question becomes "Is our first-order model good enough, or do we need a model with higher-order terms?" Maybe a higher-order model, shown by the curved dashed line, would be better.

If you examine Figure 16.4, you see that the residuals, e_{ij}, can be partitioned into two components. One component is called the "pure error," designated as ε_{ij}. Pure error is the difference between the individual observations at x_i and the mean at x_i. Recall that our assumption is that these differences are normally distributed. The other component is the "error due to lack of fit," shown as lof_i. That error is the difference between the value of y predicted by the model and the sample mean. If all the sample means were sitting on the line defined by the model, the lack-of-fit error would be 0 and the residuals would be the same as the pure error.

The sum of squares of the residuals becomes

$$\sum_{i=1}^{m} \sum_{j=1}^{n} e_{ij}^2 = \sum_{i=1}^{m} \sum_{j=1}^{n} (lof_i + \varepsilon_{ij})^2 \tag{16.30}$$

Expanding this we get

$$\sum_{i=1}^{m} \sum_{j=1}^{n} e_{ij}^2 = \sum_{i=1}^{m} \sum_{j=1}^{n} (lof_i^2 + 2 lof_i \varepsilon_{ij} + \varepsilon_{ij}^2) \tag{16.31}$$

For each sample the contribution of the middle term is 0. When you sum over all of the observations, you have a simple and useful relation for the sum of squares of the error:

$$SS_{residuals} = SS_{lof} + SS_{pe} \tag{16.32}$$

The quantity SS_{pe} is the sum of squares of the pure error. We use it when we calculate the mean-square pure error, which will be our reference for evaluating both the statistical significance of the model as well as goodness of fit.

So, using equation (16.32), you can evaluate the relative contributions of the lack of fit and the pure error to the total sum of squares of the error.

First consider the statistical significance of the lack of fit.

As you did when using ANOVA to assess the statistical significance of the model, form an F-statistic to determine the statistical significance of the lack of fit.

$$F_{lof} = \frac{MS_{lof}}{MS_{pe}} \tag{16.33}$$

To calculate the MS terms in equation (16.33), you need the degrees of freedom for each term, since

$$df(SS_{residuals}) = df(SS_{lof}) + df(SS_{pe}) \tag{16.34}$$

Since there are mn observations and one predictor variable:

$$df(SS_{residuals}) = mn - 2 \tag{16.35}$$

The degrees of freedom for the pure error sum of squares is

$$df(SS_{pe}) = m(n - 1) \tag{16.36}$$

So you can calculate the degrees of freedom for the lack-of-fit sum of squares:

$$df(SS_{lof}) = df(SS_{residuals}) - df(SS_{pe}) = m - 2 \tag{16.37}$$

Now calculate the MS values that you need for the F-statistic.

$$MS_{lof} = \frac{SS_{lof}}{m - 2} \tag{16.38}$$

$$MS_{pe} = \frac{SS_{pe}}{m(n - 1)} \tag{16.39}$$

What Do the Residuals Tell You?

The residuals are a rich source of information about the quality of the model you have built. If you have a good model, the residuals will have certain characteristics:

1. The residuals will be normally distributed and should have a mean that is close to 0. Recall that the residuals are composed of the pure error plus the lack of fit.

$$e_{ij} = \varepsilon_{ij} + e_{lof_i} \tag{16.40}$$

Ideally you would like lack of fit to be small. Since the pure error is assumed to be Gaussian, you would like the residuals to be normally distributed with a mean near 0.

2. There should be no trend when residuals are plotted across the range of the model. On average, the residuals should be about the same size for both large and small values of

the dependent variable. If there is a trend, that is a sign of an inadequate model and an associated lack of fit.

3. There should be no trend if you look at residuals versus the order in which the data were taken. If you see a trend for residuals versus order of data, that is a sign that there could be an external factor, not intentionally part of your experiment, that was changing with time. This is sometimes called a "lurking variable." An example would be if you are measuring the surface temperature of a heat sink, and ambient temperature is changing at the same time. As ambient temperature rises, the device temperature will also rise to enable continued heat transfer to ambient. Knowing the physics of the device, you could eliminate this problem by instead measuring the difference between ambient temperature and device temperature.

Let's look at an example that will illustrate how you can apply what we have discussed so far.

Example 2: Building a simple linear regression model

You have a resistor mounted on a heat sink and are interested in measuring the surface temperature as a function of current. Engineering physics tells you that the power dissipation is proportional to the square of the current. Start by fitting a simple linear model to the data and test it to see if it's a good model. You have three replicates at each of six current levels. The ambient temperature was constant during the test.

(Current, Temperature)
(0.25, 27.7), (0.25, 39.5), (0.25, 26.8)
(0.50, 43.4), (0.50, 52.0), (0.50, 52.1)
(0.75, 57.5), (0.75, 53.4), (0.75, 53.6)
(1.00, 83.2), (1.00, 77.5), (1.00, 74.2)
(1.25, 116.3), (1.25, 109.7), (1.25, 107.9)
(1.50, 147.6), (1.50, 145.7), (1.50, 136.9)

Using Minitab, fit a simple linear regression model to the data of the form

$$y = a + bx + cx^2 \tag{16.41}$$

The results of the analysis are shown below. All plots are from Minitab. Looking at the ANOVA table and the table of coefficients on the following page, you see several important results:

1. The model is highly significant, explaining over 98% of the variance in the observations as shown by the small p value for "Regression."

2. The predictor variable, current squared, and the constant are both significant. However, current is not significant as shown by the p value of 0.886.

Regression Analysis: Temp versus I and I^2

```
The regression equation is
Temp = 31.1 - 2.3 I + 51.6 I^2


Predictor    Coef  SE Coef      T      P
Constant   31.147    6.044   5.15  0.000
I           -2.31    15.82  -0.15  0.886
I^2        51.619    8.848   5.83  0.000


S = 5.85214   R-Sq = 98.1%   R-Sq(adj) = 97.9%


Analysis of Variance

Source          DF     SS     MS      F      P
Regression       2  26589  13294  388.19  0.000
Residual Error  15    514     34
Total           17  27103
```

Look at the plots of both the fitted line and the residuals and see what they tell you.

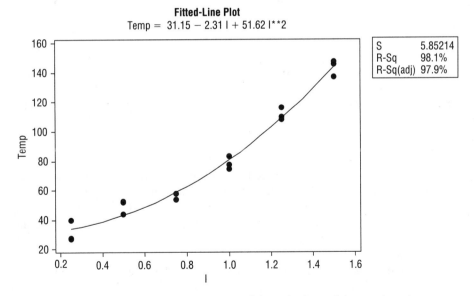

Figure 16.5 Plots of the fitted line $y = a + bx + cx^2$ from the heat sink experiment

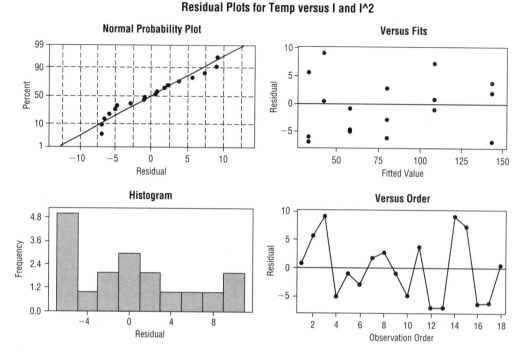

Figure 16.6   Plots of the residuals for $y = a + bx + cx^2$ from the heat sink experiment

Fitted-Line Plot

The fitted-line plot shown in Figure 16.5 looks good. This is not surprising since first principles tell you that there is a quadratic relationship between current and power. Since the rate of heat transfer (power) depends on temperature difference, this would suggest a quadratic relationship between current and temperature.

Residuals

1. The plot at the upper left of Figure 16.6 is the probability plot for the residuals. It looks acceptable, but to be certain, we stored the residuals in the Minitab worksheet and ran a test for normality which showed that the residuals are normally distributed.
2. The upper right plot shows residuals versus fits. There is no concern here because the plot has no distinctive shape, appearing to be random. This is a sign that the model is a good fit.
3. Even though the residuals are normally distributed, the lower left plot is not bell-shaped. When dealing with a small sample, this is not surprising. With only 18 residuals, you can get a variety of histograms even though the sample of 18 may come from a normal distribution.
4. The plot of residuals in observation order doesn't show anything exceptional.
5. Since the run order for the experiment was randomized, residuals versus fits do not appear highly correlated with residuals versus order.

So the quadratic model explains over 98% of the variability in the data, although it may be possible to simplify it. Apply your understanding of first principles to eliminate the first-order term from the model. You can refit the model and then review the results of the ANOVA and the plots of the residuals.

If power and temperature rise are proportional to the square of the current, the simplest model, excluding the first-order term, is

$$y = a + bx^2 \tag{16.42}$$

After fitting this model to the data and analyzing the result, Minitab gives you the following:

Regression Analysis: Temp versus I^2

```
The regression equation is
Temp = 30.3 + 50.4 I^2


Predictor    Coef  SE Coef     T      P
Constant   30.324    2.131  14.23  0.000
I^2        50.354    1.751  28.76  0.000


S = 5.67034  R-Sq = 98.1%  R-Sq(adj) = 98.0%


Analysis of Variance

Source          DF     SS     MS      F      P
Regression       1  26588  26588  826.93  0.000
Residual Error  16    514     32
  Lack of Fit    4    208     52    2.03  0.154
  Pure Error    12    307     26
Total           17  27103
```

So eliminating the first-order term from the quadratic model did not improve the model fit or prediction capability, although the model was simplified. The residuals plots, shown in Figure 16.7, are about the same as the residuals plots for the quadratic model. The fitted-line plot, which is not shown, is indistinguishable by eye from the plot for the full quadratic model.

Summary

In this chapter we have covered the basics of ANOVA and regression. Both tools are fundamental to understanding how to build empirical models. These tools used to involve hard work to complete the analyses required to gain statistical insight. With the desktop computing power now available, we are blessed with an "embarrassment of riches." What used to take days now takes minutes, allowing scientists and engineers to focus on the strategy for developing understanding rather than the tedium of number crunching.

We have shown that a simple regression model with a single predictor variable is easily built. However, when we consider the possibility of a more complex model with multiple factors,

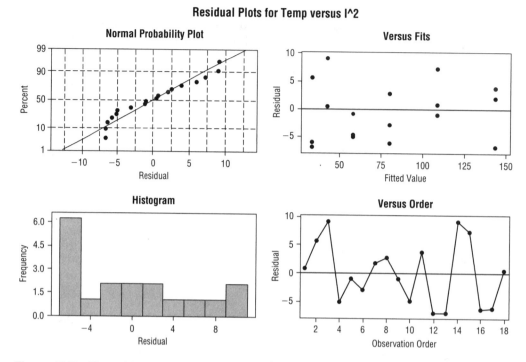

Figure 16.7   Plots of the residuals for $y = a + bx^2$ from the heat sink experiment

there are a number of considerations. The most obvious is how to vary each of the factors when running the experiment. When we have only a single independent variable, the mechanics of running the experiment are straightforward. However, if there are multiple predictor variables, also known as control factors, it's not so obvious. The challenge is to find the minimal set of data that will provide a useful model, since more data points mean more time and money.

The model-fitting approach can be generalized to higher-order models as well as to multiple predictor variables and response factors. The analysis of regressions is best left to the computer since increasing the order of the model and adding more factors make the prospects of analysis by hand daunting. Any good statistical analysis application has regression built in. We can focus on planning and running the experiment rather than on tedious calculations.

In the next chapter we discuss the design of experiments (DOE), which is the best way to build regression models with multiple factors and levels.

Key Points

1. It can be very difficult to show cause and effect using retrospective data.
2. With retrospective data there is the danger that "lurking variables" are the actual cause of the response.

3. With a planned experiment we have the opportunity to control for any suspected lurking variables, or to include them as covariates.

4. The best way to show cause and effect is with a planned experiment.

5. When planning an experiment, we need to identify the important responses and then develop a list of candidate control factors.

6. Responses in an experiment need to be linked to responses that are important to stakeholders, such as customers, end users, and the business.

7. One way ANOVA can be used is to analyze the simplest experiment, where we vary a single predictor variable across a range while measuring the response. Predictor variables can be continuous or categorical.

8. With ANOVA we calculate the mean sum of squares contribution of the factor and compare it to the "noise" or random variation in the data. The statistical approach is to use the F-statistic, discussed in Chapter 15, where we are interested in comparing the variance of two processes to see if the difference in variance is statistically significant.

9. The data in an experiment come from two distinct processes. The first process is the one that we are trying to model. It follows the underlying rules of the type of process we are working with. It represents nature's law uncontaminated by random variations. The process could be one of engineering science, life science, or a transactional business process. The second process is the random variation or Gaussian noise present in any system. It will be system-dependent.

10. The larger the F-statistic, the bigger the ratio of the variances and the less the likelihood that the variances were produced by the same process.

11. Regression is a key tool for building empirical models of processes and systems.

12. After we build a regression model, we test it to determine if it's a "good" model.

13. Good models have some important qualities:

    a. They explain a large part of the variation in the data.

    b. They are a good fit to the data.

    c. Most of the model error is the random variation from the Gaussian noise.

    d. They are "parsimonious" and would be of reduced quality if we were to add to or take away from the model.

Discussion Questions

1. What tool can be useful in understanding cause and effect?

2. Is it possible to establish cause and effect using retrospective data?

3. When evaluating the effect of a factor at three levels, why use ANOVA and not multiple t-tests?

4. Why is it correct to use the ratio of mean-square contributions when formulating the F-statistic for testing the statistical significance of factors?

5. Give some examples of Gaussian noise for some different processes.

6. When building a regression model, what is "overfitting"?

7. How can you tell if a model is a good fit to the data?

8. What are residuals and why are they important?

9. What is pure error?
10. What is lack of fit?
11. What is the ideal relationship among the residuals, pure error, and lack of fit?
12. When building an empirical model, what is the relationship between ANOVA and regression?
13. In regression, what comes first, fitting the model or ANOVA?
14. What tool is useful for planning how you will run a test to build an empirical model with multiple factors and multiple levels?

Building Empirical Models Using DOE

"The best time to design an experiment is after it is finished."
Box, Hunter, and Hunter, Statistics for Experimenters[1]

Engineering science has come a long way in its ability to build useful first-principles models. Computer-aided engineering tools such as finite element analysis (FEA) and computational fluid dynamics (CFD), along with increased computing power, have made it both possible and practical to build models of complex systems that would not have been possible in the past. In spite of all this, there still are processes that defy understanding, using first-principles models. Examples abound in business, engineering, and science. The good news is that when faced with problems that are not amenable to first-principles analysis, we have a rich array of tools to build empirical models. A useful empirical model depends on following a logical process and using the right tools. In this chapter we discuss the use of the tools of classical design of experiments (DOE). Our goals are to introduce the subject, to show the benefits of applying DOE when developing or improving systems, and to leave you interested in learning more.

Space does not permit in-depth coverage of DOE, so we cannot make you an expert. However, we want you to have an awareness of the more important concepts. It is possible to learn a lot first by reading background guidance and then by using some of the commercial software. To illustrate the tools, we use Design-Expert,[2] an excellent application for DOE.

This chapter is a foundation for the next chapter, in which we show the application of three different optimization tools to improve robustness using empirical models derived from DOE.

Overview of DOE

Design of experiments (DOE) is a tool that enables a structured process for sampling the response of a system as you systematically make changes to the settings of controllable parameters. By following the process, you can be confident that you have not created any statistical problems

---

1. George E. P. Box, William G. Hunter, and J. Stuart Hunter, *Statistics for Experimenters: An Introduction to Design, Data Analysis, and Model Building* (New York: John Wiley, 1978).

2. Design-Expert is a product of Stat-Ease (www.statease.com) in Minneapolis, MN.

because of a poor design for your experiment. As you will see, most of the opportunities for success are not in the realm of statistics. If you experience a useless result from an experiment, usually it is because you didn't do a good job of answering one or more of the important questions we cover.

Here are some of the basic characteristics of a designed experiment:

- A designed experiment is expressed in the form of a table where each row is a run specifying how the control parameters should be set for the run.
- Each unique combination of control parameter settings or levels is called a "treatment."
- Treatments can be run multiple times.
- Repeated runs of a treatment are called "replicates."
- Replicates are run to enable estimates of the pure error, the variation in response due to common cause noise.

For the topics of ANOVA and regression (Chapter 16) we discuss how to build a model and evaluate its statistical significance. In order to perform an ANOVA, you need an estimate of the error, hence the need for replicates. Regression helps you to build a model using the data that you take, while ANOVA tells you if your model is useful. The good news is that both of these tools are used in DOE. It turns out that the "front end" of a DOE is the selection of the experiment design, while the back end is the analysis using regression and ANOVA, both of which are built into most DOE applications. So in your learning, you are already partway to being able to use DOE.

This chapter is concerned mostly with the front end of the process, where planning and your understanding of the engineering science are important. The answers to important questions can improve your chances of being successful.

Some History

R. A. Fisher was a pioneer in the development and application of designed experiments. Beginning in 1919, while working at the Rothamsted Experimental Station in England, Fisher began developing the statistical basis for DOE, applying his methods to understand the importance of factors that affect agricultural yields. Factors such as soil type, seed, fertilizer, sun exposure, and water all affect the growing process. However, that process, being fundamentally one of life science, is very complex and, at that time, was not amenable to building first-principles models. Fisher made many important contributions in genetics, biology, and statistics. His success at developing and applying DOE and giving it a solid statistical foundation earned him recognition for making DOE a useful and practical tool.

However, to understand the mathematical foundations that make DOE work, we would need to study some of the early developments in the field of linear algebra, before Fisher's time. Much of this work was done during the nineteenth century by pioneers such as Hamilton, Cayley, Sylvester, and Hadamard, who were interested in pure mathematics rather than its later and at that time unforeseen practical applications. Other than a tip of the hat to those who did the heavy lifting, we won't discuss the mathematical foundations of DOE. They are far outside the scope of this book. Since Fisher's time, the body of knowledge for DOE has continued to grow, with much emphasis on modeling efficiency. For example, the increase in desktop computing power has made computer-generated designs, such as D-optimal and I-optimal, widely available and easily used.

What Do You Need to Know?

Today the newcomer to DOE is confronted by many options that can be overwhelming. Here we focus on the tools that have the most value in helping you develop robust systems with a minimum of scrapes and bruises.

While the utility of DOE depends on fundamental mathematical principles, you don't need to be a mathematician to make use of the DOE toolbox. With commercially available software, you can be confident that the design of your experiments will be mathematically sound. However, you do need a basic understanding of some statistics, especially concepts such as power, ANOVA, and linear regression. Fortunately, DOE software does these calculations, allowing you to spend more time understanding the results of the experiments. The mathematical and statistical parts of DOE are the least of your worries.

Your focus in DOE should be on planning your experiment, deciding what type of experimental strategy to follow, selecting meaningful responses and control factors, and identifying the sources of variation that can affect the performance of your system. These are the challenges. Your understanding of engineering science, rather than the foundational principles of DOE, is most important.

Why Empirical Models Using Designed Experiments?

First-principles models can be difficult to develop. They can require a great deal of time and energy, as well as a deep understanding of the underlying technology and engineering science. If there is widespread interest in a model, it will usually get worked on and often be commercialized. It may be used to solve problems having significant economic or human benefits. The first principles may be understood well and mathematically described. This has been the case in areas such as solid mechanics, field problems, heat transfer, and fluid dynamics. For these applications, the mathematical and analytical groundwork had been in place for a long time. The computational tools, both hardware and software, have been developed to the point that it is feasible to codify the general solutions, enabling the development of today's analysis systems.

However, in many areas our understanding of the underlying science and technology is incomplete, and it would require a large investment of time and resources to bring it to the point of developing useful first-principles models. In the life sciences or behavioral science, for example, systems have great complexity. The development of useful first-principles models is a substantial challenge. Economics also plays a large part. If there is not enough general interest in a problem, and if the underlying science is difficult and not well developed, it might be concluded that the benefits of developing a first-principles model for a particular problem cannot command the investment. These situations are ones for which DOE can be a cost-effective tool that yields results that you can use to improve your designs within a reasonable budget and time frame.

Screening, Modeling, and Robustness

There are three broad categories of experiments: screening, modeling, and robustness. Screening experiments often precede modeling and robustness experiments to reduce the number of factors that are included in the experiment and thereby their time and expense. Screening experiments are used to identify the important noise and control factors that determine system performance. However, they are not useful for modeling because they have some limitations, which we discuss later. While modeling experiments are focused on developing the $y = f(x)$ transfer function, robustness experiments concentrate on making your product or system less sensitive to noises that can inject variability into performance. There is no good reason to create an artificial barrier that separates modeling from the optimization of robustness when using classical DOE. It should be seamless. Ultimately the development teams want their design to be insensitive to sources of variations.

Advantages of Designed Experiments

Good empirical models can help teams develop better designs. The main benefits include these:

- An empirical model has the advantage of realism over a first-principles model. This is because the performance measured is the result of the signals, control factors, all sources of noise both known and unknown, and any "lurking variables" acting at the time of the experiment. Building a good first-principles model requires that you identify and include all the important factors in the model. Using a first-principles model, you can simulate the effects of variability and external noises by applying a tool such as Monte Carlo simulation. However, with first-principles models you have no way to include the "unknown unknowns." With an experiment, all the factors that have an effect can influence the result, even if they have not been included intentionally. They are all present either as controllable parameters or as noises. The noises are either explicitly included as factors, or implicitly as random effects that impact the response. Small effects of less important control factors that are not explicitly included get lumped in with the noises.

- Empirical models can be used to predict performance across a design space. Being able to predict performance for a range of values for design parameters helps to avoid choosing operating points on the "edge of the design cliff." By this we mean an operating point where a small change in the value of a design parameter or the addition of an external perturbation such as a noise can cause a precipitous change in the system performance.

- Empirical models can help you understand the relative importance of factors affecting performance. You can then focus on what's most important and identify the critical parameters, factors having the greatest influence on performance. As with many things, Pareto's law applies. Not all design parameters will have the same degree of effect on performance. There is also a benefit from knowing what factors have only small effects on performance, since they may represent opportunities to loosen tolerances and reduce costs. The process of understanding and controlling the important design parameters is called "critical parameter management" (see Chapter 23). It is an ongoing process and a key activity in developing robust designs.

- Good models can be used to find favorable operating conditions that will make systems robust against sources of variations. As you will see when we discuss optimization, both external noises, such as environmental factors, and internal noises can cause variable performance. In designing for robustness, we look for set points for controllable parameters that can minimize the variability transmitted to those critical responses that are important to customers.

There are significant business benefits from having a good empirical model of your system. Understanding performance across a design space and knowing what design decisions are critical to performance have direct economic consequences. The more robust you can make a product against sources of variation, the higher can be the level of customers' satisfaction that you can achieve and the lower will be your service costs. In some industries, such as medical products, there are the added considerations of regulatory agencies that can order product recalls in the event of compliance problems after market entry. Increased satisfaction of customers leads to more referrals and repeat sales. Reduced repair cost means lower costs of poor quality. There will be fewer demands for product development teams to correct problems in production rather than develop new products.

Ways of Experimenting

One Factor at a Time

Probably the most widely used alternative to DOE is the one-factor-at-a-time (OFAT) technique. Using OFAT, you start with all factors having nominal settings, maybe your best guess. You then select the factor that you think is most important and vary it across a design range. When you find the set point that optimizes the response, freeze that parameter and then vary the next factor in the same way. One by one you arrive at settings for each control factor. This method is still used by many development teams to understand system performance and to identify preferred operating points. However, the method suffers from some serious deficiencies, among which are these:

- This type of experiment is not efficient. On the other hand, a designed experiment evaluates the effects of variations in several factors simultaneously. DOE becomes even more important as the system complexity and the number of control and noise factors increases.
- The OFAT approach does not produce a model that can be used to predict performance over an operating space. It may get you close to the top of the mountain, but you won't know the topography.
- The OFAT approach will not identify interactions among factors. Interactions, especially between control and noise factors, are essential to reducing the vulnerability to noises.
- If you stop an OFAT experiment too soon, you risk suboptimization.

Random Walk Guided by Intuition

Investigators with lots of experience, deep technical understanding, and good intuition can achieve success at times by changing factors, in an unstructured way, to study the results. However, as system complexity and the number of control and response factors increase, without DOE it becomes increasingly difficult to find the "sweet spot" that gives good performance. A complicating factor is the need to find operating points that achieve both on-target performance and reduced sensitivity to sources of variation. Random methods can be chaotic and suffer from all of the deficiencies of the OFAT method.

DOE

A designed experiment can provide deeper understanding and better performance in less time and at lower cost. DOE has been used successfully across many industries to improve technologies, products, and processes. In areas as diverse as IC manufacturing and the operation of customer call centers, DOE has been used to develop superior products and more efficient processes, all leading to increases in customers' satisfaction and lower costs.

Experimentation is an area where the 80/20 rule is in effect. Most of the effort should be spent in planning the experiment and understanding the results. In the next section we list some questions that should be considered both before and after the experiment. If you do a good job with the "before" questions, you will have an easier time with the "after" questions.

Important Questions to Answer before an Experiment

In our experience, the "activity trap" can be a problem in many organizations. Management and even bystanders may assume that there's no real progress unless prototypes are being built and tests are running. Another term for this is the "rush to design." The problem with succumbing to this "Don't just stand there, do something!" pressure is that what is done is often poorly planned and does not achieve real progress. Before jumping into an experiment using the "ready, fire, aim" approach, you should be able to answer questions such as the ones in this section.

1. **What are the goals for your experiment?**
    The first question to answer is the objective for the experiment. The goal could be purely to build an empirical model, or it could be to optimize performance and robustness.

    Possible experiment goals are the following:

    a. Map performance across a range of settings for the controllable parameters in an existing system. In some cases the system operating point has been chosen, possibly without the use of DOE, but the development team wants to understand how performance varies with variations in those parameters.

    b. Minimize or maximize a functional response. For smaller-is-better or larger-is-better responses, the use of DOE can help a design team choose a set of operating points that will maximize customers' satisfaction.

    c. Make a functional response greater than or less than some value with the added requirement of minimizing its variations due to noise. From an optimization standpoint this is the same problem as co-optimizing multiple responses.

    d. Place a functional response on target and minimize response variations caused by noise.

    e. Co-optimize multiple responses. Sometimes multiple responses are affected by the same control factors. This situation requires trade-offs in performance, possibly forcing you to accept less for each response compared to an individual response being optimized. Co-optimization requires the use of an objective function rather than an individual response.

2. **What responses will be measured and how are they linked to customers' needs?**
    The next question to answer is what to measure as the response in the experiment. Responses selected should be linked to customers' satisfaction, either directly or indirectly. Often, important responses are not observable by customers but affect another functional response that is important to customers.

    Understanding what is important requires that a good job be done understanding customers' needs and translating them into technical requirements. In some situations experiments may be needed to explore the limits of acceptable performance. Often this is the case when acceptable performance is "in the eye of the beholder." An example is the requirement for an attribute of image quality for a printer, for which there is a range of acceptable performance across a population of customers for a market segment.

3. **How will the responses be measured?**
    An important part of any experiment is the measurement system that will be used. The instrumentation can add noise to the observed response, so it should be

characterized in a separate set of experiments called a Measurement System Analysis[3] (MSA). You need to ensure that the measurement system is adequate for the task. For some measurements a ruler may be fine. Others may require a laser interferometer. It depends on the differences that you are trying to resolve and the inherent variability of the measurement process. The higher the stakes in your experiment, the more pressing is the need to analyze your measurement system before taking data and presenting results.

4. **Are the responses in the experiment continuous variables?**
   Concentrate on continuous responses rather than categorical responses. Avoid qualitative responses, such as percent defective. This type of response represents an after-the-fact result of system performance. When developing a new system, process, or product, you should use continuous responses that are leading indicators of performance so that you can detect performance degradations before failures occur.

5. **Have the important control factors been identified?**
   Control factors are identified after you have selected the responses to measure. In cases where many candidate control factors are proposed and you are uncertain about their significance, you may need to run screening experiments. Tools that are useful include brainstorming, prior experience, cause-and-effect diagrams, QFD matrix, understanding of first principles, and screening experiments.

6. **What sources of external and internal noises can affect system responses?**
   Understanding the noises that affect performance is a crucial step in improving system robustness. You can use some of the same tools that you rely upon to identify control factors. Think of noises as control factors that cannot be controlled. In your $y = f(x)$ model, factors are factors. Nature makes no distinction among signal, controllable parameters, and noise factors. As is the case with control factor identification, you may need to run a noise screening experiment if there are many candidate noise factors.

7. **Which noises can be included as factors in the experiment?**
   Robustness improvement requires noises to be included in the design of the experiment. There are a couple of ways to do this. We focus on including noises as controllable factors. Even though the eventual plan is to operate in the presence of uncontrolled noises, you must control them in the experiment so that the test conditions can be replicated.

   Examples of how to do this include
   a. Environmental factors such as temperature, relative humidity, and atmospheric pressure—use an environmental test chamber
   b. Deterioration of parts—interchange aged parts with new parts as part of the experiment
   c. Variations in properties of consumable materials—use material batches with properties at their extremes

---

3. MSA is set of special-purpose designed experiments that can be used to evaluate measurement error, bias, linearity, and drift. Tools to create specific MSA experiments for your application are built into most good statistical applications such as Minitab and JMP.

   d. Out-of-the-box variations in parts dimensions—use parts from the extremes of the tolerance range or beyond

   e. Assembly variations—use setups representing the extremes of the specifications for parameters

8. What is the size of the change in the response you need to detect?

There is no point in running an experiment where across the experiment space the response varies by an amount to which you are indifferent. You want the experiment to evaluate differences in performance that can be important to your customers and perhaps to your business. Timid changes in control factors that yield unimportant changes in response are of no value. Also, the resulting small changes in response might not be statistically significant if the common cause noise is large enough. That can leave you with an unsatisfying result from your experiment. The decision of how much of a change in performance is important is a business question, not a technical one.

9. Does your planned experiment have sufficient power?

Recall the t-test and our discussion of test power. A two-sample t-test is actually a one-factor experiment with the factor at two levels. "Test power" is the probability that you will correctly reject the null hypothesis. For the experiment, the null hypothesis is that the changes in factor level settings do not cause a change in the response that is significantly different from the average response. Just as we found with the t-test, the smaller the difference that you are trying to detect and the greater the common cause noise, the lower the test power.

10. What is the minimum number of runs required to give adequate test power?

For a two-level factorial experiment you can calculate the power of the experiment. While it is a bit more complicated to calculate the power for a multifactor experiment, the idea is the same. The power and sample size calculations consider several things:

   a. The smallest difference in response that you want to detect

   b. The magnitude of the common cause noise

   c. The number of terms included in the candidate model

   d. The number of runs in the experiment

      Small experiments with few runs are limited in the number of model terms that can be estimated. In our discussion of ANOVA and regression, we mentioned the need to "save" some degrees of freedom for estimating experimental error. While it is possible to fit a quadratic function to three data points, there are no degrees of freedom left to estimate the error. Without that estimate of the error, you will not be able to judge the statistical significance of the model and thus its usefulness.

      If the changes in the experiment responses due to the changes in control factors are large compared to the variations due to noises, test power would not be a concern. However, often you are searching for smaller changes in responses. That requires enough runs to estimate the error and extract the signal from the noise.

11. Will the planned level changes in control factors be achievable and produce response changes that are significant compared to those from common cause noise?

The power of the test is a conditional probability. It's the probability you will conclude correctly that the change in a control factor level caused a difference in the response equal to or greater than a specified difference, if that difference really exists. The test power tells you nothing about the amount of change in a control factor level that is required to cause a change in response large enough to be statistically significant. That question depends on the nature of the system transfer function and the "gain" for each of the factors. So deciding on control factor levels requires that you use your past experience, perhaps first principles, and maybe some single-factor experiments or a screening experiment to get a sense of how much change in a response will be caused by turning the control factor knobs.

When setting control factor levels in the experiment, you need to ensure that you will be able to accomplish all the runs. If the response for a treatment is too small, you may have set factor levels too low and missed an opportunity for a statistically significant response. At the other extreme, you don't want the system to blow up. So you need to ensure that you can run all of the required combinations of control factors. There is nothing more frustrating than discovering on the last run of an experiment that the control factor settings required for the treatment are not achievable because of equipment or safety limitations. While it is possible to analyze the results of a "botched" experiment having missing data, it will have a negative effect on the quality of the results. This is to be avoided. Before selecting the experiment design, if there are known limitations in the settings of control factors, it is possible to include these constraints in the design of the experiment.

12. **Are there any factors that should be blocked?**

Blocking is an experiment technique used when you have "nuisance factors." Anything that could potentially change during the experiment should be considered for blocking. Examples include environment, operators, instruments, machines, raw materials, and measuring devices.

Suppose you are running an experiment with a production process and, to save time, you plan to use two identical pieces of production equipment. By simultaneously making half the runs on each machine, you can complete the experiment in half the clock time. Although the machines are assumed to be the same, slight differences could cause differences in performance. A blocked experiment design assigns the nuisance factor from different machines to the column that represents the highest-order interaction. So the nuisance factor of the machine is aliased or mixed with the highest-order interaction. The result is that the contribution of the nuisance factor is removed from the noise and assigned to the block, effectively increasing the power of the experiment. Had you not blocked, the contribution of the block would have been included in the noise. Larger noises mean that larger responses are needed to be judged statistically significant.

13. **Can the experiment runs be randomized?**

Conventional wisdom, which is sound thinking, is that you should randomize the run order in the experiment. This goes for everything, including replicates and center points. However, there are times when randomization can be time-consuming, difficult, or costly. Here are some examples:

a. You are running an experiment using a large industrial oven. Oven temperature is a control factor in the experiment. When you change the oven temperature, it takes several hours for the temperature to change levels and stabilize. You have limited time for the experiment because you are "stealing" time from production to get it done.

b. You are running an experiment on a manufacturing process. One of the setups required to change a particular control factor level is difficult and time-consuming.

In cases like these you need to weigh the benefits of randomization against the disadvantages. One strategy is to assign factors that are difficult or time-consuming to change to columns that change the fewest times when the experiment is run in its standard order. Then all runs that have the same "difficult" level setting can be randomized.

14. **Are important interactions expected? Can anything be done to minimize them?**
Sometimes an interaction can be a surprise, other times not surprising at all. Past experience and first principles can help you identify potential interactions and, if you are lucky, maybe eliminate them. A classic example from photography is the interaction between the lens opening and the shutter speed. Let's assume that you want to explore the optical density of an image as a function of the camera settings. What is important is the number of photons that hit the image sensor or film. This determines the lightness of the image. More photons hitting a given area of the film result in more exposure and a lighter image. If you analyze an experiment that has lens opening and shutter speed as two independent control factors, you would find an interaction between the factors. The reason is that you can get the same exposure with a small lens opening and slow shutter speed as you would with a large lens opening and fast shutter speed. The control factors interact. By knowing the relationship between lens opening and light intensity at the image sensor or film plane, and properly combining those factors into a compound factor, you can eliminate the interaction.

15. **Is a linear model adequate or is a higher-order model needed?**
Answering this question requires you to consider the questions of practical importance and statistical significance. A good way to assess the statistical significance of the nonlinearity in a process is by including center points in the factorial model. Center points are used to estimate the significance of process nonlinearity as well as its error or process noise. Center points will tell you if nonlinearity, also called curvature, is a problem. However, they will not tell you which factor is the one causing the nonlinear behavior.

Important Questions to Answer after Running the Experiment

1. **Is the empirical model a good fit to the data?**
In the preceding chapter we discussed goodness of fit when we analyzed a simple regression model. You have the same issue when you fit a multifactor model to experimental data. As part of the ANOVA, the significance of lack of fit will be calculated if there are sufficient degrees of freedom available. If either curvature or lack of fit is highly significant, you will have an inadequate model and may need to add higher-order terms to make it a better predictor of performance across the experiment range.

2. **Which control factors are statistically significant?**
This is derived from the results of the ANOVA. Factors having a p-value below the chosen alpha error threshold are statistically significant. They are candidates to be included in the model. Remember, the p-value is the probability that you will get a certain test

statistic, if the null hypothesis is true. The null asserts that, compared to the residuals, the model does not explain enough of the variability to be significant.

Remember that statistical significance is not absolute but depends on the size of the response, the number of samples or replicates, and the level of common cause noise in the process. If a factor is statistically significant, it is not necessarily practically important.

3. **Which control factors are practically important?**

While the assessment of statistical significance is straightforward, there are a few subtleties in judging the practical importance. You need to think of practical importance in the context of activities that are important in product development, such as

a. Identifying the control factors that are important in determining nominal performance and choosing the design set point for each control factor

b. Identifying the control factors that can reduce variation and choosing set points for them that reduce the effect of noise on performance

c. Identifying the factors that are likely to degrade performance during your product's useful life

It is logical to think that any control factor having a large effect on nominal performance would be a candidate for causing changes in performance if it varies. This is true. However, it is important to understand the potential for variation. If a given factor is chosen, and found to be stable after its initial setting with a high manufacturing capability, it will be less of a problem in production than a factor with high sensitivity and low manufacturing capability.

4. **How much uncertainty is there in the predictive ability of the model?**

After fitting a model to the data, you are usually interested in using the model to estimate performance across a range of variations in the control factors. There is some level of uncertainty in the predictions from the model, depending on the amount of noise present when the experiment is run, the number of replicates, and the goodness of fit for the model. Most good DOE applications generate estimates of the prediction variance that can be used to understand this.

5. **Is your system sufficiently robust against the noises that will affect it?**

The answer to this question can be derived from the results of the experiment, if you have included noises in the model. If noises have not been explicitly included, you can subject the system to a stress test, also known as a "ruggedness test." Set all control factors at their nominal values. For the experiment, you subject the system to an array of different noises using a Resolution III screening experiment, such as a Plackett-Burman design. The analysis of the data can tell you how large are the effects of the noises. If a stress test shows that the noises have little effect on the response, you have reason to celebrate.

6. **Can the desired performance goals be achieved with the system design, or are improvements required?**

This is the big question you have to answer as you develop a new product or system. If the robustness experiment shows that your system is sensitive to noise and difficult to desensitize, you should consider that to be a shot across the bow. If you cannot make the system robust, you should seriously consider rethinking your design concept. Moving ahead with a design that is very sensitive invites ongoing problems in production and exposes you to the risks associated with unhappy customers.

Some Important Considerations

Purpose of an Experiment

In classical DOE, we can put experiments into two distinct categories: screening experiments and modeling experiments.

Screening Experiments

The purpose of a screening experiment is to identify the more important control factors and noise factors. Often, before the experiment is run, you do not know the more important control factors with certainty. A knowledgeable team might develop a long list of candidate factors during a brainstorming session. However, the list is almost always longer than you have either the time or money to include in your modeling experiment. The good news is that many of the candidate factors will not be important enough to worry about. So your challenge in the screening experiment is to eliminate those factors that are not a threat to on-target performance. This will enable your modeling experiment to be more manageable and able to be completed within a reasonable budget and time period. The screening experiment is an efficient way to do this. We usually choose a low-resolution experiment that gives results that can be used to identify the noise and control factors that have the largest effect on system performance.

Modeling Experiments

The modeling experiment is used to develop the $y = f(x)$ relationships between the control factors and the responses. If you intend to use the empirical model to improve robustness, the model must be a good representation of the system and include both control and noise factors.

Some experiment designs are not suitable for modeling, including these:

- Plackett-Burman designs are Resolution III experiments having main effects and two-way interactions aliased. They are useful for screening, but they are not good for modeling if there are important two-way interactions.
- Fractional factorial Resolution III designs have the same drawbacks as Plackett-Burman designs.
- Two-level factorial experiments should not be used for modeling systems that have significant curvature in their response. You can begin with a two-level experiment, but it should be augmented if curvature is significant.

Your goal for a modeling experiment is to define the transfer function within the least time and expense. This means that an efficient experiment strategy should be selected.

Type of Model Required

We can classify empirical models in two broad categories: linear and nonlinear. The importance is that it dictates the type of experiment that must be run in order to build a useful empirical model. Linear models generally require fewer runs since there are fewer unknown coefficients to determine. Nonlinear models, which can have second-order and higher terms, can represent processes having quadratic and higher-order transfer functions. The other distinguishing charac-teristic is the absence or presence of interactions that are basically cross-product terms involv-ing the control factors and, if included, the noise factors.

Linear

Linear systems can be modeled with two-level factorial experiments. The number of runs in a factorial experiment depends on the number of control factors that are included and how many interactions you are trying to identify. An example of a transfer function that could be derived from a factorial experiment is

$$y(x) = c_0 + c_1 x_1 + c_2 x_2 + c_{12} x_1 x_2 \tag{17.1}$$

There are four coefficients in this transfer function. The constant c_0 is the grand mean of all the observations in the experiment. The term with the cross-product $x_1 x_2$ is the two-way interaction. You can look at some plots that will help you understand what a simple linear transfer function looks like and the effect of the interaction term.

Nonlinear

We use *nonlinear* in the sense of a transfer function having higher-order terms rather than in the statistically rigorous sense that we discussed when describing linear and nonlinear regression. Experiments with nonlinear systems require more than two levels for any factors having a nonlin-ear relationship with the functional response. While there are other experiment designs that can be used to derive a nonlinear transfer function, the "response surface" designs are probably the most useful. They have been applied to many types of problems. An example of a transfer func-tion derived from a response surface experiment is

$$y(x) = c_0 + c_1 x_1 + c_2 x_2 + c_{12} x_1 x_2 + c_{11} x_1^2 + c_{22} x_2^2 \tag{17.2}$$

In this transfer function, there are six coefficients. The first four describe the linear and interaction terms, while the two squared terms cause curvature of the response surface. The two commonly used response surface designs are the central composite design (CCD) and the Box-Behnken. Either design can produce a transfer function with all first-order terms, all squared terms, and all two-way interactions.

Principle of Parsimony

The principle of parsimony[4] is a rule of thumb that holds that the simplest explanation for a phe-nomenon is the best. While there are exceptions, in experiment design it's a blessing because it can simplify test plans, allowing the use of smaller experiments.

The implication of the principle of parsimony for experiment design is that most systems are well described by models having two-way interactions, at most. Usually, interactions that involve three or more control factors are not important. Of course, you can find exceptions to the rule,

---

4. Also known as Occam's razor, named for William of Ockham, a thirteenth-century Franciscan friar.

especially in chemistry and life sciences, but it is true in many situations. This means that full factorial experiments are seldom needed to model linear systems with numerous factors. For nonlinear systems, the result is that response surface designs have become the workhorse, since they can provide transfer functions that include all two-way interactions in addition to the quadratic terms.

The applicability of parsimony to the behavior of physical systems is based only on experience. There is no fundamental law that makes it so. The principle of parsimony is supported by empirical observations when you are trying to uncover nature's secrets. It is a reason to embrace the sequential approach to experimentation. In the long run it will save you time and money.

Aliasing

When you want to estimate the coefficients in an empirical model, you need data in the form of the response measured for each of the treatments required by the experiment design. For a given number of unique treatments there will be limitations on how many model coefficients can be estimated from the analysis of the data.

Aliasing means that some effects cannot be separated from others when the data are analyzed. This is a mathematical outcome of the selected experiment design. The alias structure of an experiment shows how factor effects are mixed together. If main effects are aliased with three-way and higher interactions, thanks to parsimony you may not have a problem. However, main effects aliased with two-way interactions could be trouble, suggesting the choice of a different test design. While some experiment designs can have complex alias structures, most computer applications for DOE can give the alias structure for a given design.

Experiment resolution is a useful mnemonic for understanding alias structure at a basic level.

Resolution Required for Modeling

When planning a modeling experiment, you need to think about what level of interactions you plan to identify. You can use resolution as a concept to guide your choice of the design for the experiment. If two-way interactions are important, Resolution V experiments are probably a safe bet. The two-way effects are aliased with three-ways. Parsimony suggests that important three-way interactions are rare. Although some alias structures are very complex, these are the important aliases that occur:

Resolution	**Alias Structure**
III	Main effects are aliased with two-way interactions.
IV	Main effects are aliased with three-way interactions; two-way interactions are aliased with other two-way interactions.
V	Main effects are aliased with four-way interactions; two-ways are aliased with three-ways.
VI	Main effects are aliased with five-way interactions; two-ways are aliased with four-ways; three-ways are aliased with other three-ways.

Table 17.1 shows the resolution of various factorial experiments and the reason why modeling experiments require more runs than screening experiments.

Some applications offer more efficient alternatives. Design-Expert, for example, has a selection of Minimum-Run Resolution IV and Resolution V designs. The Resolution V designs allow the estimation of main effects and two-way interactions aliased with three-ways. Just to compare, the Minimum-Run Resolution V for 8 factors requires 38 runs compared to the standard 2^{8-3} (one-eighth) fraction requiring 64 runs shown in Table 17.1. That is a significant savings.

Table 17.1 Experiment resolution versus number of factors and runs

		Number of Factors													
Number of Runs		**2**	**3**	**4**	**5**	**6**	**7**	**8**	**9**	**10**	**11**	**12**	**13**	**14**	**15**
	4	Full	III												
	8		Full	IV	III	III	III								
	16			Full	V	IV	IV	IV	III	III	III	III	III	III	III
	32				Full	VI	IV	IV	IV	IV	IV	IV	IV	IV	IV
	64					Full	VII	V	IV	IV	IV	IV	IV	IV	IV
	128						Full	VIII	VI	V	V	IV	IV	IV	IV
	256							Full	IX	VI	VI	VI	V	V	V
	512								Full	X	VII	VI	VI	VI	VI

Orthogonality

The concept of orthogonality is fundamental to DOE. If you consider the standard full and fractional factorial designs, note that the vector dot product of any two different columns is identically zero. This means that they are orthogonal vectors.

Orthogonality and independence of the factors enable you to separate the effects of each factor in the experiment and to do it efficiently with a minimum number of runs. There are experiment designs that are not orthogonal. We discuss these when we cover optimal experiments, which are often used when there are constraints on the experiment space that may prevent some factor levels from covering the full range, resulting in non-orthogonality.

Types of Experiments

Factorial Experiments

Factorial experiments are the simplest type of experiments. In our discussion, because they have broad applications, we concentrate on two-level factorial experiments. There are other factorial variants, including mixed-level designs, but two-level factorials are used most often, by far. They are appropriate for any system having a transfer function with only linear and interaction terms.

Full Factorial Designs

Full factorial experiments can be very large, depending on the number of factors. Thanks to parsimony, they are usually unnecessary. You can calculate the number of runs required for a full factorial experiment as follows:

$$N = l^k \tag{17.3}$$

where
l = number of levels for the factors
k = number of factors.

So every time you add a factor to a two-level full factorial experiment, you double the number of runs. If you want to model curvature, you can use a three-level factorial design. However, the number of runs gets even larger, faster, with three-level factorials. Every factor added to a three-level factorial increases the number of runs by a factor of three. For this reason, again because of parsimony, three-level factorials are seldom used. Response surface methods are the tool of choice for systems where the response surface has curvature, requiring a higher-order model.

Fractional Factorial Designs

Among the factorial designs, fractional factorials are used the most, since it's almost never necessary to identify all interactions, especially as the number of factors increases. Using equation (17.4), you can calculate the number of runs for a two-level fractional factorial experiment.

$$N = 2^{k-p} \tag{17.4}$$

The additional constant in equation (17.4), p, is the size of the fraction used. So $p = 1$ indicates a half fraction, $p = 2$ a quarter fraction, and so on. If there are no interactions in a model, the smallest fraction will suffice and the response is a linear combination of all the main effects. For example, if you have seven factors, a full factorial would require 128 runs. However with no interactions, you could build the empirical model with only 8 runs (1/16 fraction). The transfer function would consist of a constant term plus the sum of seven additional terms due to the control factors, assuming that all main effects were important.

Plackett-Burman Designs

The Plackett-Burman experiments are a family of factorial designs derived from Hadamard matrices. They are very useful as screening experiments but not for modeling, since they are Resolution III. They have very complex alias structures, and each of the main effects is aliased with fractions of many different two-way and higher-order interactions.

Response Surface Methods (RSM)

The response surface experiment is the most widely used design for modeling systems or processes with transfer functions having higher-order terms. The utility of the response surface design is a direct result of Taylor's theorem. Taylor's theorem holds that a polynomial can approximate any continuously differentiable function over a sufficiently small region. So no matter how complicated the function describing a process, a polynomial is a good approximation if that approximation is used over a small enough design space.

Response surface experiments come in two flavors: central composite designs (CCDs) and Box-Behnken designs. Using the sequential approach, you can evolve a fractional factorial experiment to a CCD. However, if the Box-Behnken design is used, it must be the chosen design from the beginning. It will not result from following the sequential approach.

The CCD and Box-Behnken designs differ in point selection. As shown in Figure 17.1, the CCD has points at the corners of the cube defining the "area of interest" as well as axial points. The area of interest is usually taken to be the volume within the cube. It is the candidate space for the set points for the control factors. On the other hand, the Box-Behnken design has points on each edge of the cube, but no axial points. The model derived from the CCD is a better predictor than the Box-Behnken at the corners of the cube. This is not

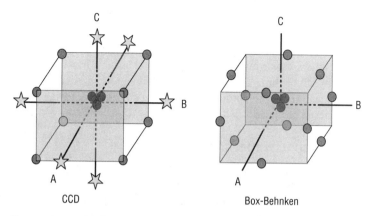

C

C

B

B

A

A

CCD

Box-Behnken

Figure 17.1   Comparison of CCD and Box-Behnken designs

surprising, considering the location of points for each design. The advantage of the Box-Behnken is that none of the required factor levels exceeds the bounds of the cube. The star points of the CCD, however, are outside the cube. This means that you must achieve higher-level settings relative to the area of interest and might be more likely to risk running up against maximum allowable operating levels.

Both the CCD and the Box-Behnken designs are able to provide full quadratic models as well as all two-way interactions.

Computer-Generated Optimal Experiments

Suppose that you need a model to improve the robustness of a system with numerous factors. It could be either a linear system, which can be modeled using a two-level factorial experiment, or one with curvature, best modeled by a response surface design such as a CCD or Box-Behnken design. Selecting an experiment design, such as a full factorial, one of the standard fractions, or a standard RSM design, will lock you into a set number of runs and treatments. The advantage when selecting a standard design is that you can be confident that the experiment is balanced and orthogonal, two characteristics determined by how the experiment design is generated. However, if you have reason to believe that there will be a limited number of important terms in the model, and you have some idea what they are, a standard design may have more runs than are actually needed. You may also have a constrained design space where, because of equipment limitations or safety concerns, there are constraints on some factor level settings, and you are not able to run all of the treatments specified in the standard design. In that case, an optimal experiment can be an excellent choice for the experiment design.

Optimal designs, also known as algorithmic designs, "bend the rules" on balance and orthogonality because they are generated with a specific objective in mind. Two commonly used optimal designs are the D-optimal and the I-optimal.

When creating a D-optimal design, the goal is to minimize the covariance of the model coefficients.

The objective of the I-optimal is the minimization of the average variance of the predicted response.

To create an optimal design, you start with the "deluxe" experiment. For example, if you are planning an optimal factorial experiment, you start with the full factorial, which is balanced and orthogonal. After you choose which terms are to be in the model and specify any constraints on the design space, the software will select a subset of all the possible treatments using the appropriate optimality algorithm. The resulting experiment will usually be smaller than the "deluxe" version and will meet the optimality criterion of minimum variance for the number of treatments included in the design.

Sequential Approach to Experimentation

The sequential approach to experimentation takes advantage of the implications of the principle of parsimony. Rather than assume that a system model has the highest level of complexity, you assume the opposite, until proven otherwise. The sequential approach to building an experiment design usually starts with a fractional factorial experiment that has center points. If the center points show that nonlinearity or curvature is not significant, there is no need to identify nonlinear terms in the transfer function. The question then becomes whether the factorial experiment has sufficient resolution to identify the significant interactions. You should have adequate resolution if the experiment is Resolution V. Again, the principle of parsimony suggests that three-way interactions are rare. Assuming that curvature is not significant, a linear model with two-way interactions is likely to be adequate.

The steps in the sequential approach are as follows:

1. Develop a list of candidate control factors and noise factors.
2. Run screening experiments, if needed, to identify the more important factors.
3. Run a factorial experiment with center points to see if curvature is significant.
4. If curvature is not significant, modify the experiment to model the curvature. Your concerns will focus on whether interactions are important.
5. If you see significant curvature, add treatments and additional runs to model the curvature. This will involve upgrading the factorial experiment to an RSM design, such as CCD. Then you can identify linear and second-order terms as well as two-way interactions.

You have to use some common sense when deciding when to follow the sequential process. With few factors and easily run experiments, the "just do it" approach can develop a model without "busting the budget." For example, with three factors there is limited value in hand-wringing over which experiment to do. In eight runs you can get all the model coefficients for a linear model. However, as the number of factors and the required resources increase, you need to consider seriously the sequential approach, starting with the appropriate screening experiments. It is difficult to give hard and fast rules, so each situation deserves thought.

A Simple Example

Suppose you are interested in optimizing a machining process.[5] The control factors are feed, speed, and depth of cut. There are two responses. The first response, "Delta," is the difference

5. Example derived from Patrick J. Whitcomb and Mark J. Anderson, "Robust Design—Reducing Transmitted Variation," 50th Annual Quality Congress (1996).

between the target dimension for the part and the measured dimension after machining. The second response, "Productivity," is derived from the product of the three control factors. It is proportional to the volumetric rate of the removal of material. We use it as a measure of the productivity of the machining process.

We have decided to follow the sequential process and have selected a factorial experiment as the starting point. To determine if curvature (nonlinearity) is significant, we add center points to the design. Table 17.2 shows the experiment design in standard order and the response data taken.

Table 17.3 shows the analysis after running the experiment. Looking at the ANOVA, you can see that the model and all of its coefficients are significant. You also see that curvature is significant, since it also has a p-value of less than 0.05. This means that a linear model with only main effects and two-way interactions is insufficient. So you need to augment your two-level factorial experiment by adding the axial points required to make the factorial experiment a CCD. This will include the squared terms in the model.

We can look at response plots for the data to get some insight into what the linear response surface looks like. Figure 17.2 shows the linear response surface.

The response surface of Figure 17.2 is characteristic of systems with significant main effects and interactions. If the system had no interactions, the response planes would be flat with no twist, and the slope along any axis would indicate the strength of the contribution of the particular factor (main effect). You can see that the response surface is twisted because of the AB interaction. Note also that lines parallel to either the A- or B-axis are straight. This is because the model we have chosen has no terms to model curvature. If you travel in any direction other than parallel to an axis, the plane will be curved.

Table 17.2 Full factorial experiment and response data for lathe machining experiment

Std	Run	Block	A: Feed	B: Speed	C: Depth	Delta	Prod.
1	6	Block 1	0.010	330.00	0.050	−0.403	0.165
2	1	Block 1	0.022	330.00	0.050	−0.557	0.363
3	7	Block 1	0.010	700.00	0.050	−0.549	0.350
4	8	Block 1	0.022	700.00	0.050	0.508	0.770
5	5	Block 1	0.010	330.00	0.100	0.415	0.330
6	3	Block 1	0.022	330.00	0.100	−0.225	0.726
7	2	Block 1	0.010	700.00	0.100	0.125	0.700
8	4	Block 1	0.022	700.00	0.100	0.525	1.540
9	9	Block 1	0.016	515.00	0.075	−0.132	0.618
10	10	Block 1	0.016	515.00	0.075	−0.193	0.618
11	11	Block 1	0.016	515.00	0.075	−0.081	0.618
12	12	Block 1	0.016	515.00	0.075	−0.052	0.618
13	13	Block 1	0.016	515.00	0.075	−0.117	0.618

Table 17.3 Response 1 Delta ANOVA for two-level factorial model analysis of variance table

Source	Sum of Squares	df	Mean Square	F Value	p-value Prob > F
Model	1.54	6	0.26	84.62	**< 0.0001** *significant*
A: Feed	0.055	1	0.055	18.11	0.0080
B: Speed	0.24	1	0.24	78.39	0.0003
C: Depth	0.42	1	0.42	139.69	< 0.0001
AB	0.63	1	0.63	208.88	< 0.0001
AC	0.16	1	0.16	53.99	0.0007
BC	0.026	1	0.026	8.65	0.0322
Curvature	0.028	1	0.028	9.14	**0.0293** *significant*
Residual	0.015	5	0.003032		
Lack of fit	0.00363	1	0.003638	1.26	0.3240 *not significant*
Pure error	0.012	4	0.002881		
Total	1.5812				

We return to this example in the next chapter to show how you can augment the experiment and use it to improve the robustness of the machining process.

Key Points

1. Many of the challenges in DOE are in the planning phase.
2. Most of the effort in planning an experiment involves having good responses to key questions before running the experiments.
3. The principle of parsimony can help you get more value from experiments where value is defined as (amount of useful information) divided by (required effort).
4. Never assume complexity unless there are compelling reasons. If in doubt, assume simplicity, until proven otherwise.
5. Use the sequential approach to experimentation unless past experience or first principles tell you to assume that a more complex model is required.
6. Consider using optimal experiment designs, especially when there are many factors remaining after the screening experiment or if the design space is constrained where certain factor level settings are not possible because of safety or equipment limitations.
7. Small or easily run experiments may not justify efforts to make them highly efficient. If planning time swamps run time, you may want to just "go for it."

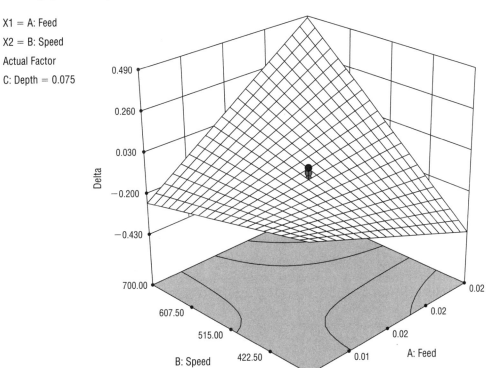

Delta

● Design points above predicted value

◐ Design points below predicted value

X1 = A: Feed

X2 = B: Speed

Actual Factor

C: Depth = 0.075

Figure 17.2 Response surface from factorial experiment for Delta versus Speed and Feed

8. Screening experiments can save a lot of time when there are many candidate control factors and noise factors, making it cost- or time-prohibitive to include them all in the modeling experiment.

Discussion Questions

1. Does your organization use designed experiments or a one-factor-at-a-time approach to experimentation?
2. Is the use of DOE encouraged to facilitate product development in your organization?
3. Within your organization, which groups make use of DOE: Research? Technology development? Manufacturing process development?
4. Describe both a successful and an unsuccessful experience with DOE. To what do you attribute the outcome in each case?
5. Should experiment runs always be randomized? Why or why not?
6. Describe two experiment techniques that can help protect against "lurking variables."

7. Describe desirable characteristics for the residuals when fitting an empirical model using DOE.

8. In what ways is an empirical model from a designed experiment a more realistic representation than a first-principles model?

9. When fitting a model to a data set, what tools can be used to assess a model's usefulness?

Developing System Robustness Using DOE

In this chapter we show how you can use traditional modeling experiments to improve system robustness. To do this you have to consider the effect of variations in the control factors and any external noise factors that you may have included in the experiment. Once you have a model that includes both control factors and noises, you can select favorable settings for the control factors to place performance on target with minimum variability.

Using System Models and Noise to Improve Robustness

First we cover the role of noises in your experiment strategy, building on the ideas presented in Chapter 7. To make a system more robust, you use noises to intentionally stress your system. This is normally done by including noises as factors in the robustness experiments.

At a high level, any system can be described by a block diagram, as shown in Figure 18.1. In the most general case, a system can have multiple inputs and outputs with a transfer function given in matrix form by

$$Y = f(X,Z) + \varepsilon \tag{18.1}$$

where

Y = a vector of responses
X = a vector of control factors
Z = a vector of controlled noises
ε = a vector of common cause noises

Assume that each y in the vector of responses is a polynomial function of x and z.

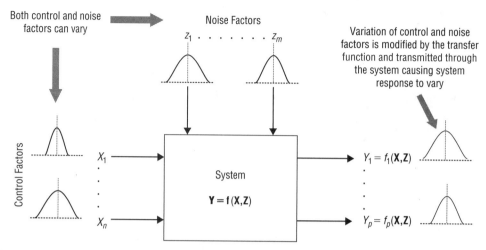

Figure 18.1 General system block diagram

There are many noise factors that can affect system performance. At a high level we can separate them into three fundamental groups:

1. **Within production processes:** These include all the factors that can make the performance of one product different from that of others. Batch-to-batch variations and those within batches can cause variations in parts, components, and materials. Factors in this group include contributors to manufacturing variability that can translate to variations in the product's control factors. These factors usually cause out-of-box performance variability. Noises within shipping, handling, storage, maintenance, and repair processes can also contribute to unit-to-unit variations.
2. **Internal to each product:** These are all of the factors within a product that contribute to performance variability, including deterioration that affects subsystem performance and negative interactions among subsystems in the product.
3. **External in the use of the product:** Factors of this type are stresses in customers' usage, stresses from environmental factors such as temperature and relative humidity, and stresses such as contamination and noises from electromagnetic fields, such as EMI/RFI.

When you build an empirical model, you have to keep in mind some important characteristics of noises:

1. A common cause noise, by definition, is assumed to be Gaussian and not subject to control. It is estimated using the data from the experiment.
2. Noises that you think can have an important influence on the response should not be ignored when planning the experiment. Otherwise they can be "lurking variables." If important, they will affect the response, but you will have no way to evaluate their functional relationship with the response. The result is that you will lose the ability to forecast their effect on performance and will be unable to make the system robust against them.
3. Where possible, important noises should be included as factors and controlled during the experiment. Historical data are good sources for estimates of the variability of each noise.

4. Every control factor is also a potential noise factor. Control factor variations will be transmitted by the system and cause variations in the system output response. The more significant and important a control factor is, the more threatening are its variations to system performance. Control factors with high significance and importance, combined with poor manufacturing capabilities, can jeopardize on-target performance.

5. If it is not possible to control an important noise, it can be included in the experiment as a covariate. You can use the General Linear Model (GLM) to analyze the data and include the noise in the empirical model.

What to Do about Noises

In Chapter 7 we discussed options for dealing with noises. Here we just recap the highlights and suggest a review of the original material for additional detail. We have several options to deal with noise or stresses that can affect a system:

1. Eliminate or control the sources of the stress.
   * Can be very expensive.
   * Can limit the application of the product.
   * Need to understand costs versus benefits before choosing this option.
2. Inspect the product prior to shipment.
   * Can be very expensive.
   * Is not a guaranteed 100% solution.
3. Depend on additional service, preventive maintenance, or customer support to compensate for manufacturing variability.
   * Can be expensive.
   * Increases service frequency and cost.
4. Depend on feedback from use in customers' applications to define the "real" problems that need to be corrected.
   * Will catch high-frequency, high-impact problems, but not until significant consequences have been endured in the field.
   * Product failure is necessary to generate data.
   * Delays availability of a reliable product.
   * Solution implementation may require expensive field upgrades
5. Incorporate feedback or feed-forward control systems to compensate for the consequences that stresses have on the functional performance.
   * Adds cost.
   * Must understand additional product cost versus reduction in service cost.
   * Additional components can be sources of failures.
6. Incorporate redundant features or increased safety margins.
   * Adds cost.
   * Added features can be another source of failure.
7. Optimize the design by developing it to be robust.
   * Lower overall cost.
   * Higher customer satisfaction sooner.
   * Increased competitive advantage.

Before deciding what to do about noise, you need to understand its effects on the system. If, after running a noise screening experiment, you find that the system is insensitive to noise, as long as you have not overlooked any important noises, you may not need to reduce system sensitivity. If this happens, the planets must be aligned for great things to happen and it might be time to purchase a lottery ticket!

Often you have no control over some of the noises, such as the environment and customers' usage. So noise elimination in actual use may not be an option. Use of noise barriers entails expense and additional complexity and should be considered a last resort. Usually when designing a new product or system, teams are not lucky enough for their designs to be unaffected. Their best choice can be to reduce the vulnerability of their designs to noises. Sometimes this can be accomplished by choosing set points for control factors that minimize the response to noise. We will look at some examples that illustrate this.

How Noises Affect Responses and What Can Be Done about It

Consider a general system having both control factors and external noise factors. We assume that the system transfer function can be approximated by a polynomial function of all factors. Knowing that reduced variability means increased customer satisfaction, we are interested in selecting an operating point for the design parameters that will align the system response with its requirement and minimize the effect of noise on the variation of that response.

In order to make a system robust against internal variations or external noises, it is necessary to have interactions between control and noise factors. For example, consider the following two transfer functions where x_1 and x_2 are control factors and z_3 is a noise that was included as a factor in the experiment:

$$y_1 = c_0 + c_1x_1 + c_2x_2 + b_3z_3 \tag{18.2}$$

$$y_2 = c_0 + c_1x_1 + c_2x_2 + c_{12}x_1x_2 + b_{23}x_2z_3 \tag{18.3}$$

There is a big difference between these two transfer functions in the potential for making the system robust against the external noise z_3. For the system having the transfer function described by equation (18.2), there is no opportunity to reduce sensitivity to noise factor z_3 by selecting a particular setting for either of the control factors x_1 or x_2. This is because there are no interactions between the control and noise factors. However, with the transfer function of equation (18.3), you could reduce the effect of noise z_3 on the response by choosing an appropriate level for x_2. For example, if you set $x_2 = 0$, you can cancel the effect of z_3. So z_3, which is an external noise, could vary over a wide range without affecting the system. You could also make the system less sensitive to internal variations since there is an interaction between x_1 and x_2. This would leave x_1 as a tuning factor to put the mean on target. This concept is fundamental to improving robustness by using favorable settings for control factors. Often, interactions can be problematic, making their elimination desirable. However, this is one situation where interactions can help.

Minimizing Error Transmission to Improve Robustness

For the general system transfer function given in equation (18.1), the standard deviation of the response is given by equation (18.4). It expresses the standard deviation of the response as a function of the standard deviations for the control and noise factors.

$$\sigma_y = \sqrt{\sum_{i=1}^{n}\left(\frac{\partial y}{\partial x_i}\sigma_{x_i}\right)^2 + \sum_{i=1}^{m}\left(\frac{\partial y}{\partial z_i}\sigma_{z_i}\right)^2 + \sigma_{error}^2} \tag{18.4}$$

In equation (18.4) y is the response and x_i and z_i are control factors and controllable noise factors, respectively, with their standard deviations. The first two terms account for the variations in the response caused by variations in the control factors and controllable noise factors. The last term is common cause noise. The objective is to search for an operating point that is on a plateau rather than on a steep slope. In three dimensions you can visualize a response "mountain." You want to avoid pitching your tent on the edge of a cliff.

Once you have the empirical model for the system, the computer can use equation (18.4) to generate a response function for the variance. Using an optimization tool, you can co-optimize response and variance. To do this you have to include the important external noise factors in the experiment. Those that are important can be identified by a noise screening experiment.

The expression for the standard deviation of the response due to factor variation is commonly called the "error transmission."[1] In Design-Expert it is called the "propagation of error" (POE).

Next we discuss the application of optimization to our example machining problem, by minimizing error transmission. First we improve the model by adding axial points that will enable modeling curvature.

Robustness Optimization of the Machining Process

Augmenting the Model for the Machining Example

Let's return to our machining example where we started to build an empirical model to predict machining accuracy as a function of speed, feed, and depth of cut.

When we analyze the factorial experiment with center points, we find that curvature is significant. Following the sequential process for experimentation, the next logical step is to augment the factorial experiment with axial points to enable a quadratic model. Let's assume that we run the factorial on the first day, analyze the results, and run the CCD on the second day. If this is the case, it is a good idea to put the additional axial points in a second block. That way, if there is an effect of having run the experiment on two different days, we have covered ourselves by blocking. The "second-day" effect will be removed from the noise, increasing the power of the experiment.

Figure 18.2 shows the additional axial points that have been added to the original factorial experiment. We now have a CCD capable of giving us a full quadratic model. Assuming that the three-way interaction is probably not important, we should emerge with a useful model that we can use to optimize performance.

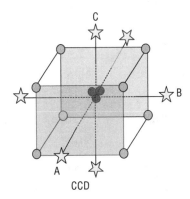

Figure 18.2   Factorial experiment augmented with axial points to make it a CCD

---

1. Raymond H. Myers and Douglas C. Montgomery, *Response Surface Methodology* (New York: John Wiley, 2002).

Table 18.1 shows the original factorial experiment, with treatments designated as Block 1. The additional treatments for the axial or star points are Block 2. With the addition of the axial points we now have 19 runs versus 13 for the factorial experiment with center points.

The analysis of the results, with the axial points added, shows that we have a better model. Table 18.2 shows the ANOVA table. By adding the axial points, we can model curvature and it no longer appears in the ANOVA table. The lack of fit is not significant.

We have chosen a threshold of 0.05 for α error. Examination of the ANOVA table shows that the linear terms are all significant as are the two-way interactions. The only significant squared term is A^2. Since the B^2 and C^2 terms have p-values > 0.05, they are not significant and should not be included in the model.

If we remove the nonsignificant squared terms, we get the model shown here for uncoded factors:

Table 18.1   CCD experiment and response data for lathe machining experiment

Std	Run	Block	A: Feed	B: Speed	C: Depth	Delta	Prod.
1	6	Block 1	0.010	330.00	0.050	−0.403	0.165
2	1	Block 1	0.022	330.00	0.050	−0.557	0.363
3	7	Block 1	0.010	700.00	0.050	−0.549	0.350
4	8	Block 1	0.022	700.00	0.050	0.508	0.770
5	5	Block 1	0.010	330.00	0.100	0.415	0.330
6	3	Block 1	0.022	330.00	0.100	−0.225	0.726
7	2	Block 1	0.010	700.00	0.100	0.125	0.700
8	4	Block 1	0.022	700.00	0.100	0.525	1.540
9	9	Block 1	0.016	515.00	0.075	−0.132	0.618
10	10	Block 1	0.016	515.00	0.075	−0.193	0.618
11	11	Block 1	0.016	515.00	0.075	−0.081	0.618
12	12	Block 1	0.016	515.00	0.075	−0.052	0.618
13	13	Block 1	0.016	515.00	0.075	−0.117	0.618
14	19	Block 2	0.006	515.00	0.075	−0.023	0.228
15	17	Block 2	0.026	515.00	0.075	0.213	1.008
16	16	Block 2	0.016	203.87	0.075	−0.378	0.245
17	18	Block 2	0.016	826.13	0.075	0.166	0.991
18	14	Block 2	0.016	515.00	0.033	−0.376	0.272
19	15	Block 2	0.016	515.00	0.117	0.254	0.964

Table 18.2 Response 1 Delta ANOVA for CCD quadratic model analysis of variance table

Source	Sum of Squares	df	Mean Square	F Value	p-value Prob > F
Block	4.326E-003	1	0.00433		
Model	1.98	9	0.22	79.46	**<0.001** **significant**
A-Feed	0.082	1	0.082	29.74	0.0006
B-Speed	0.39	1	0.39	139.25	<0.0001
C-Depth	0.62	1	0.62	222.56	<0.0001
AB	0.63	1	0.63	228.88	<0.0001
AC	0.16	1	0.16	59.16	<0.0001
BC	0.026	1	0.026	9.48	0.0151
A^2	0.070	1	0.070	25.20	0.0010
B^2	9.542E-005	1	0.000095	0.034	0.8573
C^2	4.414E-003	1	0.00441	1.59	0.2422
Residual	0.022	8	0.00277		
Lack of fit	0.011	4	0.00253	0.92	**0.5307** **not significant**
Pure error	0.012	4	0.00288		
Total	2.01	18			

$$Delta = -0.441 - 108.6A - .0022B + 30.12C + 0.253AB - 952.5AC$$
$$- 0.012BC + 1950A^2 \tag{18.5}$$

We will use this model to make the machining process robust against the sources of internal variation.

Using the System Model to Optimize Robustness

Equation (18.5) can now be used to optimize performance. We have several responses that we can co-optimize, including

- Delta
- Error transmission for Delta
- Productivity
- Error transmission for Productivity

Let's look at the response surface with the addition of terms for curvature. Figure 18.3 shows the response surface for the augmented factorial experiment. By adding the axial points, we can model the curvature. If we compare this response surface to the one shown in Figure 17.2 in the previous chapter, we see strong similarities. The response surface remains twisted and inclined in the direction of the A- and B-axes. The main difference is that elements of the plane parallel to the A-axis are now curved instead of straight. This is caused by the addition of the A^2 term to the model. Since there is no significant B^2 contribution, lines in the response surface parallel to the B-axis remain straight.

Before starting the optimization process, you have to choose the optimization constraints. For the control factors of speed, feed, and depth in the lathe example, we can allow any values in the range within the experiment cube. Our goal is to put Delta on target (Delta = 0), minimize POE (Delta), and maximize Productivity.

When deciding what constraints to choose and how to weight them, you need to remember that to improve on one objective probably means giving up some performance on another. It is a process of making trade-offs. It is not possible to maximize or minimize all objectives simultaneously. Table 18.3 shows the optimization constraints selected for our problem.

Desirability, which can range from 0 to 1, is a measure of how well we have maximized the objective function. It is derived using the optimization constraints, weightings, and assigned importance.

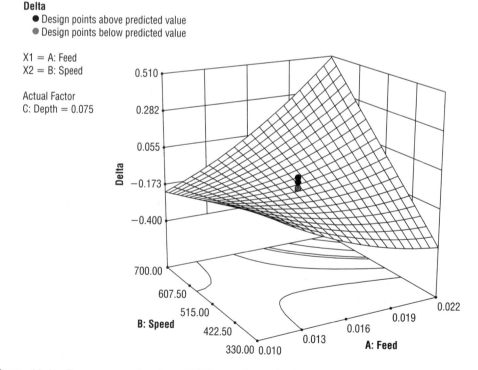

Figure 18.3   Response surface from CCD experiment for Delta versus Speed and Feed

Table 18.3 Constraints for POE optimization in Design-Expert

Name	Goal	Lower Limit	Upper Limit	Lower Weight	Upper Weight	Importance
Feed	is in range	0.01	0.022	1	1	3
Speed	is in range	330	700	1	1	3
Depth	is in range	0.05	0.1	1	1	3
Delta	is target = 0	−0.5566	0.525	1	1	3
POE (Delta)	minimize	0.0918393	0.325637	1	1	3
Productivity	maximize	0.165	1.54	1	1	3

Table 18.4 Solutions for POE optimization in Design-Expert

Number	Feed	Speed	Depth	Delta	POE (Delta)	Productivity	Desirability
1	0.022	457.38	0.100	0.000	0.069	1.006	0.849
2	0.022	458.38	0.099	−0.000	0.068	1.002	0.848
3	0.022	458.78	0.099	0.000	0.068	1.001	0.847
4	0.022	459.14	0.099	0.000	0.068	0.999	0.847
5	0.022	457.17	0.100	0.000	0.070	0.998	0.846
6	0.022	457.09	0.100	0.000	0.071	0.995	0.845
7	0.022	468.79	0.100	0.024	0.067	1.031	0.844
8	0.021	456.39	0.100	0.000	0.074	0.977	0.839
9	0.022	466.89	0.100	0.020	0.070	1.009	0.839
10	0.022	466.92	0.094	0.000	0.070	0.966	0.835

In the optimization we have put Delta on target, while minimizing the variability of Delta due to the control factor variation. In addition, we have selected an operating point that will give us improved productivity.

After completing the optimization, Design-Expert will offer a number of alternative solutions in order of decending desirability, as shown in Table 18.4. It is up to the development team to decide which is best. In the optimization constraints we have weighted the objectives Delta, POE (Delta), and Productivity equally.

The error transmission function, given by equation (18.4), describes the response surface for the standard deviation of the response we are optimizing. In order to generate this function, we have to specify the standard deviation for each control factor and controllable noise factor. This requires knowledge of process capability for each of the control factor settings, as well as some understanding of the expected range of variation for any external noises included in the model.

We have seen how minimizing error transmission is a way to reduce the variability of system responses when impacted by either internal or external noise factors. Having the option of using the propagation of error analysis enhances Design-Expert as a tool for robustness development.

Additional Options for Optimization

If you do not have access to a DOE application that has the ability to minimize error transmission, there are other options. Each requires you to have the model for the system transfer function. If you intend to use an empirical model, you still need to plan, run, and analyze a designed experiment.

We can now describe an alternative method to enhance robustness and thereby reduce performance variability.

Special-Purpose Optimization Programs

Optimization Process

In order to optimize a process, you need several important ingredients:

- **Transfer function for the process:** As long as you have a transfer function that defines the relationship between important responses and control factors, you can perform an optimization. The transfer function could be the result of either a first-principles analysis or an empirical model derived from a designed experiment.
- **Important responses:** Important responses are outputs from the system that affect customers' satisfaction.
- **Understanding of constraints:** Examples of constraints might include limits on control factor values and relationships among control factors and between responses. There are many possible forms of constraints that might be specified.

Examples of Things You Might Want to Optimize

To optimize a single response, you may be interested in taking one of several approaches:

- Maximize a response.
- Minimize a response.
- Put a response on target.

Doing each of these is relatively straightforward. You can be guaranteed that in the case of the maximization and minimization you will find a solution. When you want to put a response on target, as long as the target you specify is within the range of the transfer function, you will also find a solution. Keep in mind that whenever you use an empirical model for optimization, you should limit the search to be within the experiment space. When optimization is based on an empirical model, using optimal values outside the experiment space is risky and should be avoided.

Objective Functions

Objective functions are necessary when you have to co-optimize multiple responses. For example, you might want to put one response on target and at the same time minimize its variation. This is what we chose to do in our lathe example. An objective function is some combination of the desired outputs you want to optimize. Depending on how you set up the objective function, you will either maximize or minimize it.

When using any optimization program, it will be necessary to formulate an objective function in order to co-optimize multiple responses. This raises the question of the weighting of the responses. While two responses may be of equal importance, they may have different magnitudes. So you need to consider this when formulating the objective function. Some of the responses you might choose to optimize include

- Mean response and standard deviation
- Response capability (Cpk)
- Upper or lower end of the confidence interval for the response and standard deviation

Lathe Example Revisited

Not all DOE applications have the optimization capabilities of Design-Expert. If you are working with an application that has limitations, an alternative is to use a stand-alone optimization program. You still need the system transfer function, developed using either first-principles analysis or a DOE application.

For our lathe example we start with the empirical model developed using the CCD experiment that came from augmenting the original factorial experiment. Figure 18.4 shows the system block diagram from VarTran,[2] a flexible and easy-to-use application capable of both tolerance design and response optimization.

Our three control factors and ranges are

A: Feed 0.010–0.022 in
B: Speed 330–700 fpm
C: Depth 0.050–0.10 in

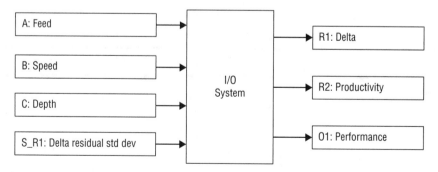

Figure 18.4   Block diagram of system from VarTran model

2. VarTran is a software product of Taylor Enterprises (www.variation.com) in Libertyville, IL.

An additional input to the model is the variation of the residuals for the response Delta, a measure of the common cause variation while performing the experiment.

The outputs of the process are

1. R1: Delta—the difference between the required machined dimension and the actual dimension.
2. R2: Productivity—derived response proportional to the rate of material removal during machining.
3. O1: Performance—an objective function that must be specified by the user. In this example we want to put the response, Delta, on a target of 0 and minimize its variation.
4. We are also interested in maximizing productivity.

There are many options that we could choose when formulating our objective function. For this case, choosing to maximize the capability for Delta would put Delta on target, satisfying our requirement. For Delta, the capability is defined as

$$Cpk = Min\left[\frac{(\mu - LSL)}{3\sigma}, \frac{(USL - \mu)}{3\sigma}\right]_{R1} \tag{18.6}$$

We need to specify values for LSL and USL. If we arbitrarily let $LSL = -0.1$ and $USL = 0.1$, the maximum value of Cpk will occur with $R1 = 0$ and a minimal value for σ.

VarTran allows easy calculations of capability. For our objective function we choose the product of capability for Delta ($R1$) and the mean of productivity ($R2$).

$$O1 = Cpk\mu_{R2} \tag{18.7}$$

So if the optimizer searches for values that maximize the objective function, both capability and productivity will be driven in the right direction. We should find an advantageous operating point that meets our performance goals. Table 18.5 shows the comparison of optimization results using Design-Expert and VarTran. The results for each are very close.

A caution to keep in mind when formulating an objective function is that there may not always be an obvious best choice. It takes some thought and in some cases a little experimentation to see what choice is most effective. When optimizing multiple responses, you should have performance requirements for each response, or at least minimum or maximum acceptable values for the responses of interest.

Table 18.5 Comparison of Design-Expert and VarTran optimization results

	VarTran	Design-Expert
Delta: R1		
Mean	0.000	0.000
Standard deviation	0.068	0.069
Productivity: R2		
Mean	1.004	1.006
Standard deviation	0.186	0.186

What Should You Optimize?

Let's consider a couple of alternatives for optimization. A traditional goal is to put the mean response on target while minimizing the standard deviation. This is a good objective, but is it the best? A better strategy may be to focus on Cpk. If we maximize Cpk, we will minimize defects per unit (DPU), the fraction of components or systems having characteristics outside the specification limits.

Reliability DOE

Another powerful application of DOE is in the optimization of the life of a component. In Chapter 21 we discuss the application of the log-likelihood function in determining the distribution parameters for a set of life data. An important assumption in the following approach is that component life is a function of a linear combination of the main effects and two-way interactions. In the life model, the log of the distribution scale factor for the ith treatment is assumed to be:

$$\ln(\eta_i) = c_0 + c_1 A_i + c_2 B_i + \cdots + c_{12} A_i B_i + \cdots \tag{18.8}$$

where A_i and B_i are control factor settings for the ith treatment and the c are unknown coefficients to be determined. For the reliability experiment, the life data can be a mix of failures and suspensions. The purpose of the experiment is to determine the design parameter values (control factor settings) that maximize time to failure. The parameters can be any component characteristic affecting time to failure. The process is as follows:

Select control factors—At this stage you need to rely on your technical understanding and use whatever historical data you have.

Design the experiment—Design a two-level Resolution V factorial experiment with the component characteristics as control factors and life as the response.

Run the experiment—It is not necessary to run all components to failure since suspensions can be included using the log-likelihood function.

Select a candidate life distribution—The candidate distributions most likely to be useful are exponential, Weibull, and lognormal.

Formulate the log-likelihood function—Equation (18.9) assumes a two-parameter Weibull for the life distribution, and a mix of failures and suspensions.

$$\ln L = \ln[f(x_i,\beta,\eta_i)] + \cdots \ln[f(x_m,\beta,\eta_m)] \\ + \ln[1 - F(y_j,\beta,\eta_j)] \cdots + \ln[1 - F(y_n,\beta,\eta_n)] \tag{18.9}$$

In equation (18.9) the x_i are the times to failure and the y_j are the times to suspension; β is the distribution shape factor and η_i is the distribution scale factor for the ith treatment.

Find the optimal values for the unknown control factor coefficients and shape factor—For a given life distribution, the optimal values for these parameters will maximize the log-likelihood function. Finding the optimal values is easily done using stochastic optimization.

Identify the important control factors—This is done using the likelihood ratio to understand the significance of each control factor.

Maximize the distribution scale factor—Reduce the life model to include only the significant control factors and find the control factor settings that maximize the distribution scale factor.
Choose the best-fit life distribution—Go through the process with several candidate life distributions. The best fit life distribution will maximize the log-likelihood function.

Use of the log-likelihood function is discussed more in Chapter 22 and will make more sense after reading about reliability functions and life distributions in Chapter 21. For many details, omitted here because of space constraints, see the references[3] below.

Key Points

1. Classical DOE can be used to improve the robustness of systems and processes.
2. Improving robustness using DOE can be a seamless process if noises are included when you build the empirical model.
3. If you plan to include noises in your empirical model, you have to be able to control them when running the experiment.
4. Every control factor is also a potential noise factor.
5. The importance of variation of a specific control factor depends on the relative size of its contribution to the response.
6. Interactions between control factors and noise factors are necessary to find control factor settings that will reduce the sensitivity to noise.
7. Improving robustness usually requires that you put performance on target and reduce its variability. However, sometimes the best performance is when some critical response is either maximized or minimized.
8. Multi-response optimization requires that you consider the relative importance of the responses.
9. When you must co-optimize multiple responses, you have to formulate an objective function.
10. Determining a good objective function requires thought and sometimes experimentation.

Discussion Questions

1. What are some metrics that can be used for optimizing robustness?
2. What methods have you used for optimizing the robustness of your products? How successful have you been?
3. What is co-optimization?
4. Is it possible to maximize a functional response and at the same time minimize its standard deviation?
5. Can you translate increased robustness into an increase in system MTBF?
6. What are some ways to reduce the effect of external noises or stresses on system performance? Which have been successful in your organization?
7. What is a way to strike a balance when two responses must be optimized?
8. Is it necessary to run a designed experiment to do performance optimization?

---

3. ReliaSoft Corporation, *Experiment Design and Analysis Reference*, ReliaSoft Publishing, Tucson, AZ, 2008, Wu and Hamada, *Experiments: Planning, Analysis, and Parameter Design Optimization.*

Robustness Optimization Using Taguchi's Methods

There are choices of tools to achieve robustness, including classical DOE, covered in Chapters 17 and 18, and Taguchi's methods. In the discussions of quality loss in Chapter 2 and of robustness development in Chapter 7 we introduced some of the concepts that are central to Taguchi's quality and robustness improvement philosophy. In this chapter we concentrate on Taguchi's use of designed experiments to improve system robustness.

Both Taguchi's methods and classical DOE have advantages and disadvantages. Unfortunately the preferences have been polarized. The debate over Taguchi versus classical DOE has been a battle between Taguchi advocates, who say it works, and statisticians, who say it's not always statistically rigorous. The truth is that they're both right. Most of the controversy centers on Taguchi's use of signal-to-noise ratio as a response, the treatment of interactions, the use of large outer noise arrays, and the use of orthogonal arrays as the experiment designs of choice.

Many users of Taguchi's methods have enjoyed great success in improving both products and processes. For an excellent book with a comprehensive treatment of Taguchi's methods, see Fowlkes and Creveling.[1] In this chapter we cover Taguchi's methods and show how they can help you achieve the goal of improving system or product robustness. We discuss some of the differences between classical DOE and Taguchi's methods.

Taguchi's Contributions to Robust Design

Dr. Genichi Taguchi has made many contributions to quality engineering. While he did not invent or discover all of the following, he actively and effectively promoted their use.

1. The principle that product designers and builders should work to put performance on target rather than just within specification, a notion embodied in the quality loss function

---

1. Fowlkes and Creveling, *Engineering Methods for Robust Product Design.*

Before Taguchi's influence, being within specification was considered to be good enough. This was a legacy of the quality inspection process, which judged compliance with specifications and either accepted or rejected a material, part, or assembly. This, being a binary decision, drove a view of quality as having two possible states, good or bad. Known as the "goalpost" approach to quality, just inside the spec was considered to be acceptable, while just outside was not acceptable.

Intuitively we know this to be wrong. If system performance is a continuous variable, quality is also a continuous variable, peaking at the point where performance is on target. Of course, this all depends on having a set of performance specifications closely coupled and aligned to the satisfaction of customers' needs.

2. The concepts of quality loss and the quality loss function

These concepts provide valuable insights. They recognize that quality is often in the eye of the beholder. Certain customers and market applications of a product are more sensitive to off-target performance than others. These more sensitive customers complain when product performance drifts off target even if it is within the specification.

It is arguable that quality loss is never zero since some customers complain about on-target performance. This is likely to happen when a product has been designed for a market with great diversity of customers' expectations. Since it is difficult to completely satisfy every customer in this situation, some complaints may arise from customers who may have made a buying decision without completely understanding the product's performance map.

A logical outcome of the quality loss function is that performance that is just outside the specification is not materially different from that which is just inside the specification.

Taguchi also broadly defined quality loss as the total loss to society that occurs when products have off-target performance. It includes the losses to the producer, to consumers, and to any others affected by off-target performance or outright failure of a product. This idea was ahead of its time. It resonates with today's ethos of sustainability that has growing influence in customers' buying decisions.

3. The principle that strong interactions among control factors should be avoided where possible

This principle is unarguable but unfortunately cannot always be followed. Some systems have interactions that are fundamental and not created by an unfortunate choice of responses or control factors, or by a design concept that is prone to interactions.

4. The recognition that an orthogonal array is an efficient tool that can be used to develop robustness

It also can be an inspection tool for evaluating the quality of a design. This is one application for which Taguchi has been criticized, especially by statisticians. Their position is that the understanding of interactions flows from a well-designed modeling experiment. Taguchi's position is that strong interactions will have a negative effect on product quality and should be minimized or, if possible, eliminated by using our knowledge of engineering science.

5. Exploiting interactions among control factors and noise factors to improve robustness against external noises

As you saw in Chapter 18, noise–control factor interactions are essential to increasing robustness against noises or stresses.

6. Exploiting nonlinearities in the system transfer function to find locations in the design space with lower slope, where the system response is less affected by variations in control factors

    This is the foundation of reducing error transmission, discussed in Chapter 18 when we used the propagation-of-error tool in Design-Expert to find advantageous control factor settings.

7. The use of stresses or noises as factors in designed experiments to find advantageous operating points that can reduce the variability of critical responses

    We discussed this in detail in Chapter 18. Noises that are ignored and not included in the experiment design can become "lurking variables," surfacing at the worst possible time to cause performance problems, often in front of customers.

8. The idea that it is essential to use and build upon our understanding of engineering science when planning experiments, rather than to treat the system as a black box that we know nothing about

    Taguchi believes that when planning and analyzing experiments, it is essential to leverage our subject matter expertise. Whether we are scientists or engineers, our knowledge of the system we are studying is a key element of success.

Building a Detailed Model versus Improving System Robustness

It is important to understand that when using Taguchi's methods the objective is to improve robustness and reliability, not to build a detailed empirical model. This is a distinction that is sometimes overlooked. The question then becomes whether it is possible to improve system robustness without having an empirical model that includes all significant factors and interaction terms. The answer is yes, as long as the missing terms are not important. In this context, importance is defined by the percentage of the total variability explained by a term in the empirical model. While a term in the empirical model may be statistically significant, it is not necessarily practically important.

Interactions

Interactions among Control Factors

Interactions among control factors drive a number of results that can make robustness optimization more difficult. What are the negative effects of interactions?

- Increased system complexity—never a good thing
- Increased variance of the response
- Increased complexity when using control factors to adjust performance
- Reduced manufacturing quality
- Reduced capabilities of field service to make adjustments efficiently
- Difficulties if future design changes are made without regard to the effects of interactions on performance—especially troublesome if your organization has a short institutional memory and is weak in documenting the history of design development

Interactions cannot always be eliminated. However, there are several ways that you can reduce their consequences:

- Choice of product concept
- Choice of control factors
- Choice of functional responses

In each of these activities you have opportunities to reduce and possibly eliminate some interactions.

In the design of electromechanical systems and the planning of experiments, you may be able to eliminate interactions of your own making. However, some processes such as chemical and life sciences have interactions present at a very basic level. In these situations, you probably cannot eliminate important interactions and must deal with them in your process of optimization and robustness improvement.

One of the criticisms statisticians level at Taguchi is that the process for treating interactions has a logical dilemma. Orthogonal arrays can include interactions if you make the appropriate column assignment for the control factors. Statisticians argue that you cannot tell which interactions are important if you have not yet run the experiment. Some of the criticism flows from the belief that an experiment is part of the route taken on a voyage of discovery, which is true in a research environment. In the early stages of developing a new technology you may lack the understanding required to identify all of the important interactions. However, as you move further downstream into the design process, your understanding grows. If your early modeling experiments in the research phase identify interactions, you can include them if you choose to use orthogonal arrays when developing robustness.

Interactions between Control and Noise Factors

The main reason to include noises in the experiment design is to find interactions between control and noise factors. As we discussed in Chapter 18, interactions between control factors and external noises are the "good" interactions. The existence of these interactions enables the selection of advantaged set points that can minimize sensitivity to external noises. The general strategy for dealing with interactions is to eliminate interactions among control factors, or reduce their effects, and take advantage of interactions between control and noise factors.

Additivity

The assumption of additivity is fundamental to the use of orthogonal arrays in developing design robustness. The implication of additivity is that the important system responses can be approximated by a linear combination of all the main effects. This does not mean that you assume that there are no interactions, only that they are not dominant and are less important than the main effects. If the interactions are strong, the assumption of additivity is probably not valid. It is then possible that, if interactions are not included in your robustness experiment, the confirmation experiment will not confirm. If you know that an interaction is important, you can include it in the experiment.

Inner and Outer Arrays and Compound Noise Factors

When Taguchi first published papers and articles describing his methods, he advocated the use of experiment designs with inner and outer arrays. The inner array specifies how the control factors are varied, while the outer array specifies the settings for each noise factor for each run. The use of large outer arrays results in larger experiments. This has been a subject of discussion and a focus of some of the criticism of his methods. Over time he began the use of compound noise factors, which can greatly reduce the total number of runs required. The cost of developing the compound noise factors is another experiment, called a "noise screening experiment." In a noise screening experiment, you set the system control factors at nominal levels and examine the effect of the external noises on the response. Using main effects plots, you then select how the individual noises should be set to drive the system response either high or low. This allows you to create two compound noise vectors that give you a way to stress the system and calculate the variance in the response, required to form the S/N. Figure 19.1 shows the system block diagram shown before in Chapter 18. By formulating a compound noise factor, you reduce the total number of noise factors to two noise vectors that essentially point in opposite directions. Of course, if there is only one important noise factor, you do not need to worry about compounding.

The result is a marked reduction in the number of required runs. For example, if you are planning to run an L9 inner array with seven noise factors, the outer array could be an L8, for a total of 72 runs. However, if you screened the noises using the L8 to develop the compound noise factors, you would need a total of only 26 runs, consisting of 8 runs for the noise screening experiment and 18 runs for the L9.

While compound noise factors can reduce the number of runs in the experiment, as with compound control factors, they also result in the loss of some information.[2]

Compound Control Factors

Compound control factors are used for two main reasons:

1. To reduce the number of control factors and the size of the experiment
2. To eliminate interactions

In the example we use to illustrate Taguchi's experimental approach, we show how to use a compound control factor. A concern, as mentioned previously, is the loss of information.

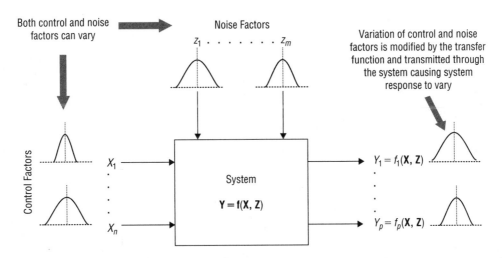

Figure 19.1   System having both control and noise factors.

---

2. C. F. Jeff Wu and Michael Hamada, *Experiments: Planning, Analysis, and Parameter Design Optimization* (New York: John Wiley, 2000).

Orthogonal Arrays

Taguchi's Robust Design methods employ orthogonal array experiments drawn from a category of designs known as Latin Squares. These test plans are efficient at evaluating the unknown effects of control parameters on system performance and at defining better set points for them. They require an investment by management in training to develop these capabilities in their organizations to get the most benefit from the approach. Also, managers need to be patient with the important test planning process, resisting the urge to demand premature testing as a sign of progress. The benefits are in the achievement of better solutions in less overall time.

An orthogonal array is a designed experiment that selects combinations of parameters in a particular order to provide information that is sufficient for the estimation of main effects by the analysis of means. The term *orthogonal* indicates that there is balance among the combinations of parameters so that no biases exist toward particular combinations. Also, the evaluation of the effect of each parameter on the output response is independent of the other parameters.

There are many alternative designs for orthogonal arrays. Figure 19.2 illustrates one that requires nine runs, hence the name L9. L stands for "Latin," a reference to the early origins of this type of statistically designed experiment. The L9 example evaluates the effects of four parameters, each with three alternative values intended to identify curvature in the design space. The selection of the most appropriate matrix design is a trade-off among the number of alternatives to be evaluated, the amount of information to be gathered, and the time and resources required for the experiment.

Many other orthogonal array designs have been cataloged for use. A few examples are

L8   Up to seven factors at two levels, or four factors at one level and one factor at four levels
L16 Up to 15 factors at two levels
L18 Up to seven factors at three levels and one factor at two levels, or up to six factors at three levels and one at six levels

Design Parameter	Levels		
	1	2	3
Material Elasticity	Low	Nominal	High
Material Thickness	Low	Nominal	High
Vibration Dampening	Low	Nominal	High
Vibration Isolation	Low	Nominal	High

With four parameters and three levels for each, a full factorial experiment would require $3^4 = 81$ runs.

An L9 orthogonal array collects information that may be sufficient for this scenario with only 9 runs.

Each combination of parameters occurs an equal number of times, e.g., once for an L9 test matrix.

Run #	Assignment of Parameter Levels			
	Elasticity	Thickness	Dampening	Isolation
1	Low	Low	Low	Low
2	Low	Nominal	Nominal	Nominal
3	Low	High	High	High
4	Nominal	Low	Nominal	High
5	Nominal	Nominal	High	Low
6	Nominal	High	Low	Nominal
7	High	Low	High	Nominal
8	High	Nominal	Low	High
9	High	High	Nominal	Low

Example of an L9 orthogonal array.

Figure 19.2   A fractional factorial experiment, such as an orthogonal array, can be very efficient in collecting data toward selecting the critical design parameters and their set points.

If you had to describe the primary advantage of orthogonal arrays, it is their ability to accommodate many control factors with mixed levels of both continuous and categorical variables. This is possible because they are Resolution III designs. This great flexibility cuts both ways and requires a thoughtful approach. When using orthogonal arrays, key to success is the confirmation experiment that is used to measure performance improvements realized by the improved control factor set points.

When selecting which orthogonal array should be used, you should err on the side of assuming curvature and including more than a minimal set of control factors. This advice is different from what you would do when using DOE to build an empirical model. Standard orthogonal arrays are built into many of the DOE computer applications that can handle mixed levels and multiple factors efficiently.

Importance of Confirmation Experiments

When using Taguchi's approach, it is always prudent to run a confirmation experiment. The reason is that with an orthogonal array, you sample only a small part of the total experiment space. In our example we experiment with a system having four factors each at three levels. The total number of possible factor combinations is 81. However, our experiment includes only 9 of all possible combinations. What if the confirmation experiment does not confirm?

Taguchi chose to use orthogonal arrays that are "highly saturated designs." So it is unlikely that the best combination of factor level settings was one of the treatments run in the robustness experiment. If the optimum set points do not correspond to a treatment that was run, you have to run a confirmation experiment.

If the confirmation experiment does not confirm, you need to understand why. Likely it is because of a strong interaction among control factors aliased with a main effect. This is a warning that you have a design that has a good chance of being problematic in production and later in the market. Probably you will need to run an experiment that can accommodate at least two-way interactions.

Signal-to-Noise Ratio (S/N)

The S/N is the metric for robustness. It evaluates the relationship between the power in the output response and the power in its variations. In this context, *signal* refers to the output response of the function and *noise* represents the unwanted variations in that response. Reduced noise, a stronger signal, or both can provide a higher S/N. That is the objective. Given that the stressful conditions are chosen because they are effective at forcing variations, those variations are a function's response to an artificial, extreme stimulation. So the S/N is relevant to the conditions of the test.

The objective is a significant increase in S/N for the same stressful conditions. That objective is not zero variation. It would not be practical. Also, the objective is not to predict field reliability. If a product design has a higher S/N than its competitive product under the same stressful laboratory conditions, it can be expected to have a higher reliability in the market under whatever conditions are involved in its handling and use.

In a product system, the function being developed is part of a system of linked functions, as shown in Figure 2.6 of Chapter 2. The output of the function is an input "signal" to the next function downstream. Variation in the output response is an input "noise" to the downstream function. Likewise, its input signal is actually an output from its upstream function, with its variations. So the terminology *signal-to-noise* makes sense from the viewpoint of the linkages among functions being signals.

Types of S/N

Nominal-Is-Best Type I

This S/N is used for responses where being on target is best. Many functional outputs fall into this category, such as

- Color reproduction by a printer
- Voltage output from a power supply
- Volume of blood drawn by an evacuated blood collection tube

The S/N takes different forms for different scenarios, such as whether or not the function is static or dynamic, and how its requirement is reflected in its quality loss function. For the static nominal-is-best scenario, as an example, the S/N is defined as

$$S/N_{NB} = 10 \log_{10}\left(\frac{\mu^2}{\sigma^2}\right) \tag{19.1}$$

where

μ = the mean

σ = the standard deviation of the variation in the function's output response

The base-10 logarithm gives the S/N decibel units (dB). Increasing the S/N is equivalent to reducing the vulnerability to stressful factors. The S/N employs the square of the mean and standard deviation and is quadratic, as is the quality loss function. Then, increasing the S/N is equivalent to reducing the quality loss, particularly with the mean being adjusted to its target. Table 19.1 shows that small increases in S/N, less than 0.3 dB, do not represent much of an improvement, while increases over 1 dB provide major benefits. The effect of increasing signal-to-noise can be calculated using equation (19.2):

$$\frac{\sigma_2^2}{\sigma_1^2} = 10^{-\Delta(S/N)/10} \tag{19.2}$$

where σ_1^2 and σ_2^2 are the variances before and after the robustness improvements respectively.

The nominal-is-best type I S/N is not suitable for a target value of 0. For a zero target, the nominal-is-best type II should be used (equation (19.4)).

Figure 19.3 shows the probability distribution for a critical functional response superimposed on the quality loss function for that response. The cost of field repair, related to customers' tolerance (Δ_C)

Table 19.1   Reduction in quality loss versus increase in S/N

Increase in S/N	Decrease in Standard Deviation (σ)	Reduction in Quality Loss (μ on target)
0.3	3%	7%
1.0	11%	26%
3.0	29%	50%
6.0	50%	75%

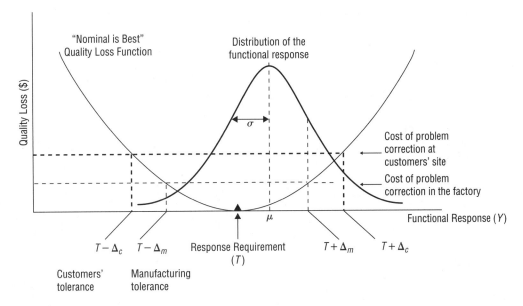

Figure 19.3   The distribution of response variations defines the probability of occurrence of the quality loss. It contributes to decisions about process and design improvements.

for deviations of the function's response, determines the shape of the curve. The cost of factory repair is one way to set the tolerance (Δ_M) for deviations detected during manufacturing. The distribution of variations in the response is shown to have significant variability with its mean not aligned with its requirement.

For a particular product whose performance variations interact with the quality loss function, the quality loss can be determined by the equation

$$QL = k[\sigma^2 + (\mu - T)^2] \tag{19.3}$$

where k is the constant that describes the shape of the quality loss function, converting performance deviations into economic terms. We described this in Chapter 2. The calculation can contribute to decisions about how much to invest in corrections to the performance distribution, such as developing the manufacturing process to be less variable and finding ways to adjust the mean of performance to its target requirement.

The nominal-is-best type I S/N factor is the one that you encounter most often in product development, since the majority of product requirements are stated as target values, usually centered within an acceptable range.

Nominal-Is-Best Type II

The nominal-is-best type II S/N is used when the standard deviation of the response does not scale with the mean. The acceptable mean response can range from $-\infty$ to $+\infty$. Some examples of responses for which this S/N is appropriate are

- Dimensional errors in machining and molding parts
- Difference between oven temperature and set point
- Tracking variation in the lateral position of a web or belt

The nominal-is-best type II S/N is given by

$$S/N_{NB-II} = -10 \log_{10}(\sigma^2) \tag{19.4}$$

Smaller-Is-Better

The smaller-is-better S/N is appropriate when you wish to minimize a response, such as

- Defects per unit area on a silicon wafer
- Parasitic losses such as friction
- Stopping distance for a vehicle

For the smaller-is-better response the S/N is given by

$$S/N_{SB} = -10 \log_{10}\left[\frac{1}{n}\sum_{i=1}^{n} y_i^2\right] \tag{19.5}$$

Larger-Is-Better

The larger-is-better S/N is used for responses we wish to maximize. Examples of larger-is-better responses include

- Gas mileage
- Strength of an adhesive bond
- Maximum data rate for a communications channel
- Risk-adjusted return on an investment portfolio

$$S/N_{LB} = -10 \log_{10}\left[\frac{1}{n}\sum_{i=1}^{n} \frac{1}{y_i^2}\right] \tag{19.6}$$

Effect of Changing the S/N

If you know the variance of the response for two different sets of control factor settings, you can calculate the change in S/N.

$$\Delta(S/N) = 10 \log\left(\frac{\sigma_1^2}{\sigma_2^2}\right) \tag{19.7}$$

The ratio of the "before" and "after" variances can be calculated when you know the change in S/N, given by equation (19.2).

Example Problem

In this example we use Taguchi's methods to improve the robustness of one of the original weapons of mass destruction, the catapult, a standard tool in teaching DOE. We then take another look at the problem using a modeling experiment to see how the results compare.

The purpose of the catapult is to throw projectiles accurately and consistently. Several adjustments affect the distance thrown. In Figure 19.4 we show the catapult and the various adjustment factors that can be used to affect throwing distance.

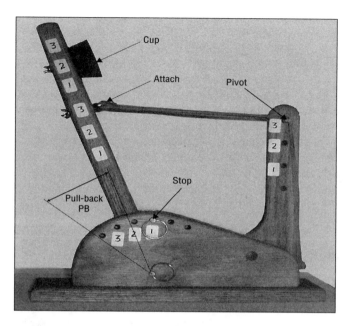

Figure 19.4   Catapult showing control factor adjustments.

The control factors are

PB   Pull-back angle for the throwing arm, measured in degrees relative to the stop pin
Cup   Position of the cup on the throwing arm
Stop   Stop pin position
Energy   Compound factor composed of the point of attachment of the elastic band to the throwing arm (attach point) and the position of the pin on the vertical beam (pivot point). The attachment and the pivot are changed together with each set at the same level to give three energy levels.

The benefit of using a compound control factor, such as the energy factor, is the reduction in the number of control factors. The disadvantage is that by changing attach and pivot together, we may fail to explore combinations of the attach and pivot factors that would occur if we treated them as separate factors.

In this example, variability in the mass of the projectile is an external noise. Our goal is to find settings for the control factors that can minimize the variability of distance thrown when using projectiles having a range of weights. The extremes of the noise are represented by a light plastic ball and a heavier golf ball.

For our optimization experiment the specific goal is to find factor level settings that will achieve a target distance of 72 inches for both plastic balls and golf balls, with minimum variability around that target. We know that we cannot achieve the same performance with the same factor level settings for both balls. Ballistics would allow this possibility if we were able to launch straight up, but the thrown distance would be zero—not very useful.

Orthogonal Array Experiment

By combining the attachment point and pivot point as the energy factor, we can use the L9 orthogonal array for the experiment design. The L9 allows us to explore a system having four control factors each at three levels. If we had not combined the attachment and pivot settings as the energy factor, we would have needed a design for five factors, requiring a larger experiment.

Using Taguchi's approach, Robust Design is a two-step process. First we reduce the variability and then adjust the mean to its target. The ideal situation is to have separate factors that can be used to independently reduce variability and adjust the mean.

We have two performance metrics: signal-to-noise and mean response. Since our goal is to hit a target while minimizing variability, we use the nominal-is-best type I signal-to-noise factor, as given by equation (19.1).

Table 19.2 shows the L9 design and the resulting data from Minitab. Each treatment is run twice, once with the plastic ball and once with the golf ball. The two runs allow us to calculate a mean and standard deviation for each treatment. The ball is a noise factor.

Main Effects Plots

The plots of the main effects from the analysis of means for the S/N and the mean are shown in Figures 19.5 and 19.6. We hope to find control factors that we can use to maximize signal-to-noise and put the mean on target. We see in Figure 19.5 that the two factors having the largest effect on signal-to-noise are cup and stop position. Pull-back angle and energy have very little effect.

However, we see a different situation in Figure 19.6. For the mean response, pull-back angle is the factor having the largest effect. This is good because it means we have a factor that can be used to put the mean on target without having much effect on signal-to-noise.

Optimal Settings

We can use the main effects plots to select control factor set points to improve robustness. If we choose factors strictly for maximizing robustness, we would use the following settings:

Stop = 1
Cup = 1

Table 19.2   L9 experiment design and data

PB	Energy	Stop	Cup	Plastic	Golf	SNRA1	STDE1	MEAN1
15	1	1	1	31.25	29.25	26.60	1.41	30.25
15	2	2	2	44.19	37.94	19.36	4.42	41.04
15	3	3	3	59.75	45.13	14.11	10.33	52.45
30	1	2	3	64.56	50.63	15.33	9.86	57.60
30	2	3	1	67.31	59.94	21.64	5.27	63.66
30	3	1	2	53.50	47.94	22.20	3.94	50.72
45	1	3	2	84.00	68.94	17.09	10.69	76.50
45	2	1	3	76.13	62.88	17.41	9.37	69.51
45	3	2	1	88.31	81.63	25.18	4.68	84.94

Energy = 3

PB = any setting within the range of 15 deg to 45 deg

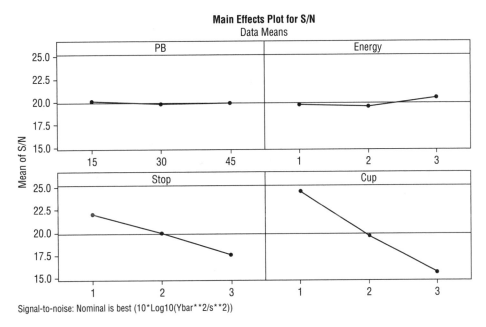

Figure 19.5   Main effects plots for S/Ns from Minitab

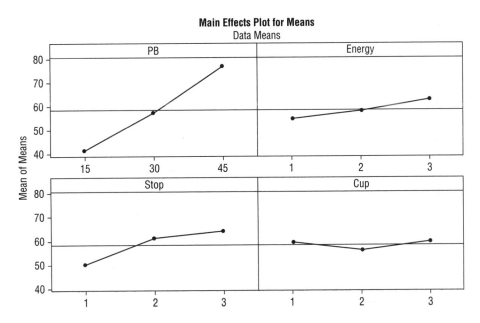

Figure 19.6   Main effects plots for means from Minitab

As it turns out, there is not enough energy to achieve the target distance of 72 inches with these settings. We will have to sacrifice some robustness to achieve the desired throwing distance. Looking at the main effects plots for signal-to-noise, we see that cup is the dominant factor for maximizing signal-to-noise. Stop position is next in importance. However, we can increase throwing distance by moving the stop to level 2. So a compromise strategy for improving robustness, while meeting the throw distance requirement, is

Stop = 2
Cup = 1

If we were depending on an adjustment factor to tune the distance, we would not depend totally on a model. Instead we would run a small series of runs varying pull-back, with the other factors set at the most favorable levels to increase robustness and achieve the distance required. This result is shown in Figure 19.7. We can use the fitted-line model from Minitab to calculate the required pull-back angle. So a compromise that will enable increased robustness while achieving the distance required is to set

Stop = 2
Cup = 1
Energy = 3
Pull-back to achieve adequate distance

With the following settings we get an average throw distance of approximately 72 inches:

Stop = 2
Cup = 1

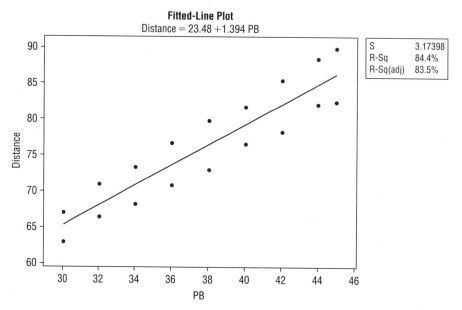

Figure 19.7   Regression model for distance versus pull-back angle

Energy = 3

PB = 34.8 deg

So by giving up a little on signal-to-noise, we are able to meet the distance requirement and still improve its robustness. You can see that using Taguchi's methods with an orthogonal array experiment requires thought and care but can pay dividends in products and systems with improved robustness.

Next we use a response surface experiment on the same problem and compare the processes and results.

Response Surface Experiment

In order to develop a more detailed understanding of the important factors affecting catapult performance, we ran a response surface experiment. The response surface design was a D-optimal with 31 runs. We chose this design because it was easy to set up a mixed-level experiment with all control factors at three levels, the same as the L9. The ball, a noise factor, is set at two levels as a factor in the experiment. Table 19.3 shows the factor level settings for the D-optimal RSM experiment.

The design will create a quadratic model that includes linear and square terms as well as all two-way interactions. By using this model to improve robustness, we will be able to compare the results to those from the robustness experiment using the L9 orthogonal array.

The response surface experiment produced by the D-optimal does not have the structure of the traditional response surface design. This is because the algorithm usually starts with a standard design having a large number of runs and selects a subset of all the treatments from the starting experiment. In this case we specify a quadratic model having linear, square, and two-way interaction terms. The goal is to minimize the covariance of the model coefficients for the specified model. So the end result is an experiment with 31 runs and with all factors except the ball at three levels. We let Design-Expert select the significant model coefficients using the backward elimination process to build the model. The results of the ANOVA for the response surface design are shown in Table 19.4.

As you can see, all of the main effects are significant. The top three factors in terms of significance are the pull-back angle, the stop position, and the energy level. This makes sense because they all affect the amount of potential energy stored in the elastic band before release. We also see numerous interactions. There are four significant interactions among control factors, AB, AC, AD, and BC. Recall from Chapters 14 and 15 that any factor with a p-value below the chosen threshold (often 0.05) is statistically significant. Also recall that statistically significant does not necessarily mean practically important.

Although smaller in magnitude than the control factor interactions, the most interesting and useful interactions are those involving the ball and the control factors. The ball is a noise factor, and interactions between the ball and the system control factors present opportunities to increase the robustness of the system. The major interactions involving the ball are

AE – pull-back angle and ball

CE – stop and ball

DE – cup and ball

Table 19.3 D-optimal RSM experiment for catapult

Std	Run	Factor 1 A: PB	Factor 2 B: Energy	Factor 3 C: Stop	Factor 4 D: Cup	Factor 5 E: Ball	Response 1 Distance
1	29	15	1	2	3	1	42.25
2	25	45	3	3	3	1	134.38
3	19	45	1	2	3	2	69.13
4	20	45	3	3	1	2	99.38
5	1	15	1	3	1	2	26.00
6	4	45	3	1	1	1	61.00
7	9	15	3	3	3	2	44.81
8	24	15	2	3	3	1	48.06
9	31	30	1	3	3	2	52.25
10	30	15	3	1	3	1	39.31
11	16	45	3	1	3	2	69.50
12	15	15	3	1	1	2	30.93
13	28	30	3	3	1	1	78.13
14	6	45	1	3	2	1	81.25
15	7	15	3	3	2	1	53.63
16	10	45	2	1	3	1	74.00
17	27	15	1	1	1	1	25.88
18	21	30	1	1	3	1	52.50
19	2	45	1	1	1	2	45.00
20	5	45	1	2	1	1	66.38
21	8	15	1	1	3	2	29.25
22	14	15	3	1	1	1	31.88
23	18	15	1	3	1	1	33.75
24	11	15	1	1	1	2	29.06
25	17	45	3	1	1	2	56.75
26	26	45	1	3	1	2	60.50
27	12	15	2	3	3	1	52.25
28	3	30	3	3	1	1	75.88
29	22	45	1	2	3	2	69.50
30	23	45	1	3	1	2	58.13
31	13	45	3	1	1	2	57.31

Figure 19.8 shows the distance thrown versus the type of ball and the cup position. Figure 19.10 shows the distance thrown versus the ball type and the pin stop position.

The regression model from the RSM is

$$Distance = 1.86 - 1.57A - 8.2B + 15.9C + 6.1D + 13.15E + 0.26AB + 0.41AC$$

$$+ 0.23AD - 0.19AE + 4.78BC - 3.38CE - 4.97DE - 5.52C^2 + \varepsilon \quad (19.8)$$

Table 19.4 ANOVA for response surface reduced quadratic model for throw distance versus control factors

Source	Sum of Squares	df	Mean Square	F Value	p-value Prob > F	
Model	16217.90	13	1247.53	172.91	< 0.0001	significant
A-PB	9489.00	1	9489.00	1315.23	< 0.0001	
B-Energy	1801.03	1	1801.03	249.63	< 0.0001	
C-Stop	2747.63	1	2747.63	380.84	< 0.0001	
D-Cup	768.20	1	768.20	106.48	< 0.0001	
E-Ball	575.88	1	575.88	79.82	< 0.0001	
AB	314.83	1	314.83	43.64	< 0.0001	
AC	722.56	1	722.56	100.15	< 0.0001	
AD	49.86	1	249.86	34.63	< 0.0001	
AE	46.11	1	46.11	6.39	0.0217	
BC	471.21	1	471.21	65.31	< 0.0001	
CE	67.12	1	67.12	9.30	0.0072	
DE	147.52	1	147.52	20.45	0.0003	
C^2	70.03	1	70.03	9.71	0.0063	
Residual	122.65	17	7.21			
Lack of fit	108.31	12	9.03	3.15	0.1072	not significant
Pure error	14.34	5	2.87			
Total	16340.55	30				

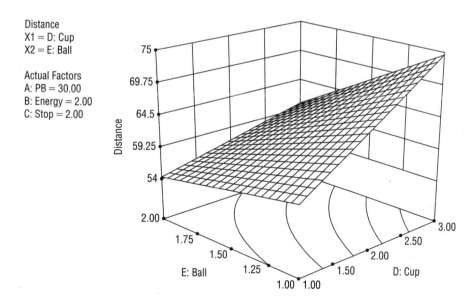

Distance
X1 = D: Cup
X2 = E: Ball

Actual Factors
A: PB = 30.00
B: Energy = 2.00
C: Stop = 2.00

Figure 19.8 Distance versus ball and cup

In Figures 19.5 and 19.6 note that in the L9 experiment the factors having the largest effect on robustness are cup position and stop position. The cup-ball interaction is important since both cup position and ball affect the rotational inertia of the arm-ball system. With the cup at level 3, the ball and cup have more impact on this rotational inertia since their contribution is proportional to the square of the distance from the cup to the arm's axis of rotation. If we convert the same amount of strain energy to rotational kinetic energy of the ball-arm-cup system, the greater rotational inertia reduces angular velocity when the arm impacts the stop pin. Recall from projectile ballistics that the three determinants of projectile distance are the initial projectile velocity, the initial angle of the velocity vector with respect to the horizontal plane, and the initial height above the horizontal plane.

Remember that in this problem, we want to minimize the difference in throwing distance for the plastic ball and golf ball. With the cup position at level 1, changing the ball from level 1 (plastic) to level 2 (golf) has a smaller effect on distance than if the cup is set at level 3. The best setting for the stop is level 1.

We have taken a liberty in the design of this experiment by assuming that ball type is a surrogate for ball weight, which can vary continuously between the heavier golf ball and the lighter plastic ball. This may be a good assumption if it is possible to have a range of projectile weights bracketed by the heavy and light ball, and weight is the only characteristic of the ball that's important. However, if there were only two possible levels for the ball, and, in addition to weight, the ball had other important physical characteristics affecting distance, the response surfaces shown in Figures 19.8 and 19.10 would not make sense. In that case, a better approach would be to treat the ball as a categorical factor that has only two possible levels, golf ball and plastic ball.

Figure 19.9 shows the traditional interaction plot. We see the near and far edges of the response surface shown in Figure 19.10, looking in the direction of the "cup" axis. The plane is twisted and sloped in two directions.

Figure 19.10 shows the plot of distance versus ball and stop. The third significant interaction involving the ball is the ball–pull-back interaction. This is the smallest of the interactions involving the ball and control factor settings.

Now that we have the system model given by equation (19.8), we can optimize performance. There is an optimizer in Design-Expert that we used in the lathe optimization problem in Chapters 17 and 18. Design-Expert treats variables defined as continuous variables during model creation as being continuous during optimization. However, for the catapult design all factor settings are discrete, except pull-back angle. They are set by placing a pin in the appropriate hole. So we use a different optimization tool that easily handles discrete variables.

Stochastic Optimization Using the RSM Model

As we have already discussed, any process can be characterized by its transfer function, which defines the relationship between the system inputs and outputs. When system inputs or outputs have a component of random variation, it becomes attractive to use Monte Carlo simulation to determine optimal settings for control factors. "Stochastic optimization" is the name for the process, and there are a number of cost-effective tools available to choose from. OptQuest, part of the Crystal Ball[3] suite, is a convenient and user-friendly application for the optimization used in our example.

---

3. Crystal Ball is a product of Oracle (www.oracle.com/crystalball).

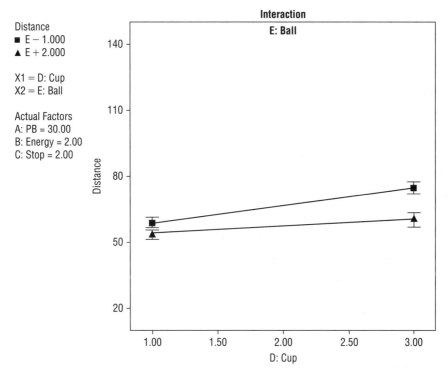

Figure 19.9 Ball-cup interaction plot

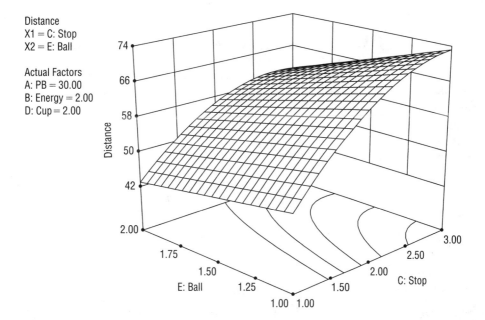

Figure 19.10 Distance versus ball and stop

There are three types of variables in an OptQuest model: assumptions, decision variables, and forecasts. An assumption variable is described by a probability distribution. While the simulation runs, values for each assumption are sampled from the appropriate distribution. Decision variables are factors that are varied across a range of values during the simulation. They can be either continuous or discrete. They can also have a random component sampled from the appropriate probability distribution. The objective of the simulation is to find settings for the decision variables, derived from the simulation, that satisfy both the objectives and the requirements for the forecasts. In our case, the decision variables are pull-back angle, cup, stop, and energy. The assumption variable is ball type, and the forecast variable is ball throw distance. A Monte Carlo simulation is run for each set of decision variable settings. The output is a statistical distribution for the forecast variables. In this case the forecast is the throw distance. Values for the assumption are sampled, the forecast is repeatedly recalculated, and the values are saved. Using the simulation results, the mean and variance of the response are calculated and settings for the decision variable are selected that meet the specified criteria for putting the mean on target while minimizing the variance. The important intelligence in the optimization algorithm is how to efficiently change the decision variables to speed the convergence to settings that will achieve a global rather than local optimum without having to perform an exhaustive search. In order to prevent convergence on a local optimum, OptQuest uses three different algorithms for sampling the operating space.

Table 19.5 shows the acceptable solutions found during the optimization. The solutions are ranked in order of increasing variance. There is a large difference between the best and worst acceptable solutions, with a worst-case variance of 95.3 and best case of 10.6.

In OptQuest, a requirement must be satisfied. Our requirement is a mean throwing distance of 72 inches when using both the plastic ball and the golf ball. An objective is the best that can be achieved for a given performance metric while meeting the requirement. Our objective is to minimize the variance.

Figures 19.11 and 19.12 show the results of the simulation using the best and worst acceptable settings. There are two contributors to the variability observed in the simulation: the ball and the residuals. Figure 19.11 shows the best-case settings that minimize the difference between the plastic and golf balls. In Figure 19.12, which is the worst case from the simulation, we can see the separate contributions of the ball and the residuals. The difference in the masses of the plastic ball and the golf ball moves the individual distributions for each ball farther apart. The individual distributions are caused by the residuals. The distribution of higher values is for the plastic ball, while the one of lower values is for the heavier golf ball.

If we look at the difference in robustness, comparing best case to worst case, the difference in terms of S/N is given by

$$\Delta(S/N) = 10 \log \left(\frac{\sigma_w^2}{\sigma_B^2} \right) = 10 \log \left(\frac{95.3}{10.6} \right) \tag{19.9}$$

This is a difference of 9.5 dB, which translates to a reduction in quality loss of nearly 89%!

Conclusions

Comparing Orthogonal Array Results to RSM

In the catapult example, we see that the orthogonal array got us to the same set of control factor settings as did the response surface experiment and stochastic optimization. Even though there were significant control factor interactions, they were not large enough to lead us to wrong

Table 19.5 Acceptable solutions to stochastic optimization using OptQuest

| | | Objective | Requirements | Decision Variables | | | |
| | | Minimize Variance of | Mean Distance between 71.50 | | | | |
Rank	Solution #	Distance	and 72.50	Cup	Energy	PB	Stop
1	417	10.59	71.90	1.00	3.00	34.50	2.00
2	535	15.65	71.69	1.00	2.00	43.50	2.00
3	208	15.97	72.25	1.00	2.00	44.00	2.00
4	520	17.29	71.80	1.00	3.00	28.00	3.00
5	1193	24.18	72.19	1.00	2.00	37.00	3.00
6	884	25.86	71.98	2.00	3.00	30.50	2.00
7	1553	32.45	72.17	2.00	2.00	37.50	2.00
8	1118	38.76	71.81	2.00	3.00	25.50	3.00
9	934	45.70	71.79	3.00	3.00	41.00	1.00
10	131	47.57	72.26	2.00	2.00	33.00	3.00
11	267	51.31	72.08	3.00	3.00	27.50	2.00
12	480	58.69	72.27	3.00	2.00	33.00	2.00
13	1446	60.15	71.80	2.00	1.00	42.50	3.00
14	252	68.86	71.58	3.00	1.00	40.00	2.00
15	1285	69.62	72.24	3.00	1.00	40.50	2.00
16	133	71.12	71.79	3.00	3.00	23.50	3.00
17	852	81.50	72.50	3.00	2.00	30.00	3.00
18	264	94.41	71.53	3.00	1.00	37.50	3.00
19	7	95.31	72.40	3.00	1.00	38.00	3.00

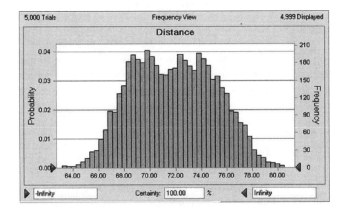

Figure 19.11 Distribution of distance for best acceptable control factor settings, solution 417

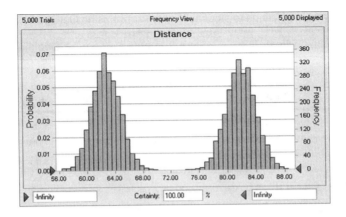

Figure 19.12   Distribution of distance for worst acceptable control factor settings, solution 7

conclusions. However, this will not always be the case. In some problems, interactions can play a much larger role, so we must be cautious.

Control Factor Settings Must Be Chosen with Care

For any system subject to noises or stresses, there are two responses we must consider when deciding on control factor settings. Selecting control factor settings with the sole objective of putting the mean response on target is almost a guarantee of suboptimal performance. If our only concern is on-target performance, we are likely to settle for too much variability in the response. From the stochastic optimization results, we see almost an order of magnitude difference between best- and worst-case variance for the same mean response. Our conclusion should be that for the small cost of an experiment our customers can enjoy much-improved performance for every product produced, throughout its life. The result will be reduced quality loss and improved customer satisfaction with the financial benefits accruing over time.

Alternative Solutions Might Offer Value

One advantage of stochastic optimization using OptQuest is that we are presented with multiple options for control factor level settings. This can be useful in helping us choose how we can improve robustness. Having choices might have value when we have to make decisions on the implementation details for a product design concept. It is possible that certain alternative control factor level settings might have cost, schedule, or other advantages.

Key Points

1. Taguchi's philosophy of using noises to improve robustness is foundational to robust engineering, regardless of whether we are using classical DOE or orthogonal arrays.
2. Orthogonal arrays can be used to improve system robustness.
3. Using orthogonal arrays requires an understanding of the system and care in planning the experiment. Some things we must consider when planning robustness experiments include
   a. Number of control factors
   b. Likelihood of curvature in the response

    c. Important noises affecting system response

    d. Nature of any important interactions

       Points a, b, and c are also key to success in planning a response surface experiment. However, with orthogonal arrays you need to consider important interactions when you plan the experiment and assign the columns. Consequently, you need some prior knowledge of important interactions since in the standard Resolution III orthogonal array the main effects are aliased with the two-way interactions.

4. Important interactions among control factors that are not included in the experiment can cause problems, resulting in the confirmation experiment not confirming.

5. Taguchi's position that strong control factor interactions should be avoided is correct. Product development will be less complicated if you can do it. However, there are situations where interactions are unavoidable. They must be considered by making appropriate column assignments when planning robustness improvement experiments using orthogonal arrays.

6. When using orthogonal arrays, if the optimal settings are not in one of the treatments run during the robustness experiment, the confirmation experiment is very important.

7. If interactions are not explicitly included in the experiment, the size of the interactions, relative to the main effects with which they are confounded, is important in order to determine whether or not the orthogonal array will yield a useful result.

8. The decision of whether to choose an orthogonal array or a classical optimization experiment should be made on the basis of the facts of each situation. The decision should be based on productivity rather than whose statistics have more powerful "mojo." It is not possible to make a blanket statement that one choice or the other will always have fewer runs and thus better productivity. Here are some of the things you should consider:

    a. Presence of interactions: Either approach will work if interactions and curvature are considered and included if appropriate. If you have prior knowledge of whether there are important control factor interactions, you can handle them with orthogonal arrays. This would be the case if modeling experiments had been done earlier during technology or product development. If important interactions were present, you should include them in your robustness experiment if you choose to use orthogonal arrays. If you know that there are no important control factor interactions, Resolution III orthogonal arrays will work well. If there are important interactions and you use an orthogonal array but do not plan for them, it is possible that your confirmation experiment will not confirm, requiring more work and probably another experiment.

    b. Curvature: If you have evidence of curvature in the system transfer function, based on experience, previous experiments, or first-principles analyses, you should choose an orthogonal array that allows control factors at three levels.

    c. Total runs: If a response surface design or a Resolution V factorial design can be done in fewer total runs, it is a better choice than a Resolution III orthogonal array since either design can model two-way interactions.

    d. Large external arrays of noise factors will result in a large number of runs. A more productive strategy is to use compound noise factors derived from a noise screening experiment. As with most things, there are probably just a few important noise factors. You can save yourself time and money if you eliminate the trivial many before embarking on your robustness experiment.

Discussion Questions

1. Compare the use of orthogonal arrays and response surface experiments for robustness improvement.
2. If a system has no interactions and no curvature, what experiment designs would be effective for robustness optimization if there are three control factors and one noise factor?
3. What interactions are desirable when trying to improve robustness?
4. What are the implications of no interactions among noise factors and control factors?
5. How can you improve system robustness if there are no advantageous settings for control factors that will reduce system sensitivity to noise?
6. Can you use orthogonal arrays for robustness improvement if there are strong control factor interactions?
7. What are the ways to include noise in a robustness experiment?
8. How can you identify noises that can be effective at forcing variations during laboratory experiments?
9. What is a compound noise factor?
10. How do you formulate a compound noise factor?
11. How do you decide which signal-to-noise factor is appropriate to use in a robustness experiment?

CHAPTER 20

Tolerance Optimization

The design of a tolerance is, in part, a trade-off between the costs of reducing variability in production and the economic consequences of that variability. The decision criterion is minimum total cost.

The process of specifying a controllable design parameter has three fundamental steps:

1. *With tolerances relaxed to economically practical levels, increase the robustness of the designs to reduce their vulnerability to variations in their control parameters.*
2. *Increase the robustness of manufacturing and service processes to reduce the probability of unacceptable variations in replicated control parameters.*
3. *Tighten the tolerances of those control parameters having the most influence over system performance to the degree that the increases in production costs are less than the savings in quality loss. Where there is an opportunity, loosen a tolerance if the reduction in production cost exceeds the increase in quality loss.*

This chapter is not about the language of tolerancing or about tolerancing best practices as described by standards for geometric dimensioning and tolerancing (GD&T). Instead we consider the design of tolerances at a relatively high level, describing two main activities in tolerance design: the analysis of tolerance accumulations (stacks) and the allocation of tolerances among multiple control parameters.

The Tolerance Design Problem

Tolerances are limits. They apply to operational functions throughout the value delivery system over which the producer has some control. The implication is that organizations within that value stream are required to take appropriate actions to prevent "out-of-spec" parameters or responses from imposing consequences on customers. So the specification of tolerances should be based on what is required for customers' satisfaction, at the full system level.

If a variation occurs in manufacturing, a preplanned corrective action would prevent the defect from escaping to the market. If the variation occurs in service, a remedial action would protect customers from the consequences of the variation. The actions triggered by a tolerance being exceeded increase the costs of delivering stable performance to customers.

Efforts to reduce the probability of these variations are investments. Resources can be applied to increase the robustness of product designs and that of manufacturing and service processes. Secondary processes might be required to achieve tolerances that are tighter than the capabilities of the initial production processes. Procedures can be imposed to control processes more tightly or to apply inspections to prevent "out-of-spec" parts from entering the supply chain.

In Chapter 2 we discussed the concept of quality loss, the economic consequence of variations in a performance response. The quality loss function increases exponentially as the magnitude of a variation in the functional response increases. An unacceptable variation in a design parameter does not necessarily mean that the system-level performance will be unacceptable to customers. It may depend on the simultaneous variations in other parameters that contribute to a particular performance response. It can also depend on the customers' applications and their sensitivity to the relevant performance variations.

The objective of tolerance design is to

Determine the allowable variation in each control parameter, x_i, to meet the requirement for allowable variation in the functional response, y, while minimizing total costs.

This is an optimization problem. It requires the simultaneous calculation of the economic consequences of variations in a functional response and the additional production costs associated with delivering tighter tolerances as they are allocated in the product design.

The problem can be approached in a couple of ways. In the most general case, teams start with an understanding of the quality loss function for the functional response and of the increased production cost of tightening each of the tolerances. The allowable variation in the functional response, *y*, and the allocation of tolerances among the x_i can be derived as an outcome of the optimization. A second approach is to assume that the amount of allowable variation in the functional response is known. In this case, assign tolerances to the controllable parameters with the objective of minimizing the cumulative cost increases associated with the allocation of the system tolerance. Solving this type of problem can involve intensive calculations, requiring the use of the computer.

What Is a Design Tolerance?

In product design, a tolerance is a specified limit to acceptable variation in a functional response or in a controllable design parameter that influences a functional response.

A design parameter can be the dimension of a mechanical part, the value of an electronic component, a behavioral property of a material, the composition of a chemical mixture, or one of many other possibilities. We refer to the design parameters used to control functional responses as control factors, consistent with our description of the parameter diagram (Figure 7.2) and the nomenclature for DOE. The control factors can be dimensions or other characteristics replicated by manufacturing processes or tooling. They may be adjustments determined by human procedures with fixtures and instrumentation. Excessive variations in these critical factors can cause degradations in system-level performance responses and thereby affect customers' perception of the product quality. That's quality loss for customers. To the extent that customers decide to spend their money for competitors' products, or not at all, that is quality loss for the producer—in other words, lost revenues.

When you design a tolerance, it should have the objective of consistently meeting the needs of both sets of stakeholders, the customers and your business. Insights from the quality loss function indicate that a tolerance should be specified so that the increased cost of delivering that constrained performance variation is less than the savings in the quality loss for that level of deviation from target.

Fundamentally there are three ways to specify a tolerance for a parameter. Each is consistent with Taguchi's concept of S/N:

1. Most common are tolerances specified by upper and lower limits around a target value. Being on target is better than being at either of the extremes of the specification limit.
2. There is the "larger-is-better" situation. Its tolerance is described as a minimum acceptable value. The target is infinity, never realizable.
3. There is the "smaller-is-better" situation. Its tolerance is described as a maximum acceptable value. The target is zero.

We focus here on the first case since it represents the majority of the tolerance design scenarios.

Truth in Tolerance Design: The Need for Communication and Trust

The specifying of tolerances has, at times, had psychological elements similar to bluffing in a poker game. Design engineers, concerned about additional deviations of parameters in customers' applications, would ask for tolerances to be tighter than were actually required for the desired functionality. Often design teams really didn't know the limits of unacceptable performance or parameter variations. Possibly the design was leveraged from work done in research using rough prototypes. Perhaps tolerance requirements never had a solid foundation based on the system-level consequences.

It's easy to recall conversations such as this:

Question by the manufacturing engineer: "What tolerance is required?"

Unspoken thought of the design engineer: "I don't know what's required. The research people delivered a working lab prototype, but we never had a discussion about tolerance requirements. If I ask for the best they can do, maybe I'll be okay in production."

Once we heard this:
"I specify tight tolerances as a way to set expectations for high-quality work in manufacturing."

That viewpoint neglects the additional manufacturing costs applied to design elements that have little contribution to customers' value. It's an avoidable waste of money, resources, and time!

Tolerance specifications should be based on trade-offs among their consequences for system functionality, production process capabilities, and production costs. Cross-functional teamwork that develops those trade-offs starts early in a development project and continues well into production. The process consequences and costs should be agreed to by those reproducing and maintaining the product with a full understanding of the rationale. The thought and analyses that establish tolerances may even begin during technology development as the manufacturability of critical functional parameters is considered. The capabilities and economics of replicating the tolerances should influence the business case before the product transitions into high-volume manufacturing.

There are times when the progress to establish rational tolerances stalls. A supplier, unable to satisfy the tolerance requirements, may continue to deliver out-of-tolerance components. Design teams may agree to "waivers" that allow the use of those materials that do not meet specifications, hoping that they will work anyway. In many cases there can be realistic uncertainties about whether or not the system-level performance will be unacceptable to customers. This process can become a standoff. Suppliers can claim that "if you want products, you takes what you gets," while customers' feedback may not indicate whether or not there is a recognizable system problem. Without the hard work of establishing the "real requirements," the direction forward can be very cloudy.

As a general rule, everyone involved in product development should have a common goal for tolerance design: Make the parts and drawings agree. The tooling, fixtures, and procedures determine the actual dimensions. If the drawings correlate with "reality," the analyses of tolerances can provide value. Specified tolerances can be economically based on what's actually needed, the supply chain can produce what is required consistently, and all stakeholders can understand the logic behind the requirements.

Controlling Variability

In earlier chapters on quality loss and robustness development we discussed the sources of variations that can affect the alignment and stability of product performance. Included in these "noises" are degrading forces suffered during production, in manufacturing, assembly, shipping, storage, installation, and service. Tolerance design is a process to limit the consequences of these stresses in a cost-effective manner. Its costs are production costs.

What can cause unit-to-unit or batch-to-batch variations?

Supply chain: The primary sources of variation in product parameters may be in the manufacturing processes of contributing suppliers. Unique to a supplier's process, the distribution of a specified parameter may not be centered on its target, or it may have a standard deviation causing a significant probability of the parameter exceeding its tolerance. This perspective is represented in the manufacturing process capability:

$$Cpk = \frac{Minimum\{(USL - \mu)\}, \{(\mu - LSL)\}}{3\sigma} \tag{20.1}$$

This version of Cpk includes the effect of the mean being off target. Variations of material properties for consumables, such as the fuel for an engine, are similar examples of variability from sources in the supply chain.

Assembly: Product designs may provide adjustable control parameters that require setting during final assembly. Regardless of whether these adjustments are mechanical or electrical, usually they are selected because they can shift the mean performance to its target. They tend to compensate for the inability of fixed parameters, replicated by tooling, for example, to achieve the specification. Human errors, or calibration problems with instrumentation or fixtures, can cause variations in specified product responses.

Service: Products requiring field service are vulnerable to variations in adjustments of critical design parameters as well as to manufacturing variations in replaceable parts and components. Off-brand consumables, for example, may not be produced with the same quality metrics as expected from the equipment manufacturer.

Environment: Changing environmental factors can interact with certain product parameters to cause performance variations. This can happen in both production and in the product's use. Component dimensions can suffer thermal expansion. Material electrical properties or stiffness may be vulnerable to humidity. Air-handling devices may be sensitive to changes in atmospheric pressure.

How Do Tolerances Affect Product Performance?

Your new design is expected to perform in two ways. Most important, it is intended to deliver value to your customers, that is, needed benefits compared to their cost of ownership. If few people are willing to spend their money on your product, there is no further discussion and no opportunity for improvement. A product is also expected to return value to your business through a dependable stream of revenues, offset by a necessary stream of costs.

Things Customers See

Benefits delivered by your products can be thought of in categories.

Functionality

"It does what I need it to do." This customer need is highly dependent on the product applications. Whether you are designing a motor vehicle or an insulin pump, there are specific requirements that must be met for customers to be satisfied. Obviously, tolerance specifications must be derived from the acceptable range of performance. How can the functionality of your product be perceived as being better than that of its competitors?

Nonfunctional Requirements

Also called constraints, nonfunctional requirements describe what the product *is* rather than what it *does*. Satisfaction of these requirements can also be affected by tolerances on control factors. Some examples of nonfunctional requirements are

- Life: "It lasts a long time."
    The usable life of a product depends on the life of its components. In some cases, tolerances can affect component life. A good example is a rolling element bearing: Tolerances on its parts can affect the life of the component.
- Aesthetics: "It looks good!"
    "Pretty" can be important. When advantageous, development teams invest in effective human factors and industrial design that can contribute value in their markets. Often, in a side-by-side comparison, the product that is more attractive wins the sale in both consumer and industrial segments. For example, the uniformity of the gaps between panels on a machine or instrument enclosure or among car panels is a design requirement for which effective tolerances directly contribute to perceived value. Consumers easily notice unevenness and react swiftly with disapproval.

- Price: "The price is right."

  In theory, the marketplace sets prices while the development teams determine costs. Higher production costs create pressure to increase prices, especially in non-commodity markets where products have a fair amount of differentiation and price inelasticity. As the differentiation in benefits to a market increases, it becomes more difficult to assess value, since the benchmark comparisons become less clear. Also, specific features and functions may deliver different values in different market segments. When there is reduced price pressure from the marketplace, higher costs can push prices up while producers strive to maintain their margins. However, higher margins attract competition. If your competitors have lower cost structures, because of superior product designs or manufacturing processes with higher capabilities and lower costs, the pricing of your new product can be vulnerable.

Not Seen by Customers, but Important to Your Business

Manufacturability, Process Capability, and Costs

There are several business concerns that are affected by tolerances. They have to be addressed continually by cross-functional development teams as a design moves from concept to production.

When assessing the reasonableness of tolerances, teams need to ask questions such as these:

- What do all the pieces cost?
- Do any elements of the design require tolerances that are tight relative to the normal variations in economically efficient manufacturing processes and in the materials selected?
- Note that for "three-sigma quality" there should be a spacing of three standard deviations between the target value and either specification limit. "Six Sigma quality" requires a spacing of six standard deviations.
- Can the design be changed to reduce the need for tighter manufacturing tolerances?
- Can secondary manufacturing operations be reduced or eliminated?
- What is the expected manufacturing process capability for a fabricated part or material, given the selected supplier, process, and material?
- How do process capabilities vary among candidate suppliers?
- What is the expected process capability for each purchased component and assembled subsystem?
- Can your operations assemble the product without human errors, rework, and excessive costs?
- Do parts or assemblies require iterative adjustments?
- Will the quality for the stream of products be consistent over time?
- What are the expected manufacturing first-pass yields for both piece parts and assemblies?

Tighter tolerances drive up the production costs of components. When the acceptable limits for process variability are less than the normal process variability, there are consequences for yields, scrap, and rework costs as well as for increased process control and for investments to improve the robustness of the process. Tighter tolerances for assembly may require instrumentation, fixtures, adjustment procedures, break-in tests, and even additional training, further increasing costs.

How Do You Identify Tolerances That Are Tight?

Development teams have to determine the width of the proposed tolerance band relative to the expected variability of the manufacturing process. This variability is a function of the configuration of the design, the material chosen for the part, the condition of the tooling, and methods practiced by the supplier.

Recall the definition of process capability when the mean is on target:

$$Cp = \frac{USL - LSL}{6\sigma} \tag{20.2}$$

where

USL = upper specification limit

LSL = lower specification limit

σ = standard deviation of the process variability

With knowledge of the process capability, the first-pass process yield can be estimated. That is the fraction of parts produced that can be used as made. Parts not meeting this criterion must be reworked, matched with parts of opposite variation, or scrapped if rework is not economically practical.

Equation (20.2) is the process capability for a centered process. It does not account for the mean drifting off target, as would be expected over time from tooling wear, for example. This process capability is as good as it gets. Of course, the calculation of the capability relevant to a real production environment should include the expectation that the process will not necessarily be centered on the target requirement or stay that way over time.

In high-volume production of parts and components for which deviation is not acceptable, a desirable process capability for manufacturing a component is $Cp = 2$. With 6σ between the target and the nearest specification limit, there is considerable room for process drift before the specification is exceeded.

Often products are manufactured with much lower process capabilities. It may not be economically practical to develop many improvements. If there is great pressure to get to market on time, management may decide to absorb the cost of poor quality associated with rework. It might also be preferable to absorb the higher costs in manufacturing rather than depend on field service to correct escaping defects. A choice to rework rather than to delay shipment might be a good business decision if there is a path forward to improved process capabilities through follow-on development work. If there is no credible path to improve process capabilities, such a decision is a tacit acceptance of reduced profitability. The business case should reflect that. In our experience, shipping products with marginal quality, robustness, and reliability can easily dilute the value delivered to customers and reduce the value returned to the business.

Steps for Tolerance Design

In general, the more complex the product, the more parameters there are that can affect system performance. There are several key steps to tolerance design:

1. **Identify the more critical functional responses delivered by the product.** The required functional responses are derived directly from understanding customers' needs (Chapter 9). Normally, they are identified early in development using a mapping method such as QFD (Chapter 10). As the requirements are deployed from the system level to lower levels in the architecture, teams identify those technical requirements that describe measurable functional responses.

2. **Establish the acceptable range of variation for each of the critical functional responses.** Ideally this is based on customers' inputs. With VOC processes, teams have opportunities to understand customers' expectations more deeply and to determine how sensitive their applications are to variations in performance responses. Understanding when customers can be expected take action to remedy off-target performance is valuable information. It's a parameter of the quality loss function.

   You should not expect all customers in a market segment to have the same sensitivity to off-target situations. Their applications may differ and the consequences to their activities or business may vary. Certain customers may object to "just barely noticeable" deviations, while others may have applications that are not so critical. Some customers may correct problems themselves, while others may require service by the producer.

   Gathering this information may be as easy as asking good questions. It may require some experiments. For example, establishing thresholds for image quality characteristics in the output of a printer may require psychophysical experiments to understand the sensitivities to variations in specific image parameters. There can be many, such as background density, line width, solid area density, and peculiar artifacts. Customers in certain markets can have very high expectations. Understandably, others may not. Who are your chosen customers? What are their quality expectations?

3. **Develop the transfer function for each critical functional response.** Developing the transfer function may require experiments. Certain transfer functions are simple, but others can be complex with many critical parameters affecting the functional response.

4. **Rank the design parameters in order of their effect on functional responses.** This can be calculated directly from the transfer function or by using ANOVA. Criteria can include, for example, the sensitivity of the functional response to changes in the parameter, or similar effects on the S/N.

5. **Allocate tolerances among the controllable parameters to minimize the total cost.** The allocation of tolerances among several contributing parameters is an economic trade-off. When there are multiple parameters affecting a functional response, teams have to allocate the total system tolerance among the contributing parameters so that each increase to manufacturing costs is justified by a greater savings in quality loss. Increases in S/N can be surrogates for savings in quality loss.

Error Transmission

We discussed error transmission (propagation of error) and robustness in Chapter 17. Equation (20.3) defines the relationships among the variability of a functional response and the control parameters and noise factors that determine it.

$$\sigma_y = \sqrt{\sum_{i=1}^{n}\left(\frac{\partial y}{\partial x_i}\sigma_{x_i}\right)^2 + \sum_{i=1}^{m}\left(\frac{\partial y}{\partial z_i}\sigma_{z_i}\right)^2 + \sigma_{error}^2} \tag{20.3}$$

In equation (20.3) we have several variables:

y = functional response

σ_y = standard deviation of the functional response

x_i = ith controllable parameter affecting the functional response

σ_{x_i} = standard deviation of the controllable parameter, x_i

z_i = ith noise factor affecting the functional response

σ_{z_i} = standard deviation of the noise factor, z_i

σ_{error} = standard deviation of the experimental error due to common cause noise

The relationships require you to understand the transfer function for the system. Guidance from the principles of "Design for Manufacturing and Assembly" for selected materials and processes can provide useful estimates of the variability of many manufacturing processes. Since it may vary among suppliers, the costs and process capabilities that are demonstrated will be more credible. With this knowledge you have what you need to co-optimize performance variability and production costs.

The Effect of Error Transmission on Tolerances

An interesting insight from equation (20.3) is that the variation of the functional response caused by variability in both the control parameters and the noise factors must be considered when setting tolerances for the controllable parameters. As the effect of the variability in the noise factors on the functional response increases, you have to tighten the tolerances on the controllable parameters to compensate, to limit the resulting variation of the functional response. If the effect of uncontrolled noise is large enough, it may not be possible with tolerances alone to keep the total variation of the system response within specification.

This is the reason to improve the robustness of the system first. When set points for the controllable parameters that minimize error transmission are selected, the sensitivity to the noises is reduced. The search for parameter set points that improve robustness is driven by the objective to make the magnitude of both $\partial y/\partial x_i$ and $\partial y/\partial z_i$ as small as possible. In your search for the best set points, look for parameter settings that place the functional response on its target but with a minimal slope of the response surface. This reduces the variation in y due to variations in both control parameters and uncontrolled noises.

Tolerances should not be assigned until after the system has been developed to be robust. As a result, wider tolerance bands will be acceptable because of the increased latitude in the design for parameter variations. This translates directly into improved manufacturing process capabilities and reduced production costs.

Tolerances and the Cost of Quality

Quality Loss and Manufacturing Cost

It is useful to consider how the total cost of quality (cost of poor quality) is affected by tolerances. Assume for the moment that your product delivers a single functional response. There is a required target for that output. When the response moves off target, your customers will be increasingly dissatisfied and, at some level of deviation, act to resolve that condition. Your customers may return the product for a replacement or demand repair, depending on their circumstances.

We can use the quality loss function and the economics of the tolerance design to build a conceptual model of how tolerances and the cost of quality are related. It is reasonable to assume that, as production tolerances are tightened, the probability of the response being unacceptable in customers' use is reduced. So the cost of quality associated with a functional response being off target should diminish. At the same time, the production costs of the product should increase because of the extra efforts required to achieve those tighter tolerances. For a survey of cost of quality models, see Schiffauerova and Thomson.[1]

Cost of Quality and Optimal Tolerances for a Simple Case

A relationship can be defined between tolerances and the total cost of quality for this simple case. Assume that the relationship between variation in response and quality loss is understood. Recall our earlier discussions of the quality loss function (Chapter 2). The average quality loss for the nominal-is-best case can be expressed as

$$QL = k[(T - \mu)^2 + \sigma^2] \qquad (20.4)$$

where

k = economic proportionality constant

T = target for the response

μ = mean response across the population

σ = standard deviation of the population response

Equation (20.5) shows a general relationship between tolerance specifications and production costs. The reasoning is that as the tolerance becomes very small, product costs increase dramatically.

---

1. Andrea Schiffauerova and Vince Thomson, "A Review of Research on Cost of Quality Models and Best Practices," *International Journal of Quality and Reliability Management* 23, no. 4 (2006).

At the high end of a tolerance, its cost approaches an asymptote where a further relaxation of the tolerance would have much less effect on cost.

$$Cost_p = C_0 + \frac{C_1}{Tol} \tag{20.5}$$

You can combine production cost and quality loss to express the total cost as

$$Cost_t = k[(T - \mu)^2 + \sigma^2] + C_0 + \frac{C_1}{Tol} \tag{20.6}$$

Now assume that there is a proportional relationship between the tolerance and the standard deviation of the response for the population.

$$Tol = C_2\sigma \tag{20.7}$$

This means that the tolerance is specified to be some multiple of the standard deviation of the variation in the manufacturing process. We could argue that tolerances and process variability are unrelated, but in this strategy for tolerance design they are not. Essentially this assumption means that when performance requirements permit, tolerances can be relaxed and greater process variability can be acceptable. This would be the case if the quality loss constant, k, were small, meaning that relatively large deviations from target would result in small economic loss. If you specify a small tolerance range, you have to limit the inherent variability in the process or be willing to accept increased scrap and rework. Either approach costs additional money.

Substituting equation (20.7) into equation (20.6) yields an expression for total cost as a function of the tolerance:

$$Cost_t = k\left[(T - \mu)^2 + \left(\frac{Tol}{C_2}\right)^2\right] + C_0 + \frac{C_1}{Tol} \tag{20.8}$$

Figure 20.1 shows how a plot of total cost might look, assuming the mean of the response variation to be on target. The economically optimal tolerance is determined by the deviation with minimum total cost. If that level of deviation from target is not acceptable to your market, your incremental production costs are too high.

Toward the objective of specifying a tolerance that minimizes total cost, equation (20.8) can be differentiated with respect to Tol to get the following relationship, after setting the result equal to 0:

$$Tol = \sqrt[3]{\frac{C_1 C_2^2}{2k}} \tag{20.9}$$

An examination of equation (20.9) shows the optimal tolerance to depend on three factors:

1. k is the constant of economic proportionality relating quality loss to the total squared deviation of all members of the population. The total squared deviation has two components: the square of the difference between the mean of the population and its target, and the standard deviation of the population, for which the tolerance is a surrogate.

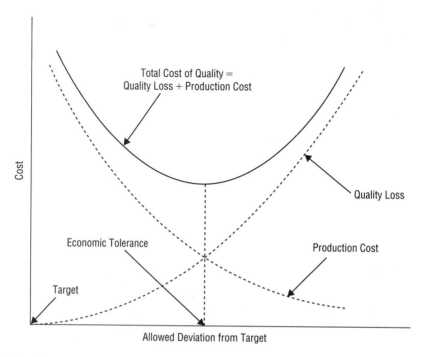

Figure 20.1   Total cost of quality versus tolerance

2. C_1 is the constant of proportionality relating the inverse of the tolerance and product cost.

3. C_2 is the constant of proportionality relating the standard deviation and the tolerance.

Interestingly, the value of the tolerance that minimizes total cost depends neither on whether the mean response is on target nor on the asymptote that production cost approaches as tolerances are relaxed.

The relationships expressed by equations (20.8) and (20.9) are useful as models for thinking about how tolerances are related to product cost and quality loss. Whether this relationship is applicable in building a quantitative model depends on how well the relationship of production costs to tolerances is understood, and whether the quality loss for the product can be quantified. Developing an understanding of these concepts requires working with customers and both internal and external supply chain partners to identify those associated additional costs.

Analysis of Tolerance Stacks

Types of Tolerance Design Problems

There are three basic approaches to analyzing the tolerances in a system of dimensions that stack. Each has its advantages and limitations.

1. The worst-case addition of upper and lower limits is simple, but it is very conservative and ignores the statistical nature of the individual variations.

2. The root sum of squares (RSS) approach is relatively easy and employs the standard deviations of the variations in each contributing dimension. However, it assumes that the means of the variations for each dimension are centered on their targets.
3. Monte Carlo simulation takes advantage of the randomness of individual variations but requires a computer application. The Monte Carlo approach is a valuable way to analyze the allocation of tolerances among contributing parameters.

So the individual tolerances are a given, and you have to estimate how they will add to determine a resulting gap or critical position. The tolerance stack analysis does not require you to be concerned with how the individual tolerances are allocated across the contributing dimensions.

Analysis of Tolerance Stacks: Simple Cases versus the Not-So-Simple

Linear Stacks: One-Dimensional Tolerance Chains

The simplest scenario for a tolerance addition is a linear tolerance stack. The analysis requires the definition of the individual upper and lower tolerance limits for each part in the stack. The objective is to calculate how they add to determine the potential variation in a resulting gap or critical dimension.

Mathematically, all tolerance stacks are defined by a transfer function. The simple linear stack has a simple transfer function, where the dimensions are summed to calculate the tolerance of the critical dimension. Figure 20.2 is an example.

Worst-Case Analysis

For a worst-case analysis, start at a reference point. Treating each dimension as a vector, make a loop and sum all of the dimensions. Assume that vectors pointing to the right (A, B, C, and G) are positive, while vectors pointing to the left (D) are negative. For Figure 20.2, the equation summing all the dimensions around the loop is

$$A + B + C + G - D = 0 \qquad (20.10)$$

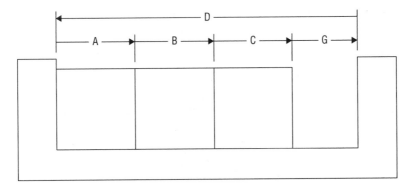

Figure 20.2   Simple one-dimensional tolerance stack

Note that to add the vectors, there is no preferred starting point. The important step is to start somewhere, make a loop, and return to the starting point. The gap, G, is

$$G = D - (A + B + C)$$ (20.11)

In worst-case analysis, the interest is in understanding the extremes of the tolerance range. The maximum and minimum dimension for the gap, G, is the concern. You know that the maximum gap will occur when A, B, and C are at their minimum and D is at its maximum. Likewise, the minimum gap will occur when A, B, and C are at their minimum and D is at its maximum. So the equations for the maximum and minimum values of G are

$$G_{max} = D_{max} - (A_{min} + B_{min} + C_{min})$$ (20.12)

$$G_{min} = D_{min} - (A_{max} + B_{max} + C_{max})$$ (20.13)

Using equations (20.12) and (20.13), the tolerance for G can be stated as

$$G = G_{nom} \pm \frac{1}{2}[(D_{max} - D_{min}) + (A_{max} - A_{min}) + (C_{max} - C_{min}) + (D_{max} - D_{min})]$$ (20.14)

$$G = G_{nom} \pm [Tol_A + Tol_B + Tol_C + Tol_D]$$ (20.15)

So the tolerance on the gap turns out to be the sum of the tolerances on all of the components comprising the stack. This is the essence of worst-case tolerance design.

When Can Worst-Case Analysis Be Misleading?

The advantage of a worst-case tolerance design is its simplicity. The calculation is straightforward and, if the worst-case result is favorable, you should have no problem. Right? It turns out that it's not quite that simple.

There is only one situation where a worst-case analysis can prove, absolutely, that there will be no cases of tolerance stacks outside the calculated worst-case limits. If the production process includes 100% inspection with the removal of all components that have dimensions outside their individual tolerance limits, you can be certain that the tolerance stacks for all assemblies will be within the worst-case limits for the assembly.

You might wonder how you could have components outside the tolerance limits. The outputs of manufacturing processes vary. Statistical process control (SPC) can be applied to monitor the variations and trigger corrective actions when needed. However, unless 100% inspection is applied, there is some probability that parts with dimensions exceeding their tolerance limits could enter an assembly. Few processes can justify the costs of 100% inspection, except where the consequences clearly are not acceptable, such as for an implanted cardiac pacemaker. With process control, the objective is a stable process having some acceptable common cause variation, with a mean on target, rather than catching individual defects.

This brings us to another weakness of the worst-case approach. When there are many contributors to a tolerance stack, a worst-case analysis will be very conservative. Having all component dimensions at their extreme dimensions simultaneously is highly unlikely, particularly if the characteristics of interest vary independently with statistical distributions such as the normal distribution. The more independent contributors there are to the tolerance stack, the more likely it becomes that the accumulated variation will have a normal distribution. If there are dependencies, the problem becomes more complicated because of the possible covariance between some dimensions.

While the worst-case analysis may be too conservative, a stack with only a few components may be too liberal, resulting in an estimate of variation that may have a higher probability of exceeding desired limits.

Worst-case analysis does not address the statistical nature of the tolerance stack. It is more of a go/no-go analysis. The RSS approach, however, does recognize the statistics of the tolerance analysis and addresses some of the weaknesses of worst-case analysis.

Root Sum of Squares (RSS)

Recall three characteristics of random variables:

1. If a set of random variables, x_1, x_2, x_3, and x_n, are all normally distributed, then y, as given by equation (20.16), is also normally distributed:

$$y = x_1 + x_2 + x_3 + \cdots + x_n \tag{20.16}$$

It turns out that y will approach a normal distribution if it is a function of many x_i, even though the individual x_i may not be normally distributed. This is a consequence of the central limit theorem, discussed in Chapter 14 when we outlined some of the reasons for the normal distribution to be experienced so often.

2. The mean of y is the sum of the means of the x_i, the individual random variables:

$$\bar{y} = \bar{x}_1 + \bar{x}_2 + \bar{x}_3 + \cdots \bar{x}_n \tag{20.17}$$

3. The variance of y is the sum of the variances of the x_i:

$$\sigma_y^2 = \sum_{i=1}^{n} \sigma_{x_i}^2 \tag{20.18}$$

These three relationships are useful in the process for RSS tolerance design. If you apply these relationships to the calculation of the gap dimension, G, given by equation (20.11), you get the following relationships:

- The nominal, or average, gap dimension is

$$G_{nom} = D_{nom} - (A_{nom} + B_{nom} + C_{nom}) \tag{20.19}$$

- The variance of the gap dimension is given by

$$\sigma_G^2 = \sigma_A^2 + \sigma_B^2 + \sigma_C^2 + \sigma_D^2 \tag{20.20}$$

With equations (20.19) and (20.20), the tolerance analysis has a statistical footing. Now you can calculate the probability of the gap dimension being greater than or less than a specified amount.

Example

Calculate the confidence interval for a gap that includes 99.9% of the possible outcomes. By using a table or Minitab, you can see that it requires ±3.291 standard deviations from the mean to capture 99.9% of the possible outcomes. The 99.9% confidence interval for the gap is given by

$$G_{nom} - 3.291\sigma_G < G < G_{nom} + 3.291\sigma_G \tag{20.21}$$

where G_{nom} and σ_G are given by equations (20.19) and (20.20) respectively.

The RSS analysis enables the calculation of the confidence interval for a tolerance stack, given the desired probability of being within a particular range. Alternatively, you can solve the inverse problem, to calculate the probability of the tolerance stack being within a specified range.

Limitations of RSS Analysis

You have seen that an RSS analysis puts the tolerance design on a statistical basis. However, there are some concerns to keep in mind:

1. RSS assumes normality of the distribution of dimensions or characteristics of interest when analyzing a tolerance stack. You can waive the normality requirement if there are many contributors to the tolerance stack.
2. RSS assumes that all dimensions or functional responses considered in the tolerance analysis are centered and on target. This will not be the case when there is drift in the process causing the mean to move off center.
3. RSS requires knowledge of the variance of the manufacturing process for each part.
4. RSS assumes the independence of all of the random variables comprising the stack. The practical meaning of this is that in the manufacturing processes, the dimension or characteristic of each variable is determined to be independent of all others. There are opportunities for this independence requirement to be violated. Suppose that the components are created by the same manufacturing process. When there is drift in a given direction, it is reasonable to expect a correlation among the properties of other parts or materials delivered by that process. Another example is when materials are manufactured in large batches. It is reasonable to expect a correlation of material properties within each batch. For many manufacturing processes you can estimate the dimension or characteristic of the next part, based on that of the previous part.

    If parts from many lots are mixed in the parts bin, possibly the assumption of normality is reasonable. So whether or not RSS will yield meaningful results is dependent on the details of the origin of the parts or materials and how they were distributed in the manufacturing process.

In the next section we discuss Monte Carlo simulation, a valuable tool for analyzing tolerances. The method is applicable to most tolerance design problems and can be applied to those that are not amenable to analysis using either worst-case or RSS methods. Beyond tolerance analysis, Monte Carlo simulation is an integral part of the tolerance allocation process using stochastic optimization.

Monte Carlo Simulation

Monte Carlo simulation became a very useful tool once computing power became widely available. However, it was first used long before computers. One of its earliest applications was Buffon's needle, a famous problem in probability first considered by Georges-Louis Leclerc, Comte de Buffon, an eighteenth-century French polymath. Buffon had broad interests and made contributions in a diverse range of subjects, including mathematics and the natural sciences. By Darwin's estimation he was the first to recognize and write about the likelihood of evolution and its effects on the development of species.

Buffon's needle problem consists of determining the probability that a needle of length l will intersect a crack between uniformly spaced floorboards of width w, when tossed at random over one's shoulder. Buffon used calculus to derive an analytical solution in which the probability depends on needle length, floorboard width, and pi.

Buffon also performed experiments, making many needle-tossing trials and observing the frequency of successful outcomes. The end result was an analytical relationship as well as empirical data that could be used to estimate pi.

The most famous early use of Monte Carlo simulation, which really launched the method as a modern simulation tool, was by Stanislaw Ulam and John von Neumann on the Manhattan Project during World War II. They applied it to solving problems in nuclear fission that, at the time, were considered intractable using the tools available.

Today, Monte Carlo simulation is a useful analytical tool whenever the performance of a system design or a technical concept is vulnerable to the effects of variability and uncertainty. The quality that makes it so useful is its flexibility.

Some of the advantages of Monte Carlo simulation are these:

1. You are not limited to variables that are continuous and normally distributed.
2. You can tackle problems of greater complexity.
3. There can be complex relationships among the variables, relaxing the requirement that all variables be independently distributed.

The availability of affordable desktop applications, such as Crystal Ball and @Risk, makes Monte Carlo simulation a tool that belongs in the toolbox of every engineer and scientist.

Stochastic Optimization of Tolerance Allocation

The process followed in tolerance analysis using Monte Carlo simulation is to select randomly a sample value for each variable from its statistical distribution and calculate the result. For our example problem shown in Figure 20.2, you would sample values for A, B, C, and D and then, using equation (20.11), calculate G, saving all the values to see how they are distributed and to compute statistical parameters for G such as the standard deviation. This is an important part of stochastic optimization that we will discuss now.

Consider the simple one-dimensional tolerance stack shown in Figure 20.2. The expression for the gap, repeated here, is

$$G = D - (A + B + C) \tag{20.22}$$

The initial assumption is to make the tolerance for each part the same multiple of the manufacturing process variance for that part. Suppose you learn that the objectives have been restated, requiring the maintenance of the quality level for the gap, G, and at the same time reducing parts cost. Data relating part cost to part tolerance can be developed by working with suppliers.

For example, the goals might be

1. Maintain a nominal gap dimension of $G = 1.5$.
2. Maintain quality level as measured by the standard deviation of the gap dimension.
3. Reduce the cost of parts.

In order to minimize the cost of parts, the part costs need to be described as a function of the tolerance for each part. It is likely that changing the tolerances on parts will not have the same effect on the cost for each part. It can depend on many factors, such as

• The materials and manufacturing processes used
• The suppliers' methods
• Whether or not the processes can be improved
• The cost of process improvements
• How the costs of process improvements are amortized over the life of production
• If a process cannot be improved, the existing process capability and the cost of rework or scrap for out-of-specification parts
• The decision criteria for rework versus scrap and associated costs

So the decision algorithm for what to do when parts fall outside their specified tolerance range can become complex with a possible multitiered cost structure.

For simplicity, assume that the only cost affected by the tolerances is parts cost. Also assume that the gap requirement is a given and that you do not need to include quality loss in your analysis. A simple relationship between part cost and tolerance is shown by equation (20.23). Instead of expressing cost in terms of the standard deviation of the critical dimension for each part, assume that the cost is given by the following relationship:

$$Cost_i = C_{0i} + C_i Q_i^{n_i} \tag{20.23}$$

There are a number of constants in this cost relationship, as defined here:

C_{0i} = base cost for a part with quality requirements relaxed

C_i = constant of proportionality

Q_i = the "sigma" quality level, also expressed as the ratio of a part's tolerance to its standard deviation. The initial assumption in this example is three-sigma quality. With this assumption, you can easily see which tolerance has been tightened or loosened after the optimization process has completed.

n_i = exponent defining how part cost varies with quality level

This cost model is fairly general and can handle a range of possible tolerance-versus-cost relationships. The functional tolerance-versus-cost relationship for an individual part will be specific to the part and dependent on the production process and materials applied.

Monte Carlo simulation can analyze such a problem and allocate tolerances for each of the parts. An input is an assumed statistical distribution for each of the dimensions. Crystal Ball can then be used to calculate the tolerance stack. The vector relationships in the design configuration are derived from the CAD models.

The optimization problem has the goal to minimize the total parts cost, where total cost is

$$Cost = \sum_i (C_{0i} + C_i Q_i^n) \tag{20.24}$$

Minimization of cost is subject to the requirement that the standard deviation of the gap after optimization is less than or equal to its initial value. The standard deviation of the gap is

$$\sigma_G = \sqrt{\sum_i \left(\frac{Tol_i}{Q_i}\right)^2} \tag{20.25}$$

For this case assume $n = 2$, meaning that part cost increases with the square of quality level. This is the same as the cost increasing inversely with the square of the standard deviation (the variance) of the part manufacturing process.

Keep in mind that the constants in equation (20.24) could be different for each part, since the cost of each part could have a different functional relationship with the standard deviation of the part's manufacturing process. The only independent variable in the optimization is the quality level for each part. The initial tolerances from the RSS analysis and the cost for each part are shown in Table 20.1 along with the results of the optimization using Crystal Ball/OptQuest.

This stochastic optimization process is straightforward. It involves a search for values of the decision variables that will achieve the goals set for the optimization. In the case of our example, the decision variables are the quality levels for each of the parts.

During the optimization, each of the quality levels is varied following several different algorithms. This is done for a couple of important reasons. One reason is to ensure that the optimization does not stop too early by converging on a local optimum. The other is to make the process as efficient as possible without requiring an exhaustive search.

For various combinations of decision variable settings, a Monte Carlo simulation is performed to understand the variability of system performance. The winning combination of set

Table 20.1   Results of optimization

	Nominal	Initial σ	$Tol = \pm 3\sigma$	Initial Q	Optimum Q	Optimum σ	C_{0i}	C_i	Initial Cost	Optimum Cost
A	2.00	0.002	0.006	3.00	2.00	0.003	3.50	0.50	8.00	5.50
B	2.00	0.003	0.009	3.00	3.00	0.003	4.75	0.25	7.00	7.00
C	2.00	0.005	0.015	3.00	5.00	0.003	2.50	0.10	3.40	5.00
D	7.50	0.003	0.009	3.00	2.00	0.0045	5.00	1.00	14.00	9.00
G	1.50	0.006856	0.021			0.006874	Total Cost		32.40	26.50

points is the one with minimum total cost that satisfies the quality requirements. In the example, the part costs were reduced while maintaining the quality level for the assembly.

The initial quality is shown in the column labeled "Initial Q." The column "Optimum Q" shows the quality level after the optimization. The quality level was increased for part C, decreased for parts A and D, and remained the same for part B. After optimization the standard deviation of the gap is still about the same, while the cost has been reduced by 5.90 units, about 18%.

A big advantage of stochastic optimization is its flexibility. The previous example shows one approach on one type of problem. Another approach to this problem could be to set a target for DPU while minimizing cost. This would require setting upper and lower limits for the G dimension before running the simulation.

Optimization of Functional Responses

The preceding example illustrates tolerance optimization for the simplest case, the one-dimensional tolerance stack. While tolerance stacks are an important class of problems considered in design, there are other golden opportunities to use tolerance optimization.

Reconsider the example of the lathe, discussed in Chapters 17 and 18. The goal is to produce machined parts with a minimal amount of deviation from their specification while having acceptable variability of the critical control process parameters. The important lathe functional response, *Delta*, is the difference between the specified part dimension and the dimension of the machined part. The transfer function for this response is

$$Delta = -0.441 - 108.6A - 0.0022B + 30.12C + 0.253AB - 952.5AC$$
$$- 0.012BC + 1950.6A^2 + \varepsilon \tag{20.26}$$

Table 20.2 shows the control parameter ranges, standard deviations and initial tolerance levels.

The objective for this problem is to minimize the total cost of quality where the cost of quality includes all costs associated with improving the process or relaxing the performance requirements, plus the quality loss associated with a given level of performance. Look for opportunities to improve performance and reduce quality loss, while at the same time spending as little as possible on design or process improvements. For this problem assume that the relationship between performance and quality loss is understood. The appropriate quality loss function is the nominal-is-best model given by

$$QL = k[(\mu - T)^2 + \sigma^2] \tag{20.27}$$

where

　　　k = a constant of economic proportionality

　　　μ = process mean

　　　T = target for the functional response

　　　σ = process standard deviation

Table 20.2 Lathe optimization control factors

Factor	Range	Standard Deviation	Tolerance	Initial Quality Level, Q	Units
A: Feed	0.010–0.022	0.003	0.009	3	in/revolution
B: Speed	330–700	5	15	3	ft/min
C: Depth	0.05–0.10	0.0125	0.375	3	in

The only way to build the quality loss function is by understanding customers' sensitivity to off-target performance. You can gain this understanding by talking to customers and by reviewing the costs of service interventions.

The cost of quality can be reduced by improving the lathe subsystems that control speed, feed, and depth of cut. If you understand how subsystem costs are affected by subsystem performance requirements, you can perform an optimization with the goal to minimize the total cost of quality. To keep things simple, include only quality loss and subsystem cost in the total cost of quality. Therefore, total cost of quality is given by

$$COQ = k[(\mu - T)^2 + \sigma^2] + \sum_i (C_{0i} + C_i Q_i^{n_i}) \tag{20.28}$$

This model assumes that the cost of quality is a continuous function of Q_i, the quality level. However, this does not have to be the case. Subsystem costs could increase stepwise as the quality level is increased, or they could be tabulated data described in a lookup table to be used for the optimization runs. The key to a successful optimization is having a credible description for the cost of quality that depends on the quality level for each important system element.

In this optimization, the strategy is to search for a combination of the quality levels for each subsystem that minimizes the cost of quality. The standard deviation of the response, *Delta*, depends on both the nominal setting for the control factors and the tolerance for each setting. In the last example, the requirement was to limit the standard deviation of the gap to be no larger than the original design, with the objective to minimize the cost of the parts. Quality loss was not included explicitly in that example. The approach for this problem could have been the same as in the first example. Instead it included quality loss in *COQ* and allowed variability to be an outcome.

Remember that the model for the cost of quality can be as complex as you want to make it. Given enough time and thought, you could include many other factors in the model. However, a better strategy is to include just those factors that are most critical to the functional response and closely coupled to the design of the system tolerances. If you minimize those costs, it is likely that you will also minimize many secondary and tertiary costs without having to include them explicitly in the *COQ* function. The farther away you are from the important contributors to *COQ*, the fuzzier will be your understanding. Larger models are not necessarily better models.

Natural Tolerances

Natural tolerances for manufacturing processes are those that are economically reasonable, that yield the most consistent outputs without extraordinary efforts. Tolerances tighter than economically efficient tolerances will drive reduced yields and higher costs. This definition is imprecise and recognizes that the quality of a process, as measured by its process capability, will vary among businesses and suppliers. If a good quality process is in place, and tools such as SPC and MSA are part of the fabric of your organization, your products can be delivered with more consistency and higher quality. To answer the question "Is the quality that is delivered adequate?" you have to compare the quality requirements, as driven by customer needs, and manufacturing process capabilities. Capability shortfalls are opportunities to improve and likely to reduce the total cost of quality. Remember, the total cost of quality includes the quality losses for all stakeholders, including customers, suppliers, and your own business.

In many cases, process variability may increase over time. This can be expected as a natural result of the deterioration of tooling, such as wear of a mold for making injection molded parts or of a die for making stampings. This is yet another reason to monitor manufacturing processes using tools such as SPC.

Accommodate Looser Tolerances with Assembly Adjustments

In some cases you may need to maintain a dimensional relationship that would be impractical to achieve with fabricated tolerances alone. In such a case you should first answer the question "Do we have a good design concept and associated production implementation?"

A reasonable and cost-effective alternative may be to provide adjustments during product assembly. However, adjustments are additional sources of errors. In general they involve fixtures, instrumentation, and training, so they are another step to the manufacturing process with associated costs.

Let's Fix It in Software

There are situations for which critical design parameters can drift over time with customers' usage. This deterioration can be caused by contamination or environmental factors, degradation of parameters, or interactions among components. An example is a property of a consumable material that degrades or a replaceable component that suffers wear in a copier/printer. In the design of such systems, it is common practice to make compensating adjustments in certain process parameters. This approach, known as process control, has enabled manufacturers to improve product availability by extending the interval between service interventions. So if you are designing a serviceable product and parameter drift over time is a major concern, the system architecture can include the instrumentation and algorithms for process control required to maintain on-target performance and thereby extend the mean time between service calls.

Implementing such a strategy requires the product to have the built-in intelligence and sensors that enable it. These add cost and complexity to the product and more sources of failure. The additional product cost could be offset by the savings in service costs as well as the savings in quality loss. Your market may perceive a product that does not need internal process control as being inherently more robust, a more efficient design.

Tolerances and Repeatable, High-Precision Positioning

At times the functionality of a system requires precise and repeatable positioning. An example is the design of a demountable lens for a camera. The lens assembly must be removable, and when remounted, it must be in the same location. Depending only on tight tolerances in this situation would be costly and ineffective. Exact constraint design[2] is a technique to consider for this kind

---

2. Douglass L. Blanding, *Exact Constraint: Machine Design Using Kinematic Principles* (New York: ASME Press, 1999).

of situation. Although outside the scope of this book, it is a better choice than tight tolerances to meet certain types of design requirements.

Key Points

1. Sources of variability affecting product performance can include
   a. Unit-to-unit variations due to stresses in the supply chain, assembly, storage, distribution, installation, repair, and maintenance
   b. Internal deterioration during the use of the product
   c. Environmental effects in the product's storage, distribution, installation, and use
   d. Stressful elements in users' applications, including customer abuse
2. A tolerance is an acceptable range of variation in a critical functional response or in a critical design parameter affecting that functional response.
3. Tolerance design is not a poker game. It is a process requiring information sharing, trade-offs, and trust among cross-functional development teams and partners in the production supply chain.
4. First make the system robust, and then design the tolerances. Increased robustness will enable relaxed parameter tolerances for a given amount of variability in a critical functional response.
5. Tolerances should be derived from criteria for customers' satisfaction, either directly or indirectly.
6. The stakeholders for tolerance design include customers, suppliers, and your business.
7. To be useful in tolerance design, quality loss requires an understanding of the economic consequences of off-target performance.
8. A higher manufacturing process capability can be caused by a smaller variation in a replicated parameter, a wider latitude for that variation, or both.
9. There are two kinds of tolerance design problems that you can analyze. The simple tolerance stack determines how an assembly of parts fits together. The more complex tolerancing problem requires a transfer function to determine how the tolerances for system elements affect expected variations in a critical functional response for the system.
10. Tolerances can affect some important factors including parts costs, ease of assembly, serviceability, and customers' satisfaction. It is important to understand the effects of tolerances on each of these when making tolerancing decisions.

Discussion Questions

1. Why should you make a system robust before assigning tolerances?
2. How can you justify relaxed tolerances?
3. What elements must change to loosen tolerances?
4. How would you go about determining the quality loss function for a system functional response?
5. What information is needed to perform tolerance optimization?

6. Which functions in an organization have the most input to tolerance design decisions?
7. What are the negative consequences of tightening tolerances?
8. What should be done if the supply chain is unable to deliver components with the required tolerances at high yield?
9. Are there any manufacturing alternatives to tighter tolerances?
10. At what point in a product development project should tolerances be considered?
11. What is the risk of waiting to address tolerances until late in a development project?
12. What is the risk of concluding that tolerances are acceptable based on alternate process parts that often are used before the final tooled parts are available?

Reliability Functions and Life Distributions

Reliability and probabilities are closely coupled. In fact, reliability is a probability. Recall our definition early in the book:

> **Reliability** *is the probability that a system or component will perform its intended function for a specified duration under the expected design operating conditions.*

Of course, reliability is much more than a probability. So far, we have focused on the proactive steps you can take to develop reliability. Understanding the measures of reliability is very important for the design of a new product. Design teams may need to estimate service or warranty costs, or focus their efforts to improve customers' satisfaction. Because of variability in both the strength of designs and the deteriorating loads or stresses that are imposed, failures of systems or components are expected to be distributed over time. In this chapter we discuss fundamental reliability functions and some very useful reliability distributions.

Fundamental Reliability Functions

This chapter is an introduction to the basic concepts of reliability functions, although not a comprehensive in-depth treatment, which would be beyond the scope of the book. The body of knowledge for reliability is broad, deep, and, in many cases, very mathematical. There are many excellent references[1] for those interested in greater detail. The good news is that developing a working knowledge of reliability fundamentals is within the grasp of anyone with a basic understanding of calculus. If you are comfortable with derivatives and integrals, you are in good shape to absorb the chapter. If your calculus skills are rusty, read on anyway. Chances are that the concepts will still make sense.

---

1. Marvin Rausand, and Arnljot Høyland, *System Reliability Theory: Models, Statistical Methods, and Applications, 2nd ed.* (Hoboken, NJ: John Wiley, 2002); Elsayed A. Elsayed, *Reliability Engineering* (Reading, MA: Addison Wesley Longman, 1996).

Modeling Reliability

There are several reasons to fit a life distribution to failure data:

1. To make forecasts of expected system reliability and to predict service and warranty costs

   If you have a life distribution based on historical life data or on the results of accelerated testing, you can make predictions about early-life failures. A credible forecast of related warranty costs is an important part of the business plan for any new product.

2. To understand if failures are caused by quality problems, random events, or wear-out due to deterioration

   Our later discussion of the Weibull distribution points out that the shape of the life distribution gives you insight into whether failures are expected early in life, randomly, or at the end of the product's life. Failures in the early life of the product will point you toward improving the manufactured quality. If early-life failures occur in prototypes, it's expected that design quality also could be a factor. Random failures tell you that your system may be affected by external stresses in operation, such as customers' usage, power surges, or environment changes. Wear-out failures, depending on when they occur, might indicate the need for more durable components.

3. To set priorities for improvements

   Usually, in complex designs, there will be a few critical elements that determine the system reliability. Once you have identified them, you will have a good start to making effective improvements. The Pareto principle (80/20 rule) often applies.

In the Kano Model, reliability is a product characteristic that most customers view as a basic need. It is assumed that a product will be reliable. If it is not, your customers can be expected to take their business elsewhere.

Fundamental Reliability Functions and Relationships among Them

There are four fundamental reliability functions:

1. Life distribution, $f(t)$; also known as the probability density function (PDF)
2. Cumulative distribution function, $F(t)$; also known as the CDF
3. Reliability or survival function, $R(t)$
4. Hazard function, $h(t)$

Since you usually observe failures and then fit a life distribution to the data, we start with the life distribution and show how the other functions can be derived from it. In fact, you can start with any one of them and derive the others.

Consider the situation where you have a sample of components from a large population and test them to failure, tracking when each component fails. You are interested in questions such as these:

- How are the failures distributed across time? Are they clustered with a central tendency or spread out over a long period of use?
- At any given time after starting the test, what percentage of the total components tested has failed? What percentage of the population tested is still surviving?
- Over time, at what rate are failures occurring?
- Is the failure rate increasing, decreasing, or remaining constant over time?

Life Distribution (PDF)

The life distribution is a mathematical description of how failures are distributed across time or actuations, or whatever metric you have chosen as a measure of usage life. It is especially important to be able to predict early failures because of their consequences for customers' satisfaction and service costs. Sometimes analysts start with a histogram to show the distribution of failures over time. A histogram is a non-parametric display of the data, a model that does not assume that system or component life is describable by a probability distribution. When you take the next step to fit a life distribution to the data, you then have a parametric model that can be used to forecast expected system reliability. Most life models used in reliability analyses are parametric, because usually a parametric model can be fit to data, making analyses and predictions more efficient. In Figure 21.1 you can see that the histogram and life distribution have similar shapes.

A life distribution is smooth and continuous while a histogram is stepped and irregular. The figure shows a histogram for 100 times to failure for a component, with the fitted distribution labeled $f(t)$. The major advantage of the life distribution, over the histogram, is that it can be used to estimate reliability, especially early-life failures. The life distribution is scaled to make the area under the curve equal to 1. So the units for the life distribution are the fraction failed per unit of time or per number of actuations. The life distribution is usually designated as $f(t)$ and is also called the probability density function or probability distribution function (PDF). The area under the life distribution between two points in time is the probability of failure in that time interval. If there are no life distributions that are a good fit to the data set, there are non-parametric methods that can be used to analyze the data. This is often the case in the life sciences, where tools such as Kaplan-Meier are used.

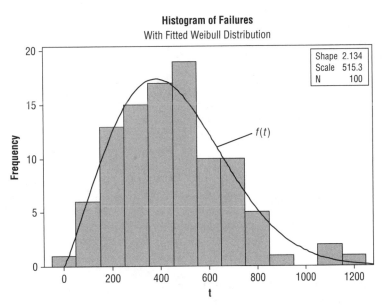

Figure 21.1   Histogram of 100 data points and a fitted Weibull distribution

Cumulative Distribution Function (CDF)

The CDF shows the fraction failed versus time or actuations and is the cumulative area under the curve that describes the life distribution. It can be derived from the life distribution:

$$F(t) = \int_0^t f(t)\,dt \tag{21.1}$$

So starting with the CDF, we can get the life distribution by differentiating equation (21.1):

$$f(t) = \frac{d}{dt}F(t) \tag{21.2}$$

Reliability or Survival Function

The reliability function expresses the fraction surviving at time t. It is given by

$$R(t) = 1 - F(t) \tag{21.3}$$

Equation (21.3) is true because there are only two possible states for a component, surviving or failed, so at any time during the process,

$$R(t) + F(t) = 1 \tag{21.4}$$

Using equations (21.2) and (21.3), you can get the following useful relationship:

$$f(t) = -\frac{d}{dt}R(t) \tag{21.5}$$

Hazard Function

The hazard function is the failure rate over time. It is defined as

$$h(t) = \frac{\text{failures per unit time}}{\text{number of survivors}} \tag{21.6}$$

You can express this as equation (21.7):

$$h(t) = \frac{\frac{d}{dt}F(t)}{R(t)} \tag{21.7}$$

Using equation (21.3), you get

$$h(t) = \frac{-\frac{d}{dt}R(t)}{R(t)} \tag{21.8}$$

By integrating equation (21.8), you get the relationship

$$\ln[R(t)] = -\int_0^t h(t)\,dt \tag{21.9}$$

This leads to a relationship that is fundamental and independent of any particular life distribution:

$$R(t) = \exp\left[-\int_0^t h(t)\,dt \right] \tag{21.10}$$

Using equations (21.5) and (21.8), you can also get the relationship

$$f(t) = h(t)R(t) \tag{21.11}$$

Using these relationships, it's possible to derive the reliability functions for life distributions, such as the Weibull and exponential models.

Reliability Metrics

Mean Time between Failures (MTBF)

The MTBF is defined as the total operating time divided by the total number of failures. It is not tied to any particular distribution.

$$MTBF = \frac{\text{total operating time}}{\text{total number of failures}} \tag{21.12}$$

Mean Time to Repair (MTTR)

MTTR is the factor that makes system availability less than 100%. It includes both response time and service time. Both the time between failures and the time to repair can be described by different statistical distributions.

Availability

Another metric, useful for repairable systems, is availability, defined as

$$Availability = \frac{MTBF}{MTBF + MTTR} \tag{21.13}$$

Availability is improved by increasing MTBF and decreasing MTTR.

Useful Life Distributions

For reasons discussed in Chapter 14, many natural and man-made phenomena are well described by the normal distribution. Curiously, the normal distribution is used rarely, if ever, to describe the distribution of times to failure. While the normal distribution is bell-shaped and symmetrical, the distributions most useful in reliability tend to be skewed, often having a long tail extending out in time.

The following distributions are among the most useful for the parametric analysis of life data:

- Exponential
- Weibull
- Lognormal
- Poisson
- Binomial

If you plot a histogram of your survival data and see a shape that is not characteristic of the normal distribution, you may be dealing with one of these distributions.

Exponential Distribution

The exponential is a continuous, one-parameter distribution. It is highly skewed, covering a range from zero to infinity. The exponential distribution is a good reliability model in a couple of situations:

1. For complex repairable systems, it is a good model for the time between failures at the system level, although not necessarily at the component level. Large systems with many components have failures arriving at an average rate over time. Even if all components in a system fail by wearing out, if there are enough of them, at the system level the life distribution will have a close resemblance to the exponential distribution.
2. Failures caused by random external stress factors will arrive at some average failure rate over time, with the time between failures distributed exponentially.

The exponential distribution can be derived by making the simple assumption of a constant failure rate. If the number of failures per unit time is a constant fraction of the number of survivors, the equation describing the process is

$$\frac{dn}{dt} = -\lambda n \tag{21.14}$$

where

n = the number of survivors at any time t

λ = the failure rate

Equation (21.14) can be integrated to get the relationship

$$\frac{n}{n_0} = e^{-\lambda t} \tag{21.15}$$

where n_0 is the number of components at the start of the process.

Equation (21.15) is the reliability function. Using equation (21.5), you can get the PDF for the exponential distribution, shown as equation (21.16):

$$f(t) = \lambda e^{-\lambda t} \tag{21.16}$$

The exponential distribution for a system with MTBF = 500 hours is plotted in Figure 21.2.

For the exponential distribution, the cumulative distribution function (CDF), the fraction failed, is given by

$$F(t) = \int \lambda e^{-\lambda t} dt = 1 - e^{-\lambda t} \tag{21.17}$$

The reliability or survival function is given by

$$R(t) = e^{-\lambda t} \tag{21.18}$$

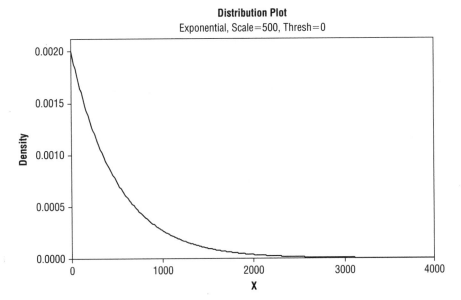

Distribution Plot
Exponential, Scale=500, Thresh=0

Figure 21.2   Exponential distribution

Substituting equation (21.18) into equation (21.8), the hazard function for the exponential distribution is

$$h(t) = \lambda \qquad (21.19)$$

If a hazard function is constant, it means that a constant percentage of the survivors fail per unit time. If you know the MTBF, the failure rate is just the inverse:

$$\lambda = \frac{1}{MTBF} \qquad (21.20)$$

Remember that failures do not arrive with clockwork regularity. Over time they arrive at an average rate, but the time between arrivals will be exponentially distributed, appearing to be random and unpredictable. This means that some failures will occur in close proximity while occasionally there will be relatively long intervals between failures. In Chapter 22 we discuss the memory-less property of the exponential distribution and its practical meaning.

Examples

1. A product has an MTBF of 1500 hours, and its time between failures is exponentially distributed. How long will it take for 10% of products placed into service to fail?

   You know that if 10% have failed, 90% are surviving, so the reliability R = 0.9. Using equation (21.18), you can solve for the time in terms of reliability and MTBF.

$$t = -MTBF \ln(R) \qquad (21.21)$$

Solving equation (21.21) yields

$$t = -1500 \ln(0.9) = 158 \text{ hrs} \qquad (21.22)$$

2. A disk drive manufacturer claims that a hard drive has an MTTF = 500,000 hours. Assuming an "on-time" of 100%, what fraction will have failed by one year?

It is reasonable to assume an exponential life distribution, since disk drives are complex electromechanical devices. Using equation (21.18), you can calculate the reliability for one year, 8760 hours of operation.

$$R = \exp\left(-\frac{8760}{500,000}\right) = 0.9826 \qquad (21.23)$$

So the fraction failed by the end of one year is expected to be

$$F(t) = 1 - R = .0174 \qquad (21.24)$$

As we mentioned, this assumes a constant failure rate. If the failure rate is not constant, the exponential distribution is not appropriate and another life distribution should be used.

This example raises questions about the quoted MTTF in product descriptions, such as

- Is the drive performing its function consistently?
- With what duty cycle is the drive operating?
- How many read/write cycles are being experienced?

When reliability is quoted as part of a marketing presentation or in sales literature, it may be advisable to be skeptical until the basis for the claims is understood. The assessment of reliability claims can include comparisons of data for similar products under the same conditions, and the results of accelerated life testing. Based on conclusions from studies[2] of actual field data, the claim of 500,000 hours is probably not representative of real-world reliability in the hands of customers for this type of electromechanical device.

Weibull Distribution

The Weibull distribution is a continuous, two-parameter distribution. It is named after Waloddi Weibull, a Swedish engineer who published an important paper[3] leading to widespread use of his methods. It is a versatile model for a range of failure behaviors, used often in the analysis of reliability life data.

The PDF for the Weibull distribution is

$$f(t) = \frac{\beta}{\eta}\left(\frac{t}{\eta}\right)^{\beta-1} \exp\left[-\left(\frac{t}{\eta}\right)^{\beta}\right] \qquad (21.25)$$

2. Bianca Schroeder and Garth Gibson, "Disk Failures in the Real World: What Does an MTTF of 1,000,000 Hours Mean to You?" Fast '07: 5th USENIX Conference on File and Storage Technologies (2007).

3. Waloddi Weibull, "A Statistical Distribution Function of Wide Applicability," *ASME Journal of Applied Mechanics,* September 1951, 293–97.

The reliability or survival function is

$$R(t) = \exp\left[-\left(\frac{t}{\eta}\right)^\beta\right]$$ (21.26)

The hazard function is

$$h(t) = \frac{\beta}{\eta}\left(\frac{t}{\eta}\right)^{\beta-1}$$ (21.27)

Going back to our earlier discussion, if you assume a two-parameter hazard function, as given by equation (21.27), using equations (21.10) and (21.11) you can derive the Weibull PDF given by equation (21.25).

The two distribution parameters for the Weibull are the shape factor and the scale factor. The shape factor, β, provides valuable information for the analysis of failure data.

Values of β less than 1 indicate early-life failures which, in a mature product may indicate quality problems in the supply chain. During new product development, early-life failures can also be caused by a lack of developed robustness in new technical concepts or design configurations, as well as inadequate implementation of DFMA guidelines.

$\beta = 1$ indicates that the time between failures is exponentially distributed. That means either that failures are caused by external events or that the failure data represent a complex system having many elements that can fail.

Values of β greater than 1 mean that it is likely that wear-out is a contributor to failures.

As β increases, the Weibull distribution has a stronger central tendency, with more failures clustered near the mean.

The Weibull scale factor, η, is related to the distribution mean, as shown here:

$$\mu = \eta\Gamma\left(1 + \frac{1}{\beta}\right)$$ (21.28)

where $\Gamma()$ is the Gamma function, which can be calculated using applications such as Excel or Mathcad.

Although the Weibull scale factor is a parameter of the Weibull distribution, people normally think about the mean life when considering reliability. For the exponential distribution where $\beta = 1$, the mean life is equal to the scale factor. For other values of β they are not equal.

If $\beta < 1$, $\eta < \mu$.   for $\beta = 0.5$, $\mu/\eta = 2.0$

If $\beta = 1$, $\eta = \mu$.   for $\beta = 1.0$, $\mu/\eta = 1.0$

If $\beta > 1$, $\eta > \mu$.   for $\beta = 3.0$, $\mu/\eta = 0.89$

The various functions for a Weibull life distribution fit to a data set are plotted in Figure 21.3.

The form of the hazard function is determined by the physics of failure. Most components that fail from deterioration degrade monotonically during use. At some point failure occurs when degradation has progressed to the point that the normal operating stresses exceed the strength of the component. Fatigue of metals is an example of this type of behavior. After many cycles of

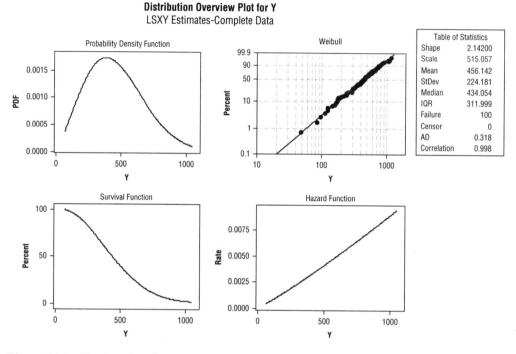

Figure 21.3   Key functions for Weibull distribution; data are sampled from a distribution with shape factor = 2, scale factor = 500.

stress, a crack may initiate at some point determined by the stress field and the presence of micro defects. The crack will then propagate in a direction perpendicular to the direction of the maximum stress, until failure occurs. Other types of failure might involve parts seizing because of changes in friction properties due to mechanisms such as galling. The common theme is a gradual degradation, eventually ending in a failure to perform its function.

In the context of a large population of samples being tested, it is reasonable to expect a statistical distribution of time or cycles to failure. As an increasing number of samples approach the critical conditions for failure, the failure rate is expected to increase. It makes intuitive sense that the failure rate and hence the shape of the life distribution is determined by the physics of failure. There are many references[4] that discuss how the physics of a failure determines the shape of the life distribution. In order to effectively analyze life data and fit a distribution, you should be aware of which distributions have been found to be most useful in modeling the different processes of degradation and failure. However, it is not necessary to understand how micro-level behaviors determine the hazard function for a process.

Figure 21.4 shows the hazard function for various values of β.

---

4. Meeker and Escobar, *Statistical Methods for Reliability Data*.

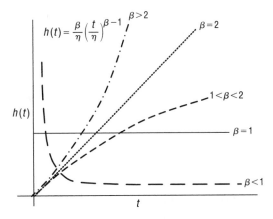

$$h(t) = \frac{\beta}{\eta}\left(\frac{t}{\eta}\right)^{\beta-1}$$

Figure 21.4 Hazard function versus time for various values of the shape factor

Table 21.1 Relationship of Weibull shape factor and failure causes

Value of β	Failure Mode	Likely Causes
$\beta < 1$	Early-life failures	Look for supply chain quality problems.
		Examples:
		• Failure of electronics due to weak solder joints
		• Product being DOA due to incorrect wiring
		Look for possible design quality problems when early failures occur during product development.
$\beta = 1$	Random events	Look for external factors causing stress to exceed strength.
		Examples:
		• Electronics failing from a voltage spike
		• Tire blowout from a road hazard
$\beta > 1$	Wear-out	Look for mechanisms such as fatigue, corrosion, diffusion, migration, abrasion, wear.
		Examples:
		• Fatigue of parts
		• Pitting of gear teeth surfaces
		• Wear of tire tread

When analyzing failure data, the Weibull distribution can be used to model early-life failures, random failures, and end-of-life wear-out. As summarized in Table 21.1, the distribution shape factor tells you which type of failure you are observing. Whenever you analyze failure data, it is important to ensure that all failures of a component included in the data set have the same root cause. In reality there can be multiple failure modes for a given component. However, when you fit a distribution to a data set, it must be for a single failure mode. We discuss this further in the section on mixtures of distributions.

The Weibull distribution is covered in many references[5] on reliability engineering. There is also an excellent reference[6] that is focused exclusively on the Weibull distribution. It's a good source for anyone who has failure analysis as a primary responsibility.

Lognormal Distribution

The lognormal distribution occurs in the analysis of many types of life data. Often it is a good fit to life data that come from a degrading process such as migration, diffusion, or fatigue. It's a skewed distribution with a long tail. If your data set has some very long times to failure and is not symmetric about its mean, it might be lognormally distributed. If the natural logs of the data are normally distributed, the data are lognormally distributed. The lognormal distribution is plotted in Figure 21.5.

The PDF for the lognormal distribution is

$$f(t) = \frac{1}{t\sigma_T \sqrt{2\pi}} \exp\left[-\frac{1}{2}\left(\frac{T-\mu}{\sigma_T}\right)^2\right] \tag{21.29}$$

where

$T = \ln(t)$, where the t values are the times to failure

μ = the location parameter, the mean of the natural logarithm of times to failure

σ_T = the scale parameter, the standard deviation of the natural logarithm of the times to failure

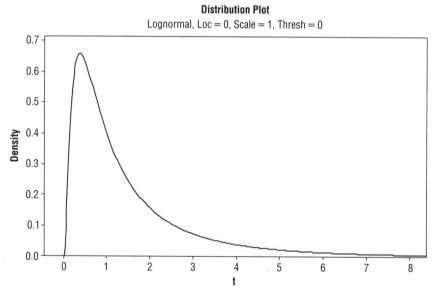

Figure 21.5   Lognormal distribution

5. O'Connor, *Practical Reliability Engineering, 3rd ed.*

6. Robert B. Abernethy, *The New Weibull Handbook, 4th ed.* (North Palm Beach, FL: Robert B. Abernethy, 2000).

Since the lognormal distribution cannot be integrated analytically, statistical analysis applications, such as Minitab or JMP, must be used to compute results such as the cumulative fraction failed and the probability of failure.

Poisson Distribution

A homogeneous Poisson process is one in which events occur at some average rate and the inter-arrival times between events are exponentially distributed. The Poisson distribution, which is fundamental to queuing processes,[7] can be derived from the binomial distribution.[8] It is a discrete distribution that is also closely related to the exponential distribution.

$$P(k) = \frac{(\lambda t)^k e^{-\lambda t}}{k!} \tag{21.30}$$

There are many diverse examples of the Poisson process, such as

- Arrival of motor vehicles at a tollbooth
- Calls to 911
- Arrival of customers at a bank drive-in window
- Arrival of cosmic rays at a detector
- Occurrence of failures for a complex system

For some of the examples involving man-made processes, the arrival rate may change over time. You might expect that the rate of arrival of motor vehicles at a tollbooth would be higher during rush hour than late at night. However, if you consider a suitably short time period with small changes in the average rate, the homogeneous Poisson process is a good representation.

With the Poisson distribution, given by equation (21.30), you can calculate the probability of an event occurring k times within a time period t, given that it occurs with an average rate λ. You can apply it to reliability in order to calculate the probability of having k failures within a time period t. The probability of having k or fewer failures is calculated by summing all of the probabilities for $n \leq k$. This is the Poisson CDF.

So the cumulative probability for having k or fewer failures within a time period t is

$$P(n \leq k) = \sum_{n=0}^{k} \frac{(\lambda t)^n e^{-\lambda t}}{n!} \tag{21.31}$$

The probability of having more than k failures within a time period t is

$$P(n > k) = 1 - \sum_{n=0}^{k} \frac{(\lambda t)^n e^{-\lambda t}}{n!} \tag{21.32}$$

Example for a Poisson Distribution

A complex machine has an MTBF of 1000 hours. How can you answer the question "What is the probability of having three or fewer failures in 2000 hours of operation?"

---

7. Alberto Leon-Garcia, *Probability and Random Processes for Electrical Engineering* (Reading, MA: Addison-Wesley, 1989).

8. Freund, *Mathematical Statistics, 5th ed.*

You know that the expected (average) number of failures for 2000 hours of operation is

$$\lambda t = \frac{1}{MTBF} t = 0.001 \times 2000 = 2 \tag{21.33}$$

The probability of having three or fewer failures in 2000 hours is

$$P(n \le 3) = \left[\frac{(2)^0}{0!} + \frac{(2)^1}{1!} + \frac{(2)^2}{2!} + \frac{(2)^3}{3!} \right] e^{-2} = 0.857 \tag{21.34}$$

What is the probability of having one or more failures in 1000 hours? You can calculate this probability by using equation (21.35):

$$P(\ge 1) = 1 - P(0) \tag{21.35}$$

For this case, $\lambda t = 0.001 \times 1000 = 1$, so

$$P(n \ge 1) = 1 - P(0) = 1 - \left[\frac{(0)^1}{0!} \right] e^{-1} = 0.632 \tag{21.36}$$

Another way to do the calculation is to use the reliability function for the exponential distribution and calculate the probability of surviving 1000 hours. That calculation is

$$P(1000) = e^{-\lambda t} = e^{-1} = 0.368 \tag{21.37}$$

So for a complex system, where the exponential distribution is appropriate, the chance of survival for as long as the MTBF is only about 37%.

Binomial Distribution

The binomial distribution is useful when calculating probabilities for combinations of events having one of two possible outcomes. If you consider the two possible outcomes as "success" and "failure," you can apply the binomial distribution to many types of problems, including the reliability of redundant systems and of quality inspection.

The binomial distribution is defined as follows:

$$P(r) = \frac{n!}{r!(n-r)!} p^r (1-p)^{(n-r)} \tag{21.38}$$

where

p = probability of success

n = number of trials

r = number of successes

$P(r)$ = probability of r successes in n trials

You can factor equation (21.38) into the product of two terms:

The term $n!/[r!(n-r)!]$ is the number of ways you can have r successes in n trials.

The term $p^r(1-p)^{(n-r)}$ is the probability of having r successes and $n-r$ failures.

Example for a Binomial Distribution

An airplane has four engines. Because two operating engines are required for the plane to remain airborne, you are interested in the probability that at least two engines will survive a ten-hour flight across the Pacific Ocean.

Assume for the calculation that the reliability of a single engine is 0.95 for ten hours of operation (it's actually much better than that). That includes one takeoff and one landing. The probability of at least two engines remaining operational is

$$P(\geq 2\ surviving) = P(2\ surviving) + P(3\ surviving) + P(4\ surviving) \qquad (21.39)$$

Using equation (21.40), calculate the probability of at least m survivors out of n engines:

$$P(r \geq m) = \sum_{r=m}^{n} \frac{n!}{r!(n-r)!} p^r (1-p)^{(n-r)} \qquad (21.40)$$

By substituting the applicable values into equation (21.40), you get

$$P(r \geq 2) = \frac{4!}{2!(4-2)!} 0.95^2 (0.05)^2 + \frac{4!}{3!(4-3)!} 0.95^3 (0.05)^1$$

$$+ \frac{4!}{4!(4-4)!} 0.95^4 (0.05)^0 = 0.999519 \qquad (21.41)$$

You see that the binomial distribution is useful for calculating the reliability of redundant systems. For this problem, the tough part is determining the reliability of an individual engine. That takes testing, lots of operational field data, and the use of engineering analysis using tools such as finite element analysis (FEA), computational fluid dynamics (CFD), and probabilistic fracture mechanics.[9] After all of the design work has been completed, you have to combine the results of analyses and field data to estimate the engine reliability using the tools of reliability engineering, such as FMECA, FTA, and reliability models. The easy part is using the binomial distribution to calculate the reliability of the redundant system.

Key Points

1. The modeling of component and system reliability helps to understand the costs of quality associated with failures for the population of products and to estimate important parameters of the business case, such as warranty expenses and service costs.
2. The fundamental functions of reliability are simple and easy to use.
3. The hazard rate over time is determined by the physics of failure.
4. The distribution that is the best fit for a set of life data depends upon the mechanism causing the degradation and failures.
5. Early-life failures in production are symptomatic of supply chain quality problems.
6. The exponential distribution is a good model for the distribution of failures for complex systems, or for failures due to random external stress factors.

9. Jack D. Mattingly, William H. Heiser, and David T. Pratt, *Aircraft Engine Design, 2nd ed.* (Reston, VA: AIAA, 2002).

7. The Weibull distribution is valuable because it can model a range of behaviors over the three stages of product life: early-life failures, failures caused by random events, and end-of-life wear-out.

8. The lognormal distribution is a good fit for failure data from degrading processes such as fatigue, diffusion, and migration. It is a good life distribution for simple solid-state electronic devices.

9. The Poisson distribution is used to calculate the probability of a given number of events occurring during a specified time period for a Poisson process.

10. The binomial distribution is useful for calculating the reliability of redundant systems.

Discussion Questions

1. What options do you have if you find that a component essential to your new product design has poor reliability?

2. After fitting a Weibull distribution to a set of life data, you find that the distribution shape factor is 0.8. What should be done to improve reliability? What factors should be considered before formulating plans for improvement?

3. When fitting two candidate distributions to a small set of life data, you often find that each is a good fit. How can you decide which is the better choice?

4. What is a homogeneous Poisson process?

5. Why is it important to fit a probability distribution to life data?

6. Do you think there is a parametric distribution that is a good fit to mortality data for humans?

7. What options exist for analyzing life data if you are unable to fit a life distribution to the data?

8. If you are given the hazard function for a component, can you calculate the PDF and CDF?

9. If you took the natural logarithm of the t values for the lognormal distribution shown in Figure 21.5, what distribution would be the best fit? What are the parameters of the distribution?

Life Data and Their Implications

In Chapter 8 we discussed HALT, HASS, and reliability growth testing. These tests are powerful tools for improving the robustness and durability of new product designs. Here we extend the discussion to the analysis of life data, a key part of the reliability growth process. When failure analyses are combined with corrective actions, sustainable improvements in reliability are enabled. Certainly there is an element of "keeping score." Life data are available only after you have tested prototypes of the production-intent design. Some data will be derived from controlled tests of components or subsystems as the development project unfolds. They may confirm your expectations or push you to make improvements. Other data may be obtained as feedback from customers' use of current products or their testing of new prototypes. These data may be collected by service databases tracking repairs or warranty costs. Once you have a set of life data for a particular component, you can fit a distribution to it and draw some conclusions about predicted reliability and service costs. Sometimes the failure of a component has more than one root cause, making failure analysis critical to the process of analyzing life data. In this chapter we discuss some important steps for the analysis of life data.

Sources of Life Data

Life Testing

Testing with Nominal Stresses

It is reasonable to test the life of a component under "nominal" conditions, with the mechanical or electrical stresses being representative of expected operating conditions. The life requirement for such a component must specify adequate durability. That may depend on a number of factors, such as

- Intended usage life of the product
- Stress levels expected to be experienced by the component during the product's use
- System-level consequences of the component failure
- Customers' expectations for related performance
- Whether or not the component is serviceable
- Component manufacturing cost
- Benchmark comparisons to similar products

Of all the factors affecting a component's life, the applied stresses are the most important. Generally there is reciprocity between stress and life, where higher stress means shorter life. If life tests under nominal conditions have demonstrated that the durability requirements have been satisfied, it may indicate that longer test durations are required to accumulate enough failures to fit a distribution and to specify the distribution parameters with acceptable uncertainty.

As an alternative to testing under nominal stress conditions, many development teams have embraced accelerated life testing, since it yields useful results in less testing time.

Accelerated Life Testing (ALT)

The goal of accelerated life testing is to learn more in less time for lower costs. The strategy contributes to "lean" product development. There are two ways to accomplish this. One is to apply higher-than-normal stresses to the component under test. The other is to test devices that fail due to actuations by increasing their actuation rate. Test plans can combine these approaches. Both cause failures to happen more quickly.

The objective of ALT is to accelerate existing failure modes, not to create new ones. Consequently, the stresses applied must not be so severe that new failure modes are stimulated, ones that will never be experienced in realistic operation. ALT is discussed in more detail later in this chapter.

Field Data

Analysis of field data is often challenging. While laboratory testing is structured and controlled, products in the field are exposed to combinations, levels, and frequencies of stresses that are neither controlled nor predictable. The most significant challenge for failure analysis is the assignment of the root cause of a failure. In the laboratory, simple component tests can be performed to ensure that a single root cause is responsible for a failure. When products are in customers' use, operating conditions and stresses can combine in uncontrolled and unpredictable patterns.

The implication is that failure analysts must have access not only to the parts that failed but also to relevant knowledge of the conditions under which the failures occurred. Often this can be a major challenge to the logistics of collecting data among customers and service organizations, which may not have incentives to contribute useful inputs to new product development.

Competing Failure Modes, Mixtures of Life Distributions, and the Importance of Failure Analysis

Most systems are affected by many different stresses that can cause failures. Some components may have multiple failure modes with different root causes. Analyses of their failures must link a specific root cause to each failure. This is essential if you intend to fit a life distribution to the data and use it to predict early failures and associated warranty costs.

Consider the tires on a vehicle. Do you remember the bathtub curve in Figure 2.1? There are three main types of failure: early-life failures due to manufacturing defects, random events causing failures at any time, and end-of-life wear-out failures. Tires leaving the factory have two populations: defective tires and those free of defects. Each has a different intrinsic life distribution. Once the tires are in use, there is a race to see which root cause is guilty of a failure in a particular tire. A defective tire could fail by accelerated deterioration at the site of the defect. It could fail by a random event, such as a road hazard or sharp debris, triggering failure in an area weakened by the defect. If its defect is not severe, a tire could fail by wear-out rather than by a premature end-of-life scenario.

Defect-free tires are still subject to two of the three types of failure. Random events can be hazardous, depending on the operating environment. Tread wear depends on many factors such as tire inflation, tire rotation (preventive maintenance), vehicle load, acceleration and braking patterns, road conditions, and, of course, rubber formulation and tire design.

So you have three competing failure modes with very different life distributions. Over time, the failures due to road hazards arrive at some average rate, so you would expect that the time between failures caused by road hazards is exponentially distributed. If you were to record all failures, without regard for their root causes, you would have trouble fitting a useful life distribution to the data. This is because the early-life failures, random failures, and wear-out failures are described by different life distributions. Even in this relatively simple case, with few root causes of failure, the problem can become complicated.

Whenever you have failures of a component, either in the market or in laboratory tests, it is extremely important to perform a thorough failure analysis. That would include a complete physical examination of the part with correlated knowledge of the operating conditions to understand the specific root cause.

Although some reliability analysis software will attempt to separate the failures into different groups, most experienced reliability analysts do not recommend this approach. Instead they make a convincing argument that after a careful examination of the part, the assignment of root cause by an experienced person is much preferred.[1] Applications such as Minitab allow the analysis of life data from mixtures of distributions. This capability requires the user to classify failures by their root cause. To see the result of distribution mixtures, consider the system reliability diagrams shown in Figures 22.1 and 22.2.

The reliability block diagram in Figure 22.1 shows a system with two components, each with reliability following the exponential distribution. The system reliability is given by

$$R = R_1 R_2 = \exp[-(\lambda_1 + \lambda_2)t] \tag{22.1}$$

The probability density function (PDF), described in Chapter 21, is obtained by differentiating the reliability function:

$$f(t) = -\frac{dR(t)}{dt} = (\lambda_1 + \lambda_2)\exp[-(\lambda_1 + \lambda_2)t] \tag{22.2}$$

In this case, the system PDF is the exponential distribution and the failure rate is the sum of the individual failure rates. If you were to lump all of the failures together without doing a failure analysis, you could correctly fit an exponential distribution to the data. However, without failure analysis you would not know that there were two distinct failure modes.

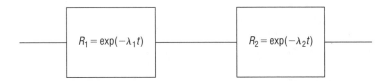

Figure 22.1   System with component failures following exponential distribution.

---

1. Abernethy, *The New Weibull Handbook*.

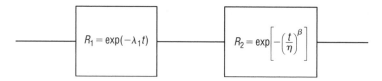

Figure 22.2   System with component failures following exponential and Weibull distribution.

If the mixture consists of different distributions, the analysis is not as simple. Consider the case of a system that has a reliability block diagram with two model elements, illustrated in Figure 22.2. Assume that one block has a constant failure rate, and the time between failures is exponentially distributed. The other block models a component whose wear-out is modeled by the Weibull distribution.

The system reliability for a series system is the product of the individual component reliabilities.

$$R = R_1R_2 = \exp\left[-\lambda_1 t - \left(\frac{t}{\eta}\right)^\beta\right] \tag{22.3}$$

Again, the PDF is obtained by differentiating the reliability:

$$f(t) = -\frac{dR(t)}{dt} = \left[\lambda_1 + \frac{\beta}{\eta}\left(\frac{t}{\eta}\right)^{\beta-1}\right]\exp\left[-\lambda_1 t - \left(\frac{t}{\eta}\right)^\beta\right] \tag{22.4}$$

Equation (22.4) shows the PDF to be neither Weibull nor exponential. Attempts to fit a single life distribution to the data would not be successful. Because the data are from a mixture of distributions, you need to assign a cause to each failure and fit a separate distribution to each. In this case there would be wear-out failures and randomly occurring failures. In the tire example, you would have to differentiate the random "blowout" from tread wear.

The PDF in equation (22.4) can be plotted and the function used to calculate the distribution of failures over time. However, by putting all failures in the same group, you actually lose information. By analyzing the failure and assigning root causes, you learn more and are able to formulate corrective action plans aimed at fixing specific problems.

Later in the chapter we discuss how to fit life distributions to a mixture of failures caused by competing failure modes.

Mixtures of Distributions and the Bathtub Curve

As a last example of distribution mixtures, consider the bathtub curve that we introduced in Chapter 2. Shown in Figure 22.3, the bathtub curve is a plot of the failure rate over the life of a population of devices. As described for tires, when products leave the factory they can have two distinct populations: those that are defect-free and those with defects escaping to the market. The defective group includes products having significant weaknesses, prone to early failures because they are not able to tolerate the stresses of normal operation. The stronger population has no latent manufacturing defects. Strong products surviving failures from random events will start to fail as deterioration accumulates with use, leading to end-of-life failures. The bathtub curve represents both multiple populations and failure modes.

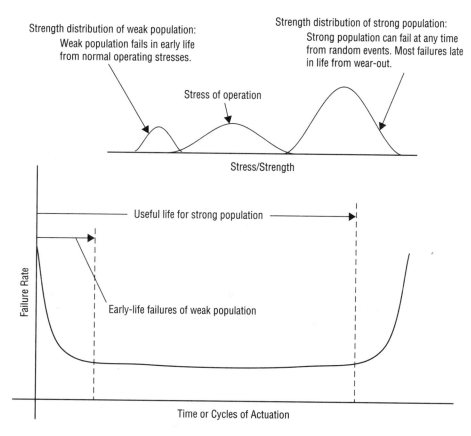

Figure 22.3 Bathtub curve and stress-strength relationship for weak and strong populations

Preventive Maintenance of Repairable Systems

Components Failing from Random Events

The exponential distribution has a "memory-less" property. This means that while a system is operating, the probability of failure during some arbitrary time period in the future is unchanged. It is not dependent on cumulative use. It's common to hear people say, "It's been a while since we have had a failure, so we are due for one." This perspective has some validity when considering a component that fails because of wear-out. However, it is not true for complex systems having exponentially distributed times between failures.

Suppose a system follows the exponential distribution and you are interested in answering the question "What is the probability of the system surviving to time $T + t$, if it has already survived until T without failing?"

The probability of surviving until time T is given by

$$p(T) = e^{-\lambda T} \tag{22.5}$$

The probability of surviving until $T + t$ is

$$p(T + t) = p(T)p[(T + t)|T] \tag{22.6}$$

The expression $p[(T + t)|T]$ is the conditional probability that the system will survive until $T + t$, given that it has survived already until T without failure. After we rearrange equation (22.6), the conditional probability becomes

$$p[(T + t)|T] = \frac{p(T + t)}{p(T)} = \frac{e^{-\lambda(T+t)}}{e^{-\lambda T}} = e^{-\lambda t} \tag{22.7}$$

So for the exponential distribution, the time since the last failure is not useful as a predictor of the probability of survival for a given future time period. The conclusion is that preventive maintenance is not effective in preventing randomly occurring system failures. This is the case if the system is a "black box" and you do not have detailed knowledge of what is inside. All you can do is repair the product when it fails. So your service strategy can be summarized as "If it's not broken, don't fix it."

Components Failing from Wear-out

Almost always, complex systems will have components that fail because of wear-out. They should not be treated as a "black box" that generates failures. If there are components that fail by wear-out, it is reasonable to use preventive maintenance (PM) to keep those failures from affecting customers. This is because the conditional probability of survival for a specified time decreases as the hazard rate increases.

You can see this with the Weibull distribution, for which the conditional probability of survival is given by

$$p[(T + t)|T] = \exp\left[\left(\frac{T}{\eta}\right)^{\beta} - \left(\frac{T + t}{\eta}\right)^{\beta} \right] \tag{22.8}$$

As described in Chapter 21, β is the shape factor for the distribution. For $\beta = 1$ you get the same result as you do with the exponential distribution. However, for $\beta > 1$ the conditional probability of survival decreases with T. As the hazard rate for a component increases, so does its deterioration and its probability of failure.

The implication is that PM will be effective in reducing the frequency of wear-out failures. This means that the designed system must have enough built-in intelligence to track operating hours and actuation cycles, so that the service person can know when to perform PM and on which subsystems. For an automobile, engine oil and tires are examples for which PM is effective.

PM is not effective for a component with a constant failure rate, such as engine computers and electronics in general. In fact, at the micro level, electronic devices also fail from deterioration.[2] However, the high level of integration and the resulting complexity of modern electronics (e.g., microprocessors with a billion transistors, a number that will only increase in the future) make the exponential distribution a reasonable life distribution for electronic systems.

For PM programs, the challenge is how to decide when replacement is appropriate. It is possible to calculate an optimal replacement interval for components that fail from wear-out and are part of a complex system.[3] While this is an interesting mathematical exercise, experience has shown that the results are often not that effective in minimizing cost. In the real

---

2. Milton Ohring, *Reliability and Failure of Electronic Materials and Devices* (San Diego: Academic Press, 1998).

3. John P. King and John R. Thompson, "Using Maintenance Strategy to Improve the Availability of Complex Systems," *Proceedings of NIP 17*, International Conference on Digital Printing Technologies (2001).

world, many serviceable systems operate in an environment where external factors can affect system performance. Another factor is that a sample of components, over time, may exhibit a life distribution in operation that often will be less peaked and much broader than one would expect based on laboratory life testing.

Reliability-Centered Maintenance

A better approach is to use a process called Reliability-Centered Maintenance[4] (RCM). Instead of being totally schedule-driven, it relies on a more holistic approach that includes the monitoring of conditions to detect deterioration and trigger PM. Examples of this are checking for wear and abrasion of mechanical parts or monitoring the vibration amplitude and frequency spectrum for rotating machinery.

When maintenance triggered by a failure is avoided, the higher costs of unscheduled service to both customers and to your company are replaced with more acceptable costs of PM that can be anticipated and scheduled with fewer consequences. The cost for servicing an unexpected failure is almost always more than the cost for sensibly timed PM. For example, chemical processing plants, with miles of heat exchanger tubing and multitudes of pumps and valves, are environments where it is extremely disruptive and costly when unscheduled maintenance is required to fix broken equipment. When safety is involved, the potential liabilities of failures can be enormous. So wear-out failures that can be anticipated must be prevented wherever possible. Another example is aircraft, where airframes and engines are subject to regular inspection and maintenance. The foregoing discussions apply to repairable systems where it is economically attractive and technically practical to repair a product rather than to consign it to the dumpster. Usually this is the case for expensive capital assets with long useful lives. The situation is quite different for low-cost consumer products. High-volume manufacturing has driven prices down and brought higher standards of living and better lives to many. Think about the implications of not having a refrigerator—daily trips to the market for fresh ingredients and no leftovers (possibly that's not all bad). At the same time it has created a throwaway culture, especially with lower-cost products where economics drives decisions that are often troubling and definitely not "green." Parts may not be available and products may not be designed to be repaired. Although the economics may tell us to dispose of a failed product, to many consumers, especially children of the Great Depression and those who have embraced green philosophies, this action may not seem right.

Fitting a Life Distribution to a Data Set

Complete Data

In a life test that runs all components to failure, you have "complete data." The analysis of complete data is straightforward. The process of choosing the best-fit life distribution has four steps:

1. Choose a candidate distribution.
2. Transform the selected distribution to yield a linear model for $F(t)$ versus time.
3. Perform a linear least-squares regression on the observed data against the chosen model to determine the unknown distribution parameters.
4. Decide if the chosen distribution has an acceptable goodness of fit.

---

4. John Moubray, *Reliability-Centred Maintenance, 2nd ed.* (Oxford, UK: Butterworth-Heinmann, 1999).

Suppose you have a goal to construct a probability plot of data for the Weibull distribution. Probability plots have the *x-y* axes scaled appropriately to yield a straight-line plot. The closer the data points are to the straight line, the better the candidate distribution fits the data. The probability plot for the Weibull distribution has a log scale for the *x*-axis (time or actuations) and a log-log scale for the *y*-axis (cumulative distribution function, or CDF). The CDF for the Weibull distribution is

$$F(t) = 1 - \exp\left[-\left(\frac{t}{\eta}\right)^{\beta}\right] \tag{22.9}$$

Rearranging equation (22.9) and taking the log of both sides twice yields

$$\ln\left[-\ln\left[1 - F(t)\right]\right] = -\beta\ln(\eta) + \beta\ln(t) \tag{22.10}$$

This has the form $y = a + bx$ where $a = -\beta\ln(\eta)$ and $b = \beta$. The only other information you need for the probability plot is $F(t)$. You can approximate $F(t)$ by using a device called "Bernard's approximation" applied to the adjusted median ranks. With the point-wise approximation for $F(t)$, apply the methods of least-squares regression (discussed in Chapter 16) to get a best fit of the straight-line relationship of equation (22.10) to the data. There is no need to worry about the tedium of doing this by hand, since there are good statistics applications to do the hard work, such as Minitab and JMP. Fitting a distribution to complete data is simpler than fitting it to censored data, which we discuss in the next section.

Censored Data

"Censored data" are a data set that includes a mix of failures and suspensions. "Suspensions" come from stopping a life test before all members of the population have failed. This is also the case when analyzing field data having a mix of failed components and components still in service.

The gathering of life data for a component must include the effect of suspensions when fitting a distribution. With "complete data," there are no suspensions, making the previous least-squares approach workable. As you saw with complete data, after transforming the data to a linear relationship you can use linear regression to determine the unknown distribution parameters.

The problem is a bit more complicated with censored data. Common sense suggests that an estimate of product life should not ignore survivors or suspensions. However, the previous method for complete data is not a good way to include them. A better way is to use the "likelihood function," invented by statisticians to handle this type of situation.

As an example, the likelihood function can be formulated for the Weibull distribution. The Weibull distribution is a two-parameter distribution, so the PDF and CDF will each be functions of the two parameters. The likelihood function is formed by taking the product of the PDF for each failure with what is essentially the reliability for each suspension.

$$L = \prod_{i=1}^{m} f_i(x_i, \beta, \eta) \prod_{j=1}^{n}\left[1 - F_j(y_j, \beta, \eta)\right] \tag{22.11}$$

After forming this likelihood function, you can make the process more efficient by taking its natural log. The resulting function, called the "log-likelihood," consists of the sum of the log of the terms in the product. The log-likelihood function is

$$\ln L = \ln[f(x_1, \beta, \eta)] + \cdots \ln[f(x_m, \beta, \eta)] + \ln[1 - F(y_1, \beta, \eta] \cdots + \ln[1 - F(y_n, \beta, \eta] \quad (22.12)$$

The next step is to maximize the function by differentiating the function with respect to β and η and setting the result to 0.

$$\frac{\partial(\ln L)}{\partial \beta} = \frac{\partial(\ln L)}{\partial \eta} = 0 \quad (22.13)$$

The results are two equations and two unknowns. They can be solved for the distribution shape factor (β) and scale factor (η). If you are unsure about which distribution to fit to the data, go through this process for several candidate distributions, perform goodness-of-fit tests, and pick the distribution that has the best fit. Clearly this task should not be done without a good statistics application to analyze the life data. This is especially true for distributions with more parameters as well as for distributions not amenable to closed-form analysis, such as the normal and lognormal. Applications such as Minitab and JMP are well equipped to analyze data sets having suspensions.

Competing Failure Modes and Mixtures of Distributions

When life data for a component represent more than one failure mode, it is essential to perform failure analyses and assign a cause for each failure. Without this step, two problems will be created. First, you will not be able to fit a single life distribution to the data, making it more difficult to make useful predictions of reliability and the costs of service and warranties. Second, you will miss the opportunity to perform corrective action on each failure mode.

Example of Analysis of a Mixture of Distributions

When analyzing data with a mixture of failure modes, you have to account for each failure mode. For example, assume that the data represent two failure modes and compromise a set of complete data. Each component was run to failure and each failed from one of the two root causes. It turns out that when you analyze the data set, you will have to consider failures for one failure mode as suspensions for the other failure mode.

The process is the following:

1. Put each failure in the appropriate category, failure mode A or B.
2. Analyze the data for failure mode A and consider the failure times for failure mode B as suspensions for failure mode A.
3. Analyze the data for failure mode B and consider the failure times for failure mode A as suspensions for failure mode B.

Why does this make sense? It's a "race to failure" between two competing failure modes, with several factors determining which failure mode occurs earlier. Among them are

- The proximity of the means of the life distributions for each failure mode. If they are close, both failure modes have a reasonable chance to occur and you will see both well represented in the data. If they are far apart, one failure mode will dominate and the other will seldom be seen.
- The spread of the life distributions. A larger spread makes it more likely to have overlap. Both failure modes will be seen in the data analysis.

- The relative strength of each of the stresses causing failure. If one stress is much stronger than the other, the stronger stress will have an advantage in the race to failure.
- The relative strength of each of the components against each of the failure modes.

Figure 22.4 shows two cases with different life distributions. The example on top shows two life distributions that are relatively close. In this case you expect to see both failure modes with the one on the left dominating because of its lower mean. The example on the bottom shows two life distributions farther apart with a smaller spread. In this situation most, if not all, of the failures would be generated by the failure mode on the left.

Figure 22.5 shows probability plots for a data set composed of two different life distributions. One distribution is for a wear-out mode of failure and the second is for a random failure mode with a much higher mean life than the wear-out mode. If you do not do failure analysis, but instead assume that if the data were all generated by a single failure mode, you would reach an incorrect conclusion. If you look at the best fit of the four candidate distributions, you might conclude that the normal distribution is the best choice. Of course, this is not correct.

With failure analysis, you can assign each failure to the appropriate failure mode. When the data are analyzed, it is possible to determine the best life distribution for each failure. You do have to make an informed guess at which distributions would be best. Figure 22.6 shows the result if you perform failure analyses and make the right assumptions about the fit of distributions. Making a good guess requires that you understand the root cause of the failure as well as a bit about each of the useful reliability distributions. You need to know which distributions have proven to be useful for describing life data for various types of degradation processes.

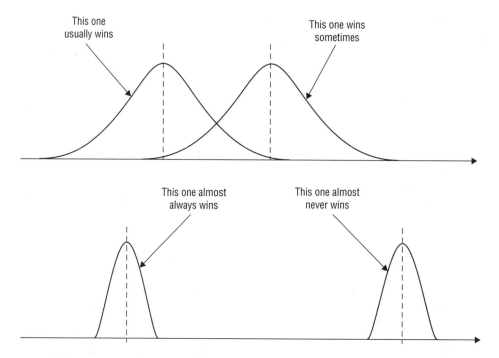

Figure 22.4   Competing failure modes

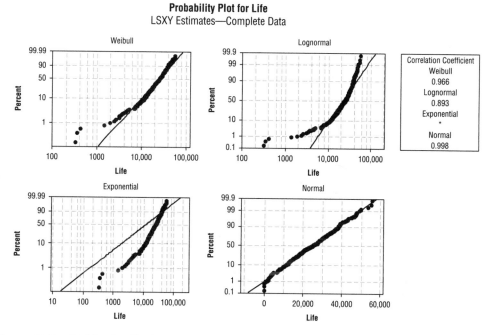

Figure 22.5   Distribution ID plots assuming a single failure mode

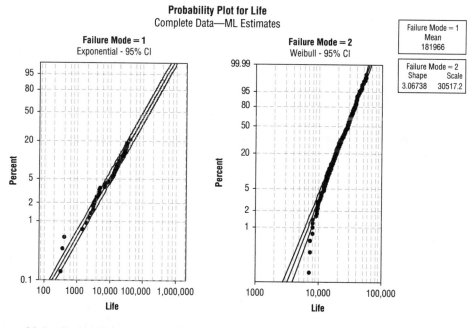

Figure 22.6   Probability plots of combined data set for two failure modes

When the data are analyzed, each failure is assigned to the distribution for a particular failure mode. Then the analysis applies the method of "maximum likelihood." As we discussed earlier, failures for failure mode B are treated as suspensions for failure mode A, and vice versa. Fortunately the analysis can be done easily using applications such as Minitab, JMP, or other statistical applications.

Accelerated Life Testing (ALT)

ALT is used to force failures to occur in less time by applying increased stresses to the component under test. By fitting the appropriate accelerated life model to the data, you can build a prediction tool that will enable you to forecast the expected life under nominal stress conditions. This can shorten the time to understand how component life can affect reliability and the costs for service and warranty repairs. For a comprehensive reference on accelerated testing, see Nelson.[5]

Accelerated Life Models

The result of ALT is a fitted model that can be used to make projections of life under nominal operating conditions, based on demonstrated life under elevated stress conditions. ALT is a process that produces quantitative results. We discuss two models, the Arrhenius model and the inverse power law model.

Arrhenius Model

The Arrhenius model, named after Swedish chemist Svante Arrhenius,[6] expresses the effect of temperature on the reaction rate for chemical processes. Because many degradation processes are chemical in nature, the rate for such a process has a strong dependence on temperature. Since the rate increases with temperature, it's reasonable to expect that time to failure will decrease with temperature. The Arrhenius model, given by equation (22.14), has broad applications in ALT.

$$t_f = A \exp\left[\frac{\Delta H}{kT}\right] \tag{22.14}$$

In this equation,

t_f = time to failure

A = a proportionality constant that drops out when calculating the ratio of times to failure

ΔH = the activation energy for the process (eV)

k = Boltzmann's constant (8.617×10^{-5} eVK^{-1})

T = the absolute temperature in degrees Kelvin

The ratio of time to failure for two different temperatures can be calculated using equation (22.15):

$$\frac{t_1}{t_2} = \exp\left[\frac{\Delta H}{k}\left(\frac{1}{T_1} - \frac{1}{T_2}\right)\right] \tag{22.15}$$

5. Wayne Nelson, *Accelerated Testing: Statistical Models, Test Plans, and Data Analysis* (New York: John Wiley, 1990).

6. Arrhenius was awarded the Nobel Prize in chemistry in 1903.

If the activation energy is unknown, it is possible to test at two different elevated temperatures, solve for ΔH, and calculate time to failure under nominal conditions using equation (22.15).

Examples of where the Arrhenius model has been applied successfully include

- Electrical insulation and dielectrics
- Solid-state devices
- Battery cells
- Lubricants
- Plastics
- Lamp filaments

Inverse Power Law Model

The inverse power law relates the number of cycles to failure under two different levels of stress to some power of the inverse ratio of the stress. It is given by equation (22.16):

$$\frac{n_1}{n_2} = \left(\frac{s_2}{s_1} \right)^N \tag{22.16}$$

where

n_1 = cycles to failure at stress level s_1
n_2 = cycles to failure at stress level s_2
N = an unknown exponent that can be derived from testing at two stress levels

If you know the usage life at two different stress levels, N is given by

$$N = \frac{\ln\left(\dfrac{n_1}{n_2} \right)}{\ln\left(\dfrac{s_2}{s_1} \right)} \tag{22.17}$$

Areas where the inverse power law has been applied successfully include

- Electrical insulation and dielectrics in voltage endurance tests
- Ball and roller bearings
- Incandescent lamps
- Flash lamps
- Metal fatigue from mechanical loading or thermal cycling

Fitting a Model and Distribution to ALT Data

Example Problem

The following classroom experiment is a nice illustration of the use of ALT. The method is simple: the bending of wire samples, cut from paper clips, through two different angles until failure. The idea is to perform the test at two larger angles and then use the result to predict life at a smaller bend angle. The samples are gripped in a small collet at each end and then bent through the chosen angle until failure. The bend angle is measured with a protractor that is part of a fixture used in the experiment. The experiment would never win awards for precision or accuracy. However, it turns out that, if you are careful and relatively consistent, the results are good enough to illustrate some of the important points about running an accelerated life test. The nice thing is that the equipment is affordable and the experiment can be done in a classroom with a large group.

Team 1

Bend angle: 135 deg	Cycles to failure: 5.50, 5.50, 6.25, 5.50, 5.25, 5.25, 4.75, 5.75, 5.50, 5.25
Bend angle: 180 deg	Cycles to failure: 3.75, 3.25, 3.75, 3.75, 3.75, 3.50, 3.25, 3.50, 3.75, 4.50

Team 2

Bend angle: 135 deg	Cycles to failure: 5.75, 5.75, 6.00, 5.50, 4.25, 6.75, 6.75, 5.00, 6.00, 5.75
Bend angle: 180 deg	Cycles to failure: 2.75, 4.00, 3.50, 3.50, 3.25, 3.25, 3.50, 3.50, 3.75, 2.75

Probability Plots

A probability plot of these data, with angle as a variable, gives the plots shown in Figure 22.7. We have selected four candidate distributions to fit the data: Weibull, lognormal, exponential, and normal. We look for parallel probability plots for the two different bend angles and want the data points to hew closely to the probability plot line. While the Weibull and lognormal distributions both look reasonable, the normal and exponential models are not good choices. The normal model is not a good choice because the probability plots at the high and low bend angles are not parallel.

This brings us to an important point: How should you decide which distribution to use when there are multiple candidates that look reasonable? Usually this is a problem when the data set is small. In this case we have 20 data points for each bend angle, not a very large sample for the identification of a distribution. What you can do for guidance is to consider the type of physical process causing failure. In this case it is metal fatigue caused by the strain from large bend angles. Experience indicates that the lognormal distribution has been useful for this type of problem. The next question is which accelerated life model to use. Again, experience suggests the power law model. These are good choices until proven otherwise. They are recommended by experience and fit the data well. As more data accumulate, you can revisit these choices and see if the weight of the experimental evidence is still on their side. You shouldn't think of your choices at this stage as being irrevocable. They are subject to future review and change, as dictated by the data.

Life Model

As discussed in the previous section, the power law model has been useful for ALT data involving metal fatigue. That model is fairly simple. For the lognormal distribution the model is

$$Y_p = \beta_0 + \beta_1 X + \sigma \varepsilon_p \tag{22.18}$$

where

$Y_p = p$th percentile of the log (cycles to failure) distribution. If cycles to failure have a lognormal distribution, the log of cycles to failure will be normally distributed.

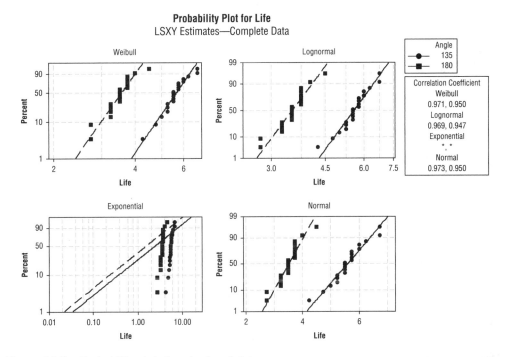

Figure 22.7   Probability plots for wire bend data

β_0 = y-intercept of the fiftieth percentile line for the log of cycles-to-failure distribution. It is the value of the log-life at $X = 1$.

β_1 = slope of regression line

X = natural log of bend angle

σ = scale parameter, the standard deviation of the natural log of the cycles to failure

ε_p = pth percentile of the error distribution. For the lognormal distribution plotted in log coordinates, it is the standard normal distribution.

$\beta_0 + \beta_1 X$ = fiftieth percentile line given by the regression line fitted to the data in log coordinates.

The fitted model is shown in Figure 22.8, with some explanatory annotation followed by the regression table from Minitab.

The full model describing the location of the mean and distribution of life for each level of stress is given by equation (22.19), which is equation (22.18) with the appropriate constants inserted:

$$\ln(cycles\text{-}to\text{-}failure) = 9.62 - 1.61[\ln(BendAngle)] + 0.109\varepsilon_p \qquad (22.19)$$

The first two terms on the right-hand side of equation (22.19) are the intercept and slope of the regression line for the fiftieth percentile line as a function of bend angle (stress). At any position along the fiftieth percentile line there is a normal distribution describing the life distribution at that particular stress level. To locate any chosen percentile line, use the additional term, $0.019\varepsilon_p$ in equation (22.19). It defines a normal distribution with a mean of 0 and standard deviation of 0.019, which is the scale factor given in the regression table following Figure 22.8.

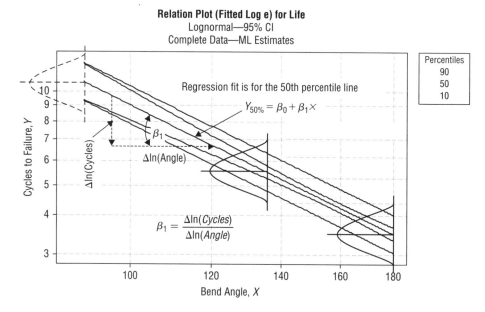

Figure 22.8   Fitted power law life model for wire bend data

Regression Table

Predictor	Coef	Standard Error	Z	P	95.0% Normal CI Lower	Upper
Intercept	9.62006	0.602826	15.96	0.000	8.43854	10.8016
Angle	-1.61108	0.119344	-13.50	0.000	-1.84499	-1.37717
Scale	0.108571	0.0121386			0.0872060	0.135170

Mixed Failures from Time and Actuations

In many systems there will be components for which the power-on time is the main driver of failure. Electronics are examples of this. Other components may fail because deterioration occurs from usage actuations. A mechanical component such as a bearing or drive belt is an example. Copiers and printers are examples of this behavior. Systems remain in a standby condition waiting for someone on the network to send a print job, or for a customer to use the machine locally to make copies. So usage is a mix of powered-on waiting combined with actuations accumulated while running to produce printed or copied output. System reliability in this situation depends on both time and actuations. The question is how to measure system reliability: Should you use time or the number of actuations? What difference does it make?

The following analysis assumes that you are dealing with a complex system. There are two main drivers of failure: power-on time and cycles of actuations.

Failures over time and actuation cycles can be expressed as failures from power-on time:

$$N_T = \lambda T \tag{22.20}$$

or failures from cycles of actuations:

$$N_A = \omega C \qquad (22.21)$$

where

T = power-on time

C = cycles of operation occurring in time T

λ and ω = are the respective failure rates for time and cycles

Total failures can be expressed in terms of time, given by summing failures due to on-time and failures due to cycles of operation:

$$Failures = N_T + N_A = \lambda T + \omega RDT \qquad (22.22)$$

where

D = duty cycle, between 0 and 100%

R = rate at which actuations accumulate, per unit time, when the machine is on and producing

Equation (22.22) gives accumulated failures at any time.

By dividing equation (22.22) by time (T), you can solve for *MTBF*:

$$MTBF = \frac{1}{\lambda + \omega RD} \qquad (22.23)$$

You can also arrange the expression in terms of actuation cycles by dividing equation (22.22) by cycles, C, and solving for mean cycles between failures (*MCBF*):

$$MCBF = \frac{1}{\dfrac{\lambda}{RD} + \omega} \qquad (22.24)$$

So depending on whether you measure system reliability in terms of time or cycles between failures, you can see a decrease or increase in reliability with increasing duty cycle. In equations (22.23) and (22.24), MTBF decreases and MCBF increases with increasing duty cycle. This behavior makes sense. If you are measuring reliability by MTBF, for a given time period a high-duty-cycle machine will accumulate more failures from cycles of operation than will a low-duty-cycle machine. If you measure reliability in terms of MCBF, a high-duty-cycle machine will take less time to accumulate a given number of cycles and therefore have fewer time-related failures for a given number of cycles. Is this a real change in reliability with duty cycle? Not really. It is an artifact of bookkeeping, the choice of how you keep score. Since many systems, especially electromechanical ones, have components that are affected by cycles and time, you should choose the metric that is more useful for measuring reliability. The choice can be somewhat arbitrary, but you should consider which measure is more meaningful to your business.

Key Points

1. There are two main sources of life data: testing and field experience.
2. Life testing performed in the laboratory is usually more structured and controlled than are the experiences of products in customers' applications.
3. Life testing can be done under nominal stress levels or with elevated stresses.
4. Life testing with elevated stresses will shorten the time to failure and improve the productivity of the test program.
5. Failure analysis is an essential part of any testing program. It is especially important when analyzing field data.
6. Competing failure modes cause a mixture of distributions in the life data, as illustrated by the bathtub curve.
7. Preventive maintenance is not effective for components that fail from random events. PM strategies are effective when components fail from predictable or detectable wear-out.
8. It is possible to estimate the optimal replacement interval for components that wear out.
9. A life distribution that has been fitted to a data set can be useful for estimating service and warranty costs.
10. When analyzing life data to fit a distribution and to estimate distribution parameters, you must include the effect of both failures and suspensions.
11. Accelerated life testing can be used to develop predictive models with less test time than is required under nominal stress conditions.
12. Some systems have failures caused by both time and actuations. If reliability is measured in terms of MTBF and mean actuations between failures, the results will look different. Actually they are the same.

Discussion Questions

1. What are the advantages and disadvantages of laboratory test data and field data?
2. Why is accelerated life testing a productivity improvement tool?
3. Describe one of the accelerated life models and its applicable situations.
4. What are the reasons that failure analysis must be done?
5. Why will life testing always be required when developing new products?
6. How are mixtures of distributions and the bathtub curve related?
7. What are some actions that will reduce early-life failures for a product?
8. What are some ways to reduce the failure rate from random events?
9. How well does your organization make use of field data when developing new products?
10. Has your product development organization made productive use of accelerated life testing?
11. Are field data readily accessible to teams doing product development?
12. What improvements can you suggest to make better use of both field and laboratory data in your product development process?

SECTION IV

Integration of Framework and Methods

Chapter 23: Critical Parameter Management

Once the performance of a design concept has been optimized and specified, its controllable parameters must be recorded in a manner that is readily accessible to everyone involved in maintaining the design, reproducing it in production, and leveraging it in future products. An effective system for doing so imposes analysis and discipline on the management of design changes and the resulting product configuration. It is a companion to product data management systems designed to manage the bills of materials.

Chapter 24: An Illustration of Process Methods and Tools

As a way to show the integration of several of the practices described in the previous chapters, a hypothetical example takes the reader through their application to the design of a product familiar to many.

Chapter 25: Balancing Schedules, Budgets, and Business Risks

The strategies and principles of product development continue to evolve as companies strive to achieve shorter times to market without compromising production costs or imposing more risks on market entry schedules. The wisdom in this book should provide a good foundation for further improvements of your own processes.

Glossary

Bibliography

This reading list is a collection of those references identified in the chapters.

Critical Parameter Management

After customers' needs are understood and translated into technical requirements, the backbone of any product development project becomes critical parameter management (CPM). CPM is intended to ensure that all product characteristics affecting attributes important to customers are characterized, specified, and in compliance with requirements. We could argue that everything discussed so far is aimed at understanding behaviors of product characteristics that are really important. You can then evaluate risks. Will the product be easy to manufacture and service with the likelihood of few customer complaints? Or will it be a test of the organization with lots of bumps in the road? CPM is a process that can help you to gain that understanding sooner in the development process. We discuss the subject of CPM late in the book because it integrates many concepts covered earlier. It is a way of thinking and acting that should be used early in any development project, whether you are developing a new product or process.

CPM and the Domains of Product Development

What Is a Critical Parameter?

Recall our discussion of axiomatic design in Chapter 12. The four domains of product development, shown again in Figure 23.1, describe categories of activities that lead to knowledge development. Usually knowledge and information are developed from left to right, starting in the Customer Domain, where customers' needs are identified, and ending in the Process Domain, where the supporting manufacturing and service capabilities are developed.

As development teams move from one domain to the next, critical parameters emerge that support the requirements of the previous domain. The mapping process is a result of applying tools such as QFD, axiomatic design, concept selection, DOE, Robust Design, and Tolerance Design during product development.

A critical parameter is any measurable product characteristic that affects customers' satisfaction and any controllable factor that can have an important influence on that characteristic. The definition is broad, and can include factors that are not controllable. To identify critical parameters you have to understand which design, manufacturing, or service parameters control performance and where deterioration can cause a failure. In Chapter 2 we described two ways that a product can fail: hard failures, where the product can no longer operate, and soft failures, where an important output of the product is off target.

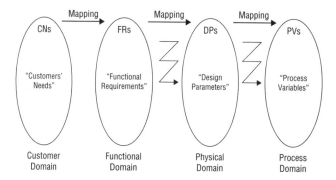

Figure 23.1   The four domains of product development

Usually, experimentation is required to develop a quantitative understanding of the relationship between critical parameters and customers' satisfaction. The focus is on the transfer function, which relates design parameters to system performance, measured by critical functional responses (CFRs). The system design parameters determine the alignment and stability of product performance metrics, which determine the product's useful life.

Categories of Critical Parameters

Functional Domain

Response Factors

A critical functional response (CFR) is the measurable result of a function that the product performs. The response factors that concern us are those that are important to customers. An important response factor either is seen directly by customers or affects attributes that are important to customers' satisfaction. Examples of response factors for familiar products are

- Fuel consumption for a motor vehicle
- Print quality for a printer
- Energy consumption for a refrigerator
- Response time for 911 calls for emergency medical services

Physical Domain

Control Factors

Any control factor that affects a functional response that is important to customers is a critical parameter. Identifying control factors could be simple, requiring the use of common sense combined with scientific or engineering knowledge and a cause-and-effect analysis. To understand the quantitative relationships and their relative importance probably requires a designed experiment. Typically, control factors are design parameters. Their set points are the result of design decisions. However, control factors are not necessarily attributes that can be set while the product is operating. In the spirit of DOE, a control factor is any product characteristic that has an important influence on product performance. Examples of control factors include

- Fatigue strength of a part subject to cyclic loads
- Number of ambulances or crews available to respond to a 911 dispatcher

- Particle size of toner for a copier or printer
- Interference fit for a hub shrunk on a shaft

Observable Quality Metrics

These are attributes such as fit and finish that are seen by customers. If they are off target, they can give the impression of poor quality and cause dissatisfaction among customers, possibly the loss of their business. Specifications for observable quality metrics have to be established by engaging your customers and by understanding the history of complaints for similar metrics on similar products. These are requirements for which "beauty is in the eye of the beholder." It may be that psychophysical experiments are required to establish thresholds of acceptability, which can be specific to a market segment. Examples of observable quality metrics are

- Smoothness of paint on a vehicle body panel
- Gap consistency among vehicle body panels
- Amount of flash on injection molded parts
- Maximum defect size and defects per unit area in chrome plating

Process Domain

Manufacturing Process Factors

Manufacturing process factors are parameters that specify setups, adjustments, or tooling characteristics that enable the production processes to replicate the design's critical parameters. If a manufacturing process factor has an important effect on a product's performance, stability, or durability, the manufacturing process factor is a critical parameter. Manufacturing process factors can also affect observable quality metrics such as fit and finish. Examples of manufacturing process factors can include

- Pressure, temperature, and flow rate for an injection molding process
- Rivet temperature for a hot forming process
- Air pressure setting for a spray-painting process
- Bath chemistry for an electroforming process

Service Factors

Service factors are elements of the system available for service interventions such as adjustments or parts replacement procedures. Service factors are similar to manufacturing process factors. They can affect product performance and durability, but, instead of being the responsibility of manufacturing, they are managed by the service organization. With service interventions, a documented process is required to re-establish on-target operation with minimized variability. Inherently they compensate for inadequate robustness in the product's design. They add costs and opportunities for human mistakes, so it would be better if they were not needed. Justify them by answering these questions: Is it possible to eliminate them by design? What are their economic consequences? Examples of service factors are

- Drive belt tension
- Tire pressure
- Wheel alignment settings for a motor vehicle
- Cleanliness of the lens in an imaging system

External Noise Factors

Critical parameters can include noise factors that have important and direct effects on the functional response or an indirect effect by affecting control factors, manufacturing process factors, or durability. Although you cannot control external noises, you must understand their behaviors and, in a sense, manage them by improving the robustness of product designs and production processes. Common external noises are

- Ambient temperature and relative humidity
- Pressure from altitude or water depth
- Customers' use or abuse
- Variability in properties of consumable materials
- Variations in power sources such as line voltage surges

Durability and Deterioration

Durability

Factors affecting durability are component or system characteristics that have a direct effect on the useful life of the system. Although factors that affect durability do not necessarily affect system response, they can affect product life as measured by mean time or actuations until a fatal failure.

Requirements for durability must be driven by the system reliability requirements, as well as by the desired useful life of the product and by its service strategy. Life testing is one of the tools that are useful to determine component life requirements. When making design decisions affecting durability, you have to consider both nominal stresses as well as random events, where the latter can subject the product to much higher stresses.

Examples of factors affecting durability include

- Endurance strength of a material that is subject to fatigue failure
- Strength of adhesive bonds that are subject to force
- Strength of rubber used in a tire
- Hardness of steel parts that are subject to abrasion

Deterioration

Any design control factors, manufacturing process factors, or service factors that can degrade by physical deterioration may affect product performance, durability, or customers' satisfaction. In order to manage those critical parameters, you have to identify the noises that cause the deterioration and the mechanisms for that deterioration. This generally requires testing or the analysis of failure data. Control factors can be monitored for deterioration in serviceable products. This is called "condition monitoring," requiring periodic inspection and assessment. An automotive example is the checking of tire tread wear when the vehicle is serviced. Any manufacturing process factor that can deteriorate and introduce variability into the product can be monitored indirectly using tools such as SPC. It is also possible to monitor deterioration directly. An example is the tracking of the wear of a tool used in an injection molding process.

In some cases deterioration affects an observable quality metric. Although product functionality and durability may not be affected, the satisfaction of customers can suffer because of the perception of poor quality.

Examples of deterioration are

- Wear of sliding surfaces
- Crack propagation in an airframe
- Color shift or fade of painted surfaces
- Corrosion of metal surfaces

Linking Customer Satisfaction and Critical Parameters

Figure 23.2, discussed earlier in Chapter 2, illustrates ways that a system can fail. An important part of CPM is to understand the linkage between product performance and the decisions in design, manufacturing, and service that define the product during and after product development.

Soft Failures: Performance That Is Out of Specification

Most products have response outputs that are important to customers. Performance can be off target "out of the box" or it can drift because of deterioration. Recall our discussion of quality loss in Chapter 2. Examples of soft failures include

- Decrease in fuel economy for a motor vehicle
- Decline in image quality for a copier/printer
- Drift in accuracy or precision for a measurement system
- Deterioration in parameters of consumable medical products during shelf life

Hard Failures: Things That Break

This category includes everything that fails in ways that prevent customers' continued use of the product. Hard failures can occur at any time during product life. Examples include

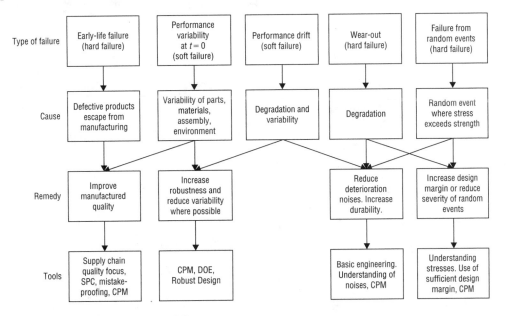

Figure 23.2   How systems can fail

- Burnout of an incandescent light bulb
- Flat tire on a motor vehicle
- Failure of a laptop computer power supply
- Fatigue of a drive belt

How Do You Decide What Is Critical?

Empirical Models

An empirical model can identify control factors that have important effects on attributes affecting customers' satisfaction. They are developed best by using DOE, discussed in Chapter 17. With an empirical model, the statistically significant factors can be identified and ranked in order of importance with the analysis of the transmission of error covered in Chapter 18.

First Principles/Common Sense

A good first-principles model can both identify and rank critical parameters in order of importance. Of course, it is necessary to validate the model with experimental data.

At times it will be difficult to build a good predictive first-principles model. However, with your technical knowledge and some common sense you may be able to identify those parameters that are important to performance, reliability, and durability.

Figure 23.3, similar to the P-diagram discussed earlier, shows a general model for the relationships in a system. Many systems are simpler.

The product design, manufacturing process, and service factors are specified by development teams. Although external noise factors cannot be controlled, they must be anticipated, and their

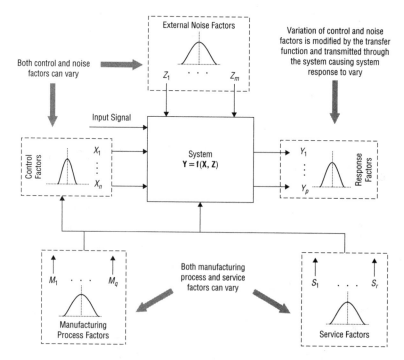

Figure 23.3   A system and its critical parameters

consequences managed. Only then can you use robustness development to make your system less vulnerable to them.

Critical parameters always exist in the context of a hierarchy, shown in Figure 23.4. The product's feature and function flow-downs are decomposed from the value proposition at the top of the hierarchy. Using the critical parameter tree, you can show how fulfilling the value proposition depends on meeting the specifications at lower levels in the tree.

There are transfer function relationships among levels in the hierarchy. They have to be defined in order to have quantitative descriptions of the effects of critical parameter variations and their relative importance. Usually these relationships are understood through designed experiments or first-principles analyses. Remember that variations at the bottom of the CPM tree can

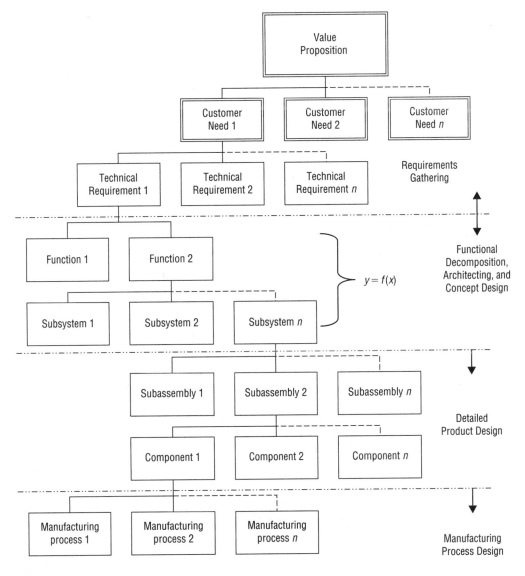

Figure 23.4 Critical parameter "tree"

be transmitted to the top and affect the fulfillment of the value proposition. Requirements are deployed from the top down, while performance is transmitted from the bottom up.

While it may be easy to achieve the requirements for certain critical parameters, others may be difficult. To evaluate the "difficulty" metric, there are both qualitative and quantitative factors to consider. In the following section we discuss both types.

Assessing Difficulty

Before discussing the critical parameters related to design and manufacturing, consider some qualitative factors that can affect a product or system. These factors are not quantitative, as are the transfer function and the design capability. However, they can be helpful in identifying risks.

Qualitative Factors to Consider

Early in a project, often before you have any data, there are a few things you should consider when trying to evaluate the difficulties facing you. These factors should not necessarily discourage you from proceeding. Consider them to be markers that tell you to move forward with care. They are indicators of where to focus your attention and resources.

Experience

There is comfort in knowing that something has already been done. A design concept may be commercialized in an existing product. This does not necessarily mean that it can be done easily, only that it can be done. How well has it succeeded in production? Is it a mature and robust design or one plagued with reliability and cost problems? Should it be leveraged into new products? What difficulties have been experienced? How would you adapt the concept yet develop improvements that correct its shortcomings? What about a new concept expected from technology development? It is natural to want to "go where no one has gone before." It can be exciting and interesting with great payoffs. The risks must be managed.

The novelty of your challenge should be considered in the context of your organizational competency. If you are planning a foray into a technical space that will be new to your organization, you have to understand its risks. Study the experience of others.[1] We remember a design concept that looked promising. Unfortunately, after much development, we found that it was neither superior to alternatives nor robust against stresses in its applications. Worse yet, our chief competitor had already proven that the concept was inferior. We paid a high price for that knowledge.

Although we have expressed concern about potential pitfalls, some wonderful advances have been driven by champions who did not know that "it can't be done." Just be aware of the odds that you are facing, learn from the lessons of others, and manage the risks.

Complexity

In recent years there has been much focus on the reduction of complexity in business processes and designed systems. There is ample evidence that businesses that are less complex are easier to manage, tend to have lower costs and higher profitability, and often offer more growth

---

1. Al Ries, *Focus: The Future of Your Company Depends on It* (New York: Harper Business, 1997).

opportunities.[2] In physical systems we understand that "less can be more."[3] Fewer system elements translate into fewer potential critical parameters and thus fewer opportunities for failure. Is the product concept or its supporting technology simple or highly complex? Are there a few critical factors, or are there many important parameters?

A toxic mix that can threaten the success of a project is the combination of many new technical concepts and a complex system architecture. Throw in the build-test-fix approach, ignore the lack of inherent robustness, and you have a nightmare project. We have seen this approach cause great difficulties. Often this type of project suffers repeated schedule delays and a premature product launch. The damage to your business can be substantial.

Supply Chain Issues

Will production manufacturing be managed internally or by supply chain partners? What are the competencies of manufacturing partners? Reproducing a design that is beyond current capabilities can be risky. Outsourcing it can aggravate that risk, unless the supplier has superior capabilities. Certain parameters can become critical just because of the weak capabilities of the supplier, given the selected manufacturing process and materials. The comparison of alternative supply chain sources makes the process capability a relative rather than an absolute metric.

An absolute measurement of supplier competency is the manufacturing process capability (Cpk) for the characteristic of a part or material. High capability means that the manufacturing process is under control and will replicate the parameter at its specification with small variability compared to the parameter tolerance.

Service Issues

Although at the far end of the value chain, service can be a make-or-break function for products needing it. Consider motor vehicles. Once the buyer chooses a model, the product becomes a commodity available from numerous dealers. Service quality and buyers' experience can be important factors in selecting the dealer for both the initial purchase and the ongoing maintenance and repair.

How costly will service be for the producer during the warranty period and for the buyer after warranty? Is the service of the product easy or difficult? Are there nagging reliability problems that will affect both warranty expenses and customers' satisfaction? Is the producer prepared to service all of the units that will be sold?

The service strategy should provide inputs to the product requirements. This should, in turn, drive concept selections and design decisions affecting robustness, durability, and serviceability and thereby affect many critical parameters. The service organization should have critical parameters of its own, requiring inputs from product design and manufacturing.

Quantitative Measures of Difficulty
Manufacturing Process Capability (Cpk)

Manufacturing process capability is a measure of your supply chain's ability to reproduce design parameters that are on target with their specifications, with a limited amount of variation. A low

---

2. Michael L. George and Stephen A. Wilson, *Conquering Complexity in Your Business* (New York: McGraw-Hill, 2004).

3. Appropriated by minimalist architect Ludwig Mies van der Rohe from Christoph Martin Wieland, 1733–1813.

process capability indicates either a shift of the population mean toward a specification limit, or excessive variations, or perhaps both. Low process capabilities are associated with excessively tight tolerances where the specification limits start to encroach on the natural variations of the process. The implication is that the design concept is not tolerant of expected manufacturing variations. Usually process drift, monitored by SPC, is easier to correct than the inherent process variability. Manufacturing process capability is calculated using equation (23.1):

$$Cpk = Min\left(\frac{USL - \mu_x}{3\sigma_x}, \frac{\mu_x - LSL}{3\sigma_x}\right) \tag{23.1}$$

where

USL = the upper specification limit

LSL = the lower specification limit

μ_x = the mean

σ_x = the standard deviation

When the manufacturing process capability is low, it is more difficult to reproduce a design parameter with competitive yield costs. The difficulties in achieving better manufacturing capabilities and lower costs can persist throughout the life of the product. The corrective approaches are either to relax the tolerances or to increase the robustness of the manufacturing process. Reducing process variability requires investments in process improvement.

Design Capability

Determining the importance of quantitative factors, such as control parameters, requires that you understand the system transfer function. Recall Chapter 18, where we discussed the transfer function and how variability from both internal variations and external noises in customers' applications is transmitted to functional responses of the system. Equation (23.2) repeats the general form for the transfer function:

$$\mathbf{Y} = \mathbf{f}(\mathbf{X}, \mathbf{Z}) + \varepsilon \tag{23.2}$$

Once you have the transfer function, you can optimize the robustness of the system. With an understanding of the differences in costs associated with loosening or tightening tolerances, you can allocate the tolerances among the control factors. Knowing the effect of each control factor and its variability, you can determine the importance of each factor and calculate the "design capability." Recall that the standard deviation of the response to variations in control and noise factors can be calculated using equation (23.3):

$$\sigma_y = \sqrt{\sum_{i=1}^{n}\left(\frac{\partial y}{\partial x_i}\sigma_{x_i}\right)^2 + \sum_{i=1}^{m}\left(\frac{\partial y}{\partial z_i}\sigma_{z_i}\right)^2 + \sigma_{error}^2} \tag{23.3}$$

For a response with an upper and lower specification limit, the design capability is given by

$$Cpk = Min\left(\frac{USL - \mu_y}{3\sigma_y}, \frac{\mu_y - LSL}{3\sigma_y}\right) \tag{23.4}$$

The calculation assumes the normality of the response. If y is not normally distributed, it must be transformed into normal form, or an appropriate distribution must be fit to the data before the calculation. Keep in mind that if the data are transformed to normal form, the specification limits must also be transformed.

If the response is centered between the USL and the LSL, the design capability is

$$Cp = \frac{USL - LSL}{6\sigma_y} \tag{23.5}$$

To calculate the design capability, first determine the optimal set points for the control factors, placing the response on target while minimizing the standard deviation. Then, knowing the mean and standard deviation of the expected production population and the specification limits for the response, you can calculate the design capability using equation (23.4). If there are many control factors, it is easier to perform a Monte Carlo simulation.

Note that the calculated variation in the response depends on the slope of the response surface. It depends on the control factors and noise factors at the operating point of interest, as well as on the standard deviation of both the noise and control factors.

Service Process Capability

Service interventions, such as adjustments, can be characterized using the capability metric given by equation (23.4). It is a surrogate for service difficulty and service costs. Highly "capable" adjustments are, on average, closer to target, less variable, and less costly. Of course, it is important to understand the importance of the adjustment being made. Does the adjustment affect a response that is very important to customers? If yes, then it is very important to have good adjustment capability.

Assessing Importance

To understand the importance of each of the control factors, go back to the transfer function and the calculation of the variations in the response due to variations of the control factors and external noises. For many problems, the tool of choice will be DOE. The relative contribution of each factor to variation is a measure of importance. With the transfer function, the calculation is more easily done using a tool such as a Monte Carlo simulation.

Recall our DOE example for lathe machining error in Chapter 18. The transfer function for the error is given by equation (23.6):

$$Delta = -0.441 - 108.6A - .0022B + 30.12C + 0.253AB - 952.5AC$$

$$-0.012BC + 1950.6A^2 + \varepsilon \tag{23.6}$$

Using this transfer function, we optimized the control factor settings to minimize the error transmission. In this case, external noises were not included in the model, just the variations of the control factors. In order to complete the optimization we had to specify the standard deviation for each control factor. We also needed the standard deviations when we calculated the relative importance. The standard deviation for each of the factors is given here:

Feed: $\sigma_A = 0.005$
Speed: $\sigma_B = 5.00$
Depth: $\sigma_C = 0.0125$

The optimum solution for control factor set points and the resulting variation of the response is shown in Table 23.1. The key calculation is of the standard deviation of the response. The contribution of each control factor to the variance of the response is

$$contribution\ of\ variation\ in\ x\ to\ the\ variance\ of\ y = \left(\frac{\partial y}{\partial x}\sigma_x\right)^2 \qquad (23.7)$$

So there are two factors that are important: the slope $(2y/2x)$ of the response surface in the direction of each control factor and the standard deviation of the control factor. It's the product of the two that determines the control factor variations that are transmitted to the response, as shown in Table 23.2 for each control factor.

Because the control factor settings were already optimized to minimize the transmitted variations, reducing the slope is not a useful option to reduce the transmitted variation. The remaining workable option is to reduce the variation of the control factors. However, the product is a measure of the contribution of the control factor variations to CFR variations. Once you determine the relative importance of each factor's contribution, you will know better where to focus your efforts for improvement and your ongoing quality monitoring.

The standard deviation of the response for the optimum control factor settings is given by

$$\sigma_y = \sqrt{(0.000846)^2 + (0.010)^2 + (0.0440)^2 + (0.05)^2} = 0.0679 \qquad (23.8)$$

You can see from the analysis that the most important control factor is depth of cut, followed by speed and, last, by feed. Both make much smaller contributions to variability. The last

Table 23.1   Optimum solution for minimizing transmitted error

Number	A: Feed	B: Speed	C: Depth	y: Delta	σ_y
1	0.022	457.4	0.100	0.000	0.069

Table 23.2   Factors contributing to variation of the response

Factor		A: Feed x_1	B: Speed x_2	C: Depth x_3
Slope	$\partial y / \partial x_i$	0.282	0.002	3.52
Standard deviation	σ_{x_i}	0.003	5.0	0.0125
Slope × standard deviation	$(\partial y / \partial x)\sigma_{x_i}$	0.000846	0.010	0.0440
Quality loss fraction	$QL_i = [(\partial y / \partial x)\sigma_{x_i}]^2 / \sigma_y^2$	0.0002	0.022	0.43

component given by equation (23.8) is the error (0.05) that contributes over half of the total variation in this example.

The results of this analysis can be combined with the Kano Model to assess the risk generated by variations in any of the control parameters.

Quality Loss—A Measure of Importance

Recall the expression for quality loss that was discussed in Chapters 2 and 19:

$$QL = k[\sigma^2 + (T - \mu)^2] \tag{23.9}$$

Equation (23.9) defines the relationship between the variability of a CFR and its contribution to the quality loss.

k = the proportionality constant between performance variation and monetary loss that should be derived from customers' interviews and field data

σ = the standard deviation of the CFR for the population

T = the target value for the CFR

μ = the population mean for the CFR

Using equations (23.7) and (23.9), you can calculate the contribution of each factor to quality loss. In making the calculation, remember that the observed variance of a CFR is the sum of the contributions of the control and noise factors as well as the error, as shown in equation (23.10):

$$\sigma_y^2 = \sum_{i=1}^{n} \left(\frac{\partial y}{\partial x_i} \sigma_{x_i} \right)^2 + \sum_{i=1}^{m} \left(\frac{\partial y}{\partial z_i} \sigma_{z_i} \right)^2 + \sigma_{error}^2 \tag{23.10}$$

Assuming that design optimization has put the CFR on target, using equation (23.10) you can calculate the fraction of the total quality loss, for a given CFR, that is assignable to each of the control factors.

$$QL_i = \frac{\left(\frac{\partial y}{\partial x_i} \sigma_{x_i} \right)^2}{\sigma_y^2} \tag{23.11}$$

The value of QL_i depends on the control factor settings, which affect the values of both $\partial y/\partial x_i$ and $\partial y/\partial z_i$. For our lathe machining example, the last row of Table 23.2 gives the fraction of quality loss assignable to each control factor. The control factor variation does not account for the entire quality loss, since the error makes a significant contribution. In the lathe example, no noise factors were included in the experiment design, so variations due to external noises are part of the error.

You can see that the fraction of the quality loss for a particular CFR that is assignable to a particular control factor is identical to the fraction of the total variance of the CFR assignable to the control factor. This gives a useful working definition for the relative importance of each control factor.

Critical Parameters and Risk

Risk Assessment

Different types of risk can affect schedule, product performance, product costs, and development costs. In this section we focus on identifying the risks associated with shortfalls in product performance.

Early in a development project it may not be possible to have a good understanding of design capabilities. The longer your lack of understanding persists, the higher the risk you are assuming. For a project with a fixed delivery date (most projects), problems must be understood and addressed early. Usually a poor design capability is rooted in either a weak design concept or a poor manufacturing process capability. So risk mitigation might require either design changes or manufacturing process improvements. They require time and resources to understand trade-offs and develop solutions.

Risk assessments enable risk mitigation. If certain of the CFRs and supporting control factors are risky, you need to know it. It may make perfect business sense to assume risks in going forward with a project. What never makes sense is not to understand the risks. To not look is to not understand. It can have serious consequences when problems surface later in a project. A common definition of risk includes two important factors: the probability of occurrence of an event and the consequences of the event if it does occur. So risk can be defined as shown here:

$$Risk = P(event) \times Consequences(event) \qquad (23.12)$$

In this case our interest is in evaluating those business risks associated with the product failing to satisfy customers. It can happen in a number of ways, as shown in Figure 23.1.

The following are some considerations when identifying the risks associated with a design:

1. A critical parameter that has a poor manufacturing process capability is more likely to be off target. When the specification limits are tight, expect a lower first-pass manufacturing yield. It is more likely that deterioration will cause performance to drift off target and out of specification quickly.
2. A critical parameter that is important to putting a CFR on target has a proportionally larger effect on variability than do less important factors. The fraction of the variance of the CFR assignable to a particular control factor is the same as the fraction of the quality loss assignable to that control factor.
3. A critical parameter having an effect on a basic need, as defined by the Kano Model, presents a higher potential risk. Failure to satisfy a Basic need has greater consequences than failing to achieve the target for a Satisfier or Delighter.
4. A control factor for which the slope of the response surface is high presents a potential risk to product performance, even if you believe that you can handle its potential variations. If you underestimate the variations of the control factor, or if something happens to cause a larger-than-expected variation, the impact of that control factor on performance will be greater than expected. If there is the possibility of deterioration

causing set point drift over time, the effect of the set point change on a CFR will be magnified by the high slope of the response surface.

Using the Kano Model to Evaluate Risks

There are three Kano categories for classifying customers' needs (see Chapter 9):

1. **Basic needs, which customers expect:** Often there is an implicit assumption that Basic needs will be met. Customers tend not to mention them.
2. **Satisfiers:** An example is vehicle fuel economy. How much is needed? Generally, more is better.
3. **Delighters:** These are capabilities that can surprise and delight customers. If they are missing, there should be no impact unless a competitor comes out with a product having those Delighters.

You can use the Kano Model to assess risks when developing a new product.

Basic Needs

Be a prudent risk taker. If there are low-risk options, it is foolish to assume great risks in delivering a function that is a Basic need. It is possible that a new way of delivering a Basic need might result in a cost advantage. While success may deliver benefits to your company, failure can reduce value delivered to your customers.

Satisfiers

There can be some risks associated with this type of customer need, depending on the competition in your markets. To understand the situation, the benchmarking of competitive products is necessary. If you lag behind your competition and your goal for a particular Satisfier is to achieve parity, there are more risks in falling short than if you are already ahead and are trying to increase your lead over competitors.

Delighters

This is an area where you can take risks. Contingency plans should address the consequences of failure to implement designs for Delighter features. It would be prudent to ensure that you can back off gracefully in the event of performance shortfalls. The absence of a Delighter may do less harm than one implemented poorly.

You can apply the Kano Model along with the guidance about "importance" and "difficulty" to evaluate the risks associated with delivering the value proposition for a new product. The approach is the following:

1. For each CFR, determine the
   a. Type of customer need it supports, as defined by the Kano Model
   b. Design capability
   c. Quality loss
   d. Supporting control factors

2. For each supporting control factor, determine the
   a. Difficulty, as measured by the manufacturing process capability or qualitative factors discussed earlier
   b. Importance, as measured by the factor contribution to the variation of the CFR
   c. Fraction of the quality loss for the CFR assignable to each control factor

Figure 23.5 shows the outline of a process to understand the risks associated with the critical parameters of a system. As quantitative people, engineers and scientists like to assign numbers to characteristics that are important. However, when evaluating risks, the thought process can be more important than the assignment of numerical values. Probably it is more important to go through the process with the goal to understand the relative risks associated with delivering a product that will score high for customers' satisfaction as well as meet the needs of the business. The information in the table shown at the bottom of Figure 23.5 can be used to evaluate the risk associated with both CFRs and control factors. The assessment will be both product- and organization-dependent.

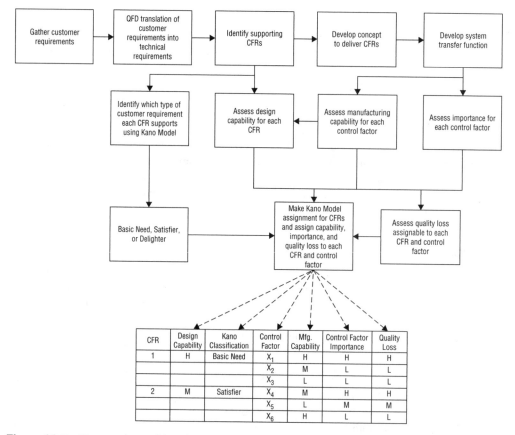

CFR	Design Capability	Kano Classification	Control Factor	Mfg. Capability	Control Factor Importance	Quality Loss
1	H	Basic Need	X_1	H	H	H
			X_2	M	L	L
			X_3	L	L	L
2	M	Satisfier	X_4	M	H	H
			X_5	L	M	M
			X_6	H	L	L

Figure 23.5 Process for ranking risks for CFRs and control factors using the transfer function and the Kano Model

Critical Parameter Measurement

The Importance of Good Data

None of us would argue against the need for data to have high integrity. If you are making measurements as part of an experiment or quality assurance activity, you are gathering data that will be used for a decision. When your measured values are different from reality, you take risks if important decisions are based on those data.

There are a couple of good principles to follow when using data to justify a recommendation:

1. Before taking data, ensure that the measurement system has sufficient capabilities for the task. Not all measurement systems are equally capable.
2. Never make a recommendation, especially a controversial one, based on measurements taken using a system that has not been characterized beforehand.

Measurement System Analysis

There is a tendency to think of the measurement system just as the equipment. It is more than that. The gauge, the operators and procedures, the software, and the measurement environment are part of that system. There are several performance metrics for the measurement system that are important:

1. **Linearity:** If you measure a collection of parts that have known characteristics, you want a straight-line relationship in the plot of actual versus measured values.
2. **Accuracy:** Accuracy is the closeness of the measured value to the actual value.
3. **Precision:** Precision is a measure of the variability in repeated measurements for the same parameter.
4. **Stability:** Stable measurement systems show little or no drift over time. If a stable system is characterized at two different points in time, the results will be similar.

Measurement System Analysis (MSA) is a process to evaluate the performance of a measurement system. Performance should be evaluated for tasks for which the measurement system is intended. Since every measurement system has limitations, it will not perform well on tasks beyond its inherent capability. Using a properly designed experiment, it is possible to determine the performance of any measurement system. An MSA is done by measuring parts covering the range of dimensions or characteristics that are expected when the system is being used. The variation in the resulting measurements can be broken into two contributions: the parts and the measurement system. It is the variation due to the measurement system that is undesirable.

Most good statistical applications[4] have MSA tools that can be used to evaluate the performance of your measurement system. A common MSA report is the Gage R&R, which gives the study results in a standard format that shows the contribution of the measurement system to the observed measurement variability. An excellent overview of MSA is given by Pyzdek.[5]

---

4. Minitab and JMP are two widely used examples.

5. Thomas Pyzdek, *The Six Sigma Handbook* (New York: McGraw-Hill, 2003).

CPM: A Process for Creating, Using, and Maintaining Information

The main users of CPM information are the organizations for technology development, design engineering, materials management, manufacturing, and service. Many critical parameters are identified during product development. They are refined as the project moves toward market entry, production, and service.

A system for CPM includes the following:

1. **Critical parameter data**
   a. **Performance specifications:** Definition of targets and tolerances for performance outputs (CFRs) to be measured in production or service to ensure that the product satisfies its requirements; where appropriate, the measurement techniques (instrument, procedure) are included.
   b. **Component and material specifications:** Targets and tolerances for all critical parameters that affect CFRs or measurable quality metrics.
   c. **Final adjustment specifications:** Definition of targets and tolerances for critical adjustment parameters to be set in production or maintained in service to ensure that the performance specifications are achieved; where appropriate, the measurement techniques (instrument, procedure) are included.
   d. **Manufacturing process specifications:** Definition of manufacturing process set points used to reproduce any components or materials that affect CFRs or measurable quality metrics that are important to customers. Whether you are producing a consumable material or a hardware component, the recipe has to be documented and maintained.
2. **Process for monitoring critical parameters**
   Critical parameters are characteristics of materials and components that are manufactured. Consequently, it is important to characterize the consistency and stability of the processes that generate those parameters. A key tool is SPC. Using SPC, you can evaluate process variability and detect process drifts. Root cause analyses and corrective actions have to be responsive to variations in process outputs that exceed common cause variation.
3. **Product data management system**
   Effective CPM requires a database that specifies the critical parameters (fixed and adjustable) with editorial controls and a disciplined change management process. The actual format might include documents, CAD files, and software files. The data include design specifications, manufacturing assembly and adjustment parameters, manufacturing process parameters for components and materials, purchase specifications for components, and service parameters. As changes are implemented in production for the product design and production processes, the data management system must provide configuration control with ready access for those who need the accurate information.

Key Points

1. A critical parameter is any product characteristic that is important to customers as well as any characteristic that affects attributes that are important to customers.
2. Critical parameters must be identified and managed proactively.

3. There are several important things to understand about critical parameters:
   a. Importance
   b. Difficulty
   c. Quality loss
   d. Kano Model considerations
4. "Importance" is assessed using the system transfer function. It should be calculated after robustness is optimized.
5. "Difficulty" is assessed using manufacturing process capability, design capability, and qualitative factors.
6. Quality loss, due to control factor variations, is determined after the development of robustness, which yields the optimal settings for control factors. The quality loss can be calculated for each CFR and allocated to the control factors.
7. After you understand these things, you are positioned to evaluate risks associated with product performance variability.

Discussion Questions

1. Define *risk*.
2. Is there a danger in classifying CFRs using customers' ranking of needs, rather than the Kano Model?
3. What is the difference between statistical significance and importance when evaluating control factors in an empirical model?
4. In the QFD matrix you evaluate the strength of the relationships between technical requirements and customer needs. How is strength similar to importance?
5. A particular control factor has a high manufacturing process capability (low difficulty) and also a high importance to a CFR that is a basic need. What is the nature of the risk?
6. When might you want to carry forward alternate technologies or concepts in development?
7. What is a common unit of measurement for all risks in business?
8. What are some tools for CPM during product development? In production?

An Illustration of Process Methods and Tools

This story is written to illustrate many of the cross-functional engineering methods that we've described. We'll not extend it into the domains of marketing, project management, portfolio management, or business management since those topics are out of scope for the book. The product concept may be familiar to those of you who have used similar gear for your leisure or professional activities.

Several years ago we worked with a product development project in a midsize company specializing in protective clothing and equipment for emergency responders and outdoor adventurers. Their products included boots, coats, gloves, helmets, goggles, and other safety apparel. The target market was defined as volunteers and professionals working in fire departments, ambulance corps, ski patrols, search-and-rescue squads, and similar activities that expose people to harsh weather and terrain, often at night. Some of their customers were also extreme adventurers such as backcountry skiers, snowmobile riders, backpackers, rock or ice climbers, and spelunkers. The company's business was growing as the market recognized their products as being superior to alternatives available in similar price ranges. Higher-priced gear was out of reach for those paying for their own equipment. Conclusions from strategic planning suggested that there were several opportunities to extend the product offerings, particularly to improve the range of product applications and the convenience of their use.

This story is focused on a development project that was chartered by that refreshment of the product development portfolio. Our purpose is to illustrate the application of some of the methods that we have described in earlier chapters.

The company was located in a town near vast forests, mountains, and lakes. Many of the employees were outdoor enthusiasts, volunteering as local emergency responders or being deeply involved in extreme adventures. As part of their culture, they used their own products, as well as those of their competitors. It was a behavior model that provided valuable insights into the competitive market and its business opportunities. By being actively involved in the market, they were positioned to know a variety of people using this category of products, to understand their activities, and to appreciate how the company's products could provide additional value to people in those activities. Some of the engineers and marketing people routinely went on adventures together, just as good friends who shared a passion. They did the same activities but tended to see things slightly differently, as their functional disciplines would suggest. By being involved participants, they could recognize opportunities to modify existing products or develop new ones that would fit the company's business strategy and would be recognized as delivering more value to customers in the market segment.

The company had been doing fairly well with its product line, although management sensed that they benefited from a lack of aggressive competition. The market was recognized to be changing. A rise in the public's enthusiasm for outdoor activities, even wilderness adventures, had created a demand for products that were also useful for the emergency responders. These two user groups were merging into the same market segment. High-tech fabrics were showing up in gloves, clothing, and packs. Lighter-weight and more rugged plastics were now in the designs for helmets and tools. Equipment was increasingly robust against extreme weather conditions. Where did the product planners see new opportunities?

Advanced Product Planning

An established planning group was charged with the task to monitor the market and identify new opportunities to improve the company's growth. It was recognized that products developed for outdoor adventurers and emergency personnel may very well appeal to people working on home projects or around farms or forests, as well as people interested in hunting, fishing, camping, hiking, and climbing. Possibly even snowmobile riders, skiers, and others active in winter sports would find benefits. Spelunkers who explore caves might also find these products valuable, particularly because of the wet and dark conditions of their environment. So the chosen market segment might have fairly fuzzy edges.

As a starting point, the planning group decided to tap the company's internal wisdom. They had developed a list of employees who were known to be participants in the market segment. These folks had firsthand experiences to contribute, particularly with the pros and cons of features and functions of products under extreme conditions. They could describe stressful conditions and the consequences of product malfunctions or inadequacies. They tended to know other people to talk with outside the company. These contacts could be found in sporting goods stores, activity clubs, relevant volunteer organizations, and adventure guide services. They received publications supporting these interests, many of which wrote of trends, catastrophic episodes, and lessons learned.

One of the suggestions that arose was that the lanterns and flashlights that the company produced were excellent at lighting a campsite or throwing a strong beam over a long distance. However, they were handheld devices that tended to be heavy and cumbersome. They pointed light where they were aimed but were difficult to position so that they could be helpful when not held by hand. In addition, they tended to have a single use, for example, to throw either a strong narrow beam or a wide soft beam. They tended to create lots of glare that made it difficult to read maps or written material. All were vulnerable to rain, cold temperatures, being dropped, momentary dunking in water, and other abuses expected in outdoor environments.

Product Concept Proposal

The collective inputs produced a project proposal to develop a hands-free headlamp, one that would be suitable for extreme applications, yet would be broadly applicable to other uses.

An example is shown in Figure 24.1. The hands-free concept was primarily a response to the frustration of the user having to devote one hand to hold existing flashlights and lanterns. The application to extreme environments reflected the target market segment, the people engaged in outdoor activities under difficult conditions of weather, lighting, terrain, and so on. Marketing liked the opportunity to take market share from conventional flashlights and lanterns as well as from existing headlamps that didn't stand up to the rigors of these applications. Engineering liked the idea of developing a lighting device that would be powerful, versatile, and robust against the stresses in these activities.

This product proposal was introduced to the portfolio of development projects, with a value proposition:

"Tough lights for tough people."

During Phase 0, management chartered a product development team (PDT) to evaluate the product concept, the availability of potential technical solutions to the preliminary requirements, and the achievability of a project to develop the product. There was open debate about its probability of success. This was focused on an initial project management plan and the viability of a supply chain for its expected components. Within a short period of time, the proposal for the product concept was deployed to potential concepts for its components and an assessment of its risks. An initial business plan was submitted, with a project management plan for Phase 1. The project was accepted as being aligned with the company's market participation strategy and having a reasonable potential to deliver improved value to these target customers.

The PDT included team leaders from mechanical and electrical disciplines in design and manufacturing engineering, industrial design, and marketing, with the assistance of a project management specialist and "as needed" representation from purchasing and finance. When the project charter was authorized, this group became the core leadership team committed for the duration of the project. With the charter being backed by Gate 0 agreements among the stakeholders, the project leader received clear empowerment to make all decisions affecting design and manufacturing, including the application of those resources, and the selection of design

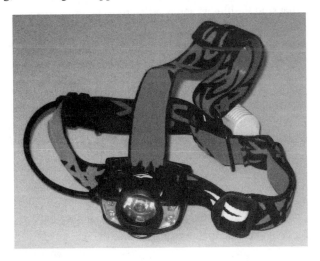

Figure 24.1   A headlamp enables both hands to be used for tasks.

concepts and their suppliers. She was also given license to exceed the budgets for development funding, within modest constraints, as long as the savings in the costs of delayed market entry were expected to be greater than the additional development costs. The target for market entry was set for late summer to coincide with the buying habits of people getting ready for the winter months.

Product Development

Initially, the concept for the product was mostly a vision. The first task of the PDT was to develop a detailed understanding of the needs of their customers so that the value proposition could be translated into a set of engineering requirements.

- What are "extreme environmental conditions" in these applications?
- What does "easy to use" mean?
- What does the light have to do?
- What does "tough" mean to these customers?
- Are spelunkers in or out of the market segment?
- Is there one product or are there several?

Voice of the Customer

The data-gathering and analysis activities were easier for this project than for more complicated product concepts or more diverse markets. Here, team members were also customers of the product. They knew product users and could ask them a wide range of probing questions to derive more insight than if they were less familiar with the market. They could identify good answers, understand the nuances in the customers' frustrations, and recognize opportunities for value-adding differentiation. The PDT decided on a VOC strategy that had four thrusts for data gathering:

- Employees who were routine users of flashlights and headlamps or who participated in activities relevant to the market segment
- Retail stores that sold flashlights and headlamps to outdoor adventurers
- Clubs of outdoor enthusiasts
- Organizations of emergency responders, such as fire departments, ambulance corps, and search-and-rescue squads

Several teams, with two people each, visited representatives of each of these categories. One person asked questions while the other recorded the essence of the inputs with as many verbatim comments as needed to capture the intent accurately. The teams had people from engineering and marketing disciplines in order to get the benefits of their range of listening frameworks. Additional "customers" were questioned until the data teams felt that they were receiving redundant inputs that they had already noted.

The "voices" of these customers, and their sources, were recorded without overt interpretation. The objective was to understand *what* the headlamp needed to do and the context of those needs, not *how* the customers wanted it done or the company thought it should be done. These were scenarios of product usage that later could be used to design tests. Since the market already was being served by both handheld and hands-free lights, it was relatively easy for customers to refer to frustrations they had experienced when using different devices for particular applications. For example, many lights that were dropped accidentally suffered broken filaments. They didn't like getting wet or very cold. Many didn't throw enough light to illuminate a trail or read a trail marker from a distance while hiking or skiing. Batteries couldn't be trusted to last all night. It was fairly easy to think about needs for this type of product.

The data gathering was coordinated out of a centralized project room that had long walls on which sticky notes could be placed and moved easily. Inputs were written as close to verbatim as was reasonable, preferably in a verb-noun-modifier format. Initially there was no deliberate organization to the arrangement. It looked something like Figure 24.2.

The team's task then was to organize the inputs into statements and hierarchies that indicated major themes and subordinate details. Each person in the data-gathering team took turns moving the sticky notes into affinity groups, collections of inputs that had a similar intent. The rearranged VOC inputs looked like Figure 24.3 with high-level headings for each group.

The teams reminded themselves of the Kano graph (Figure 9.1). Its message was that requirements were heard because they were concerns familiar to product users. Frequent comments were that they needed more light or more ruggedness. There were requirements that users tended not to mention, assuming that they were already known, such as the stability of fit or the ease of understanding how to turn the device on. And there were requirements that people did not know to talk about because they hadn't thought about the possibilities of major advances in the product's capabilities. These were the "Delighters" that could differentiate the product and be critical to its sales. The product development team would have to think creatively about those.

So the VOC teams discussed the product needs that they had collected and organized. Were they all at the same hierarchy level, or were certain ones subordinate to others? Did the teams understand the implications, or were clarifications or additional details needed? Were any need statements redundant? What was missing? A spreadsheet was developed to organize the inputs into a table (Table 24.1), the first room in the House of Quality.

Figure 24.2 VOC inputs reflected customer needs that were spoken or observed.

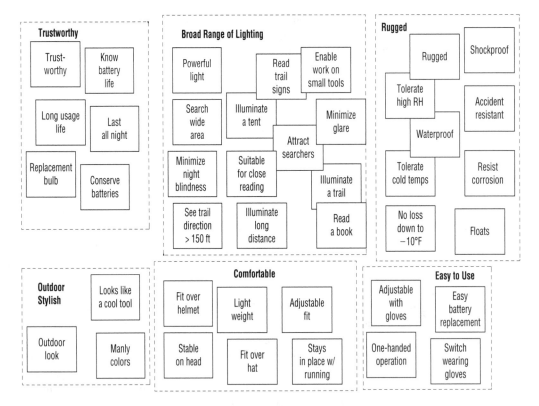

Figure 24.3   VOC inputs were placed in affinity groups with a primary-level name.

Fortunately, the company was located in a rural northern area where employees were very active, often going on camping and backpacking trips, or participating as emergency responders. Some even went frostbite camping in the winter. What a wonderful opportunity for their firsthand understanding to be enriched! Since it was early February during Phase 1 of the project, there was plenty of cold weather to experiment with the rigors of the environment. Three weekend parties were organized for PDT members:

1. The first team focused on hiking and camping, on snowshoes. They would understand the range of lighting conditions and the difficulties when wearing bulky hand coverings. Included in the adventure would be following trails at night and reading maps.
2. The second team was asked to do deliberate work on cars or trucks outside at night. This might include changing tires, batteries, or spark plugs as a way to understand the visibility needs for close work, including fine motor skills.
3. The third team, because of their existing membership in a search-and-rescue team, decided to turn their volunteer work into a VOC experiment. They were interested in how the headlamp could contribute more to the safety of the person wearing it.

The teams were to use a variety of headlamps already on the market, some of which had specific features to be tested.

Table 24.1 The VOC inputs formed the first "room" in the House of Quality

VOC for Hands-Free Headlamp		
Value Proposition: Tough lights for tough people		
Primary	**Secondary**	**Tertiary**
Easy to Use	One-handed operation	Switch wearing gloves
		Adjustable wearing gloves
	Easy to change battery	
Comfortable	Stable on head	Light weight
	Adjustable fit	Fits over hat or helmet
Safe	Safe for environment	No contaminants for disposal
	Safe for people	Not hot to touch
Proximity Lighting	Illuminate a tent	Read a book at arm's length
	Illuminate work group	
Distance Lighting	Illuminate a trail	Read trail signs
	Illuminate a long distance	See trail directions and obstacles
Rugged	Accident resistant	
	Waterproof	
	Tolerate high RH	Resist corrosion
Trustworthy	Last all night	Replacement bulb
		Conserve batteries
	Long usage life	Efficient at creating light
Outdoor stylish	Outdoor look	

Monday morning the teams showed up for the project meeting loaded with inputs and a significant level of excitement about the opportunity to provide much better solutions for their customers. None of the team members used an existing headlamp that they thought to be satisfactory for these extreme applications. Some didn't throw enough light or made the light too diffuse to illuminate a trail. A couple of batteries could not tolerate the cold weather. One with a rotating focus mechanism jammed when snow got into the device. One, made of metal, loosened as the part dimensions shrank in the extreme cold. When worn without a brimmed hat, several headlamps did not stay in place when the person was hiking vigorously or had to jump across a narrow stream. One headlamp fell into water and failed to function.

There were interesting features with a couple of devices that the team wanted to replicate, such as a battery pack that could be detached from the head straps. That allowed the battery to be placed inside a coat to keep it warmer and reduce the stress on the battery. Most stood up to being dropped in snow but not to being dropped into water. One hit a rock and died on the spot. For close repair work, those with higher light levels produced too much glare. Headlamps designed to prevent accidental switching had small or recessed "on/off" buttons that were difficult to manipulate with gloves, even worse with mittens. Some were so stiff that two hands were needed, one to hold the headlamp, the other to push the button. All the participants mentioned how much they had learned about the product that they were charged to design.

This first-cut analysis of the VOC inputs identified several important questions:

- What are the really stressful conditions for "tough" use?
- What would customers do if they knew that the batteries had little life left?
- At what depth does the "waterproof" capability need to apply?
- Does the perception of usage life apply to the battery, or light source, or both?
- To what extent would customers give up battery life to get less weight?
- How do flashlights get damaged?
- How cumbersome are gloves or mittens when trying to switch the headlamp?
- Are there concerns about accidentally turning the headlamp on or off?
- How cold can temperatures get when some lamp operation is required?
- What do customers need to do when lamp operation has deteriorated?
- Are utility linemen and roadside car mechanics part of the market segment?
- Do people wearing helmets need headlamps that are separate from the helmet?
- Are cold temperatures and high relative humidity in the same market segment?
- What styling characteristics are desired?
- What capabilities are required if the person searching for a victim also needs to be found?

A second round of conversations with target customers was then organized. This time the questions were much more probing, given the insights derived from the initial inputs and the weekend experiences. Two benefits were enjoyed: The value proposition was validated, and the VOC inputs became more detailed and more complete. The first room in the House of Quality was then refined and extended to look like Table 24.2.

Competitive Assessment

The weekend experiments enabled the VOC teams to take another step forward. They compared alternative headlamps and identified how various customer needs, if satisfied really well, could provide competitive advantages. The conclusions were recorded in the planning matrix of the House of Quality. First, prime competitors were identified with technical details that enabled the comparison, as shown in Table 24.3. The company's current product was a mid-priced LED lamp for proximity lighting.

Table 24.4 illustrates part of the House of Quality matrix, correlating the VOC details with the planning assessments for those attributes related to illumination and ruggedness. The entire QFD matrix was much larger. However, the development team subdivided it, as illustrated in the example, to make it easier to understand and manage.

You can see that the competitive assessment focused attention on the challenges to develop ruggedness in a drop and watertightness for the battery and light source. The strength of distance illumination was important, although not with the technical challenge of the other two.

System Requirements

The engineered solutions needed requirements in the technical language of the solution. The resulting requirements would then have target values that not only would satisfy customers but also, if achieved, could establish competitive differentiation for the product.

The relationship matrix in the House of Quality was a useful tool to document the results of team discussions and decisions that established those requirements as thoughtful translations of the intents of the customer needs. The cumulative scoring in the relationship matrix yielded a priority score for each requirement that was normalized and thereby guided the distribution of development

Table 24.2   The VOC room in the House of Quality was more complete and actionable

VOC for Hands-Free Headlamp		
Value Proposition: Tough lights for tough people		
Primary	**Secondary**	**Tertiary**
Easy to Use	One-handed operation	Switch wearing gloves Adjustable wearing gloves
	Easy battery change	No tools required Low force/torque to remove cover
Comfortable	Stable on head	Stays in place when running Light weight
	Adjustable fit	Fits over hat or helmet Attaches to pack, jacket Accommodates reading in sleeping bag
Safe	Safe for environment	No contaminants for disposal
	Safe for people	Not hot to touch Reflect light
Proximity Lighting	Illuminate a tent	Read a book at arm's length Enable work on small tools w/o glare
	Illuminate work group	Do not blind other people
Distance Lighting	Illuminate a trail	Read trail signs as in daylight Minimize night blindness
	Illuminate a long distance	See trail directions and obstacles to react Attract searchers
Rugged	Accident resistant	Shockproof in drop from head level Floats
	Tolerate cold temperatures	Maintains brightness through coldest night
	Waterproof	Watertight to fall in stream Battery cover won't come off accidentally
	Tolerate high RH	Resist corrosion
Trustworthy	Last all night	Know battery life remaining Replacement bulb Conserve batteries Consistent light level
	Long usage life	Efficient at creating light Lamp life very long Run throughout the night
Outdoor Stylish	Outdoor look	Unisex colors
	Attractive	Look like a cool tool

Table 24.3   The headlamps to be compared were described at the same level of detail

| Product | Primary Light Source | | | Secondary Light Source | Attributes | Price |
	Source	Light Output	Range	Source		
Our Current Product	LED-1 watt	25 lumens	140 ft	3 LEDs-1/2 watt@	Storm proof	$40
Competitor #1	Halogen	10 lumens	No data	5 LEDs	Focused; waterproof @16 ft	$110
Competitor #2	LED-1 watt	16 lumens	60 ft	1 Red LED	Waterproof @ 3 ft; −22° F; explosive environment; durable plastic	$30
Competitor #3	Xenon	20 lumens	240 ft	3 LEDs	Focused; Water resistant	$45

funds and resources. That score, if you remember, was the sum of the scores in the cells of the matrix multiplied by the normalized importance scores for each cell. Figure 24.4 illustrates a portion of the relationship matrix that the headlamp teams developed. To avoid large, cumbersome matrices, the teams worked with logical sections that enabled them to see the forest in spite of the trees and set targets for their requirements. These requirements then became the criteria for the selection of the baseline design concepts.

As examples, the customers said that they wanted to read trail signs as well as they could in daylight and see changes in trail directions or obstacles soon enough so that they could react. Engineers translated those requirements into a target of at least 150 feet in range for the primary illumination, with the insight that more than that would surpass the capabilities of competitive headlamps. An assumption was that instrumentation "range" meant the maximum distance for which, on axis, at least 0.25 lumens per square meter could be detected. That assumption cleared away some ambiguities about how well one must see objects 150 feet away. The team recognized that some additional thought and development might be needed. For example, they might modify that requirement later to have a lumens target based on illumination measured with specific instrumentation at that distance.

Functional System Analysis

With requirements determined at the system level, the work groups could think through the functions of the product that would be necessary to achieve the requirements. For example, the projection of light would need a power source, a switching function, the light source itself, and a reflector. The need for a zoom function was debatable. The functional flow diagram, illustrated in Figure 24.5, enabled the development team to make decisions about how the system should be configured, such as whether the battery was to be integrated with the lamp or attached to the strap in the back. If there was to be a strobe function, which light source would be flashed? Was there to

Table 24.4 The planning matrix identified opportunities for competitive differentiation

Primary Level VOC	Secondary Level VOC	Tertiary Level VOC	Importance to Customers	Our Current Product	Competitor #1	Competitor #2	Competitor #3	Goal for New Product	Improvement Ratio	Benefit to Sales	Raw Importance Score	Relative Importance
Proximity Lighting	Illuminate a tent	Read a book at arm's length	3	9	8	9	9	9	1	1	3.0	
		Enable work on small tools w/o glare	4	9	8	9	9	9	1	1.3	5.2	
Distance Lighting	Illuminate a trail	Read trail signs as in daylight	3	8	9	7	9	9	1.1	1.3	4.3	
		Minimize night blindness	3	8	7	9	3	8	1	1	3.0	
	Illuminate a long distance	See trail directions and obstacles to react	5	5	8	1	9	9	1.8	1.5	13.5	
		Attract searchers	5	7	3	7	5	9	1.3	1.3	8.4	
Rugged	Accident resistant	Tolerate drop from head	5	2	1	9	1	9	4.5	1.5	33.8	
		Floats	5	2	9	9	2	9	4.5	1.3	29.2	
	Tolerate cold temperatures	No loss over cold night	5	7	1	9	1	9	1.3	1.5	9.8	
	Waterproof	Watertight to fall in stream	4	2	9	5	2	9	4.5	1.5	27.0	
	Tolerate high RH	Corrosion resistant	3	4	7	9	4	5	1.2	1	3.6	

be switching of intensity levels for both the distance light source and the one for proximity lighting? Were there to be two different light sources, or just one? Was there to be an interaction between the power delivered to the light and the remaining energy left in the battery, particularly when the battery was running low?

You can imagine the discussions within the development team, probing into the intent of the customers' needs, the relevance to the value proposition, and the trade-offs among the system architecture and the capabilities of its design components. The functional flow diagram facilitated the discussions and documented their results.

You can see that the requirements for certain functions were formed by the interactions among design functions. For example, the switching of power to a lamp could either be the user's choice or be under the control of the "reduce light level function."

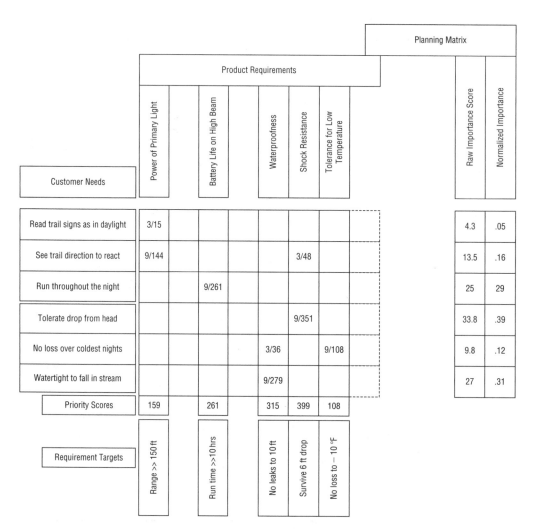

Figure 24.4   The relationship matrix in the House of Quality helped to set targets for the product requirements and guide the distribution of development funding.

Concept Generation and Selection

The engineering teams understood that they were in competition with well-known suppliers of headlamps. Those products could be found easily in outdoor outfitters, even in hardware stores. The team knew that they had to compete with their technical solutions.

By standing back from the details, they could see that they faced challenges in three fundamental categories:

1. Certain concept features were viewed as routine design tasks, not subject to development. This category included the head strap with its need to be secure, to have various styles and colors, and to enable the battery to be removed. The activities to develop them could be left to Phase 3.

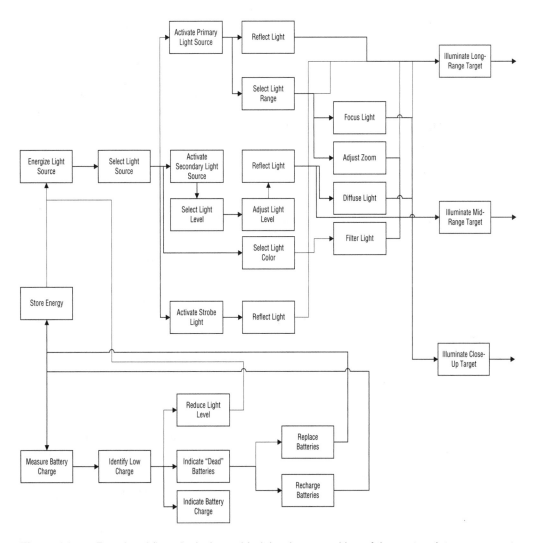

Figure 24.5 Functional flow analysis enabled the decomposition of the system into component functions for which requirements could be decomposed and solutions selected.

2. Selected attributes were inherent in the parts purchased from suppliers, such as the light sources, reflectors, and battery. The specifications of these supplied components would be deliverables for Phase 2.

3. Directly aligned with the value proposition was the need for the housings for the headlamp and its battery to be rugged and watertight. This was seen as a challenge for robustness development in Phase 1.

Price

The teams recognized that price was not a difficult constraint. People in this market segment would pay to get a product that worked well for them, but it must be better in ways that really mattered. There was a sense that the price of the headlamp should not be more than $100, but that left much room in the selection of the baseline concepts.

Light Sources

Although several studies were reviewed, the ones for the light source were particularly interesting, since the architecture of the headlamp depended on that concept selection. Table 24.5 illustrates the conclusions. Although not complete in all details, the analysis gave the illumination work group a clear sense that they needed to focus on LEDs since there would be no filament to break or burn out.

These initial assessments were seen to be not detailed enough to enable the selection of the best design concept for the light source. A little extra effort delivered a more quantitative comparison, illustrated in Table 24.6.

Table 24.5   Data gathering enabled technical comparisons of alternative design concepts

Source	Light Output	Construction	Attributes
Incandescent	Limited spectrum Biased toward red ~10% energy as light	Tungsten filament Filaments fragile	Standard flashlight Inferior to advanced bulbs Cooler than Halogen Operated close to melting point
Halogen Gas	Whiter light than Incandescent More lumens per watt than Incandescent	Tungsten filament Filament fragile	Longer life than Incandescent (~2500 hrs)
Krypton Gas		Tungsten filament Filament fragile	Longer life than Incandescent (~2000 hrs) Conducts less heat than Argon Conducts more heat than Xenon

Xenon	Clear white light
	Re. Halogen
	  – 2 × brightness
	  –1/2 power consumption
		Tungsten filament
		Filament fragile
			Longer life than Halogen
			(~5000 hrs)

High-Pressure Xenon	Clear white light
	Re. Halogen
	  – 2 × brightness
	  –1/2 power consumption
		Half size of Xenon
		Thicker glass
		Xenon pressurized
		to several
		atmospheres
			Longer life than
			Halogen (~5000 hrs)

LEDs	Very efficient at creating light Limited light output over distance Produces soft beam of gentle illumination Nichia (5 mm): 2–3 lumen @ Luxeon 1 w: 25–30 lumens @ Luxeon 3 w: 75–80 lumens@ Luxeon 5 w: ~120 lumens @	No filament to burn out Unbreakable resin construction	Environmentally friendly - no waste, no mercury Does not get hot to touch Longer lived (~100,000 hrs)

Table 24.6   The range for distance lighting was a system requirement involving the light source, number of lights, reflector design, and power source

Source	Light Output	Range	Run Time (high beam)	Attributes
Halogen		330 ft		Fragile filament
Xenon	20 lumens	150–240 ft	3–5 hrs w/3 AA (4.5v) 60 hrs w/3 AAA (4.5v)	Fragile filament
LED-5 mm	2–3 lumens			
LED-1/2 watt				Unbreakable
LED-1/2 watt × 2		33 ft	35 hrs w/2 AAA	Unbreakable
LED-1/2 watt × 3		43 ft	30 hrs w/2 AAA 80 hrs w/3 AAA 120 hrs w/3 AA	Unbreakable
LED-1/2 watt × 4		76–90 ft	80 hrs W/3 AAA 90–105 hrs w/3 AA 160 hrs w/6v Lithium	Unbreakable
LED-1 watt	25–30 lumens	43 ft		Unbreakable
LED-1 watt with technical reflector	25–30 lumens	140–165 ft	50 hrs w/3 AA??	Unbreakable
LED-3 watt	75–80 lumens	150–185 ft	8 hr w/3 AA 10 hrs w/2 Lithium CR123A (6v)	Unbreakable
LED-3 watt with technical reflector and heat sink for LEDs	75–80 lumens	185–260 ft	8 hrs w/3 AA 15 hrs w/2 Lithium CR123 (6v)	Unbreakable
LED-5 watt	120 lumens		28 hrs w/3 AAA?	Unbreakable
LED-5 watt × 6	350 lumens	396 ft (120 m)	16 hr w/2 rechargeable Lithium ion	

The company's current product already used 1/2-watt LEDs for proximity lighting. That was not a particular challenge since there were readily available alternatives. The challenge was to select the best concept for long-distance illumination. These data enabled the illumination team to see that they had to concentrate on higher-powered LEDs and the number of LEDs, with implications for the battery. Would a 1-watt LED be sufficient? How about 3-watt LEDs? Could the range be achieved by a parabolic reflector? On the market at that time, a very expensive

Table 24.7   Alternative concepts for distance lighting were described at the same level of detail

Option	Description	Light Output	Illumination Range*	Attributes	Run Time (on high beam) versus Power Source	Relative Price of Competitors' Headlamps
A	Xenon w/reflector and focus	20 lumens	150–240 ft	Fragile filament; can be focused; $2 \times$ brighter than halogen; 1/2 power consumption of halogen	3–5 hrs w/3 AA	Medium
B	LED-1 watt	25–30 lumens	40–45 ft	Unbreakable Friendly to environment	50 hrs w/3 AA	Low
C	LED-3 watts	75–80 lumens	150–185 ft	no mercury, no metal Softer light	10 hrs w/2 Lithium (6v)	Medium
D	LED-5 watt	120 lumens	No data	Efficient use of energy Tolerate high and low temperatures	No data for high beam	Medium

* = minimum 0.25 lumens per sq. meter

headlamp used six 5-watt LEDs to throw an extremely strong light. Would that be too much light? Its price was well over the $100 constraint.

To move forward in the selection of the baseline design, the illumination team chose four concepts for an initial comparison. Xenon was powerful but used a filament. The others used LEDs at three different power levels. Table 24.7 illustrates that technical comparison, at the same level of detail.

How did the capabilities of these options compare in their ability to satisfy their requirements? Remember from Chapter 11 that the question was which option was expected to be better at satisfying the requirement, not how much better it was to be. Look at the evaluation in Table 24.8. The Xenon concept was chosen as the reference, and the other approaches were compared to it. The LED options B, C, and D were judged to be better than Xenon for this application, although each was not as good for at least one requirement.

Following the process, the illumination team decided to identify additional concepts in an attempt to overcome the shortcomings of those initially compared. By introducing a technical reflector, they would either have to design it themselves or find a capable supplier to do so in collaboration with their internal teams. They could also take a brute-force approach with multiple LEDs of a particular power level. Table 24.9 shows their evaluation, which then was used in a new selection matrix shown in Table 24.10. Concepts A and B were eliminated, and concept C became the new reference for the comparison.

Table 24.8 The initial comparison of lighting systems did not identify one that was best at satisfying all criteria

Selection Requirements	Option A	Option B	Option C	Option D
Long-distance illumination	D	–	S	+
Rugged to shocks	A	+	+	+
Low-temperature capability	T	+	+	+
Efficient use of energy	U	+	+	+
Long run time	M	+	+	+
Environmentally friendly		+	+	+
Acceptable cost (< high)		+	–	–
Summary		**+ = 6** **– = 1** **S = 0**	**+ = 5** **– = 1** **S = 1**	**+ = 6** **– = 1** **S = 0**

Table 24.9 Additional concepts were identified to improve the chances of satisfying all of the important requirements

Option	Description	Light Output	Illumination Range*	Attributes	Run Time (on high beam) versus Power Source	Relative Price of Competitors' Headlamps
A	Xenon w/reflector and focus	20 lumens	150–240 ft	Fragile filament 2 × brighter than halogen 1/2 power consumption of halogen	3–5 hrs w/ 3 AA	Medium
B	LED-1 watt	25–30 lumens	40–45 ft	Unbreakable Friendly to environment no mercury, no metal	50 hrs w/3 AA	Low
E	LED-watt w/ parabolic reflector	25–30 lumens	140–165 ft	Softer light Efficient use of energy	50 hrs w/3 AA	Medium
C	LED-3 watts	75–80 lumens	150–185 ft	Tolerates high and low temperatures	10 hrs w/2 Lithium (6v)	Medium

(continues)

Table 24.9   Additional concepts were identified to improve the chances of satisfying all of the important requirements (*continued*)

Option	Description	Light Output	Illumination Range*	Attributes	Run Time (on high beam) versus Power Source	Relative Price of Competitors' Headlamps
F	LED-3 watts w/parabolic reflector and LED heat sink	75–80 lumens	185–260 ft		15 hrs w/2 Lithium (6v)	High
D	LED-5 watt	120 lumens	No data		No data for high beam	Medium
G	5 LEDs-5 watt	350 lumens	350–390 ft		16 hrs w/rechargeable Lithium ion	Very High

* = minimum 0.25 lumens per sq. meter

Table 24.10   For the second selection matrix two concepts were deleted and a new datum was chosen

Selection Requirements	Option A	Option B	Option C	Option D	Option E	Option F	Option G
Long-distance illumination			D	+	–	+	+
Rugged to shocks	D	D	A	S	S	S	S
Low-temperature capability	E	E	T	S	S	S	S
Efficient use of energy	L	L	U	S	S	S	S
Long run time	E	E	M	S	S	+	+
Environmentally friendly	T	T		S	S	S	S
Acceptable cost (<high)	E	E		–	+	S	–
Summary				+ = 1	+ = 1	+ = 2	+ = 2
				– = 1	– = 1	– = 0	– = 1
				S = 5	S = 5	S = 5	S = 4

In this comparison, concept F, the 3-watt LED with the technical reflector, was seen to have advantages over the others, particularly if its manufacturing cost could enable a street price under $100.

Baseline Design Description

The concept generation and selection process was applied to several other features of the product. The resulting choices of design approaches enabled the baseline product architecture and its design concepts to be committed to in Phase 1. Table 24.11 illustrates the description.

Table 24.11 The baseline design concepts were selected to be the best available solutions to the product requirements

Product Feature	Baseline Design Concept	Remarks
Long-distance light source	3-watt LED High-tech parabolic reflector No zoom function Two illumination levels	Depend on capabilities of reflector supplier Deleting zoom removes sealing problem with rotating parts.
Proximity light source	1-watt LED (tentative) Tilting lamp mount Swivel not necessary Two illumination levels	Existing capability Options remaining to be decided in Phase 1 are: one 1-watt LED versus four 1/2-watt LEDs, and whether or not a diffuser lens is needed.
ON-OFF switch	Large switches above surface Located on top	Easily felt through gloves or mittens No concern for accidentally turning light on or off Location should not interfere with hat or helmet brim
Battery	Two 6v Lithium batteries Option: rechargeable Lithium ion	Critical to run time requirement
Head strap	2 straps; adjustable Battery in back; removable	Removable battery eliminates obstacle behind head when reading in bed; enables battery to be kept warmer inside a parka in extremely cold weather.
Switch water seal	Rubber membrane	Existing capability
Battery water seal	Gasket in recess with screw tightening Buckle on head strap acts as screwdriver	Object for robustness development
Lamp water seal	Gasket in recess	Object for robustness development

(continues)

Table 24.11   The baseline design concepts were selected to be the best available solutions to the product requirements (*continued*)

Product Feature	Baseline Design Concept	Remarks
Ruggedness	Plastic housing Rubber trim around housing Light sources without filaments	Consider military specifications in design details; could enable marketing to military applications Object for robustness development
Low-temperature operation	LED light source Plastic housing Detachable battery with long power cord	Object for robustness development Minimum moving parts (switch and tilt)
Emergency search	Strobe function Orange and white reflectors on head band	Headband can also have styling suitable to outdoor adventures, e.g., camouflage, personalized colors/patterns
Night vision	Red night light	Red is commonly used, more so than green or blue
Battery power awareness	Battery level indicators: amber, red Reduced power when batteries are low	By knowing remaining battery power, user can change the battery or take a spare if the need is expected.

Among these selected concepts, "Delighter" features were seen to include

- Powerful light: useful illumination at a range of at least 150 feet
- More waterproof: survive submersion to a depth of at least 10 feet
- Very rugged: survive a drop from a height of at least 6 feet
- Optional strap designs: camouflaged, personalized, reflective with emergency colors
- Removable battery: accommodate cold; enable reading with head on pillow

After the selection of the baseline design concepts, different types of development work remained. For those components that were to be purchased, the requirements led to their specifications and to the subsequent work that selected the most capable suppliers, manufacturing processes, and materials.

For those requirements focused on the headlamp being rugged, waterproof, and tolerant of very low temperatures, development teams had to determine how to control the performance of those system functions that were in-house responsibilities.

Robustness Development

Design Concept

After the team generated several concepts for sealing the battery and lamp and evaluating the alternatives, the concept shown in Figure 24.6 was chosen for further development. Let's look at their thought process for the housings being waterproof, at least to be functional after being dunked in water up to 10 feet deep. No leakage can be suffered.

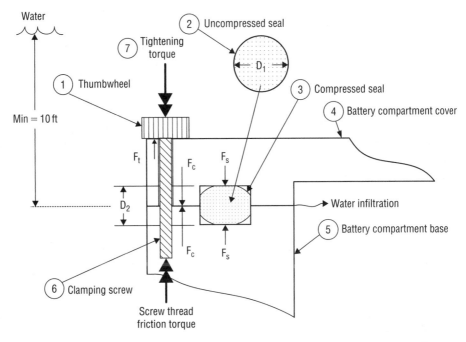

Figure 24.6   Cross section showing concept details for the battery compartment seal

Cause-and-Effect Analyses

Water Leakage

If water leakage is a potential failure mode, what could cause it? A cause-and-effect diagram was a visible way for the team to document their suspected root causes, some controllable, others not. Look at Figure 24.7. It became clear that the design of the compartment seals around both the battery and the light source were critical. The fact that the intended environment was very cold raised the concern for thermal effects on parts that contributed to the seals' effectiveness. If the seals depended on how tightly the cover was closed, human error could contribute. Discussions among the team members also focused on the material for the seals, being concerned with how well they would function when cold and less elastic.

How could the seal design work group identify the critical functional parameters? Most of the seals were static and could be designed to have high compression forces during assembly, giving zero leakage. Since those seals would not be removed during the life of the product, they did not affect the ease of use. However, the battery compartment was an exception.

Battery Change

Customers would need to remove a cover to change batteries. The battery compartment cover, ideally, should not require tools for removal, but also should not loosen unintentionally. One design concept under consideration would allow battery replacement using a thumbscrew. The concept used a standard O-ring seal that would be compressed when the thumbscrew was tightened. Figure 24.8 shows factors that would contribute to making the battery difficult to change.

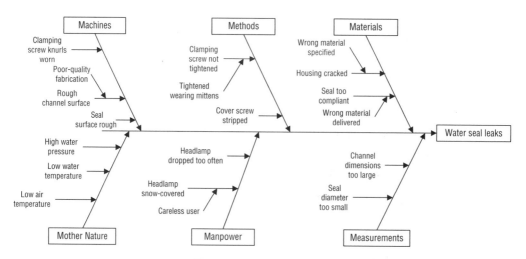

Figure 24.7   A cause-and-effect diagram showed factors that could contribute to failure of the water seal.

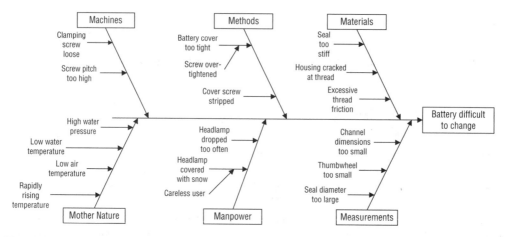

Figure 24.8   A cause-and-effect diagram showed factors that could contribute to the battery being difficult to change.

There were three design requirements for the sealing system:

1. Zero leakage of water into the battery compartment
2. Removal and replacement of the cover requiring less than a specified maximum amount of effort
3. Cover had to be removed intentionally; it must not come off accidentally during customers' use and handling

Planning the Experiment

The development team considered how the seal concept was supposed to work and used their insight to plan further analyses and experiments.

There were two critical requirements for the battery compartment:

1. There must be zero leakage of water.
2. The thumbwheel tightening torque required to fully compress the seal and close the case must be below a limit determined by human factors.

It was reasonable to expect that the no-leak performance depended on the seal compliance and the amount of seal compression. These two factors caused the interface pressure to develop between the seal and case when the seal was compressed, providing a barrier to the water.

The tightening torque was dependent on several factors, including the screw thread pitch and geometry and the friction between the screw thread and the threads in the battery compartment base. The other friction torque developed at the interface between the battery compartment cover and thumbwheel when the seal was compressed by turning the thumbwheel on the clamping screw.

Figure 24.9 shows the P-diagram that includes design parameters and noise factors affecting the two important responses, leakage pressure and thumbscrew torque required to fully compress the seal.

For the experiment design the team chose a full factorial experiment with the four factors shown in Table 24.12. The noise parameter chosen was the temperature of the device, which depended on the ambient temperature. A full factorial was chosen because the team had never developed an empirical model and was unsure of interactions. Also, since the design problem required co-optimization, it was felt that in this first experiment, it was prudent to understand the operating space with an empirical model that would establish control factor importance to meeting

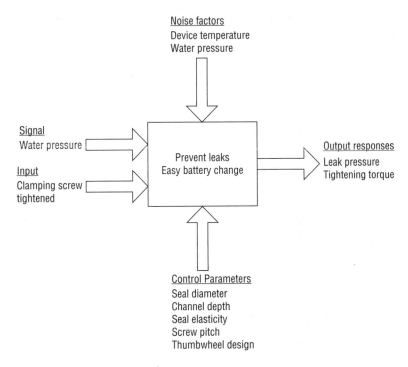

Figure 24.9 P-diagram for water seal system

the product goals. Center points were included in the experiment to test for curvature and enable a response surface design if needed.

The experiment design is shown in Table 24.13. There were two responses: the leak pressure and the tightening torque required to fully compress the seal. The process of running the experiment was to seal the case by tightening the clamping screw and submerge it in water. Then the water pressure was increased until leakage was detected. Good methods for measuring tightening torque and

Table 24.12   Parameter levels for the water seal experiment

Design Parameters	Level	
Seal material stiffness (ξ)	Low	High
Gasket diameter (D_1)	Low	High
Channel depth (D_2)	Low	High
Noise parameter		
Device temperature	Low	High

Table 24.13   Full factorial experiment design for water seal performance

| Run # | Factor Levels | | | | Responses | |
	Seal Stiffness	Seal Diameter	Channel Depth	Device Temperature	Leak Pressure	Tightening Torque
1	−1	−1	−1	−1	p_1	t_1
2	1	−1	−1	−1	p_2	t_2
3	−1	1	−1	−1	...	...
4	1	1	−1	−1	...	...
5	−1	−1	1	−1	...	...
6	1	−1	1	−1	...	...
7	−1	1	1	−1	...	...
8	1	1	1	−1	...	...
9	−1	−1	−1	1	...	...
10	1	−1	−1	1	...	...
11	−1	1	−1	1	...	...
12	1	1	−1	1	...	...
13	−1	−1	1	1	...	...
14	1	−1	1	1	...	...
15	−1	1	1	1	...	...
16	1	1	1	1	...	...
17	0	0	0	0	...	...
18	0	0	0	0	...	...
19	0	0	0	0	...	...
20	0	0	0	0	p_{20}	t_{20}
21	0	0	0	0	p_{21}	t_{21}

sensing water infiltration were required. It was expected that there would be a relationship between the two critical responses and the seal/channel geometry and seal material stiffness.

The team's goal was to find the set of factor combinations that maximized the leak pressure while having a tightening torque less than the maximum allowable torque. This required the co-optimization of the responses.

An important design task that affected the ergonomics was the design of the thumbwheel and the selection of the screw thread. The clamping screw had to allow users to compress the seal fully and to apply some additional tightening torque. It also must not loosen when used. This required some human factor experiments where a number of users were asked to tighten the clamping screw using a range of screw threads and thumbwheel designs. Then a stress test was performed to evaluate a specific design's resistance to loosening.

Further Robustness Work

After the experiment was run and control factor settings were chosen, the team planned to perform further stress tests to ensure that the design was "bulletproof" against factors such as customers' abuse and manufacturing variation in parts and materials.

Epilogue

We had the opportunity to test prototypes of the new design on a weekend cross-country ski trip, hut to hut. We chose to travel at night to see how well the headlamps functioned, particularly under difficult conditions. The nights were nasty, very cold with the falling snow being mostly sleet. Because of recent warm days, the ground was covered with slush on top of a frozen crust. It was slippery and wet.

To be objective, we brought with us examples of competitors' headlamps. The conditions were not as extreme as those in the robustness experiments, particularly the ambient temperature, which was still in the 20s. However, our clumsiness paid off. Twice headlamps fell into water, once when crossing a deep stream and once when a backpack rolled over into a puddle. In the stream incident, both our lamp and one of the competitive lamps fell in, a side-by-side test. Our lamp continued to work but theirs did not, because of water in the battery compartment. Both headlamps endured the shock of being dropped, although one example with an incandescent bulb did not.

Although these experiences were pleasing, the real stress tests were side-by-side comparisons under stressful laboratory conditions. In the lab we could drive the temperature down artificially and then subject the headlamps to the shocks of being dropped and dunked in water in measurable ways. Given the requirement to survive submersion in 10 feet of water, it was not practical to test this in a frostbite skiing trip.

From this brief experience we concluded the following:

- Our new lamp that had been subjected to robustness development did demonstrate better performance under real-life conditions.
- Laboratory tests of these products could be more extreme and determine results with higher integrity in less time.
- If a product is more rugged and lasts longer under stressful laboratory conditions, it can be expected to do so under whatever conditions are experienced in its market applications.

Balancing Schedules, Budgets, and Business Risks

Throughout the chapters of this book we have described strategies and methods that can be adapted to design new products and processes. The objective is to make them more robust to the stresses in their applications and thereby have higher reliability and durability. The work to achieve those designs is in the context of your product development process, which itself is linked to processes for portfolio balance, product planning, and technology development. The ongoing development of your company's capabilities, employees' tools and skills, business relationships and supply chains, for example, supports these processes. They are integrated to provide competitive advantages.

Over the past decades major advancements have been achieved in companies' capabilities, and further advancements are expected. The competitive business environment in which most companies find themselves demands continuous improvements with their benefits so that the businesses will thrive and renew themselves. Those benefits that are of most value vary among companies. Schedule pressures may remain, either toward shorter cycle times or more predictable delivery dates. Economic pressures may continue to force companies to seek opportunities to reduce the costs of development or the costs for the product's manufacturing, service, or support. Opportunities may remain to achieve higher levels of satisfaction for customers, as reflected in product features and functionality and in the quality and reliability with which they are implemented. Any of these attributes is subject to implementation risks, problems that may occur to reduce the achievement of the business objectives.

In this last chapter we discuss some of the ongoing initiatives that many companies are pursuing toward further improving their product development capabilities.

Over the past three decades various improvement initiatives have delivered benefits to product development projects. Some have been derived from the discipline of military programs, while others have been stimulated by the competitive pressures on commercial products. Certain contributions to good practices have come from the Far East, notably Japan, while others have grown in the United States and Europe. Forward-thinking companies have learned lessons from their own mistakes and become students of process improvements, regardless of their origins. Benchmarking has been a valuable way to learn from other companies; those who receive the Malcolm Baldrige award are obligated to contribute their wisdom to other companies' improvement efforts.

Our experience with these processes began in the 1960s with the disciplined management of requirements, the practices of systems engineering, and regular management reviews of development. In the mid-1980s we were significant contributors to initiatives that standardized the product development process with phases and gates, embracing concurrent engineering and treating the process as applied systems engineering.

Figure 25.1 illustrates that, across industries, many process strategies have become favored, fortunately not all at the same time. A company's "call to action" may focus on just a few imperatives at once. The ability to manage change processes takes both the motivation and the time to learn the new methods, to adapt them to the company's business model, and to implement them in ways that make them systemic, just the way that work gets done.

In Figure 25.1, the timelines are only approximate. We just want to give you a sense of the themes and to identify those that are relatively recent. Depending on the company, those strategies indicated through the mid-1990s tend to be reasonably well understood and practiced, while later ones may be new to your company and thereby represent opportunities. Certain methods that were learned well in the early 1990s may no longer be practiced if the workforce that learned them is no longer employed or the practices themselves have become victims of neglect. New employees

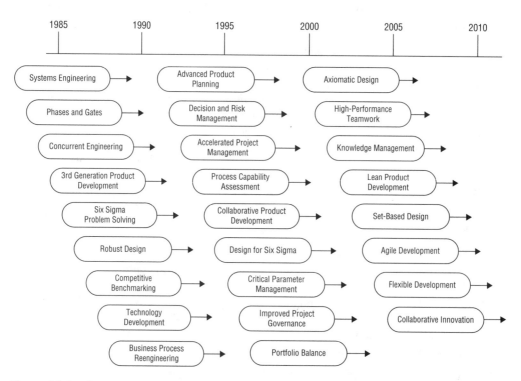

Figure 25.1   Strategies to improve product development have evolved.

may not be provided the tools or taught the methods, leaving current capabilities back where they were a decade or more in the past.

Consider the product development process itself. Project teams may view their phases-and-gates process to be rigid, bureaucratic rules that get in their way or do not seem to be relevant to the way that work needs to get done. They may then disrespect the intentions and work with more ad hoc processes. If projects treat gate reviews as an exercise in completing checklists, those reviews fall short of being value-adding decision points, critical to the management of project risks. For some companies, product development may take place in isolation from the management of a portfolio of development projects. Projects may spring to life and compete for resources without efforts to align them with the company's business strategy, capabilities, or priorities. These and other problems were resolved by the processes developed in the past, but they may once again be handicaps to companies' abilities to compete with their development process.

In the late 1980s the Six Sigma movement focused on systematic problem solving, particularly in production. Over time the principles migrated into product development in the form of Design for Six Sigma, which embraces many of the principles illustrated in Figure 25.1.

As an umbrella process, DFSS emphasizes rigorous use of engineering and team best practices, such as VOC, concept selection, and DOE, to develop robustness and prevent problems. The principles of DMAIC are incorporated to correct problems quickly as a way to grow reliability and resolve other design shortcomings. DFSS rejuvenates the attention to the intent of phases and gates. We described many of these methods in earlier chapters.

Where is your company in this evolution of product development capabilities? Which improvement initiatives are most relevant to your current business challenges?

Ongoing Attention to Product Development Capabilities

We are writing this book in the first decade of the twenty-first century, with the background of those improvements already learned, adapted, and implemented. The current hot topics, as indicated in part by the agenda of industry conferences, include four that you may find to be on target with your needs. Lean product development, flexible product development, set-based design, and agile development are well aligned with the overall theme of leanness. Certainly, other initiatives such as high-performance teamwork and collaborative innovation are very important, but their treatment is beyond our scope here. The objectives of these four can be characterized to include

- Higher satisfaction of customers
- Increased quality in manufactured products
- Shorter, more predictable cycle times for development
- Reduced technical and business risks

They build upon the focus of this book, that being to increase product reliability and usage life. Throughout the chapters we have identified some wisdom that can help, such as the following:

- Identify requirements that implement a differentiating value proposition.
- Negotiate requirements that can be achieved within the constraints of the project.
- Select the best available solutions for those requirements that directly relate to customers' perception of value.
- Develop implementation plans that adopt efficient and effective strategies.

- Prevent problems as a focus of early development work.
- Practice fast problem solving later in the project.
- Accelerate tests using stressful conditions to find problems faster and to prove solutions to be better.
- Focus on improving the robustness of designs rather than on demonstrating their reliability or usage life.
- Maintain a sense of urgency for milestones that really matter.
- Manage the prerequisite inputs that enable tasks to start on time, including the entire project.
- Establish clear acceptance criteria for project deliverables and milestones, prior to the work beginning.

Among those objectives that are directly related to the product's quality, price, and availability, we support strategies and tactics that develop capabilities in parallel, rather than compromising one to get the other two. However, for those objectives that relate more to the development project, such as its compliance with funding budgets, we do not hold such a disciplined viewpoint. For example, budgets for the investment in product development may be set to spread resources over too many projects, leaving all of them handicapped in some way. Compliance with budgets is easy to measure and usually becomes a performance metric for the project team. However, the overutilization of resources results in avoidable queues at bottlenecks that can impose delays on all projects. The overloading of people pushes them to do just the tasks that are most demanded, or to work on just the highest-priority projects. When certain projects receive the benefits and others do not, the objectives of the portfolio of projects are jeopardized. The workforce may satisfy the most pressing demands in the short run but not be able to sustain the pace over the long run.

The discussion that follows about lean product development provides another example. The investments in centralized capabilities, such as prototype shops or testing labs, often can create incentives to overload those resources to ensure that they return sufficient value. When little reserve capacity exists, the throughput of that function is vulnerable to variations in the volume and timing of the demands placed upon it. Lines form and projects are delayed. The costs of project delays can be very large, particularly if the delays are in market entry. A better strategy is derived from queuing theory.[1] As resource utilization is reduced, the costs of delay are reduced but the costs of the underutilized resources increase. The strategy is to add the costs of delay to the costs of the underutilized resources and select a resource utilization level that minimizes the total cost.

If your workforce is overloaded or your facilities overutilized, we expect that your financial models will show that the gains from the stream of revenues beginning on time and the stream of development costs ending on time would far outweigh a one-time overrun in development costs. The problem then becomes how to set aside reserves of funding and resources to be allocated to projects when needed to enable them to deliver the right quality at the right price at the right time.

For each of the "lean" topics we provide references that explain the principles comprehensively in much more detail. Lean product development is an overarching theme, with flexible and agile characteristics. Set-based design is one of the tactics that can enable your product development to be flexible and agile. Lean product development includes practices embraced by DFSS, project and risk management, queue management, high-performance teamwork, and other improvement

---

1. Reinertsen, *Managing the Design Factory.*

strategies. As with most advice derived from these initiatives, the question becomes how you can use the wisdom in the context of your business model and your specific calls to action. That can vary greatly among companies and across time. Given that qualifier, these next topics may shine some light on pathways for new attention in your organization.

Lean Product Development

The principles and many of the practices of "lean product development" are derived from Toyota's Development System. There are several excellent references[2] on the subject, so we'll concentrate here on providing a perspective.

Lean product development is focused on creating more value for customers, not just on reducing costs or shortening cycle time. When activities that are considered to provide little or no value to customers are eliminated, resources can be freed to devote more attention to value creation. That may be on the current project or on other projects in the development portfolio.

Development processes that are leaner contribute well to shorter and more predictable cycle times. A perspective that you may find enlightening is that "efficiency" can be thought of as how well a project achieves its plan. This "actual versus plan" is a common metric for projects. Leanness, on the other hand, refers to how well a project follows an ideal plan, how short timelines are, compared to how short they could be if there were no wasteful activities. That idealized plan would be one that you would have to establish for a particular project, since usually no two projects face the same challenges. The resulting leanness metric, actual versus ideal, is inherently normalized and therefore comparable across projects. Improvements in cycle time would be difficult to recognize with an absolute cycle time metric because of the differences among projects.

Lean product development emphasizes several important paradigms. Certainly, not all of these can be attributed solely to Toyota's Development System. In fact, our experience with the evolution and practice of product development can identify many practices that contribute to shorter, more predictable cycle times and to increased value for customers. However, the references identify some key points for your consideration, such as we describe in the following sections.

A. Intense Focus on Value Creation

Product development is an investment. Its expected returns are first in the value delivered to customers, which in turn generates revenues that return value to the company. The challenge is for the new product to do it better than its competition after market entry, and for the project to do it better than the next-best alternative investment.

B. Standardized Work Processes

With standardized processes, your development teams can devote their time to working on solutions rather than figuring out what work to do. Standardized but concise documentation sets expectations for content and facilitates their interpretation. It's a great advantage for new employees. There are some additional benefits worth mentioning:

- Process knowledge can be transferred from project to project, across business units and collaborative partners, with a common understanding of the expectations.

---

2. Ward, *Lean Product and Process Development*; Mascitelli, *The Lean Product Development Guidebook*; Michael N. Kennedy, *Product Development for the Lean Enterprise: Why Toyota's System Is Four Times More Productive and How You Can Implement It* (Richmond, VA: Oaklea Press, 2003); Morgan and Liker, *The Toyota Product Development System*.

- Adaptations of the standard process to a range of project characteristics can have a common basic model.
- Lessons learned from a variety of projects can contribute to continuous improvements that can assist all projects.
- Project stakeholders can provide governance to a variety of projects with a common understanding of the process, its principles and key paradigms, decision criteria, and probing questions.

C. "Pulled" Development of Information

If teams know the questions that must be answered, they can do the work necessary to deliver the required information. If they know when those questions must be answered, they can do the work when it is necessary. From a lean perspective, unnecessary or premature work can be avoided. The strategy implies reverse-engineering the project schedule to identify when key elements of information are needed in order to enable the follow-on work that leads to market entry. This "pull" process is leaner than a traditional "push" process that can foster excessive work with immature information and decisions that are made earlier than necessary.

D. Parallel Processing with Frequent Integration Events

Parallel processing is the strategy of concurrent engineering articulated for product design and manufacturing, extended to early supplier involvement. Its interpretation expands the context to the many organizational elements that should contribute to product development. Engineering and marketing have to work together, as must service, logistics, finance, and other internal functions. Collaborative partners and suppliers have to work with your internal project teams. Because these parallel resources work on behalf of common project objectives, their integration is essential. The lean advice is to integrate them frequently and to clarify any ambiguities that can get in the way of them acting as if they were on an autonomous project team. Gate reviews are minimum integration events. How about weekly project team meetings? When the level of urgency is high, daily "sunrise meetings" are integration events. So include your "partners" either physically or virtually.

E. Avoidance of Excessive Work and Unnecessary Handoffs

Building on that "pull" process, development teams need to understand clearly the information that they have to deliver to enable the next downstream activity to begin. By maintaining close communications, work groups can act as if they are on the same team. The objective is to deliver the information that is needed with the right content and integrity at the right time, and to have the receiving team trust that it will be delivered when expected.

F. Smaller Batch Sizes with Faster Learning Cycles

The focus of the advice for "small batch sizes" is on those deliverables that establish the demands for downstream work. An important example is the amount of incremental advancements in a design that is submitted for testing. The larger the amount of design changes that are bundled, the longer it takes to prepare them for testing and the longer it takes to test and analyze them. That then causes longer time periods to obtain the feedback needed to enable further work.

You may have had the experience of a prototype iteration being designed before feedback had been received from the testing of the previous iteration. It's a cause of the "hardware swamp" that Don Clausing characterizes as one of his "Ten Cash Drains."[3]

G. Managed Queues That Are Predictable

The queues that we mentioned previously can be in two categories, predictable or unexpected. If the queue is predictable, the resource utilization can be managed near a minimum total cost level. However, if the loading and timing of the demands on a centralized resource are not predictable, the resources that handle that workload need to be flexible. That flexibility can be obtained with reserve labor that has been cross-trained and maintained to be available in a crisis while doing other work. Flexibility may also be available from using backup equipment or employing contract resources, such as external test labs. The objective is to minimize the consequences on market entry for all new products in the portfolio, not just the one with the highest priority.

Lean product development is a broad and deep topic. As markets become increasingly competitive, lean methods may become major elements of your competitive advantage.

Flexible Product Development

Design changes are inevitable. Flexible product development[4] is focused on reducing the consequences of late design changes. They can be forced by external factors, such as changes in customers' tastes or needs, changes in regulatory requirements, or unexpected competitive actions. They may be forced by internal factors, such as conflicts among rapidly evolving technical solutions or immature technical functions incorporated into designs. Early design changes can improve the product with fewer schedule impacts. Late design changes, however, can have tremendous consequences for development costs, parts and tooling inventories, manufacturing costs, suppliers' delivery schedules, the product's market entry date, and the costs of poor quality. The question then is "How gracefully, with lower costs, can your organization handle these changes?"

In this context, "flexibility" in development is a function of the incremental costs of modifying a design. It's logical that the closer a product is to its market entry date, the less flexible is the process. A major benefit is that increased flexibility reduces the risks associated with errors in defining requirements or in forecasting competitive markets. Increased flexibility is particularly important when markets are changing rapidly. With more flexible processes, the product delivered to the market can have the benefits of late-arriving information and thereby be more competitive.

How can your processes have increased flexibility? Here are some suggestions:

- Develop rapid prototyping capabilities.
- Automate testing.
- Keep design options open, developing alternative solutions in parallel.
- Delay decisions until their consequences are on the critical path.
- Maintain ongoing market studies.

---

3. Clausing, *Total Quality Development*.

4. Smith, *Flexible Product Development*.

- Gain continual feedback from enlightened customers.
- Employ higher-quality people and tools.
- Follow efficient, standardized processes.
- Practice frequent iterations of redesign and refinement.

You may have many more in the context of your business model. Be careful to prioritize changes. Feature changes add value, while scope changes just add work.

Set-Based Design

Set-based design, also termed "Set-Based Concurrent Engineering," is a tactic of lean product development.[5] It is contrasted with point-based design in which a project commits to a design concept early in the process and then develops this single concept iteratively to make it work as well as possible. If the selection of the baseline design concept is flawed, development has the job of trying to make a bad concept good enough. You may have heard the observation that you "beat it into submission."

Set-based design follows a different paradigm. It keeps options open and increases the confidence in the eventual solution that is chosen. The strategy delays the commitment to a specific design concept until as late in the process as is acceptable. It develops alternative "sets" of designs solutions at both conceptual and parametric levels. When trade-offs among alternative set points for controllable parameters are developed, the design space is well understood. When alternative design concepts are developed in parallel, back-up solutions can be ready for use if needed. Once the necessary and sufficient information is available and the timeline becomes critical, select and specify the best design option for the application.

How do you make set-based design work? It doesn't come for free. Here are some suggestions:

- Invest in the development of both higher-risk and conservative design options in parallel. These options may be alternative design concepts, alternative features of a specific design concept, or alternative set points for control parameters in a concept. Devote more time in the early phases to understanding their trade-offs.
- Develop a product architecture that is not dependent on specific and risky subsystem concepts. This involves more modular designs with robust interfaces. As a result, a flawed design can be replaced by a more robust concept without jeopardizing the adjacent designs.
- Specify functional control parameters in a design after the integration of subsystems and modules into an optimized system.
- Use a process of releasing designs for tooling in phases, as the design details become more mature and stable.
- In the project management plan, decide when to decide. Allow the schedule milestones, system dependencies, and production release strategies to determine when solution details have to be frozen.

---

5. Sobek et al., "Toyota's Principles of Set-Based Concurrent Engineering"; Smith, *Flexible Product Development*.

- Maintain the documentation of trade-offs and comparisons as information changes and analyses are developed. The knowledge gained may also help the next development project.
- Commit development teams to stay within the set of design options, barring extreme circumstances, so that other work groups can rely on that information.

The benefits of set-based design can be many, including reduced costs of change, better designs, and institutionalized learning.

Agile Development

For software, agile development[6] is a conceptual framework that promotes frequent design iterations and fast learning cycles. It emphasizes working software as the major evidence of progress, planned with short cycle-time intervals. Each iteration passes through a full development cycle of requirements, analysis, design, coding, testing, and documentation. Agile development embraces the notion that design changes are natural and inevitable, so they should be desirable and easy. The question is "How can agile methods be adapted to be useful to design work other than software?"

Agile development may not be appropriate for all projects. For example, it might not be so necessary when requirements are known early and are relatively firm. The methods might not be manageable when there are complex technical challenges that are highly integrated. If the work force is relatively junior, large in number, or geographically distributed, probably more disciplined, standardized processes are necessary.

On the other hand, agile methods can be more suitable when

- Requirements and competitive actions are emergent and rapidly changing
- The focus is on adapting quickly to continuous feedback from customers
- Work is performed in a collaborative, self-organized manner by higher-skilled teams who are experienced in shipping products
- The time intervals for prototype iterations are in weeks, rather than months, with a disciplined schedule
- Engineering practices enable rapid prototyping, automated testing and analysis, and responsive corrective actions

What makes agile development work? Software is a medium that is relatively easy to change. An electromechanical design would need the same character designed into it. What else? In addition to those attributes already mentioned, here are a couple of process requirements:

- Above all, strong attention should be paid to the system architecture. In fact, because of the benefits of agility in design, there is a significant viewpoint that the selection of the product architecture is a business decision, not just one for engineering.
- Customer involvement in an agile world is face-to-face. Customers' representatives should act as part of the development teams and be available to provide feedback in rapid learning cycles.

---

6. Smith, *Flexible Product Development*.

- Risk identification, analysis, and mitigation should be routine elements of project management meetings. These concerns are too important to be left to gate reviews. You want to reward problem anticipation.
- Stakeholders need to monitor progress closely so that they can intervene with guidance and resources to ensure the success of the project.

While agile methods were derived from software development practices, such as Extreme Programming, they may find a broader range of applications and thereby deliver benefits to your entire product development project.

Flexible, set-based design and agile depend on the larger "lean" context. They are good examples of cross-fertilization among engineering disciplines and between development teams and management.

Realistic Project Planning and Management

Schedule-Driven versus Resource-Driven Plans

The project plans that you make can be a significant source of business risks. A common problem in product development is an unrealistic implementation plan that is not achievable, even before it is presented to management. Plans for developing a new product or service always affect other organizations such as marketing, sales, service, and the supply and distribution chains. The implementation of a plan in ways that cause delivery dates to be missed has negative effects on many elements of the business and eventually affects profitability.

A major factor in formulating unrealistic plans is the tendency to make them schedule-driven rather than resource-driven. Typically, the end date is taken as a given and the plan is developed working backward from the end date. There is nothing wrong with that approach. The end date may be when the business needs the new product or service. It is appropriate to "begin with the end in mind" and develop plans supporting the desired outcome. The problems arise when resource constraints are overlooked. This is aggravated when the basics of project management are ignored and wishful thinking substitutes for thoughtful planning. If the delivery date has been dictated, management may like the plan and be reluctant to ask questions that probe for achievability and risks. The presenters, pleased that management approves, are also reluctant to express concerns. As a result, there is an unspoken understanding that makes problem prevention very difficult. If resources are woefully short of those realistically required to implement the plan, trouble can surface quickly and dates for various milestones will begin to slip. Eventually, organizations burned by this approach develop more realistic plans.

The next step toward reality is to develop plans that include resource allocation. There can be many peaks and valleys in the timing of resource demands. There can be an assumption of "infinite resources," that what is needed will be available when needed. Since resources that exceed capacity are generally not available "on demand," the logical step is to smooth the resource demand curve. However, resource leveling may extend the project plan past the required market entry date.

At this point, management has to decide what to do. Their options can include the following:

- Address the problem caused by the resource demands.
- Change the project requirements so that they are achievable within the project timelines with the available resources.

- Change the project end date to enable its completion without additional resources. This is accomplished by "leveling" the resource demand to be consistent with resource availability. The time for task completion will be extended for those tasks competing for the same resources, where demand exceeds capacity.
- Cancel lower-priority activities to make certain resources more available. This requires evaluating the relative importance of projects and deciding those to delete.

A more common, but pathological, response is

- Do not be moved by a project's plea for more resources or for priority setting.
  Declare, "We want it all," and follow up by neither adding resources nor extending the end date for the plan. This is ineffective management, sometimes rooted in the belief that, above all, people just want to make things easier for themselves. It is often driven by being overwhelmed, where there may be no time or appetite for understanding. The appearance of being demanding and tough-minded is chosen over doing the hard work of addressing the capacity problems. We have seen this approach cause poor outcomes.

Deterministic versus Probabilistic Project Networks

The tools that manage variability in project planning were developed to recognize important realities. Some facts that make these tools attractive include these:

- Estimating task duration is an imprecise art heavily dependent on experience.
- Unexpected events can occur at random, affecting your ability to meet the schedule.

To address these factors, tools such as PERT and Monte Carlo simulation have been applied to project management. These tools have value in identifying the major causes of schedule variability, the first step in mitigating the schedule risks. A dilemma that results is that the project end date becomes a random variable described by a statistical distribution. This raises the question of how to manage the project. Because other stakeholders have to make concrete plans to support the project, it is not practical to communicate the end date as a confidence interval. The business requires an achievable delivery date. A method that addresses this problem is called Critical Chain Project Management (CCPM).

Critical Path versus Critical Chain

CCPM is based on work pioneered by Goldratt,[7] who developed the Theory of Constraints. Since Goldratt's original work, CCPM has become accepted as a valuable project management strategy and is now included in the Project Management Body of Knowledge (PMBOK). The CCPM approach includes the traditional tools of project management but adds some new thinking to the process of building the project plan.

While traditional project management is focused on the critical path for the project, CCPM is focused on the critical chain. The critical chain in a project is the sequence of tasks that determine project duration. It is identified after the project network is assembled and resource contention is resolved by eliminating multitasking.

---

7. Eliyahu Goldratt, *Critical Chain* (Great Barrington, MA: North River Press, 1997).

Some CCPM thinking includes the following:

- Accept that variability is inevitable, include its effects in the plan, and manage it effectively.
- Eliminate multitasking.
- Do not start tasks earlier than necessary.
- Instead of focusing on the critical path, carefully manage the critical chain in the project network.
- Rather than let everyone pad their individual schedules with contingency time, manage contingencies at the project level using schedule buffers inserted at the appropriate points in the project network.
- Manage the project buffer but don't micromanage the project.

CCPM is a rich subject, but an in-depth discussion is out of scope for this book. An excellent reference is Leach.[8]

Be a Student of the Process

The last point made about agile development points to our request that you become a student of processes, methods, and tools. Those that can deliver major benefits to your business evolve over time, driven by competitive pressures to improve. We have given you many references throughout the chapters and summarized them in a list of supplemental readings at the end. These are "trailheads" that can lead to other sources along the paths that you choose to follow. We have learned much ourselves from our years of experience with product development. However, there is much that we would never have known if we had learned only from inside our own business worlds.

Learn from others. Read. Participate in professional associations, conferences, and trade shows. Benchmark other companies. Many other companies are facing the same struggles. Many have already transformed lessons learned into competitive advantages. Respect their wisdom.

We have mentioned often that the task is to understand new principles, paradigms, methods, and what makes new practices work. Then the challenge is to adapt them to be integral to your own processes and practices, to be just the way that work gets done.

Change takes time, resources, and determination. A clear, compelling call to action is essential to channel the energies of your organization and to prioritize your focus. The benefits flow to your business, to your employees, and to your customers.

Good luck in your adventure.

---

8. Leach, *Critical Chain Project Management*.

Glossary of Terms

A

Accelerated stress testing (AST) employs stressful conditions to cause failures and thus find problems faster. There are several different tactics for this type of test, each with different goals. An improved design should demonstrate robustness and durability to be better than those of competing designs under those same conditions. That sets an expectation for reduced quality loss and longer useful life under actual conditions. See the descriptions of specific stress tests in Chapters 8 and 22.

Additive model A model in which the response is a function of main effects only. Interaction terms are not included in an additive model. The response of a system having an additive model can be adjusted by changing the control factor settings. If a transfer function has strong interaction terms, it is not additive in Taguchi's sense, and the model may overestimate or underestimate the effect of control factors on robustness.

Additivity A design concept whose critical functional parameters are either independent or interact constructively is much easier to implement in production. The effects of the parameters add to each other, rather than counteract each other's effects on the response.

Aliasing The mixing of effects in the analysis of the experimental data (also called *confounding*).

Alpha error The probability of incorrectly rejecting the null hypothesis if it is true.

ANOVA A statistical analysis that looks at the contribution of factors to the total sum of squares variation in a data set. The contribution of each factor is compared to the contribution of the noise using the F-test. Factors having an F-statistic larger than the chosen threshold for α error are significant.

Augmentation In experiment design augmentation means adding treatments to enhance the experiment and to enable actions such as the detection of curvature, or the addition of higher-order terms to the system model. Adding center points and star points are two examples of augmentation. When augmenting an experiment, it is important to consider the need for blocking, especially if the additional runs will be done on different days.

Axiom A proposition of fundamental knowledge that is generally accepted as being self-evident based on its merits, without the need for proof.

B

Balanced design In DOE a balanced design means that there are an equal number of observations taken for each treatment. Balance enables simplified analysis of the experiment results.

Basic need Defined in the Kano Model as a product requirement that is so obvious that it can remain unstated when customers describe their requirements for a product.

Bathtub curve Shows how the failure rate varies during a product's life with infant mortality problems in early life, random failures later in life, and wear-out failures at the end of life.

Beta error The probability of incorrectly failing to reject the null hypothesis if it is false.

Binomial distribution The binomial distribution can be used to describe the frequency of events for any process having two possible outcomes described as the proportion of successes or failures in a given number of trials. For example, you could calculate the probability of seven tails in ten flips of a fair coin.

Blocking An experiment design technique useful for isolating the contribution of nuisance factors that would otherwise be assigned as noise. By removing them from the noise, you effectively increase the power of the experiment.

Box-Behnken design A member of the response surface experiment designs, but has no points outside the "cube." Instead, Box-Behnken designs have their test points located on the edges of the cube defining the design space of interest.

C

Case control study An epidemiological study that focuses on a population and uses retrospective data to discover associations between chosen risk factors and the incidence of disease.

Cause-and-effect diagram A method used to identify factors that might be important in causing a certain outcome. The factors are then candidates for further evaluation using a tool such as a designed experiment or ANOVA. Also known as a *fishbone* or *Ishikawa diagram.*

Censored data Data that have a mix of complete data, where failures occurred, and suspensions, where the test was terminated before the component failed.

Central composite design (CCD) A member of the response surface family. CCD experiments include axial points outside the cube that defines the area of expected operation or design space of interest.

Chi-square distribution A statistical distribution that describes how sample variances from a normal population are distributed. The chi-square distribution can be used to estimate the likelihood that the population variance is within a specified range based on the sample variance.

Coded factors Factors can be either coded or uncoded. For example, in a two-level factorial experiment, coded factor levels are specified as -1 and $+1$ for low and high. With coded factors, the value of the model coefficient for each factor is a measure of relative importance.

Common cause noise Variability that occurs naturally in any process and has no assignable cause. Of course, every effect has a cause, but by having a category such as common cause noise you are saying that there is a level of variability in any process, and the variability for common cause noise is below the threshold of concern, difficult to reduce, and probably impossible to eliminate.

Complete data Data where all components in a life test are run to failure.

Compound control factor Combinations of control factor settings as used in the catapult experiment with the energy factor. Compound control factors can be used in any type of experiment. The advantage is the potential for fewer runs. The disadvantage is that you may miss a better set of optimal control factors. When factors are combined, there are some combinations that will not be run.

Compound noise factor Combinations of multiple noises so that all noise factors contribute to moving the system response in either the positive or negative direction. With the signs reversed on the individual noises, the response will be moved in the opposite direction. Compound noise factors are developed by running a noise screening experiment to understand how the noises affect the response.

Confidence interval (CI) A range of values for a population statistic that are calculated using sample data. A confidence interval includes an upper and lower bound as well as the probability that the population statistic is within the range. Confidence intervals can be calculated for the mean and standard deviation as well as many other statistics.

Confirmation experiment Run to verify the control factor level settings derived from a robustness experiment. The confirmation experiment should be run when the control factor level settings were not one of the treatments run in the robustness experiment. The confirmation experiment is especially important when running Resolution III robustness experiments.

Confounding factor A factor not included in a study that has an influence on the result. If confounding factors can be anticipated, they can be controlled for in a study.

Control factor A term used in DOE to describe independent variables that are controlled to determine the functional response of a process or system. See also **predictor variables**.

Corollary A proposition derived from a proven proposition, with little or no additional proof needed.

Correlation A mathematical relationship between two data sets that shows the tendency for one variable to change as another changes. Correlation coefficients can be between -1 and $+1$. A negative coefficient means that one factor increases as the other decreases. A positive coefficient means that the factors increase or decrease together.

Cost of quality The total economic loss due to a particular deviation of performance from its target and the additional production costs to keep the performance from exceeding that amount of deviation.

Covariate An uncontrolled factor that has a significant effect on the response in an experiment. An example might be ambient temperature. If impossible to control during the experiment, it can be measured and included in the experiment as a covariate. The data can then be analyzed using a tool such as the GLM and the covariate included in the empirical model. If it is not included as an important covariate in the experiment, it becomes a lurking variable.

Critical functional response (CFR) A measurable output of a product that is important to customers.

Critical parameter tree A hierarchy of customer needs that are deployed to lower levels and define the important characteristics of a product.

Cumulative distribution function (CDF) Describes the fraction failed versus time or actuations.

D

Degrees of freedom In statistics the number of degrees of freedom is the number of elements in a data set that can vary independently in the calculation of a statistical parameter.

Delighter Defined in the Kano Model as a product feature or function that is not expected by customers and thus is unstated when customers describe their requirements for a product.

Design capability A measure of a how close to being on target is a specific CFR and how much variability there is around its mean. Design capability is determined by manufacturing process capability and the system transfer function for a specific CFR.

Design of experiment (DOE) A body of knowledge with roots in linear algebra and statistics. The primary uses of DOE are building empirical models of systems and Robust Design. DOE has been successfully applied in many areas of science, engineering, and human behavior to understand causal relationships.

Deterioration Physical degradation of a component or consumable material that can affect a product's performance.

Difference testing Examples are the t-test, F-test, and chi-square test, which are used to determine if the observed difference between samples is statistically significant.

D-optimal design A computer-generated experiment design that normally starts with a full factorial or CCD. The computer selects a subset of the candidate points in the starting design with the objective of minimizing the covariance of the model coefficients. Also called an *algorithmic design*.

Durability The characteristic of a product or component that determines its useful life. Resistance to deterioration and the ability to tolerate highly stressful random events are both measures of durability.

Duty cycle For products that are powered on, it is the percentage of on-time when the product does useful work. The metric can be complicated by factors such as "sleep" modes where the product is mostly powered down but comes to life when the customer pushes a button.

E

Efficient model An efficient model is one with no significant lack of fit and no over-fitting. Adding more terms will not improve it, and taking terms away will increase lack of fit. A model in which the residuals equal the pure error is a very efficient model.

Empirical model Developed by fitting a regression model to a set of data from a designed experiment. Empirical models are essential in developing new products and processes.

Epidemiology The study of populations and the incidence of disease associated with or caused by chosen risk factors.

Equivalence testing Used to determine if the estimates of population means are within a range of indifference that has been specified. Equivalence testing is not the same as difference testing, normally done using a two-sample t-test.

Error transmission Variability in system response caused by variability in the system control factors and external noises.

Exponential distribution A one-parameter distribution of the inter-arrival times for events that occur in a homogeneous Poisson process. The exponential is a good fit for the time between failures of a complex system.

External noise factor An uncontrollable noise factor that can cause variability of system performance. External noises are managed using the tools of Robust Design.

F

Factorial Defined with an exclamation mark that indicates multiplication in a certain way, for example, $n! = (n) \times (n-1) \times (n-2) \cdots \times (2) \times (1)$ so $5! = 5 \times 4 \times 3 \times 2 \times 1 = 120$. Factorials occur in evaluating functions such as the binomial and hypergeometric distributions.

Failure analysis A process for determining the cause of failure by examining failed parts. Failure analysis can be as simple as a visual inspection and classification or can be more complex, requiring laboratory instruments to make the required measurements.

F-distribution A statistical distribution that describes how the ratio of two sample variances from a normal population are distributed. The F-distribution can be used to assess the likelihood that two samples came from populations with similar variability.

First principles The fundamental laws and equations of science and engineering.

Fisher's exact test Used to assess the probability of getting different proportions of events in two different samples. The results for each sample are typically shown in a two-by-two contingency table. Fisher's exact test calculates the total number of less likely combinations for the same marginal totals in the table and adds the probability of each to calculate the p-value.

Fractional factorials Experiment designs that are capable of identifying all main effects and some interactions depending on the resolution of the experiment.

F-test A statistical test used to make inferences about the relative importance of chosen factors based on how their variation affects the variation of a chosen response.

Full factorials Experiment designs that are capable of identifying all main effects and interactions. A two-level factorial is the most widely used experiment design.

G

Gaussian A process described as Gaussian (in honor of C. F. Gauss) is one that exhibits variability that is described by the normal distribution.

General Linear Model (GLM) A powerful model in statistical science that is a superset of the tools and methods of ANOVA and regression. The GLM can be used to analyze a variety of data sets and does not have some of the limitations of standard ANOVA, such as requiring a balanced design when taking data. The GLM can also handle covariates, which can be very useful for including uncontrolled factors that could affect the response.

Goodness of fit A measure of the dispersion of a data set around the model prediction. When the residuals approach the pure error, you have a good fit.

Grand mean The mean of all observations recorded in an experiment.

H

Hard failure The failure of a system to continue to operate because a component breaks or because performance drifts outside acceptable limits and the system shuts itself down. Hard failures are correctable only by a service intervention.

Hazard function Shows the failure rate as a fraction of survivors, plotted versus time or actuations.

Homogeneous Poisson process Generates a sequence of events. The time between events is exponentially distributed. There are many examples of Poisson processes, both natural and man-made.

Hypothesis 1. A statement made about the way you think things are, either supported by the data and accepted, or not supported and rejected. 2. A conjecture of what the facts of a situation are, subject to significance tests using statistics derived from the data set. 3. A proposition of fundamental knowledge that needs to be proven.

Hypothesis testing A structured process in which you formulate two hypotheses, the null hypothesis and the alternate hypothesis. If the data do not support the null hypothesis, you reject it and accept the alternate hypothesis.

I

Importance A measure of the contribution that variation in a control factor makes to variation of a CFR. The fraction of the variance of the CFR caused by the variance of the control factor is a good measure of importance.

Independent variable See **predictor variable**.

Inference A conclusion formed about a population based on analysis of a sample.

Inferential statistics The use of statistics to make judgments about the location and magnitude of statistical distribution parameters. Questions such as the location of the mean, the spread of the data, and the likelihood that two things were produced using equivalent processes rely on inferential statistics.

Interactions The product terms in the transfer function involving more than one control factor. For example, with three control factors, A, B, and C, there are four interaction terms, AB, AC, BC, and ABC.

I-optimal design A computer-generated experiment design that normally starts with a full factorial or CCD. The computer selects a subset of the candidate points in the starting design with the objective of minimizing the variance of the model prediction. Also called an *algorithmic design*.

L

Lack of fit Measures a portion of the deviation of the fitted model from the data that can be attributed to deficiencies in the model.

Life distribution Shows how failures are distributed across time or actuations. The life distribution is normalized so that the area under the curve is 100%. The area under the curve between any two points in time provides the fraction failed in that time interval. Also called the *probability density function* (PDF).

Linear regression A method for fitting predictive models that are linear in the coefficients but not necessarily in the independent variables.

Lognormal distribution If the natural logs of times to failure are normally distributed, the times to failure are lognormally distributed. The lognormal is closely related to the normal distribution.

Loose tolerances Specification limits that are relatively larger multiples of the process variability. For example, Cp = 2 is a looser tolerance than Cp = 1.

Lurking variable A factor that affects the system response but was not included in the experiment design. If you are lucky, a lurking variable will vary during the experiment, leaving you with a large amount of unexplained variation in the response. This should force you to rethink the experiment and probably add one or more factors to understand what is causing the unexplained variation. Bad luck would be a lurking variable that does not surface during the experiment, either because it was controlled or because it did not vary. In this case you are in for a surprise later when you may be faced with unexplained performance variation that may be difficult and expensive to correct. Generally, if you have moved further down the road toward production, both the urgency and the cost of fixing the problem are greater, sometimes much greater. See **confounding factor**.

M

Main effects The contribution to the response of each of the control factors in a two-level factorial experiment.

Manufacturing process capability (Cp, Cpk) The relationship between the variability in the process that replicates a design parameter and the latitude in the system design for those variations.

Manufacturing process factor Any process characteristic or set point that is important to hitting the target and managing the variability of a manufacturing process.

Mean cycles between failures (MCBF) A reliability metric used for devices that fail from actuations. An example is an automobile tire that fails from the miles driven, actually total rotations, similar to actuations. MCBF = total cycles/total failures.

Mean time between failures (MTBF) A reliability metric used for devices that fail from usage time. It is the arithmetic mean of the observed times between failures. MTBF is generally used for repairable systems. MTBF = total time/total failures.

Mean time to failure (MTTF)   A reliability metric used for non-repairable or "one-shot" systems or components. It is the mean time to the first and fatal failure.

Mean time to repair (MTTR)   Applies to repairable systems or components. Time to repair should be described by a statistical distribution. This helps you understand the magnitude of the long-tail problem, where some repairs are lengthy.

Measurement system analysis   A study to assess how much variation a measurement system contributes to the variability of its measurements. A measurement system includes the people, hardware, software, and process for making measurements. Measurement system evaluations are always done in the context of a particular application, often using a designed experiment to determine the fraction of measurement variation caused by the measurement system which includes the gauge, operators, software, and environment. Do you need to take measurements to the nearest inch or micron? How much variation is there in what you are trying to measure? As long as the system isn't "broken," a given system will be appropriate for some tasks and not others.

Method of maximum likelihood   A statistical method for determining unknowns such as distribution parameters by formulating the likelihood function and finding parameter values that will maximize the likelihood function.

Mixtures of distributions   These occur when there are multiple failure modes with different physics of failure that are contributing to the total failure rate.

Model   A mathematical description in the form $y = f(x)$ that specifies the relation between the control factors and the response. Models can come from either the application of first principles or the use of designed experiments.

N

Natural tolerances   Tolerances that can be produced by a given manufacturing process that are economically efficient, without heroic efforts and high levels of scrap. Different types of processes will have different natural tolerances. Different types of processes, materials, and design configurations can be expected to have different natural tolerances. They can depend also on the given organization and needs of the business. They may vary among suppliers.

Noise   Uncontrolled variability that can affect the performance of a system.

Nonlinear regression   A method for fitting predictive models which can be nonlinear in both the coefficients and the predictor variables.

O

Objective function   A mathematical expression whose value depends on the value of important system outputs. It is used to co-optimize multiple responses. It can determine the tolerance for each response as well as the relative importance of the responses.

Observable quality metric   A customer-observable product characteristic that affects customers' perceptions and their assessments of product quality. These metrics are measures of quality and appearance, not measures of critical functions that the product performs.

One-sample t-test   Used to compare a sample mean to a reference value. An example of a reference value could be a target value for a manufacturing process. So you attempt to make an inference, based on a sample, of whether you're close to or far from the target.

Optimization   A mathematical process for finding desirable set points for the parameters that control system output(s). Optimization requires a search of the experiment space, usually within the "cube of interest," which is bounded by the allowable operating limits for the factor settings. Multi-response optimization requires you to formulate an objective function that implicitly means you are willing to entertain performance trade-offs among the important system responses.

Orthogonal array   A Resolution III experiment where main effects are aliased with two-way and higher interactions. Resolution III orthogonal arrays have very complex alias structures and are not a good choice unless interactions do not dominate their main effects.

Orthogonality   In mathematics, orthogonality is a property of vectors. Vectors that are orthogonal are at "right angles." The concept can be generalized to more than three dimensions. In DOE, many experiment designs such as factorials have columns that are orthogonal vectors describing the factor level settings for each treatment. The property of orthogonality means there is no confounding of main effects and makes it possible to calculate the factor effects independently of each other using the analysis of means.

Overfitting   Fitting a model having some predictor variables that do not add value to the predictive capability of the model.

P

Pareto's law   Most of the effects in any system come from a fraction of the possible causes. Named after Italian economist Vilfredo Pareto. Example: 80% of the income is earned by 20% of the people. Says the same thing as the principle of parsimony.

Parsimony   A principle that the simplest explanation that adequately describes any phenomenon is the best. So its use hinges on the judgment of adequacy and what is "good enough." The principle of parsimony holds true in many areas ranging from the natural sciences to systems designed by humans. In DOE, parsimony tells us that if a system is well described by a model from a two-level

Resolution III experiment, upgrading to a CCD actually subtracts value.

Poisson distribution   A discrete distribution used to calculate the probability of a given number of failures within a specified time period, when the time between failures is exponentially distributed.

Population   The population of something is all things in that category that exist. A population could be dynamic—all 30-year-old males—or static—all 2005 Subaru Forester station wagons.

Power of the test   The probability that you will reject the null hypothesis if it is false.

Practical importance   Predictor variables that contribute a large enough part to the response are practically important. Note that practical importance is not a statistical characteristic.

Predictor variables   A term used in regression for the independent variables in a model that determine the response. See also **control factor**.

Preventive maintenance   The process of replacing parts or making adjustments before a failure causes an unscheduled outage and loss of use for customers. Preventive maintenance either is scheduled or is performed when a system is serviced for a failure.

Probability density function (PDF)   A statistical distribution that describes the distribution of failures over time. The vertical height of the PDF is scaled to make the area under the curve unity. The vertical height will depend on the time scale used in reporting the life data.

Prospective study   An experiment that gathers data as control factors are changed deliberately.

Pure error   The variation in the data caused by common cause noise. Pure error is the difference between an observation and the mean of the sample the observation comes from. Replicates are required to calculate pure error. By definition the sum of all the pure error is 0.

P-value   The probability that you would get the data you have if the null hypothesis were true.

Q

Quality level The "sigma" metric for the deviation of a performance response or replicated parameter from its target. It represents the ratio of the tolerance to the standard deviation of the variation from target.

Quality loss A measure of the financial consequences of a product characteristic being off target. Quality loss is applicable to both CFRs and observable quality metrics.

Quality loss function The quadratic representation of the economic loss due to the range of deviations of a performance response from it target value. The losses are those suffered by customers as well as those incurred by the producer to correct the consequences of those deviations.

R

Reliability-centered maintenance A maintenance strategy that is preventive, using tools such as condition monitoring, FMEA, and FTA to determine maintenance timing.

Reliability or survival function The fraction of the original population surviving versus time. The reliability function provides the probability that a specific population member will survive for a specified length of time.

Replication Running a particular treatment more than once. A replicate is different from a repeat. Replicates should be randomized and not run one after the other. That would be a "repeat." Usually when you replicate an experiment you run all of the treatments the same number of times. Center points would be exceptions.

Residuals The difference between the predicted and observed values of the response. The residuals can be partitioned into two parts, pure error and lack of fit.

Resolution A measure of the ability of an experiment design to identify interactions. Any fractional factorial experiment will be described as having a certain resolution. The resolution is related to the size of the fraction being run and the number of control factors.

Response The result of how the predictor variables combine to determine the value of the dependent variable in the model. It's the y in $y = f(x)$.

Response factor A measurable output of a product that is important to customers' satisfaction.

Response surface The plotted surface describing the relationship between control factors and response. You can visualize a response surface, $z = f(x, y)$, in three dimensions.

Response surface experiment An experiment design capable of identifying all main effects, two-way interactions, and square terms in a system transfer function.

Retrospective study A study that attempts to establish association between certain factors and a response by looking back in time and using historical data.

Risk The expected value of loss assignable to the occurrence of an event. Risk is the product of the probability of occurrence for an event and the consequences when it does occur.

Robustness The quality of being insensitive to sources of variation. For an engineering system there will be variability in both internal control factors and external uncontrolled noises. A system that is more robust will be less vulnerable to these sources of variability.

Root sum of squares (RSS) tolerances A tolerance design method based on the assumption that the dimensions of parts in an assembly or the characteristics of the elements in a system that affect a functional response are independent and identically distributed (i.i.d.) and, further, that they are normally distributed. As the number of elements contributing to performance variability increases, the central limit theorem enables the relaxation of the normality assumption.

R-squared The square of the correlation coefficient, where you are calculating the correlation between predicted and observed values of the response. R-squared, which is between 0 and 1, expresses the fraction of the variation of the data "explained" by the model.

R-squared adjusted   Calculated by derating or reducing R-squared as a penalty for overfitting. It is always less than R-squared.

Run   One row in the experiment design worksheet. A run could be unique, or it could represent replicates of certain treatments.

S

Sample   Consists of a small fraction of its population. You try to make judgments or inferences about the population based on characteristics of the sample.

Sample mean   The mean of an individual sample taken for a statistical purpose such as an ANOVA or a regression. There are as many sample means as there are samples.

Satisfier   A product feature or function for which increasing performance is correlated with increasing customers' satisfaction.

Service process capability   Similar to manufacturing process capability, except that it is applied to service interventions performed by a service organization.

Signal-to-noise ratio (S/N)   A term used in signal processing to specify the ratio of the power of the meaningful signal or data to the power of the noise. It has also been used to describe the statistics of any process or system where the response of interest is mixed with random variation called noise. The higher the signal-to-noise ratio, the easier it is to separate the signal or information from the noise. S/N has also been used by Taguchi in robustness development to quantify the same thing. Taguchi's S/N is

$$S/N = 10 \log\left(\frac{\mu^2}{\sigma^2}\right)$$

Significance test   Establishes the probability (p-value) that a sample statistic such as the mean or variance would have a certain value, usually determined by a sample, if the null hypothesis were true.

Soft failure   A failure caused by drifting performance that eventually becomes unacceptable to customers. The product continues to be operable but an important CFR is sufficiently off target to precipitate customers' complaints.

Standard error of the mean   Describes the spread of the sampling distribution of the mean and is equivalent to the standard deviation of a normal population.

Statistical distribution   A mathematical description of how data are distributed. Since statistical distributions are mathematical, they are very useful for both modeling and statistical inference.

Statistical significance   Measured by the p-value. Low p-values tell you that it is unlikely that you would get a given test statistic if the null hypothesis is true. Low p-values lead you to reject the null hypothesis and accept the alternative. The observed difference is not the result of chance but is significant and caused by something other than chance alone.

Stochastic optimization   An optimization technique that uses Monte Carlo simulation to find control factor settings that will improve performance. The system transfer function is used in the simulation. Control factors are often decision variables while noise factors are random variables. You are seeking optimal control factor settings that will put the response on target and minimize variability of the response caused by the noises.

Student's t-distribution   A statistical distribution that is the sampling distribution of the mean, where the parent population is normal, sample size can be small, and the population standard deviation is unknown.

Subgroup   A group within a population that might not share all population characteristics or might share them in different proportions from the general population. For example, if you look at investment bankers as a subgroup within the population of 30-year-old males, what's the proportion of Democrats and Republicans compared to all 30-year-old males? It might be different.

Suspension   A termination of a life test that leaves the components under test in a non-failed state. Data that are mixes of suspensions and failures are censored data.

T

Taguchi's methods The methods of robustness development pioneered by Genichi Taguchi. The approach is based on the use of orthogonal arrays and compound noise factors, which combined with control factors introduce response variability into the experiment. The objective is to find settings for control factors that simultaneously reduce response variability and place response on target.

Theorem A proposition that is not self-evident but can be demonstrated to be true by logical argument.

Tight tolerances Specification limits that are a relatively smaller multiple of process variability. For example Cp = 1 is a tighter tolerance than Cp = 2.

Tolerance The range of allowable variation, relative to the target requirement, of a product parameter or an important functional response delivered by the product.

Tolerance allocation The process of dividing a system tolerance into parts and assigning each system element a share of the total acceptable variability. Ideally, the tolerance allocation is driven by its economics.

Tolerance optimization The allocation of tolerances to meet quality and cost requirements. It can be a constrained optimization problem where we could choose to minimize total costs subject to achieving a quality goal, or maximize quality subject to meeting a cost goal. Assuming you can formulate a credible quality loss function, an alternative is to minimize the cost of quality.

Transfer function A mathematical description, in the form $y = f(x)$, of the functional relationship between control factors and responses for a system or process.

Treatment A unique combination of control factor settings used in both ANOVA and DOE when measuring the functional response.

Two-level factorial designs An experiment design ideal for systems that are well described by models having only main effects and interactions and no important higher-order terms.

Two-sample t-test Used to compare the means of two samples from different sources to see if the population means are comparable or not. A question may be "Can these two processes produce components that are about the same?"

Type I error The error made when you reject the null hypothesis when it is true.

Type II error The error made when you fail to reject the null hypothesis when it is false.

U

Uncoded factors The actual values for the factor levels in an experiment, such as −5 volts and +5 volts for low and high. With uncoded factors, the value of the model coefficients for each factor cannot be used as a measure of relative control factor importance.

V

Value proposition The basic reason that customers are expected to buy your product rather than the offerings of your competitors. It is an intention to deliver value in specific benefits relative to their costs. The value proposition is transformed into requirements for the product and its related services.

W

Weibull distribution A two-parameter (three if there is a location parameter) statistical distribution, one of the more useful life distributions. The Weibull distribution can model behaviors ranging from early-life failures through random failures to end-of-life wear-out. It is useful in all three regions of the bathtub curve.

Worst-case tolerances Tolerances that will occur when system elements are at their minimum or maximum allowable tolerances.

Bibliography

Abernethy, Robert B. *The New Weibull Handbook, 4th ed.* North Palm Beach, FL: Robert B. Abernethy, 2000.

Advisory Committee to the Surgeon General of the Public Health Service. *Smoking and Health*, 1964.

Altman, Douglas G. *Practical Statistics for Medical Research.* London: Chapman & Hall, 1991.

Andrews, J. D., and T. R. Moss. *Reliability and Risk Assessment, 2nd ed.* New York: ASME Press, 2002.

Anscombe, Francis J. "Graphs in Statistical Analysis." *American Statistician*, February 27, 1973.

Barabasi, Albert-Laszlo. *Linked.* New York: Plume, 2003.

Blanding, Douglass L. *Exact Constraint: Machine Design Using Kinematic Principles.* New York: ASME Press, 1999.

Box, George E. P., William G. Hunter, and J. Stuart Hunter. *Statistics for Experimenters: An Introduction to Design, Data Analysis, and Model Building.* New York: John Wiley, 1978.

Broemm, William J., Paul M. Ellner, and W. John Woodworth. *AMSAA Reliability Growth Handbook.* Aberdeen, MD: U.S. Army Materiel Systems Analysis Activity, 1999.

Burchill, Gary, and Christina Hepner Brodie. *Voices into Choices: Acting on the Voice of the Customer.* Madison, WI: Joiner Associates, 1997.

Center for Quality of Management. *Center for Quality of Management Journal* 2, no. 4 (1993).

Chan, H. Anthony, and Paul J. Englert, eds. *Accelerated Stress Testing Handbook: Guide for Achieving Quality Products.* New York: IEEE Press, 2001.

Chapman, Jonathan. *Emotionally Durable Design: Objects, Experiences, and Empathy.* London, UK: Earthscan, 2005.

Christensen, Clayton M. "Finding the Right Job for Your Product." *MIT Sloan Management Review*, Spring 2007.

Clausing, Don. *Total Quality Development: A Step-by-Step Guide to World-Class Concurrent Engineering.* New York: ASME Press, 1994.

Clausing, Don, and Stuart Pugh. "Enhanced Quality Function Deployment." Design and Productivity International Conference, Honolulu, HI, February 6–8, 1991.

Clausing, Don, and Victor Fey. *Effective Innovation: The Development of Winning Technologies.* New York: Professional Engineering Publishing, 2004.

Clemens, P. L. "Event Tree Analysis." Jacobs-Sverdrup, 2002. Available at www.fault-tree.net.

———. "Fault Tree Analysis." Jacobs-Sverdrup, 1993. Available at www.fault-tree.net.

Cochran, William G. *Sampling Techniques.* New York: John Wiley, 1977.

Cohen, Lou. "Quality Function Deployment: An Application Perspective from Digital Equipment Corporation." *National Productivity Review*, Summer 1986.

———. *Quality Function Deployment: How to Make QFD Work for You.* Englewood Cliffs, NJ: Prentice Hall, 1995.

Collins, Jim. *Good to Great.* New York: HarperCollins, 2001.

Cooper, Robert G. *Winning at New Products: Accelerating the Process from Idea to Launch,* 3rd ed. Cambridge, MA: Perseus, 2001.

Cooper, Robert G., and Scott J. Edgett. *Lean, Rapid, and Profitable New Product Development.* Canada: Product Development Institute, 2005.

Cooper, Robert G., Scott J. Edgett, and Elko J. Kleinschmidt. *Portfolio Management for New Products.* Reading, MA: Addison-Wesley, 1998.

Creveling, C. M. *Tolerance Design: A Handbook for Developing Optimal Specifications.* Reading, MA: Addison-Wesley, 1997.

Creveling, C. M., J. L. Slutsky, and D. Antis, Jr. *Design for Six Sigma in Technology and Product Development.* Upper Saddle River, NJ: Prentice Hall PTR, 2003.

Crow, Larry H. *Reliability Analysis for Complex Repairable Systems.* Aberdeen, MD: U.S. Army Materiel Systems Analysis Activity, 1975.

Damelio, Robert. *The Basics of Process Mapping.* Portland, OR: Productivity Press, 1996.

Diethelm, Pascal, and Martin McKee. "Lifting the Smokescreen: Tobacco Industry Strategy to Defeat Smoke Free Policies and Legislation." European Respiratory Society and Institut National du Cancer, 2006.

Doll, Richard. "Uncovering the Effects of Smoking: Historical Perspective." *Statistical Methods in Medical Research* 7 (1998): 87–117.

Doll, Richard, and A. Bradford Hill. "The Mortality of Doctors in Relationship to Their Smoking Habits: A Preliminary Report." *British Medical Journal* ii (1954): 1451–5.

Doty, Leonard A. *Reliability for the Technologies.* New York: Industrial Press, 1989.

Duane, J. T. "Learning Curve Approach to Reliability Monitoring." *IEEE Transactions on Aerospace* 2, no. 2 (April 1964).

El-Haik, Basem Said. *Axiomatic Quality, Integrating Axiomatic Design with Six-Sigma, Reliability, and Quality Engineering.* Hoboken, NJ: John Wiley, 2005.

El-Haik, Basem, and David M. Roy. *Service Design for Six Sigma: A Roadmap for Excellence.* Hoboken, NJ: John Wiley, 2005.

Elsayed, Elsayed A. *Reliability Engineering.* Reading, MA: Addison Wesley Longman, 1996.

Evans, Merran, Nicholas Hastings, and Brian Peacock. *Statistical Distributions, 2nd ed.* New York: John Wiley, 1993.

Fowlkes, William Y., and Clyde M. Creveling. *Engineering Methods for Robust Product Design Using Taguchi Methods in Technology and Product Development.* Upper Saddle River, NJ: Prentice Hall, 1995.

Freund, John E. *Mathematical Statistics, 5th ed.* Englewood Cliffs, NJ: Prentice Hall, 1992.

Garvin, David A. "Competing on the Eight Dimensions of Quality." *Harvard Business Review,* November–December 1987.

Geneen, Harold, and Alan Moscow. *Managing.* New York: Doubleday, 1984.

George, Michael L. *Lean Six Sigma: Combining Six Sigma Quality with Lean Speed.* New York: McGraw-Hill, 2002.

George, Michael L., David Rowlands, Mark Price, and John Maxey. *The Lean Six Sigma Pocket Toolbook.* New York: McGraw-Hill, 2005.

George, Michael L., and Stephen A. Wilson. *Conquering Complexity in Your Business.* New York: McGraw-Hill, 2004.

Ginn, Dana, Barbara Streibel, and Evelyn Varner. *The Design for Six Sigma Memory Jogger.* Salem, NH: Goal/QPC, 2004.

Goldratt, Eliyahu. *Critical Chain.* Great Barrington, MA: North River Press, 1997.

Good, Philip I., and James W. Hardin. *Common Errors in Statistics.* Hoboken, NJ: Wiley Interscience, 2006.

Gouillart, Francis J., and Frederick D. Sturdivant. "Spend a Day in the Life of Your Customers." *Harvard Business Review*, January–February 1994.

Griffin, Abbie. "Obtaining Customer Needs for Product Development." *The PDMA Handbook of New Product Development.* Hoboken, NJ: John Wiley, 2005.

Griffin, Abbie, and John R. Hauser. "The Voice of the Customer." *Marketing Science*, Winter 1993.

Hambleton, Lynne. *Treasure Chest of Six Sigma Growth Methods, Tools, and Best Practices.* Upper Saddle River, NJ: Prentice Hall, 2008.

Hamel, Gary, and C. K. Prahalad. "Seeing the Future First." *Fortune*, September 5, 1994.

Hauser, John R., and Don Clausing. "The House of Quality." *Harvard Business Review*, no. 3 (May–June 1988).

Hinckley, C. Martin. *Make No Mistake! An Outcome-Based Approach to Mistake-Proofing.* Portland, OR: Productivity Press, 2001.

Hobbs, Gregg K. *Accelerated Reliability Engineering.* Chichester, UK: John Wiley, 2000.

Hooks, Ivy F., and Kristin A. Farry. *Customer-Centered Products.* New York: AMACOM, 2001.

IEEE Standard P1233. New York: Institute of Electrical and Electronic Engineers, 1993.

Jarvis, Jeff. *What Would Google Do?* New York: HarperCollins, 2009.

Kales, Paul. *Reliability for Technology, Engineering, and Management.* Upper Saddle River, NJ: Prentice Hall, 1998.

Kaplan, Robert S., and David P. Norton. *The Balanced Scorecard: Translating Strategy into Action.* Boston: Harvard Business School Press, 1996.

Katzenbach, Jon R., and Douglas K. Smith. *The Wisdom of Teams: Creating the High-Performance Organization.* New York: HarperCollins, 2003.

Kendrick, Tom. *Identifying and Managing Project Risk.* New York: AMACOM, 2003.

Kennedy, Michael N. *Product Development for the Lean Enterprise: Why Toyota's System Is Four Times More Productive and How You Can Implement It.* Richmond, VA: Oaklea Press, 2003.

King, John P., and John R. Thompson. "Using Maintenance Strategy to Improve the Availability of Complex Systems." *Proceedings of NIP 17,* International Conference on Digital Printing Technologies, 2001.

Kotter, John P. *Leading Change.* Boston: Harvard Business School Press, 1996.

Lanning, Michael J., and Edward G. Michaels. "A Business Is a Value Delivery System." Lanning, Phillips & Associates, 1987.

Leach, Lawrence P. *Critical Chain Project Management.* Norwood, MA: Artech House, 2005.

Leon-Garcia, Alberto. *Probability and Random Processes for Electrical Engineering.* Reading, MA: Addison-Wesley, 1989.

Levin, Mark A., and Ted T. Kalal. *Improving Product Reliability: Strategies and Implementation.* West Sussex, UK: John Wiley, 2003.

Mascitelli, Ronald. *The Lean Product Development Guidebook: Everything Your Design Team Needs to Improve Efficiency and Slash Time-to-Market.* Northridge, CA: Technology Perspectives, 2007.

Mattingly, Jack D., William H. Heiser, and David T. Pratt. *Aircraft Engine Design, 2nd ed.* Reston, VA: AIAA, 2002.

McDermott, Robin E., Raymond J. Mikulak, and Michael R. Beauregard. *The Basics of FMEA.* Portland, OR: Productivity Press, 1996.

McGrath, Michael E., Michael T. Anthony, and Amram R. Shapiro. *Product Development: Success through Product and Cycle-Time Excellence.* Boston: Butterworth-Heinemann, 1992.

McQuarrie, Edward F. *Customer Visits: Building a Better Market Focus.* Thousand Oaks, CA: Sage Publications, 2008.

Meeker, William Q., and Luis A. Escobar. *Statistical Methods for Reliability Data.* New York: John Wiley, 1998.

Mello, Sheila. *Customer-centric Product Definition: The Key to Great Product Development.* New York: American Management Association, 2002.

Morgan, James M., and Jeffrey K. Liker. *The Toyota Product Development System: Integrating People, Process, and Technology.* New York: Productivity Press, 2006.

Moubray, John. *Reliability-Centred Maintenance, 2nd ed.* Oxford, UK: Butterworth-Heinemann, 1999.

Myers, Raymond H., and Douglas C. Montgomery. *Response Surface Methodology.* New York: John Wiley, 2002.

Nelson, Wayne. *Accelerated Testing: Statistical Models, Test Plans, and Data Analysis.* New York: John Wiley, 1990.

Neter, John, Michael Kutner, Christopher Nachtsheim, and William Wasserman. *Applied Linear Statistical Models.* New York: McGraw-Hill, 1996.

O'Connor, Patrick D. T. *Practical Reliability Engineering.* New York: John Wiley, 1991.

———. *Test Engineering.* Chichester, UK: John Wiley, 2000.

Ohring, Milton. *Reliability and Failure of Electronic Materials and Devices.* San Diego: Academic Press, 1998.

Ong, Elisa K., and Stanton A. Glantz. "Constructing 'Sound Science' and 'Good Epidemiology': Tobacco, Lawyers and Public Relations Firms." *American Journal of Public Health* 91, no. 11 (November 2001).

Otto, Kevin N., and Kristin L. Wood. *Product Design: Techniques in Reverse Engineering and New Product Development.* Upper Saddle River, NJ: Prentice Hall, 2001.

Pande, Peter S., Robert P. Neuman, and Roland R. Cavanagh. *The Six Sigma Way Team Fieldbook.* New York: McGraw-Hill, 2002.

Peto, Richard, Sarah Darby, Harz Deo, Paul Silcocks, Elise Whitley, and Richard Doll. "Smoking, Smoking Cessation and Lung Cancer in the UK since 1950: Combination of National Statistics with Two Case-Control Studies." *British Medical Journal* 321, no. 5 (August 2000): 323–29.

Phadke, Madhav S. *Quality Engineering Using Robust Design.* Englewood Cliffs, NJ: Prentice Hall, 1989.

Piantadosi, Steven. *Clinical Trials: A Methodologic Perspective, 2nd ed.* New York: John Wiley, 2005.

Porter, Alex. *Accelerated Testing and Validation.* Oxford, UK: Newnes-Elsevier, 2004.

Project Management Institute Standards Committee. *A Guide to the Project Management Body of Knowledge.* Newtown Square, PA: Project Management Institute, 2000.

Pugh, Stuart. *Total Design: Integrated Methods for Successful Product Engineering.* Reading, MA: Addison-Wesley, 1991.

Pyzdek, Thomas. *The Six Sigma Handbook.* New York: McGraw-Hill, 2003.

Rausand, Marvin. *System Reliability Theory.* New York: John Wiley, 2004.

Rausand, Marvin, and Arnljot Høyland. *System Reliability Theory: Models, Statistical Methods, and Applications, 2nd ed.* Hoboken, NJ: John Wiley, 2002.

Reinertsen, Donald G. *Managing the Design Factory: A Product Developer's Toolkit.* New York: Free Press, 1997.

Ries, Al. *Focus: The Future of Your Company Depends on It.* New York: Harper Business, 1997.

Ries, Al, and Jack Trout. *Positioning: The Battle for Your Mind.* New York: McGraw-Hill, 2001.

Riffenburgh, Robert H. *Statistics in Medicine.* Burlington, MA: Elsevier, 2006.

Rogers, Everett M. *Diffusion of Innovations, 5th ed.* New York: Free Press, 2003.

Rosen, Emanuel. *The Anatomy of Buzz.* New York: Currency-Doubleday, 2002.

Rother, Mike, and John Shook. *Learning to See: Value Stream Mapping to Create Value and Eliminate MUDA.* Cambridge, MA: Lean Enterprise Institute, 2003.

Schiffauerova, Andrea, and Vince Thomson. "A Review of Research on Cost of Quality Models and Best Practices." *International Journal of Quality and Reliability Management* 23, no. 4 (2006).

Schmidt, Stephen R., and Robert G. Launsby. *Understanding Industrial Designed Experiments.* Longmont, CO: CQG Ltd Printing, 1989.

Schroeder, Bianca, and Garth Gibson. "Disk Failures in the Real World: What Does an MTTF of 1,000,000 Hours Mean to You?" Fast '07: 5th USENIX Conference on File and Storage Technologies, 2007.

Sharma, Naresh K., Elizabeth A. Cudney, Kenneth M. Ragsdell, and Kloumars Paryani. "Quality Loss Function—A Common Methodology for Three Cases." *Journal of Industrial and Systems Engineering* 1, no. 3 (Fall 2007).

Shewhart, Walter A. *Economic Control of Quality of Manufactured Product/50th Anniversary Commemorative Issue.* Milwaukee, WI: ASQ/Quality Press, 1980.

Shillito, M. Larry, and David J. DeMarle. *Value: Its Measurement, Design, and Management.* New York: John Wiley, 1992.

Shook, John. *Managing to Learn: Using the A3 Management Process to Solve Problems, Gain Agreement, Mentor, and Lead.* Cambridge, MA: Lean Enterprise Institute, 2008.

Smith, Preston G. *Flexible Product Development: Building Agility for Changing Markets.* San Francisco: Jossey-Bass, 2007.

Smith, Preston G., and Guy M. Merritt, *Proactive Risk Management: Controlling Uncertainty in Product Development.* New York: Productivity Press, 2002.

Smith, Preston G., and Donald G. Reinertsen. *Developing Products in Half the Time.* New York: Van Nostrand Reinhold, 1991.

Sobek, Durward K., Allen C. Ward, and Jeffrey K. Liker. "Toyota's Principles of Set-Based Concurrent Engineering." *Sloan Management Review*, Winter 1999.

Sterne, Jonathan A. C., and George Davey Smith. "Sifting the Evidence—What's Wrong with Significance Tests?" *British Medical Journal* 322 (January 27, 2001).

Suh, Nam P. *Axiomatic Design: Advances and Applications.* New York: Oxford University Press, 2001.

Sullivan, Lawrence P. "Quality Function Deployment: A System to Assure That Customer Needs Drive the Product Design and Production Process." *Quality Progress*, June 1986.

Sundararajan, C. (Raj). *Guide to Reliability Engineering: Data, Analysis, Applications, Implementation, and Management.* New York: Van Nostrand Reinhold, 1991.

Trout, Jack, and Steve Rivkin. *Differentiate or Die.* New York: John Wiley, 2000.

Tukey, John W. *Exploratory Data Analysis.* Reading, MA: Addison-Wesley, 1977.

Ullman, Dr. David G. *Making Robust Decisions; Decision Management for Technical, Business, & Service Teams.* Victoria, BC, Canada: Trafford Publishing, 2006.

Ulrich, Karl T., and Steven D. Eppinger. *Product Design and Development.* New York: McGraw-Hill, 1995.

U.S. Department of Defense. *Procedures for Performing a Failure Mode, Effects and Criticality Analysis,* MIL-STD-1629A. Washington, DC: Department of Defense, 1980.

———. *Reliability Test Methods, Plans and Environments for Engineering, Development, Qualification, and Production,* MIL-HDBK-781A. Washington, DC: Department of Defense, 1996.

Walton, Mary. *The Deming Management Method.* New York: Perigee Books, 1986.

Ward, Allen C. *Lean Product and Process Development.* Cambridge, MA: Lean Enterprise Institute, 2007.

Weibull, Waloddi. "A Statistical Distribution Function of Wide Applicability." *ASME Journal of Applied Mechanics*, September 1951, 293–97.

Weisstein, Eric W. "Buffon's Needle Problem." From Mathworld, A Wolfram Web Resource, http://mathworld.wolfram.com/BuffonsNeedleProblem.html.

Welch, B. L. "The Generalization of Student's Problem When Several Different Population Variances Are Involved." *Biometrika*, 1947, 28–35.

Whitcomb, Patrick J., and Mark J. Anderson. "Robust Design—Reducing Transmitted Variation." 50th Annual Quality Congress, 1996.

Womack, James P., and Daniel T. Jones. *Lean Solutions: How Companies and Customers Can Create Value and Wealth Together.* New York: Free Press, 2005.

Wu, C. F. Jeff, and Michael Hamada. *Experiments: Planning, Analysis, and Parameter Design Optimization.* New York: John Wiley, 2000.

Yang, Kai, and Basem S. El-Haik. *Design for Six Sigma: A Roadmap for Product Development.* New York: McGraw-Hill, 2003.

Young, Ralph R. *Effective Requirements Practices.* Boston: Addison-Wesley, 2001.

———. *The Requirements Engineering Handbook.* Norwood, MA: Artech House, 2004.

Index

Page numbers followed by f and t indicate figures and tables

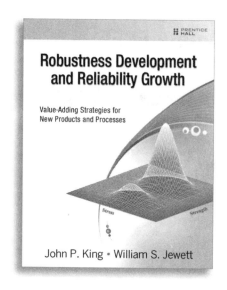

FREE Online Edition

Your purchase of **Robustness Development and Reliability Growth** includes access to a free online edition for 45 days through the Safari Books Online subscription service. Nearly every Prentice Hall book is available online through Safari Books Online, along with more than 5,000 other technical books and videos from publishers such as Addison-Wesley Professional, Cisco Press, Exam Cram, IBM Press, O'Reilly, Que, and Sams.

SAFARI BOOKS ONLINE allows you to search for a specific answer, cut and paste code, download chapters, and stay current with emerging technologies.

Activate your FREE Online Edition at www.informit.com/safarifree

> **STEP 1:** Enter the coupon code: ULTNZBI.

> **STEP 2:** New Safari users, complete the brief registration form.
> Safari subscribers, just log in.

If you have difficulty registering on Safari or accessing the online edition, please e-mail customer-service@safaribooksonline.com

 Addison Wesley   AdobePress   ALPHA   Cisco Press   FT Press FINANCIAL TIMES   IBM Press   lynda.com   Microsoft Press New Riders

 O'REILLY   Peachpit Press   que   Redbooks   SAMS   SAS Publishing   Sun microsystems   Wharton School Publishing   WILEY